Encyclopaedia of Herbal Antioxidants

The Editor

Prof. T. Pullaiah obtained his M.Sc. (1973) and Ph.D. degrees in Botany from Andhra University. He was a Post Doctoral Fellow at Moscow State University, Russia during 1976-78. He traveled widely in Europe and USA and visited Universities and Botanical Gardens in about 17 countries. Professor Pullaiah joined Sri Krishnadevaraya University as Lecturer in 1979 and became Professor in 1993. He held several positions in the University, which include Dean, Faculty of Life Sciences, Head of the Department of Botany, Chairman, BOS in Botany, Head of the Department of Sericulture, Co-ordinator and Chairman, BOS in Biotechnology, Vice Principal and Principal, S.K. University College. He retired from active service on 31ˢᵗ May 2011. He was selected by UGC as UGC-BSR Faculty Fellow and is working in Sri Krishnadevaraya University. Prof. Pullaiah has published 47 books, 250 research papers and 35 popular articles. His books include *Flora of Andhra Pradesh* (5 volumes), *Flora of Eastern Ghats* (vol. 1-4), *Biodiversity in India* (vol. 1-5), *Taxonomy of Angiosperms* (presently in 3ʳᵈ edition), *Plant Development, Plant Reproduction, Plant Tissue Culture, Flora of Guntur district, Flora of Kurnool, Flora of Anantapur district, Flora of Nizamabad, Flora of Ranga Reddi district* etc. He is Principal Investigator of 20 major Research Projects totaling more than a Crore of Rupees funded by DBT, DST, CSIR, UGC, BSI, WWF, GCC etc. Under his guidance 51 students obtained their Ph.D. degrees and 35 students their M.Phil. degrees. He is recipient of Best Teacher Award from Government of Andhra Pradesh, Prof. P. Maheswari Gold Medal and Prof. G.Panigrahi Memorial Lecture of Indian Botanical Society and Prof. Y.D.Tyagi Gold Medal of Indian Association for Angiosperm Taxonomy. He was Vice President of Indian Botanical Society. He was Member of Species Survival Commission of International Union for Conservation of Nature and Natural Resources (IUCN).

Encyclopaedia of Herbal Antioxidants

– Volume 2 –

Editor
T. Pullaiah
Department of Botany
Sri Krishnadevaraya University
Anantapur – 515 003
Andhra Pradesh

2012
REGENCY PUBLICATIONS
New Delhi - 110 002

© 2012, Editor
ISBN 9789351240167 (Set)
ISBN 9789351307273 (Volume 1) (International Edition)
ISBN 9789351307280 (Volume 2) (International Edition)
ISBN 9789351307297 (Volume 3) (International Edition)

Published by	:	**Regency Publications**
		A Division of
		Astral International Pvt. Ltd.
		– ISO 9001:2008 Certified Company –
		4760-61/23, Ansari Road, Darya Ganj
		New Delhi-110 002
		Ph. 011-43549197, 23278134
		E-mail: info@astralint.com
		Website: www.astralint.com
Laser Typesetting	:	**Classic Computer Services**
		Delhi - 110 035
Printed at	:	**Chawla Offset Printers**
		Delhi - 110 052

PRINTED IN INDIA

Preface

The role of free radicals in many disease conditions has been well established. Several biochemical reactions in our body generate reactive oxygen species and these are capable of damaging crucial bio-molecules. If they are not effectively scavenged by cellular constituents, they lead to disease conditions, *e.g.*, cerebrovascular disease, cancer, Arteriosclerosis, Atherosclerosis, Heart disease, Senelity, Ageing, Behcet's disease, Crohn's disease, Cataracts, Sunburn, Ulcers, Osteoporosis, Rheumatoid Arthritis, Diabetes Mellitus, Emphysema, Stroke, Haemorrhagic Shock, Cardiovascular Disorders, Cystic Fibrosis, Neurodegenerative Diseases (*e.g.*, Parkinsonism, Alzheimer's disease), Gastrointestinal Ulcerogenesis, AIDS and even early Senescence. Free radicals are involved in numerous skin diseases, especially inflammatory reactions and photosenescence. Antioxidants reduce blood clotting and cholesterol level. Antioxidants have antistress, immunomodulatory, cognition-facilitating, anti-inflammatory and anti-aging properties.

The interest on the role of natural antioxidants as a tool to prevent aging and degenerative diseases process is growing due to the toxicity of synthetic antioxidants such as butylated hydroxyl anisole (BHA) and butylated hydroxyl toluene (BHT) at fairly high doses which limits their therapeutic usage. In the last few years there is a flurry of research on antioxidant plants.

The book, *Encyclopaedia of Herbal Antioxidants*, gives information on antioxidant activity of different plant species. All the species are listed in alphabetical sequence. Angiosperms are given first followed by other groups of plants. Plant part used, model system studied, methods of assay and chemical constituents are given for each and every species studied.

I tried to make this book as comprehensive as possible. However for a book of this magnitude there are bound to be omissions. Had I tried for perfection I could have never completed this book. Readers are requested to bring to the notice of the author for inclusion in the revised edition.

T. Pullaiah

Contents

Abbreviations

AAPH:	2, 2 – Azobios-2-amino-propane dihydro Chloride
ABTS:	2,2'-azino-bis-3-ethyl benzothiazoline-6-sulphonic acid
ALP:	Alkaline Phosphatase
CAT:	Catalase
DMSO:	Dimethyl Sulphoxide
DPPH:	1,1,Diphenyl-2-Picryl-hydrazyl
FRAP:	Ferric Reducing and Antioxidant Power
GAE:	Gallic acid Equivalent
GPx:	Glutathione Peroxidase
GSH:	Glutathione
IC:	Inhibitory Concetration
MDA:	Malondialdehyde
ONOO:	Peroxynitrite
ORAC:	Oxygen Radical Absorbance Capacity
RNS:	Reactive Nitrogen Species
ROS:	Reactive Oxygen Species
SGOT:	Glutamate Oxaloacetate transaminase
SGPG:	Glutamate Pyruvate transaminase
SOD:	Superoxide Dismutase
TEAC:	Trolox Equivalent Antioxidant Capacity

Flemingia strobilifera (L.) R.Br. Family: Fabaceae (Leguminosae)

Madan *et al.* (2010) evaluated the antioxidant activity of various extracts of root and leaf of *Flemingia strobilifera* using different *in vitro* methods and its phytochemical analysis. The antioxidant activity was studied by DPPH radical scavenging method, nitric oxide radical inhibition assay and scavenging of hydroxyl radical by p-NDA method.

Flueggea luecopyrus Willd. Family: Euphorbiaceae

Karuppusamy *et al.* (2011) evaluated the anthocyanins, ascorbic acid, total phenolics and flavonoids and antioxidant activity of methanol extracts of fruits of *Flueggea leucopyrus*. The anthocyanins were 0.34 CGE/100g, Ascorbic acid 39.4 AAE/100g, Total phenolics 31.7GAE/100g, Total flavonoids 42.7QE/100g and antioxidant activity (DPPH) 76.8µg/ml.

Foeniculum vulgare Mill. Family: Apiaceae (Umbelliferae)

English Name: Fennel. Japanese name: Uikyou.

A study was carried out by Budinevi *et al.* (1995) on the antioxidative activity of *Foeniculum vulgare*. Using the ethanolic extracts of plant material good results were obtained.

The antioxidant activity of water and ethanol extracts of *Foeniculum vulgare* seed was evaluated by various antioxidant assays, including total antioxidant, free radical scavenging, superoxide anion radical scavenging, hydrogen peroxide scavenging, metal chelating activities and were compared to standard antioxidants such as BHA, BHT and α-tocopherol. The water and ethanol extracts showed strong antioxidant activity. 100 µg of water and ethanol extracts exhibited 99.1 per cent and 77.5 per cent inhibition of peroxidation in linoleic acid system, respectively, and greater than the same dose of α-tocopherol (36.1 per cent). Both the extracts of *Foeniculum vulgare* seeds have effective reducing power, free radical scavenging, superoxide anion radical scavenging, hydrogen peroxide scavenging, and metal chelating activities. This antioxidant property depends on concentration and increasing with increased amount of sample. In addition, total phenolic compounds in the water and ethanol extracts of fennel seeds were determined as gallic acid equivalents (Oktay *et al.,* 2003).

The antioxidant activity of the essential oil obtained from *Foeniculum vulgare* was evaluated by Ruberto *et al.* (2000) by two lipid model systems: A modified TBARS assay and a spectrophotometric detection of hydroperoxydienes from linoleic acid in a micellar system. The oil demonstrated antioxidant capacity comparable in some cases to that of α-tocopherol and BHT, used as reference antioxidants.

Cai *et al.* (2004) evaluated the antioxidant activity and phenolic compounds in traditional Chinese medicinal plants associated with anticancer. The improved ABTS.+ method was used to systematically assess the total antioxidant capacity (Trolox equivalent antioxidant capacity, TEAC) of the medicinal extracts. The TEAC values and total phenolic content for methanolic extract of fruit of *Foeniculum vulgare* were 105.9 µmol Trolox equivalent/100g dry weight (DW), and 0.60 g of gallic acid equivalent/100g DW respectively. A positive significant linear relationship between antioxidant activity and total phenolic content showed that phenolic compounds were the dominant antioxidant components in the tested herbs. Major types of phenolic compounds in *Foeniculum vulgare* include phenolic acids (cinnamic acid, vanillic acid), coumarins (6,7-dihydroxycoumarin)

Hydrodistillated extracts from Fennel were assessed for their total phenol content, and antioxidant (iron(III) reduction, inhibition of linoleic acid peroxidation, iron (II) chelation, DPPH radical-scavenging and inhibition of hydroxyl radical-mediated 2-deoxy-D-ribose degradation, site and nonsite-specific) activities by Hinneburg *et al.* (2005). The total phenols and antioxidant activity was moderate.

Maeda *et al.* (2006) examined the phenolic contents of leaf of *Foeniculum vulgare* and their radical scavenging activities by using DPPH assay. High contents of phenolic compounds and strong DPPH scavenging activities were observed in extracts from this species. They also showed radical scavenging activities for the superoxide radical and tert-butylhydroperoxyl radical (t-BuOO·).

Khalil *et al.* (2007) studied the growth, phenolic compounds and antioxidant activity of fennel grown under organic farming condition. Antioxidant activity of ethanol extract (inhibition per cent) was 76.14 at 200μl extract.

Two diglucoside stilbine trimers and a benzoisofuranone derivative were isolated by Marino *et al.* (2007) from *Foeniculum vulgare* fruit together with nine known compounds. Antioxidant activity was tested using three methods: DPPH, total antioxidant capacity and assay of lipid peroxidation.

Souri *et al.* (2008) evaluated the antioxidant activity of fruit against linoleic acid peroxidation using 1,3-diethyl-2-thiobarbituric acid as reagent. Antioxidant activity (IC_{50}) against peroxidation of linoleic acid (2mg/ml) was 8.01 and phenolic content was 165.07 mg/100g dry weight. The results of this study showed that there is no significant correlation between antioxidant activity and phenolic content of the studied plant materials and phenolic content could not be a good indicator of antioxidant capacity.

Antioxidant activities of ethanol extracts from seven Apiaceae fruits have been studied by Nickavar and Abolhasani (2009) by the DPPH radical scavenging test. All the studied extracts, including *Foeniculum vulgare*, showed antioxidant capability. For *Foeniculum vulgare* IC_{50} value of the DPPH scavenging activity was 199.20μg/ml and total flavonoid content was 16.27μg/mg. A positive correlation was found between the antioxidant potency and flavonoid content of the fractions.

The essential oils from *Foeniculum vulgare* commercial aerial parts and fruits were isolated and evaluated by Miguel *et al.* (2010) for antioxidant activity. With the DPPH method the plant oils showed better antioxidant activity than the fruits oils. With the TBARS method and at higher concentrations fennel essential oils showed a pro-oxidant activity. None of the oils showed a hydroxyl radical scavenging capacity>50 per cent, but they showed an ability to inhibit 5-lipoxygenase.

Forsythia suspensa (Thunb.) Vahl Family: Oleaceae

The antioxidant properties of twenty medicinal herbs used in the traditional Mediterranean and Chinese medicine including *Forsythia suspensa* were studied by

Schinella *et al.* (2002). Using free radical-generating systems *Forsythia suspensa* protected against enzymatic and non-enzymatic lipid peroxidation in model membranes and also showed scavenging property on the superoxide radical. All the extracts were assayed at a concentration of 100µg/ml. Most of the extracts were weak scavengers of the hydroxyl radical.

Cai *et al.* (2004) evaluated the antioxidant activity and phenolic compounds in traditional Chinese medicinal plants associated with anticancer. The improved ABTS.+ method was used to systematically assess the total antioxidant capacity (Trolox equivalent antioxidant capacity, TEAC) of the medicinal extracts. The TEAC values and total phenolic content for methanolic extract of fruit of *Forsythia suspensa* were 640.5 µmol Trolox equivalent/100g dry weight (DW), and 4.29 g of gallic acid equivalent/100g DW respectively. A positive significant linear relationship between antioxidant activity and total phenolic content showed that phenolic compounds were the dominant antioxidant components in the tested herbs. Major types of phenolic compounds in *Forsythia suspensa* include lignans (forsythigenol, forsythin), coumarins (6,7-dimethoxy-coumarin) flavonols (rutin).

Li *et al.* (2008) evaluated the antioxidant properties of 45 medicinal plants including *Forsythia suspensa* using FRAP and TEAC assays and the total phenolic contents of these plants were measured. *Forsythia suspensa* showed 153.78 µmol Fe(II)/g FRAP value, 62.59µmol Trolox/g of TEAC value and 17.69 mg GAE/g of Phenolic content. A high correlation between antioxidant capacities and their phenolic contents indicated that phenolic compounds were a major contributor of antioxidant activity of these plants.

Fragaria ananassa Duchesne Family: Rosaceae

English Name: Strawberry.

The total antioxidant activity of strawberry was measured by Wang *et al.* (1996) using automated oxygen radical absorbance capacity (ORAC) assay. On the basis of the wet weight of the fruits (edible portion), strawberry had the highest ORAC activity (15.36 micromoles of Trolox equivalents per gram). On the basis of dry weight of the fruits strawberry had the highest ORAC activity (153.36µmoles/g Trolox equivalent). The contribution of the fruit pulp fraction (extracted with acetone) to the total ORAC activity of a fruit was usually less than 10 per cent.

The antioxidative activity of phenolic extracts from fruits of *Fragaria ananassa* was examined by autoxidation of methyl linoleate by Kahkonen *et al.* (1999). Remarkable high antioxidant activity and high total phenolic content were found in strawberry. In an LDL oxidation sysem, strawberry was found to be a weaker antioxidant than blackberries, red raspberries, sweet cherries and blueberries (Heinonen *et al.*, 1998b). On the contrary, fresh strawberry extract was reported to have 15 times higher total antioxidant capacity than trolox and greater activity than many fruits, for example plum, orange, red grape, apple and tomato, in artificial peroxyl radical model system (Wang *et al.*, 1996). According to Hakkinen *et al.* (1998) strawberry has an especially high content of ellagic acid.

Fruits and leaves from different cultivars of strawberry plants were analyzed by Wang and Lin (2000) for total antioxidant capacity (ORAC) and total phenolic content.

In addition, fruits were analyzed for total anthocyanin content. Strawberries had the highest ORAC activity during the green stages. Total anthocyanin content increased with maturity of fruits. Compared with fruits, leaves were found to have higher ORAC values. These results showed a linear correlation between total phenolic content and ORAC activity of fruits and leaves. For ripe berries, a linear relationship existed between ORAC values and anthocyanin content.

Cultivar effects of phenolic contents were compared with antioxidant capacities, as measured by the total oxyradical scavenging capacity (TOSC) assay by Meyers *et al.* (2003). Free phenolic content was weakly correlate with total antioxidant activity, and flavonoid and anthocyanin content did not correlate with total antioxidant activity.

The effect of strawberry on early atherosclerosis in hamsters was investigated by Rouanet *et al.* (2010). They received an atherogenic diet at the same time a juice at a daily dose corresponding to the consumption of 275 ml by a 70kg human. After 12 weeks berry juices inhibited aortic lipid deposition by 79-96 per cent and triggered reduced activity of hepatic antioxidant enzymes, not accompanied by a lowered plasma cholesterol These findings suggest that moderate consumption of berry juices and teas can help prevent the development of early atherosclerosis.

Wang and Lin (2003) reported that compost as a soil supplement increases the level of antioxidant compounds and oxygen radical absorbance capacity in strawberries.

Fragaria vesca L. Family: Rosaceae

Agostini *et al.* (2004) carried out studies to determine the antioxidant capacity of the flavonoids in strawberries. The results indicated that the antioxidant capacity of the flavonoids of strawberries was of 0.267 equivalent μM Trolox.

Kiselova *et al.* (2006) studied the antioxidant activity of *Fragaria vesca*. Antioxidant activity was measurd by ABTS cation radical decoloration assay and the total polyphenol content was measured according to the Folin-Ciocalteu method. *Fragaria vesca* extract exhibited higher antioxidant activity than mate. A positive correlation between antioxidant activity and polyphenol content was found, suggesting that the antioxidant capacity of the aqueous plant extract was due to a greater extent to their polyphenols.

Fraxinus excelsior L. Family: Oleaceae

English Name: Ash, European ash

Middleton *et al.* (2005) evaluated the antioxidant activity of *Fraxinus excelsior*. Among the extracts, the methanol extract displayed the highest level of antioxidant activity (RC^{50} = 1.35 X 10-2 mg/ml).

Fraxinus formosana Hayata Family: Oleaceae

Lee *et al.* (2006) investigated the antioxidant activity of ethanol extract of leaves using DPPH, hydroxyl and superoxide radical scavenging and reducing power activities as well as the induction of heme oxygenase-1 (HO-1). The IC_{50} values against

DPPH radical were 13.2µg/ml, OH radical 0.58µg, superoxide radical 114.4µg/ml and total phenolics was 27.0 mg of GAE/g.

Fraxinus ornus L. Family: Oleaceae

The antioxidative action of the total ethanolic extract of *Fraxinus ornus* bark and its main hydrocoumarin constituents was investigated using kinetically pure triacylglycerols of lard (TGL) and sunflower oil (TGSO) (Marinova *et al.*, 1994). The stabilization factor (F) and the oxidation rate ratio (ORR) of the tested antioxidants were estimated. The ethanolic extract in a concentration of 0.05 per cent showed a pronounced antioxidant activity during the oxidation of TGL (F=4.8, ORR=3.6) and TGSO (F=3.6, ORR=0.60), comparable with that of butylated hydroxytoulene and butylated hydroxyanisole. The activity of fraxetin and esculetin was higher than that of the corresponding glucoside fraxin and esculetin and comparable to that of other well-known antioxidants such as caffeic acid. The data also demonstrate that fraxetin was a stronger antioxidant than esculetin, especially in the system of lower oxidizability (TGL). In the same study the presence of additional antioxidative substances was revealed by TLC analysis. Among them calceolarioside demonstrated a significant activity.

Fraxinus rhynchophylla Hance Family: Oleaceae

Cai *et al.* (2006) reported structure-radical scavenging activity relationships of phenolic compounds from traditional Chinese medicinal plants including *Fraxinus rhynchophylla*.

Li *et al.* (2008) evaluated the antioxidant capacities of 45 selected medicinal plants using ferric reducing antioxidant power (FRAP) and Trolox equivalent antioxidant capacity (TEAC) assays respectively, and the total phenolic contents of these plants were measured by the Folin-Ciocalteau method. It was found that *Fraxinus rhynchophylla* possessed the highest antioxidant capacities and thus could be potential rich resources of natural antioxidants. *Fraxinus rhynchophylla* showed 373.35 µmol Fe(II)/g FRAP value, 166.09µmol Trolox/g of TEAC value and 40.27 mg GAE/g of Phenolic content. A strong correlation between TEAC values and those obtained from FRAP assay implied that antioxidants in these plants were capable of scavenging free radicals and reducing oxidants. A high correlation between antioxidant capacities and their total phenolic contents indicated that phenolic compounds were a major contributor of antioxidant activity of these plants.

Fraxetin, phenolic coumarin, isolated from *Fraxinus rhynchophylla* possessed considerable antioxidant activities (Thuong *et al.*, 2010).

Fritillaria cirrhosa D.Don Family: Liliaceae

Antioxidant capacities of 56 selected Chinese medicinal plants, including *Fritillaria cirrhosa* were evaluated by Song *et al.* (2010) using the TEAC and FRAP assays, and their total phenolic content was measured by the Folin-Ciocaleteu method. *Fritillaria cirrhosa* showed very low antioxidant activity with TEAC value of 2.57µmol Trolox/g, FRAP value of 0.29µmol Fe^{2+}/g and phenolic content was 0.96mg GAE/g.

Fritillaria verticillata Willd. Family: Liliaceae

Antioxidant capacities of 56 selected Chinese medicinal plants, including *Fritillaria verticillata* were evaluated by Song *et al.* (2010) using the TEAC and FRAP assays, and their total phenolic content was measured by the Folin-Ciocaleteu method. *Fritillaria verticillata* showed low antioxidant activity with TEAC value of 9.83μmol Trolox/g, FRAP value of 0.91μmol Fe^{2+}/g and phenolic content was 1.07mg GAE/g.

Fumaria officinalis L. Family: Fumariaceae

Crude extract from *Fumaria officinalis* was screened by Sengul *et al.* (2009) for its *in vitro* antioxidant and antimicrobial properties. Total phenolic content of extract of this plant was also determined. β-carotene bleaching assay and Folin-Ciocalteu reagent were used to determine total antioxidant activity and total phenols of plant extract. Antioxidant activity was 78.93 per cent and the total phenolic content was 10.50 mgGAE/g DW).

Funturnea elastica Stapf Family: Apocynaceae

Oke and Hamburger (2002) studied the antioxidant activity of ethanol extract of leaf using DPPH assay. The extract showed moderate antioxidant activity.

Gaillardia megapotamica (Spreng.) Baker var. *megapotamica* Family: Asteraceae (Compositae)

Borneo *et al.* (2008) evaluated the antioxidant activity of ethanolic extract of *Gaillardia megapotamica* var. *megapotamica.* The FRAP value was 132.3μmol of Fe(II)/g and DPPH radical scavenging activity was EC_{50} = 275.7.

Galanthus reginae-olgae Orph. subsp. *vernalis* Kamari Family: Amaryllidaceae

Conforti *et al.* (2010) presented the qualitative and quantitative analysis of Amaryllidaceae-type of alkaloids in the aerial parts and bulbs of *Galanthus reginae-olgae* subsp. *vernalis*. The alkaloids galanthamine, lycorine and tazettine were identified in both the extracts while crinine and neronine were found only in bulbs. The antioxidant activity was tested by DPPH activity and the IC_{50} values were 39 and 29 mug/ml for MeOH extracts from aerial parts and bulbs, respectively. The higher activity was given by EtOAc fraction of aerial parts with IC_{50} value of 10 mug/ml. This activity is probably due to the presence of EtOAc fraction of polar compounds such as polyphenols. The fraction exhibited significant antioxidant capacity also in the β-carotene-linoleic acid test system. A higher level of antioxidant activity was observed for the EtOAc fraction from bulbs with IC_{50} of 10 mug/ml after 30 min and 9 mug/ml after 60 min of incubation.

Galeopsis speciosa Mill. Family: Lamiaceae (Labiatae)

English Name: Hemp nettle.

Matkowski and Piotrowska (2006) studied the antioxidative effects of methanolic extract with the use of three *in vitro* assays (DPPH assay, phosphomolybdenum method and lipid peroxidation). The maximum DPPH reduction was observed only after longer incubation.

Galeopsis tetrahit L. Family: Lamiaceae (Labiatae)

Matkowski *et al.* (2008) evaluated the antioxidant activity of aerial parts of *Galeopsis tetrahit* using DPPH, OH radical scavenging assays, reducing power assay and inhibition of deoxyribose degradation. EC_{50} value of DPPH radical scavenging was 26.24μg/ml for MeOH extract, 88.98μg/ml for Dichloromethane extract, 173.22μg/ml for BuOH extract and 20.66μg/ml for ethyl acetate extract.

Galium aparine L. Family: Rubiaceae

In a screening of South African indigenous food plants for antioxidant activity by testing for inhibition of lipid peroxidation the teas with the tuber of *Galium aparine* had high activity (Lindsey *et al.*, 2006).

Galium verum L. subsp. *verum* Family: Rubiaceae

Antioxidant and radical scavenging activities, reducing powers and the amount of total phenolic compounds of aqueous and methanolic extracts of *Galium verum* were studied by Mavi *et al.* (2004). The aqueous extract showed moderate activity.

Garcinia afzelii Engl. Family: Clusiaceae (Guttiferae)

Two new prenylated xanthones, afzelixanthones A and B, together with three known xanthones and two phytosterols, β-sitosterol and stigmasterol, were isolated from the CH_2Cl_2/MeOH (1:1) extract of stem bark of *Garcinia afzelii*. The antioxidant activities of the crude extracts as well as the new compounds were evaluated by Kamden Waffo *et al.* (2006).

Garcinia atroviridis Griff. ex T. Anderson Family: Clusiaceae (Guttiferae)

Kruawan and Kangsadalampai (2006) investigated the antioxidant activity and phenolic compound contents of water extract of *Garcinia atroviridis*. The extract exhibited low scavenging activity against DPPH radicals (46.52 per cent) and their FRAP value was 376.93 μmol/g. The content of phenolic compound in the extracts was determined using the Folin-Ciocalteu reagent and calculated as gallic acid equivalent (GAE). The herbal extract gave a high phenolic compound content (GAE 36.76 mg/g).

Povichit *et al.* (2010) investigated phenolic content and free radical scavenging effect of fruits of *Garcinia atroviridis* by DPPH and ABTS assays, anti-lipid peroxidation activity by TBARS and for antiglycation activity. The results revealed that the total phenolic content showed good correlation with free radical scavenging by ABTS and anti-lipid peroxidation by TBARS, but showed no correlation with antiglycation. The extract of *Garcinia atroviridis* moderate antioxidant effect and also showed a moderate antiglycation effect. The IC_{50} were 15.20mg/ml for the DPPH method, TEAC value of 0.01 mg/Trolox/mg sample for the ABTS method, IC_{50} of 18.12 mg/ml for the TBARS method and IC_{50} of 0.32 for the antiglycation method.

Garcinia hombroniana Pierre Family: Clusiaceae (Guttiferae)

The crude methanolic extract from the bark exhibited antioxidant activity (Rema *et al.*, 2007).

Garcinia indica (Dupetit-Thours) Choisy Family: Clusiaceae (Guttiferae)

English Name: Koakam.

Garcinol, a polyisoprenylated benzophenone derivative, was purified from *Garcinia indica* fruit rind, and its fee radical scavenging activity was studied using ESR spectrophotometry (Yamaguchi *et al.*, 2000a,b). In the hypoxanthine/xanthine oxidase system, emulsified garcinol suppressed superoxide anion to almost the same extent as DL-α-tocopherol by weight. In the fenton reaction system, garcinol also suppressed hydroxyl radical more strongly than DL-α-tocopherol. In the H_2O_2/NaOH/DMSO system, garcinol suppressed superoxide anion, hydroxyl radical and methyl radical.

Mishra *et al.* (2006) examined the antioxidant activity of aqueous and boiled extracts. The assays employed are ORAC, FRAP, ABTS and the ability to inhibit lipid peroxidation in rat liver mitochondria. Kokam syrup and the two aqueous extracts had significant antioxidant effects in the above assays. They have high ORAC values (29.3, 24.5 and 20.3), higher than those reported for other spices, fruits and vegetables.

Garcinia kola Family: Clusiaceae (Guttiferae)

The antioxidant and scavenging activity of *Garcinia* biflavonoid complex has been investigated in a range of established *in vitro* assays involving reactive oxygen species. The study showed that kolaviron elicited significant reducing power and dose-dependent inhibition of oxidation of linoleic acid (Farombi *et al.*, 2002). Kolaviron also significantly scavenged superoxide generated by henazine methosulfate NADH. Furthermore, kolaviron scavenged hydroxyl radicals as revealed by significant inhibition of the oxidation of deoxyribose.

Furthermore, kolaviron dose dependently inhibited the intracellular ROS production induced by H_2O_2 in HepG2 cells detected as 2272-dichlorofluorescein diacetate (DCF) fluorescence (Nwankwo *et al.*, 2000; Eddy *et al.*, 1997). Thus the ability of kolaviron to act as antioxidant in this cell underscores its role as an antioxidant and its potential role in the chemoprevention of chemically-induced genotoxicity.

Garcinia biflavonoid effectively prevented microsomal lipid peroxidation induced by iron/ascorbate in a concentration dependent manner (Farombi and Nwaokeafor, 2005). The data demonstrate that Garcinia biflavonoid protected against the oxidation of lipoprotein presumably by mechanisms involving metal chelation and antioxidant activity and as such might be of importance in relation to the development of atherosclerosis.

Garcinia mangostana L. Family: Clusiaceae (Guttiferae)

English Name: Mangosteen.

Garcinol, polyisoprenylated benzophenone derivative, was purified from *Garcinia indica* fruit rind, and its antioxidant activity, chelating activity, free radical scavenging activity, and antiglycation activity were studied by Yamaguchi *et al.* (2000a,b). Garcinol exhibited moderate antioxidative activity in the micellar linoleic acid peroxidation system and also exhibited chelating activity at almost the same level as citrate. It also showed nearly 3 times greater DPPH free radical scavenging activity than DL-α-tocopherol by weight in aqueous ethanol solution. In a phenanzine methosulfate/NADH-nitroblue tetrazolium system, garcinol exhibited superoxide anion scavenging activity and suppressed protein glycation in a bovine serum albumin/fructose system.

Chomnawang *et al.* (2007) reported that *Garcinia mangostana* was highly effective in scavenging free radicals. Antioxidant activity was determined by DPPH scavenging and NBT reduction assay.

Thirty-eight types of fruits commonly consumed in Singapore were systematically analyzed by Isabelle *et al.* (2010) for their hydrophilic oxygen radical absorbance capacity (H-ORAC), total phenolic content (TPC), ascorbic acid (AA) and various lipophilic antioxidants. Mangosteen was high in tocotrienols.

The aqueous extract of the fruits of *Garcinia mangostana* was checked for their wound healing and antioxidant potential by Nainwal *et al.* (2010). Antioxidant study was carried out by using Liquid peroxidation method. The result showed a good antioxidant potential of the drug.

Povichit *et al.* (2010) investigated phenolic content and free radical scavenging effect of *Garcinia mangostana* by DPPH and ABTS assays, anti-lipid peroxidation activity by TBARS and for antiglycation activity. The results revealed that the total phenolic content showed good correlation with free radical scavenging by ABTS and anti-lipid peroxidation by TBARS, but showed no correlation with antiglycation. The extract of *Garcinia mangostana* demonstrated moderate antioxidant effect and also showed a moderate antiglycation effect. The IC_{50} were 0.38mg/ml for the DPPH method, TEAC value of 0.52 mg/Trolox/mg sample for the ABTS method, IC_{50} of 4.30 mg/ml for the TBARS method and IC_{50} of 31.33 for the antiglycation method.

Garcinia multiflora Champ. Family: Clusiaceae (Guttiferae)

Two new xanthene derivatives, garcinones A and B, two new benzophenone derivatives were isolated from the stems of *Garcinia multiflora*. Compounds were evaluated in the brine shrimp lethality test and in the DPPH antioxidant assay (Chiang *et al.*, 2003).

Garcinia subelliptica Merr. Family: Clusiaceae (Guttiferae)

Phloroglucinols, garcinielliptone L and M isolated from the seeds of *Garcinia subelliptica* showed potent inhibitory effects on the release of β-glucuronidase and histamine, respectively, from peritoneal mast cells stimulated with *p*-methoxy-N-methylphenethylamine. These compounds also showed potent effects on NO production in culture media of RAW 264. 7cells in response to lipopolysaccharide (LPS). *Garcinielliptone* L also showed a potent effect on NO production in culture media of N9 cells in response to LPS/interferon gamma (Weng *et al.*, 2004).

Chen *et al.* (2009) studied the antioxidant activity of leaves extract towards xanthine oxidase, lipoxygenase and tyrosinase activities. The inhibitory effect of acetone extract on xanthine oxidase activity was less than 10 per cent, tyrosine activity was almost nil and lipoxygenase activity was more than 20 per cent.

Garcinia vieillardii Pierre Family: Clusiaceae (Guttiferae)

Two new xanthones, 6-O-methyl-2-deprenylrheediaxanthone B and vieillardixanthone were isolated by Hay *et al.* (2004) from the stem bark of *Garcinia vieillardii*, as were four known compounds. Each isolate was tested for its antioxidant properties based on a scavenging study using the stable DPPH free radical.

Garcinia xanthochymus Hook.f. ex T.Anders. Family: Clusiaceae (Guttiferae)

Two benzophenones, guttiferone H and gambogenone isolated from the fruits of *Garcinia xanthochymus* displayed antioxidant activity in DPPH assay (IC_{50}=64 and 38.7µM respectively) (Baggetti *et al.*, 2005).

Gardenia augusta (L.) Merr. Family: Rubiaceae

Leelapornpisid *et al.* (2007) reported antioxidant activity of the absolutes (obtained by solvent extraction) of *Gardenia augusta*.

Gardenia erubescens Stapf Family: Rubiaceae

Polyphenol content and antioxidant activity of fourteen wild edible fruits from Burkina Faso were investigated by Lamien-Meda *et al.* (2008). The data obtained show that the total phenolic and total flavonoid levels were significantly higher in the acetone than in the methanol extracts. *Gardenia erubescens* fruit had higher flavonoid content and higher antioxidant activity.

Gardenia gummifera L.f. Family: Rubiaceae

Kalim *et al.* (2010) evaluated the antioxidant activity of Methanol (50 per cent) extract of *Gardenia gummifera*. IC_{50} values for DPPH, ABTS, NO, OH, O_2 and ONOO were 82.33, 11.62, 34.33, 45.39 and 890.32µg/ml. The extract contained significant amount of polyphenols.

Gardenia jasminoides Ellis Family: Rubiaceae

Cai *et al.* (2004) evaluated the antioxidant activity and phenolic compounds in traditional Chinese medicinal plants associated with anticancer. The improved ABTS.+ method was used to systematically assess the total antioxidant capacity (Trolox equivalent antioxidant capacity, TEAC) of the medicinal extracts. The TEAC values and total phenolic content for methanolic extract of fruit of *Gardenia jasminoides* were 132.6 µmol Trolox equivalent/100g dry weight (DW), and 1.00 g of gallic acid equivalent/100g DW respectively. A positive significant linear relationship between antioxidant activity and total phenolic content showed that phenolic compounds were the dominant antioxidant components in the tested herbs. Major types of phenolic compounds in *Gardenia jasminoides* include phenolic acids (chlorogenic acid), flavones (gardenins).

Five new quinic acid derivatives – methyl 5-O-caffeoyl-3-O-sinapoylquinate, ethyl 5-O-caffeoyl-3-O-sinapoylquinate, methyl 5-O-caffeoyl-4-O-sinapoylquinate, ethyl 5-O-caffeoyl-4-O-sinapoylquinate, and methyl 3,5-di-O-caffeoyl-4-O-(3-hydroxy-3-methyl)-glutaroylquinate were isolated from the fruit of *Gardenia jasminoides* by Kim *et al.* (2006). These compounds were assessed for antioxidant activity using three different cell-free bioassay systems. Five quinic acid derivatives showed potent DPPH radical scavenging, superoxide anion scavenging, and lipid peroxidation inhibition activities.

Li *et al*. (2008) evaluated the antioxidant properties of 45 medicinal plants including *Gardenia jasminoides* using FRAP and TEAC assays and the total phenolic contents of these plants were measured. *Gardenia jasminoides* showed 89.48 μmol Fe(II)/g FRAP value, 38.92μmol Trolox/g of TEAC value and 9.47 mg GAE/g of Phenolic content. A high correlation between antioxidant capacities and their phenolic contents indicated that phenolic compounds were a major contributor of antioxidant activity of these plants.

Gastrodia elata Bl. Family: Orchidaceae

Liu and Mori (1992) investigated the antioxidant effects of *Gastrodia elata in vivo* and their free radical scavenging effects *in vitro*. Extracts significantly inhibited the increase in levels of lipid peroxide in the ipsilateral cortex. In *in vitro* experiments, the extract exhibited significant dose-dependent scavenging effects on free radicals.

Gaultheria fragrantissima Wall. Family: Ericaceae

Karuppusamy *et al*. (2011) evaluated the anthocyanins, ascorbic acid, total phenolics and flavonoids and antioxidant activity of methanol extracts of fruits of *Gaultheria fragrantissima*. The anthocyanins were 2.74 CGE/100g, Ascorbic acid 67.6 AAE/100g, Total phenolics 80.4GAE/100g, Total flavonoids 94.3QE/100g and antioxidant activity (DPPH) 240.2μg/ml.

Gaultheria shallon Pursh Family: Ericaceae

The EtOAc extract obtained from *Gaultheria shallon* was tested in the DPPH free radical assay. High antioxidant activity was obtained from the extract of fruit (Acuna *et al.*, 2002).

Gaura biennis L. Family: Onagraceae

Borchardt *et al*. (2008) evaluated the antioxidant activity of methanol extracts of seeds of *Gaura biennis* using DPPH antioxidant assay.The antioxidant value reported in μM Trolox/100g (TE) from DPPH radical scavenging activity of crude seeds extract was 44,583.

Gazania splendens Family: Asteraceae (Compositae)

The chloroform and methanolic extracts of 124 Egyptian plant species belonging to 56 families were investigated by Moussa *et al*. (2011) and compared their antioxidant activity by DPPH scavenging assay. Among the 124 plant species tested 18 exhibited extremely high antiradical activity (>80 per cent inhibition). The IC_{50} value of DPPH radical scavenging of *Gazania splendens* was 23.57μg/ml, while total phenolic and flavonoid contents were 48.06 mg Tannic acid equivalent/g extract and 208.86 mg Rutin equivalent/g extract, respectively. Correlation coefficient between DPPH radical scavenging activity and the total phenolic and flavonoid contents suggest that phenolics and flavonoids in the extracts were partly responsible for the antiradical activities.

Genista cadasonensis Valsecchi Family: Fabaceae (Papilionaceae)

The flavonoid fraction of aerial parts of *Genista cadasonensis* was examined by Serrilli *et al*. (2010). A rare flavonoid, the 6-hydroxygenistin was isolated. The antioxidant activity of dichloromethanic, ethanolic and acetonic total extracts of aerial parts was evaluated.

Genista sandrasica Hartwig et Strid Family: Fabaceae (Papilionaceae)

Orhan *et al*. (2011) evaluated the antioxidant activity of the crude methanol and hydrolyzed extracts of *Genista sandrasica* by three *in vitro* methods, namely DPPH free radical scavenging, ferrous ion-chelating and FRAP tests at 0.25, 0.50 and 1.0 mg/ml. The hydrolyzed extracts displayed greater antioxidant activity than the crude methanol extracts in all tests.

Genista tenera (Jacq. ex Murray) Kuntze Family: Fabaceae (Papilionaceae)

Rauter *et al*. (2009) evaluated the antioxidant activity of n-butanol, ethyl acetate and diethyl ether plant extracts of *Genistaa tenera* using DPPH assay. The best radical scavenging activity was observed for the ethyl acetate extract (48.7 per cent at 139.1µg/ml) which was also the most effective one at minimal concentration tested.

Genista vuralii A.Duran et H.Dural Family: Fabaceae (Papilionaceae)

Orhan *et al*. (2011) evaluated the antioxidant activity of the crude methanol and hydrolyzed extracts of *Genista vuralii* by three *in vitro* methods, namely DPPH free radical scavenging, ferrous ion-chelating and FRAP tests at 0.25, 0.50 and 1.0 mg/ml. The hydrolyzed extracts displayed greater antioxidant activity than the crude methanol extracts in all tests.

Gentiana decumbens L. Family: Gentianaceae

Myagmar and Aniya (2000) evaluated the free radical scavenging action of some medicinal herbs growing in Mongolia. The aqueous extract of nine herbs including *Gentiana decumbens* were used. The free radical scavenging action was determined *in vitro* and *ex vivo* by using ESR spectrometer and chemiluminiscence analyzer. The results showed that *Gentiana decumbens* extract possess strong scavenging action of DPPH, superoxide and hydroxyl radicals.

Gentiana lutea L. Family: Gentianaceae

English Name: Yellow gentian.

Kusar *et al*. (2006) investigated the free radical scavenging activity of methanolic extracts of yellow gentian leaves and roots in two different systems using electron spin resonance spectrometry. Assays were based on the stable free radical DPPH and the superoxide radicals generated by xanthine/xanthine oxidase system. This study proved that yellow gentian leaves and roots exhibited considerable antioxidant properties, expressed either by their capability to scavenge DPPH or superoxide radicals.

Gentiana macrophylla Pall. Family: Lardizabalaceae

Gan *et al*. (2010) evaluated the antioxidant activity of *Gentiana macrophylla* using the FRAP and TEAC assays, and its total phenolic content was measured by the

Folin-Ciocalteu method. For FRAP assay the value was 52.29 μmol Fe(II)/g dry weight. For TEAC assay, the value was 27.55μmol Trolox/g dry weight. Total phenolic content was 6.89 mg gallic acid equivalent/g dry weight of plant material.

Gentiana piasezkii Maxim. Family: Gentianaceae

An extract of *Gentiana piasezkii* afforded a new arbutin derivatives and a new flavone together with four known flavonoids. These compounds were evaluated for the antioxidant activity in the DPPH assay system (Wu *et al.*, 2006).

Gentiana scabra Bunge Family: Gentianaceae

Li *et al.* (2008) evaluated the antioxidant properties of 45 medicinal plants including *Gentiana scabra* using FRAP and TEAC assays and the total phenolic contents of these plants were measured. *Gentiana scabra* showed 41.82 μmol Fe(II)/g FRAP value, 23.40μmol Trolox/g of TEAC value and 8.99 mg GAE/g of Phenolic content. A high correlation between antioxidant capacities and their phenolic contents indicated that phenolic compounds were a major contributor of antioxidant activity of these plants.

Gentianella alborosea (Gilg) Fabris Family: Gentianaceae

Acero *et al.* (2006) tested the free radical scavenging (DPPH) activity of its methanol extract. The result showed a noticeable scavenging activity.

Gentianella austriaca (Kerner) Holub. Family: Gentianaceae

The main constituents of *Gentianella austriaca* aerial parts were gamma-pyrones and secoiridoids (Leskovac *et al.*, 2007). *Gentianella austriaca* possessed strong antioxidant properties, significantly reducing lipid peroxidation and the incidence of micronuclei while enhancing apoptosis with no perturbation of the cell cycle (Leskovac *et al.*, 2007).

Gentianella nitida Family: Gentianaceae

Lock *et al.* (2005) evaluated the *in vitro* antioxidant activity of 40 Peruvian plants using DPPH assay. *Gentianella nitida* showed highest antioxidant activity with EC_{50} value of 13.70μg/ml.

Geranium macrorrhizum L. Family: Geraniaceae

Miliauskas *et al.* (2004) investigated the radical scavenging activity of acetone, ethyl acetate and methanol extracts of leaves using DPPH and ABTS assays. They found that methanol extracts were most effective DPPH radical scavengers. Ethyl acetate and acetone extracts were considerably less effective radical scavengers. High content of phenolics in the extracts correlated with its antiradical activity, confirming that phenolic compounds were likely to contribute to the radical scavenging activity of this plant extracts.

Venskutonis *et al.* (2004) isolated seven compounds namely gallic acid, ellagic acid, 4-galloyl quinic acid, the flavonoid quercetin and three of its glycosides, quercetin-3-β-galactopyranoside and quercetin-4-β-glucopyranoside and quercetin-3-galactopyranoside and identified in the various fractions of *Geranium macrorrhizum*. Quercetin-4-β-glucopyranoside and quercetin-3-galactopyranoside showed the highest antioxidant capacity.

Geranium niveum S. Watson Family: Geraniaceae

Geranium niveum is a medicinal herb widely used by the Tarahumara Indians of Mexico. This species is rich in proanthocyanidins and other phenolics. Maldonado *et al*. (2005) evaluated the antioxidant activity of the extracts and two proanthocyanidins (geranins A and D) from the roots of *G. niveum* by using seven different assay systems, namely ABTS, DPPH, superoxide anion, hydrogen peroxide, hydroxyl radical, hypochlorous acid and singlet oxygen. The results showed that geranins A and D and the extracts were able to scavenge ABTS, DPPH, superoxide anion, hydroxyl radical and hypochlorous acid. The methanol-$CHCl_3$ (1:1) extract had a higher ability to scavenge ABTS, DPPH and O_2 radicals than the chloroform extract.

Geranium pratense L. Family: Caryophyllaceae

Myagmar and Aniya (2000) evaluated the free radical scavenging action of some medicinal herbs growing in Mongolia. The aqueous extract of nine herbs including *Geranium pratense* were used. The free radical scavenging action was determined *in vitro* and *ex vivo* by using ESR spectrometer and chemiluminiscence analyzer. The results showed that *Geranium pratense* extract possess strong scavenging action of DPPH, superoxide and hydroxyl radicals. *Geranium pratense* also depressed reactive oxygen production from polymorphonuclear leukocytes stimulated by phorbol-12-myristate *ex vivo*.

Geranium sanguineum L. Family: Geraniaceae

Murzakhmetova *et al*. (2008) investigated the antioxidant and prooxidant properties of a polyphenol-rich extract from *Geranium sanguineum in vitro* and *in vivo*. The polyphenol-rich extract protected biological membranes due to its antioxidant capacity and caused a dose-dependent decrease of the osmotic hemolysis of human erythrocytes and increased their resistance against the effect of H_2O_2; no effect on catalase activity was observed.

Geranium sibiricum L. Family: Geraniaceae

Antioxidant activity of glycoprotein isolated from *Geranium sibiricum* was carried by Shim and Lim (2009). The glycoprotein (500 µg/l) has antioxidative effects on hydroxyl and superoxide anion radicals in cell-free systems and the glycoprotein (200µg/l) significantly protected from cytotoxicity in the GO (100 mUml) treated Chang liver cells for 4 h.

The antioxidant capacity and xanthine oxidase inhibitory effects of extracts and main polyphenolic compounds of *Geranium sibiricum* were studied by Wu *et al*. (2010). The IC(50) values of the ethyl acetate fraction were 0.93, 3.32, 2.06, 2.66 and 1.64 µg/ml in the DPPH radical scavenging, superoxide radical scavenging, nitric oxide scavenging, β-carotene-linoleic acid bleaching and reducing power assays, respectively. Of the polyphenolic compounds separated from the ethyl acetate fraction, geranin showed a higher activity than corilagin and gallic acid. The IC_{50} values ranged from 0.87 to 2.53µM, which were even lower than the positive control (except for allopurinol).

Geranium wilfordii Maxim. Family: Geraniaceae

Gan *et al.* (2010) evaluated the antioxidant activity of *Geranium wilfordii* using the FRAP and TEAC assays, and its total phenolic content was measured by the Folin-Ciocalteu method. For FRAP assay the value was 347.33 µmol Fe(II)/g dry weight. For TEAC assay, the value was 215.98µmol Trolox/g dry weight. Total phenolic content was 14.98 mg gallic acid equivalent/g dry weight of plant material.

Geum quellyon Sweet Family: Rosaceae

Extract of *Geum quellyon*, containing tannins, exhibits antioxidant properties, hydrazyl radical (DPPH) and superoxide anion, inhibit xanthine oxidase activity, chelate metals and radicals (OH) and nitric oxide (NO) (Russo *et al.*, 2005).

Geum triflorum Pursh. Family: Rosaceae

Borchardt *et al.* (2008) evaluated the antioxidant activity of methanol extracts of seeds of *Geum triflorum* using DPPH antioxidant assay.The antioxidant value reported in µM Trolox/100g (TE) from DPPH radical scavenging activity of crude seeds extract was 55,723.

Gevunia avellana Mol. Family: Proteaceae

English Name: Chilean Hazelnut.

Chilean Hazelnuts (*Gevunia avellana*) meals were extracted with methanol, ethanol, acidified water, acetone, butanol, diethyl ether and ethyl acetate by Moure *et al.* (2001). The highest concentration of total polyphenols was found in the ethanolic extracts. The antioxidant activity of the extracts evaluated by the β-carotene assay and with hydrogen radical scavenging ability showed that the activity of the butanol and methanol extracts from *G. avellana* was comparable to those of synthetic antioxidants BHA and BHT. The DPPH radical scavenging activity of ethanol and water extracts was 2-3 times lower than those of BHT and BHA, respectively.

Gisekia pharnaceoides L. Family: Aizoaceae

Singh *et al.* (2009) evaluated the antioxidant activity of dichloromethane and methanolic extracts of 12 arid zone medicinal plants. *Gisekia pharnaceoides* showed appreciable antioxidant activity and per cent of inhibition of DPPH where RC_{50} (µg/ml) was 15 (Dichloromethane extract) and 7.5 (MeOH extract). per cent of inhibition of DPPH was 83.45 per cent in methanol extract of *Gisekia pharnaceoides*. Key phytochemicals include myristone, triacontane and tetracosanol.

Glaucium contortuplicatum Boiss. Family: Papaveraceae

Souri *et al.* (2004) evaluated the antioxidative activity of *Glaucium contortuplicatum* by linoleic acid peroxidation test using 1,3-diethyl-2-thiobarbituric acid as the reagent. Methanolic extract showed 55.18 per cent inhibition at 40µg concentration with IC_{50} value of 8.68µg.

Glaucium elegans Fisch. et C.A.Mey. Family: Papaveraceae

Souri *et al.* (2004) evaluated the antioxidative activity of *Glaucium elegans* by linoleic acid peroxidation test using 1,3-diethyl-2-thiobarbituric acid as the reagent.

Methanolic extract showed 61.76 per cent inhibition at 40µg concentration with IC_{50} value of 14.20µg.

Glaucium fimbrilligerum Boiss. Family: Papaveraceae

Souri *et al.* (2004) evaluated the antioxidative activity of *Glaucium fimbrilligerum* by linoleic acid peroxidation test using 1,3-diethyl-2-thiobarbituric acid as the reagent. Methanolic extract showed 89.16 per cent inhibition at 40µg concentration with IC_{50} value of 4.19µg.

Gleditsia japonica Nakai Family: Caesalpiniaceae

Methanol aqueous extract of *Gleditsia japonica* was screened by Kim *et al.* (1997) for free radical scavenging activity using the DPPH free radical generating system. The extract showed free radical scavenging activity.

The aqueous extracts of 13 oriental medicinal plants were examined for their reducing power, scavenging ability towards superoxide and hydroxyl radicals and their inhibitory effect on lipid peroxidation. The reducing power of Cortex Gleditsiae was more apparent than with other plants (Nam and Kang, 2004). All extracts tested were found to be highly active on scavenging of superoxide radicals.

Gleditsia sinensis Lam. Family: Caesalpinaceae

The aqueous extract of *Gleditsia sinensis* thorns inhibited LPS-induced NO secretion as well as inducible nitric oxide synthase expression in macrophages, without affecting cell viability (Ha *et al.*, 2008).

Antioxidant capacities of 56 selected Chinese medicinal plants, including *Gleditsia sinensis* were evaluated by Song *et al.* (2010) using the TEAC and FRAP assays, and their total phenolic content was measured by the Folin-Ciocaleteu method. *Gleditsia sinensis* showed low antioxidant activity with TEAC value of 54.14µmol Trolox/g, FRAP value of 26µmol Fe^{2+}/g and phenolic content was 6.68mg GAE/g.

Glehnia littoralis F. Schmidt Family: Apiaceae (Umbelliferae)

Ng *et al.* (2004) investigated the antioxidant effect of aqueous and organic extracts of roots of *Glehnia littoralis*. The aqueous extracts were the most potent in inhibiting erythrocyte hemolysis. The organic and aqueous extracts were potent in inhibiting lipid peroxidation.

Glinus oppositifolius (L.) A.DC. Family: Molluginaceae

Diallo *et al.* (2001) demonstrated with DPPH assay that *Glinus oppositifolius* possessed potent antioxidant activity.

Asok Kumar *et al.* (2009) investigated the free radical scavenging and antioxidant activities of *Glinus oppositifolius* using different *in vitro* assay systems which includes H-donor activity, nitric oxide scavenging, superoxide anion scavenging, reducing ability, hydroxyl radical, hydrogen peroxide scavenging, total phenolic content, total flavonoid content, total antioxidant activity by thiocyanate and phosphomolybdenum method, metal chelating and β-carotene bleaching, total peroxy radical assays. The generation of free radicals were effectively scavenged by the ethanol extract of *Glinus oppositifolius*. The antioxidant activity depends on concentration and increases with increasing amounts of the extract.

Behera *et al.* (2010) evaluated the antioxidant activity of methanol and aqueous extracts of *Glinus oppositifolius*. It was found that the extracts showed significant percentage of inhibition. In DPPH assay the extracts showed significant percentage of inhibition when compared with standard BHT.

Gliricidia sepium Jacq. Family: Fabaceae (Papilionaceae)

Ruiz-Terán *et al.* (2008) evaluated the antioxidant activity of aerial parts of *Gliricidia sepium* using DPPH and β-carotene-linoleic acid bleaching assay. The extract displayed antioxidant activity comparable to the commercial antioxidant BHA. For the methanolic extract analysed, a clear relation between the total phenolic content of the extracts and their antioxidant activity was found.

Globimetula cupulatum (DC.) Danser Family: Loranthaceae

Akinmoladun *et al.* (2010) assessed the phytochemical constituents and antioxidant and free radical scavenging activities of *Globimetula cupulatum* using seven different antioxidant assay methods. The extract had low nitric oxide radical scavenging activity and highest hydroxyl scavenging activity (63.84 per cent). The extract demonstrated high lipid peroxidation inhibitory activity. The results suggest that the methanolic extracts possess significant antioxidant and radical scavenging activity that may be due to the phytochemical content of the plant.

Globularia alypum L. Family: Globulariaceae

Khlifi *et al.* (2005) investigated the antioxidant activity of the hydromethanolic extract of aerial parts (leaves and stems) of *Globularia alypum* toward linoleic acid emulsion and human LDL peroxidation. The hydromethanolic extract of *Globularia alypum* exhibited significant antioxidant effect. There was a significant inhibition of conjugated dienes formation in copper ions mediated linoleic acid emulsion as well as human LDL peroxidation. Analysis of the plant extract revealed a high amount of polyphenols, suggesting a possible role of these compounds in the antioxidant properties.

The antioxidant activity of *Globularia alypum* phytochemicals were evaluated for their capacity to scavenge DPPH free radical and some structure-activity relationships were obtained (Es Safi *et al.*, 2007). Assay guided fractionation led to the isolation of syringin, four phenyl ethanoidss, four flavonoids and six iridoids as the main constituents of the extract and their antioxidant activity was determined. The obtained results showed that the activity towards the DPPH free radical was mainly due to the flavonoid and phenyl ethanoid constituents which were most active free radical scavengers than iridoids. Among the tested flavonoids 6-hydroxyluteolin glycosides showed the strongest activity, suggesting that the presence of the 6-hydroxy group was a favourble structural feature of flavonoids with regard to DPPH scavenging effect. The isolated phenyl ethanoid glycosides all showed potent antioxidant activity and their capacity to scavenge free DPPH radical was greater than BHT. Their high antioxidant activity could be attributed to the caffeoyl moieties contained in them, while iridoids showed moderate free radical scavenging activity.

Harzallah *et al.* (2010) investigated the antioxidant potentials of extracts of *Globularia alypum* leaves with FRAP and Reducing Power (RP) assays. Their results showed significant antioxidant capacity by the FRAP and RP methods. Aqueous extract showed the highest antioxidant capacity with a value of 8.9 mM TE, 4.5mM TE by FRAP and RP method respectively.

Glochidion wallichianum Muell.-Arg. Family: Euphorbiaceae

The phenolic compounds of five southern Thai indigenous vegetables including Mon-pu (*Glochidion wallichianum*) leaf were extracted using different solvents. The extracts were analyzed for total phenolic content using the Folin-Ciocalteu procedure, free radical scavenging capacity by using ABTS and DPPH methods, and reducing capacity by using FRAP assay. The acetone extract possessed the highest total phenolic content followed by methanolic and distilled water extracts, respectively. The acetone extract showed higher free radical scavenging capacity and reducing capacity than those of their methanolic and water extracts counterparts respectively. Among all plants tested, the extracts of Mon-pu extracted with all extracting media exhibited the highest free-radical scavenging and reducing abilities (Wu *et al.*, 2010).

Glossocardia bosvallea (L.f.) DC (Syn.: *Verbesina bosvallea* L.f.) Family: Asteraceae (Compositae)

Prasanna Anjaneya Reddy *et al.* (2011) reported that essential has shown antioxidant activity with 50 per cent radical scavenging activity in the concentration of 20µl/ml.

Glycine max Merr. Family: Fabaceae (Papilionaceae)

English Name: Soybean, black bean.

Lin *et al.* (2001) evaluated the anti-lipid peroxidation activity, free radical scavenger activity and anti-superoxide formation of hot water extracts of legumes including *Glycine max* to test the antioxidant activity. The results showed that the extract exhibited remarkable inhibition of $FeCl_2$-ascorbic acid-induced lipid peroxidation of mouse liver homogenate. Extract showed anti-lipid peroxidation activity and anti-superoxide formation.

Antioxidant activities of soybean extracts were measured by Lee *et al.* (2004) using DPPH free radical and photochemiluminescence (PCL) methods. The highest and lowest isoflavone contents were 11.75 and 4.20 µmol/g soy, respectively, while the average was 7.12 µmol/g soy. Antioxidant activities of soybean extracts ranged from 7.51 to 12.18 µmol butylated hydroxytoulene equivalent/g soy using the DPPH method. Lipid and water soluble antioxidant activities of soybean extracts ranged from 2.40 to 4.44 µmol Trolox equivalent/g soy and from 174.24 to 430.86 µmol ascorbic acid equivalent/g soy, respectively, using the PCL method.

Cho *et al.* (2008) evaluated the antioxidant activity of *Glycine max*. Water extract increased the induction of IL-6, TNF and nitric oxide on RAW 264.7 cells.

Sreeramulu *et al.* (2009) evaluated the antioxidant activity of soybean by DPPH, scavenging assay, FRAP assay and reducing power. Phenolic content of the whole seed was 100.54 mg/100g. DPPH radical scavenging activity expressed as trolox equivalent, FRAP (µmol/g) and reducing power (mg/g) were 0.97, 131 and 1.52, respectively in soybean.

Glyceria grandis S.Wats. Family: Poaceae

English Name: Reed Manna grass.

Borchardt *et al.* (2008) evaluated the antioxidant activity of methanol extracts of seeds of *Glyceria grandis* using DPPH antioxidant assay.The antioxidant value reported in µM Trolox/100g (TE) from DPPH radical scavenging activity of crude seeds extract was 70,940.

Glyceria striata S.Wats. Family: Poaceae

English Name: Fowl Manna grass.

Borchardt *et al.* (2008) evaluated the antioxidant activity of methanol extracts of seeds of *Glyceria striata* using DPPH antioxidant assay.The antioxidant value reported in µM Trolox/100g (TE) from DPPH radical scavenging activity of crude seeds extract was 67,707.

Glycyrrhiza glabra L. Family: Fabaceae (Papilionaceae)

English Name: Licorice.

From the air-dried roots of *G. glabra* five new flavonoid compounds named glucoliquiritin apioside (a flavonone bisdesmoside), phenyllicoflavone A (a bisprenylflavone), shinflavone (a prenylated pyranoflavanone), shinpterocarpin and 1-methoxyphaseollin (both pyranopterocarpans), were isolated together with 8 known saponins, 7 known flavonoid glycosides, and 11 flavonoids (Kittagawa *et al.*, 1994). The structures of the new potent antioxidant compounds have been elucidated on the basis of their chemical and physicochemical properties.

The study by Vaya *et al.* (1997) also analyzed the antioxidative properties of natural compounds from the root of licorice toward LDL oxidation. Seven constituents, with antioxidant capacity were isolated from *Glycryrrhiza glabra*. The isolated compounds were identified as the isoflavans hispaglabridin A, hispaglabridin B, glabridin, and 4'-O-methylglabridin, the two chalcones, isoprenylchalcone derivative and isoliquiritigenin, and the isoflavone, formononetin. Among these compounds, glabridin constituted the major amount in the crude extract (11.6 per cent w/w) detected by HPLC analysis. The anioxidative capacities of the isolated compounds (1-7) were tested against β-carotene destruction and LDL oxidation. These results suggest that the first 6 constituents were very potent antioxidants with glabridin being the most abundant and potent antioxidants. As LDL oxidation is a key event in the formation of the early atherosclerotic lesion, the use of these natural antioxidants may prove beneficial to attenuate atherosclerosis.

Glabridin, an isoflavan isolated from *Glycyrrhza glabra* (licorice) root, and its derivatives have been reported to inhibit the oxidation of LDL induced by copper ions or mediated by macrophages (Belinky *et al.*, 1998a). The antioxidant effect of glabridin on LDL oxidation appears to reside mainly in the 2'hydroxyl moiety of the isoglavan. The effect of the consumption of glabridin, on the susceptibility of LDL to oxidation was also studied in atherosclerotic apolipoprotein E deficient mice. The *in vivo* and *in vitro* inhibiltory activities may be related to the absorption of binding of glabridin to LDL particle and subsequent protection of LDL from oxidation and by protecting LDL associated carotenoids (Belinky *et al.*, 1998b).

Naik *et al.* (2003) evaluated the antioxidant activity of aqueous extract of *Glycyrriza glabra* for its potential as antioxidant. The antioxidant activity of these extracts was tested by studying the inhibition of radiation induced lipid peroxidation in liver microsomes at different doses in the rages of 100-600 Gy as estimated by TBARS. Naik *et al.* (2003) evaluated the antioxidant activity of aqueous extract of *Glycyrrhiza glabra* for its potential as antioxidant. The antioxidant activity of these extracts was tested by studying the inhibition of radiation induced lipid peroxidation in liver microsomes at different doses in the rages of 100-600 Gy as estimated by TBARS.

Fukai *et al.* (2003a) evaluated antinephritis and radical scavenging activities of glabridin from *Glycyrrhiza glabra*. ESR spectroscopy demonstrated that glabridin neither produced radical, nor affected the radical intensity of sodium ascorbate, suggesting the lack of correlation between the antinephritic activity and radical scavenging activity.

Biondi *et al.* (2004) isolated new dihydrostilene derivatives from the leaves of *Glycyrrhiza glabra*. They evaluated the antioxidant activity of the crude extracts and the isolated compounds.

Souri *et al.* (2004) evaluated the antioxidative activity of *Glcyrrhiza glabra* var. *glabra* by linoleic acid peroxidation test using 1,3-diethyl-2-thiobarbituric acid as the reagent. Methanolic extract showed 96.14 per cent inhibition at 40µg concentration with IC_{50} value of 2.86µg.

Hispaglabridin B, isoliquiritigenin, and paratocarpin B isolated from the roots and stolons of licorice were found to be the most potent antioxidant agents (Chin *et al.*, 2007).

Wojcikowski *et al.* (2007) evaluated the antioxidant activity of aerial parts of *Glycyrrhiza glabra*. Oxygen radical absorbance capacity of the ethyl extract, methanol extract and aqueous extracts were 196.44, 416.93 and 416.08µmol Trolox equivalent per g of dried starting material, respectively.

Glycyrrhiza inflata Batalin Family: Fabaceae (Papilionaceae)

Licochalcone A, B, C, D and echinatin, retrochalcones isolated from the roots of *Glycyrrhiza inflata* (licorice), along with an ordinary chalcone isoliquiritigenin, were assessed by Haraguchi *et al.* (1998) for their inhibitory activities on lipid peroxidation in various systems and radical scavenging activity. Among those tested, licochalcones B and D strongly inhibited superoxide anion production in the xanthine/xanthine oxidase system. These two compounds also showed potent scavenging activity on DPPH radical. Microsomal lipid peroxidation induced by Fe(III)-ADP/NADPH was inhibited almost completely by 3 micrograms/ml of licochalcones B and D. Mitochondrial lipid peroxidation induced by Fe(III)-ADP/NADPH was more sensitive to these retrochalcones; almost complete inhibition was observed at 10micrograms/ml of all retrochalcones tested. Licochalcones B and D scavenged superoxide anion in microsome. Furthermore, these retrochalcones protected red cells against oxidative hemolysis. These phenolic compounds were shown to be effective to protect biological systems against various oxidative stresses.

Antinephritic activity of prenylflavonoids similar to glabridin, isolated from *Glycyrrhiza inflata* was evaluated in mice with glomerular disease. Licochalcone A exhibited a weak scavenging activity against superoxide anion radical (Fukai *et al.*, 2003b).

Glycyrrhiza lepidota (Nutt.) Pursh. Family: Fabaceae (Papilionaceae)

English Name: Wild licorice.

Ethanolic extracts of roots of wild licorice were prepared and evaluated by Amarowicz *et al.* (2004) for their free-radical scavenging capacity and their antioxidant activity, by a number of chemical assays. Assays employed included β-carotene linoleic acid (lineolate) model system, reducing power, scavenging effect on the DPPH free radical and capacity to scavenge hydroxyl free radical (HO), by use of electron paramagnetic resonance (EPR) spectroscopy.

Glycyrrhiza uralensis Fisch. Family: Fabaceae (Papilionaceae)

Cai *et al.* (2004) evaluated the antioxidant activity and phenolic compounds in traditional Chinese medicinal plants associated with anticancer. The improved ABTS.+ method was used to systematically assess the total antioxidant capacity (Trolox equivalent antioxidant capacity, TEAC) of the medicinal extracts. The TEAC values and total phenolic content for methanolic extract of root of *Glycyrrhiza uralensis* were 545.9 µmol Trolox equivalent/100g dry weight (DW), and 2.92 g of gallic acid equivalent/100g DW respectively. A positive significant linear relationship between antioxidant activity and total phenolic content showed that phenolic compounds were the dominant antioxidant components in the tested herbs. Major types of phenolic compounds in *Glycyrrhiza uralensis* include flavonones (dihydroflavones, liquiritin, liquirigenin), licopyranocoumarins.

Antinephritis activity of prenylflavonoids similar to glabridin, isolated from *G. uralensis*, was evaluated in mice with glomerular disease. Licoricidin and licorisoflavan A showed weak scavenging activity against superoxide anion radical.

Glyphea brevis (Spreng.) Atori Family: Tiliaceae

Mamyrbekova-Bekro *et al.* (2008) evaluated the phytocompounds of the extract and assessed their antioxidant potential of *Glyphea brevis* leaves and root. The extracts showed the presence of polyphenols, flavonoids, tannins and coumarins. The extracts showed DPPH scavenging activity. This activity is probably due to the presence of the phenolic compounds, flavonoids and tannins.

Gmelina arborea Roxb. Family: Verbenaceae

Pandey *et al.* (2010) evaluated antioxidant activity of *Gmelina arborea* extracts by *in vitro* techniques. DPPH radical scavenging activity was IC_{50} = 35 µg/ml for methanolic extract and 100µg/ml for aqueous extract. IC_{50} value for ascorbic acid was 16µg/ml. Methanolic and aqueous extracts showed moderately good nitric oxide-scavenging activity. The percentage of inhibition increased with increasing concentration of the extract. IC_{50} was calculated as 60µg/ml and 140µg/ml while standard ascorbic acid has IC_{50} = 18µg/ml.

Gmelina hystrix Schult. ex Kurz. Family: Verbenaceae

Antioxidant activity of *Gmelina hystrix* and four other species have been investigated by Adib *et al.* (2010) by DPPH assay. Methanol extract of *Gmelina hystrix* was found to have the moderate antioxidant activity with an IC_{50} value of 170µg/ml.

Gnaphalium obtusifolium L. Family: Asteraceae

Borchardt *et al.* (2008) evaluated the antioxidant activity of methanol extracts of seeds of *Gnaphalium obtusifolium* using DPPH antioxidant assay.The antioxidant value reported in µM Trolox/100g (TE) from DPPH radical scavenging activity of crude seeds extract was 57,331.

Gongronema latifolium Benth. Family: Asclepiadaceae

Using streptozotocin-induced non-insulin dependent diabetes (NIDD) rat model, Ugochukwu and Babady (2002), demonstrated that aqueous and ethanolic extracts of *Gongronema latifolium* leaves harvested from Eastern Nigeria significantly increased the activity of superoxide dismutase and the level of reduced glutathione peroxidase as well as the levels of glutathione peroxidase and glucose-6-phosphate, while decreasing lipid peroxidation. Based on these observations, these authors concluded that the diabetic activities of the plant could be mediated through its antioxidant properties.

Odukoya *et al.* (2007) evaluated the antioxidant activity and Total Phenolic content of Nigerian green leafy vegetables including *Gongronema latifolium*.

Goniothalamus hookeri Thwaites Family: Annonaceae

Puvanendran *et al.* (2007) evaluated the antioxidant activity of Dichloromethane (DCM) and MeOH extract of stem bark and leaves of *Goniothalamus hookeri* using DPPH assay. The extract showed moderate antioxidant activities. Percentage of inhibition of stem bark was 12.43 (MeOH extract); leaf extract was 14.03 (MeOH extract).

Gontscharovia popovi (B.Fedsch. et Gontsch.) A.Boriss. Family: Lamiaceae (Labiatae)

Firuzi *et al.* (2010) evaluated the antioxidant activities of methanolic extracts of 24 Lamiaceae species growing in Iran using FRAP and DPPH free radical scavenging assays. *Gontscharova popovii* showed highest antioxidant activity with 50.3µM/g DW FRAP value and IC_{50} = 315.3µg/ml of DPPH radical scavenging activity. It also showed highest phenolic content of 16.2 mg CE/g. FRAP and DPPH assay results showed good correlations with the total phenolic contents of the plants, measured by the Folin-Ciocalteau assay.

Gordonia axillaris (Roxb. ex Ker.-Gawl.) Endl. Family: Theaceae

Fu *et al.* (2010) evaluated the antioxidant capacity and total phenolic contents of 56 wild fruits from South China using FRAP and TEAC assays. A high correlation between antioxidant capacity and phenolic compounds could be the main contributor to the measured antioxidant activity. The results showed that fruits of *Gordonia axillaris* possessed the highest antioxidant capacity and total phenolic contents.

Gossypium arboreum L. Family: Malvaceaae

Annan and Houghton (2008) investigated the antioxidant activity of *Gossypium arboreum*. Aqueous extracts significantly protected fibroblast cells against oxidative damage a doses up to 50µg/ml.

Graptopetalum paruguayense E. Walther Family: Crassulaceae

Chung *et al.* (2005) evaluated the antioxidant activity of water, 50 per cent ethanolic and 95 per cent ethanolic extracts of *Graptopetalum paruguyense*. The antioxidant activities, including radical scavenging effect, reducing power and antioxidant effect on Fe/ascorbate-induced lipid peroxidation in a liposome model system were studied *in vitro*. The results showed that all the extracts possessed antioxidant characteristics including radical scavenging, reducing power and lipid peroxidation inhibition. It was found that the antioxidative activities of all the extracts increased with increasing concentrations and the activities correlated with both total phenol and anthocyanin contents.

Grewia nervosa (Lour.) Panigrahi Family: Tiliaceae

Methanol extract of twig showed potent antioxidant activity (Kshirsagar and Upadhyay, 2009). The extract has the same activity profile as that of Curcumin. Its ability of scavenging radicals was measured by DPPH reduction spectrophotometric assay.

Grewia sapida Roxb. ex DC. Family: Tiliaceae

Kshirsagar and Upadhyay (2009) reported antioxidant activity of the methanol extract of twig. Its ability of scavenging radicals was measured by DPPH reduction spectrophotometric assay. The extract has the same activity profile as that of Curcumin.

Grewia tiliaefolia Vahl Family: Tiliaceae

Khadeer Ahmed *et al.* (2010) investigated *in vitro* antioxidant and *in vivo* prophylactic effects of two gamma-lactones isolated from *Grewia tiliaefolia* against hepatotoxicity in carbon tetrachloride intoxicated rats. To clarify the influence of the methanolic extract and isolated constituents on the protection of oxidative-hepatic damage, Khadeer Ahmed *et al.* (2010) examined *in vitro* antioxidant properties of the test compounds. The extract and the constituents showed significant free radical scavenging activity.

Karuppusamy *et al.* (2011) evaluated the anthocyanins, ascorbic acid, total phenolics and flavonoids and antioxidant activity of methanol extracts of fruits of *Grewia tiliaefolia*. The anthocyanins were 2.60 CGE/100g, Ascorbic acid 70.5 AAE/100g, Total phenolics 44.1GAE/100g, Total flavonoids 47.1QE/100g and antioxidant activity (DPPH) 126.2µg/ml.

Grindelia robusta Nutt. Family: Asteraceae (Compositae)

Chemical composition of the essential oil obtained from *Grindelia robusta* aerial parts was analyzed by Fraternale *et al.* (2007). The antioxidant activity of the essential oil was evaluated using the DPPH and 5-lipoxygenase tests.

Guiera senegalensis J.F.Gmel. Family: Combretaceae

The antioxidant properties of nine tannins isolated and characterized from different parts of *Guiera senegalensis* were evaluated by Bouchet *et al.* (1998). Interesting results showed that galloylquinic acids (hydrolysable tannins), resulting from a tri- or tetra-substitution of galloyl groups on the quinic acid skeleton played a crucial role in the inhibitory effect on Fe^{2+}-induced lipid peroxidation in rat liver microsomes and a radical scavenger activity in the DPPH rest. The effect of all tannins were markedly higher than that of gallic acid. Condensed tannins such as epicatechin and epigallocatechin gallate also showed fairly significant effects in both tests.

Guizotia abyssinica Cass. Family: Asteraceae (Compositae)

English Name: Niger seed

Shahidi *et al.* (2003) evaluated the antioxidant activity of defatted seeds of niger using β-carotene-lineolate and a meat model system. Three solvent systems, A [80:20 (vol/vol) ethanol : water], B (80: 20 (vol/vol) acetone : water] and C (water) were evaluated as extraction media. Extract A exhibited superior antioxidant activity, compared to extracts B and C. Phytochemical analysis revealed that the antioxidant activity, at least in part, was from the chlorogenic acid-related compounds.

Gundelia tournefortii L. Family: Asteraceae (Compositae)

Coruh *et al.* (2007) evaluated the antioxidant capacities of aerial parts and seeds of *Gundelia tournefortii* by using DPPH radical scavenging and lipid peroxidation inhibition methods. The seeds were found to have higher antioxidant potential than the aerial parts with IC_{50} values of 0.073mg/ml for DPPH scavenging and 0.146 mg/ml for lipid peroxidation inhibition capacities. In addition, total phenolic contents of the *Gundelia tournefortii* extracts, especially the seed extracts correlates to its high antioxidant activity with 105.5μg gallic acid equivalents (GAEs) per mg of seed extract.

Gymnema inodorum (Lour.) Decne (Syn.: *Gymnema tingens* Roxb. ex Spreng.) Family: Asclepiadaceae

Chanwitheesuk *et al.* (2005) investigated the antioxidant activity of *Gymnema inodorum* using a β-carotene bleaching method. The contents of plant chemicals such as vitamin C, Vitamin E, carotenoids, tannin and total phenolics, were also determined. Methanolic extract showed antioxidant activity. Highest antioxidant activity and highest amount of vitamin E was found in *Gymnema inodorum*. It has an antioxidant index of 14.8, Vitamin C 19.3 mg per cent, Vitamin E 0.0301mg per cent, Total carotenes 1.31 mg per cent, Total xanthophylls 1.07mg per cent, Tannins 11.1mg per cent and Total phenolics of 188 mg per cent.

Gymnema montanum Hook.f. Family: Asclepiadaceae

G. montanum ethanolic leaf extract possess antihyperglycemic and antiperoxidative effect. It has been reported to be rich in gymnemagenin and gymnemic acids that are responsible for antihyperglycemic effect (Murakani *et al.*, 1996). The decrease in lipid peroxides increase in reduced glutathione, ascorbic acid (vitamin C) and α-tocopherol (vitamin E). In the study by Ananthan *et al.* (2003) has shown treatment of diabetic rats with leaf extract increased the antioxidant levels.

Gymnema sylvestre (Retz.) R.Br. ex Schult. Family: Asclepiadaceae

In order to find their potential sources, the crude extracts of different parts of some selected food and medicinal plants, including *Gymnema sylvestre*, were studied by Bajpai *et al.* (2005) for total phenolic contents (TPC) and antioxidant activity (AOA). AOA was assayed by auto-oxidation of β-carotene and linoleic acid and expressed as the per cent of inhibition relative to control. TPC in leaves of *Gymnema sylvestre* was 10.8 mg/g (GAE) while antioxidant activity was 35.0 per cent.

Biswas *et al.* (2010) analyzed antioxidant activity of the aqueous extract of leaves of *Gymnema sylvestre* in different systems of assay. Aqueous extract showed Superoxide radical scavenging activity with $IC_{50} = 143.02 \mu g/ml$; Hydroxyl radical scavenging $IC_{50} = 6538.43 \mu g.ml$; Lipid peroxidation $IC_{50} = 1390.07 \mu g/ml$.

Gynostemma pentaphyllum (Thunb.) Mak. Family: Cucurbitaceae

Cai *et al.* (2004) evaluated the antioxidant activity and phenolic compounds in traditional Chinese medicinal plants associated with anticancer. The improved ABTS.+ method was used to systematically assess the total antioxidant capacity (Trolox equivalent antioxidant capacity, TEAC) of the medicinal extracts. The TEAC values and total phenolic content for methanolic extract of whole plant of *Gynostemma pentaphyllum* were 84.6 µmol Trolox equivalent/100g dry weight (DW), and 0.44 g of gallic acid equivalent/100g DW respectively. A positive significant linear relationship between antioxidant activity and total phenolic content showed that phenolic compounds were the dominant antioxidant components in the tested herbs. Major types of phenolic compounds in *Gynostemma pentaphyllum* include flavonoids (rutin).

Wang and Luo (2007) evaluated the antioxidant activity of different fractions of polysaccharide purified from *Gynostemma pentaphyllum*.

Chemical investigations of the EtOH extract of *Gynostemma pentaphyllum* led to the isolation of 3 lignans, named compounds I, II and III (Wang *et al.*, 2010). Compound III showed stronger DPPH free radical scavenging activity when compared to BHT at the same concentrations, while compound I had only very little DPPH free radical scavenging and compound II failed to scavenge DPPH free radicals within the experimental concentration range. All compounds had a weaker ability to chelate Fe^{2+} when compared to EDTA.

Gynoxys psilophylla Klatt. Family: Asteraceae (Compositae)

The aerial parts of 17 Bolivian plants including *Gynoxys psilophylla* were screened by Rosas-Romero and Saavedra (2005) to determine antioxidant activity. A methanol extract of each plant was prepared and portioned sequentially with hexane, chloroform, and ethyl acetate, leaving an aqueous solution. All extracts and their 5 fractions for a total of 102 samples, were evaluated using two techniques: an adaptation of the β-carotene bleaching technique using an emulsion of linoleic acid in water as the oxidizable substrate, and the DPPH free radical trapping technique. The results of the β-carotene bleaching technique were more discriminating and better related to the the rancidity process under normal conditions; with this assay, 11 species provided at least one fraction with highly promising antioxidant activity. All species gave good results under the DPPH technique, and in most cases they performed better than BHA, which was used as a reference antioxidant (Rosas-Romero and Saavedra, 2005)

Gynura bicolor DC. Family: Asteraceae (Compositae)

English Name: Gynura; Japanese name: Suizenjina.

Maeda *et al.* (2006) examined the phenolic contents of leaf of *Gynura bicolor* and their radical scavenging activities by using DPPH assay. High contents of phenolic compounds and strong DPPH scavenging activities were observed in extracts from this species. They also showed radical scavenging activities for the superoxide radical and tert-butylhydroperoxyl radical (t-BuOO·).

Chen *et al.* (2009) studied the antioxidant activity of leaves extract towards xanthine oxidase, lipoxygenase and tyrosinase activities. The inhibitory effect of acetone extract on xanthine oxidase activity was less than 10 per cent, tyrosine activity was almost less than 10 per cent and lipoxygenase activity was about 30 per cent.

Gynura formosana Kimura Family: Asteraceae (Compositae)

The free-radical scavenging activities of a 70 per cent aqueous acetone extract from the herb *Gynura formosana* were evaluated by Hou *et al.* (2005). Bioassay guided fractionation led to the isolation of phenolics caffeic acid (1), quercetin 3-O-rutinoside (2), kaempferol-3-O-rutinoside (3) and kaempferol 3-O-robinobioside (4). The IC_{50} values of DPPH radical scavenging activity for compounds 1-4 were 6.7, 7.7, 300.3 and 286.7µM, respectively, and for superoxide radical scavenging activity, they were 187.3, 25.8, 55.3 and 87.4µM, respectively. Using a spin trapping ESR method, caffeic acid (1) and quercetin-3-O-rutinoside (2) exhibited good hydroxyl radical activity.

Gynura procumbens Merr. Family: Asteraceae (Compositae)

The effect of extraction temperature on total phenolic contents and free radical scavenging activity of *Gynura procumbens* leaf extract was investigated by Akowuah *et al.* (2009). The extracts obtained at lower temperature exhibited significant free radical scavenging activity compared to extraction at higher temperatures.

Gypsophila eriocalyx Boiss. Family: Caryophyllaceae

Serteser *et al.* (2009) studied the antioxidant activity of 50 per cent aqueous methanol extracts by various antioxidant assays, including free radical scavenging, hydrogen peroxide scavenging and metal (Fe^{2+}) chelating activities. The DPPH radical scavenging effects of leaves extract was 0.272. Fe^{2+} chelating activity (per cent) of leaves extract was 17.34 and Hydrogen peroxide inhibition activity of leaves extract was 18.64 per cent.

Gypsophila parva Barkoudah Family: Caryophyllaceae

Serteser *et al.* (2009) studied the antioxidant activity of 50 per cent aqueous methanol extracts by various antioxidant assays, including free radical scavenging, hydrogen peroxide scavenging and metal (Fe^{2+}) chelating activities. The DPPH radical scavenging effects of leaves extract was 0.290. Fe^{2+} chelating activity (per cent) of leaves extract was 19.32 and Hydrogen peroxide inhibition activity of leaves extract was 25.43 per cent.

Gypsophila perfoliata L. Family: Caryophyllaceae

Serteser *et al.* (2009) studied the antioxidant activity of 50 per cent aqueous methanol extracts by various antioxidant assays, including free radical scavenging, hydrogen peroxide scavenging and metal (Fe^{2+}) chelating activities. The methanolic extract examined in the assay showed the strongest activities. The DPPH radical scavenging effects of leaves extract was 0.320. Fe^{2+} chelating activity (per cent) of leaves extract was 20.32 and Hydrogen peroxide inhibition activity of leaves extract was 32.65 per cent.

Gypsophila pilosa Hudson Family: Caryophyllaceae

Serteser *et al.* (2009) studied the antioxidant activity of 50 per cent aqueous methanol extracts by various antioxidant assays, including free radical scavenging, hydrogen peroxide scavenging and metal (Fe^{2+}) chelating activities. The methanolic extract examined in the assay showed the strongest activities. The DPPH radical scavenging effects of leaves extract was 0.258. Fe^{2+} chelating activity (per cent) of leaves extract was 16.34 and Hydrogen peroxide inhibition activity of leaves extract was 21.54 per cent.

Gypsophila tubulosa Boiss. Family: Caryophyllaceae

Serteser *et al.* (2009) studied the antioxidant activity of 50 per cent aqueous methanol extracts by various antioxidant assays, including free radical scavenging, hydrogen peroxide scavenging and metal (Fe^{2+}) chelating activities. The methanolic extract examined in the assay showed the strongest activities. The DPPH radical scavenging effects of leaves extract was 0.323. Fe^{2+} chelating activity (per cent) of leaves extract was 18.51 and Hydrogen peroxide inhibition activity of leaves extract was 24.27 per cent.

Gyrocarpus jathrophifolius Domin. Family: Hernandiaceae

Ruiz-Terán *et al.* (2008) evaluated the antioxidant activity of aerial parts of *Gyrocarpus jathrophifolius* using DPPH and β-carotene-linoleic acid bleaching assay. The extract displayed antioxidant activity comparable to the commercial antioxidant BHA. For the methanolic extract analysed, a clear relation between the total phenolic content of the extracts and their antioxidant activity was found.

Habenaria repens Family: Orchidaceae

Johnson *et al.* (1999) reported a phenolic antioxidant from *Habenaria repens*.

Halacsya sendtneri (Boiss.) Dorfl. Family: Boraginaceae

Niiforovi *et al.* (2010) examined the antioxidant activity of *Halacsya sendtneri* including total antioxidant capacity, DPPH free-radical scavenging, the inhibitory activity toward lipid peroxidation, Fe(3+)-reducing power, Fe(2+)-chelating ability and hydroxyl scavenging activity. Total phenolic and flavonoid contents were also determined for each alcoholic extract. Alcoholic extract of *Halacsya sendtneri* showed the highest total antioxidant capacity (231 mg AA/g dry extract) as well as DPPH free-radical scavenging (IC_{50} = 99μg/ml), inhibitory activity toward lipid peroxidation (IC_{50} = 17μg/ml) and reducing power. *Halacsya sendtneri* also showed greatest hydroxyl radical scavenging activity, as well as ferrous ion chelating ability.

Halenia elliptica Family: Gentianaceae

Huang *et al.* (2010) evaluated the hepatoprotective and antioxidant activity of *Halenia elliptica* against experimentally induced liver injury. The antioxidant property of methanolic extract (ME) of *Halenia elliptica* was investigated by employing various established *in vitro* systems. The ME possessed strong antioxidant activity *in vitro*. The results of CCl(4)-induced liver toxicity experiment showed that rats treated with the ME of *Halenia elliptica* (100 mg/kg and 200 mg/kg), and also the standard treatment, silymarin (50 mg/kg), showed a significant decrease in ALT, AST, ALP, and total bilirubin levels, which were all elevated in the CCl(4) group.

Hamamelis virginiana L. Family: Hamamelidaceae

Witch hazel (*Hamamelis virginiana*) bark is a rich source of both condensed and hydrolizable oligomeric tannins. From a polyphenolic extract soluble in both ethyl acetate and water, Tourino *et al.* (2008) have generated fractions rich in pyrogallol-containing polyphenols (proanthocyanidins, gallotannins, and gallates). The mixtures were highly active as free radical scavengers against ABTS, DPPH (hydrogen donation and electron transfer), and HNTTM (electron transfer). They were also able to reduce the newly introduced TNPTM radical, meaning that they included some highly reactive components. Witch hazel phenolics protected red blood cells from free radical-induced hemolysis and were mildly cytotoxic to 3T3 fibroblasts and HaCat keratinocytes. They also inhibited the proliferation of tumoral SK-Mel 28 melanoma cells at lower concentrations than grape and pine procyanidins. The high content in pyrogallol moieties may be behind the effect of witch hazel phenolics on skin cells. Because the most cytotoxic and antiproliferative mixtures were also the most efficient as electron transfer agents, they hypothesized that the final putative antioxidant effect of polyphenols may be in part attributed to the stimulation of defense systems by mild prooxidant challenges provided by reactive oxygen species generated through redox cycling.

Hamelia patens Jacq. Family: Rubiaceae

Ruiz-Terán *et al.* (2008) evaluated the antioxidant activity of aerial parts of *Hamelia patens* using DPPH and β-carotene-linoleic acid bleaching assay. The extract displayed antioxidant activity comparable to the commercial antioxidant BHA. For the methanolic extract analysed, a clear relation between the total phenolic content of the extracts and their antioxidant activity was found.

Haplopappus multifolius Phil. ex Reiche Family: Asteraceae (Compositae)

The antioxidant activity of eight coumarins and two flavonols isolated from *Haplopappus multifolius* was studied with the DPPH radical method (Torres *et al.*, 2006). Results showed that a high concentration of phenolic coumarins and the presence of quercetin and rhamnetin in the exudates could account for the proetection of the plant against oxidative stress.

Harpagophytum procumbens (Burch.) DC. Family: Pedaliaceae

English Name: Devil's claw, Grapple plant.

Georgiev *et al.* (2010) investigated the antioxidant activity of devil's claw cell biomass and its active constituents. The antioxidant activities of total methanol extracts, collected fractions and specific active constituents were then evaluated in DPPH and superoxide anion radical scavenging and oxygen radical absorbance capacity (ORAC) assays. The most active compounds were found to be β-OH-verbascodide (in DPPH and superoxide radical scavenging assays) and leucosceptoside A (in ORAC assays). The phenylethanoid fraction may be attractive for various commercial purposes since it displayed significant antioxidant activity.

Harpephyllum caffrum Bernh. Family: Anacardiaceae

Total phenolic content, proanthocyanidins, gallotannins, flavonoids and antioxidant activities of *Harpephyllum caffrum* methanolic extracts were evaluated by Moyo *et al.* (2010) using *in vitro* assays. *H. caffrum* stem bark extract had the highest content of proanthocyanidins (1.47 per cent). The EC_{50} values of the extracts in the DPPH free radical scavenging assay ranged from 4.26 to 6.92µg/ml, compared to 6.86µg/ml for ascorbic acid. A dose dependent linear curve was obtained for all extracts in the ferric-reducing power assay. All extracts exhibited high antioxidant activity comparable to butylated hydroxytoulene based on the rate of β-carotene bleaching (84.1-93.9 per cent).

Harungana madagascariensis Lam. ex Poir. Family: Clusiaceae (Guttiferae)

Biapa *et al.* (2007) investigated the phytochemical and antioxidant properties of four medicinal plants used in Cameroon, including *Harunga madagascariensis*. Four extracts (methanol, hydro-ethanolic, aqueous and hydrolysed) from each of the plants were prepared and analysed. Folin Ciocalteu, FRAP and DPPH assays were used to determine the antioxidant potential. The highest antioxidant activites were observed in the hydrolysed extracts of each plant, while the aqueous extract showed the least activity irrespective of the method used.

Hedeoma drummondii Benth. Family: Lamiaceae (Labiatae)

The methanolic extracts of aerial parts of *Hedeoma drummondii* exhibited a strong antioxidant effect measured by scavenging of the free DPPH radical (Viveros-Valdez *et al.*, 2008). Assay-guided fractionation of the crude methanolic extract allowed the identification of three major active constituents, chlorogenic, caffeic and rosmarinic acid, as well as sideriteflavone derivatives and simple phenolics. The high content of caffeic acid and rosmarinic acid relates to the antioxidant activity of *Hedeoma drummondii*.

Hedera colchica Koch. Family: Araliaceae

The antioxidant activities of saponins hederacolchisides-E and F from *Hedera colchica* were investigated by Gülçin *et al.* (2004) using different antioxidant tests.

458 | *Encyclopaedia of Herbal Antioxidants*

Hederacolchisides-E and F exhibited a strong total antioxidant activity. At the concentration of 75µg/ml, these saponins showed 88 and 75 per cent inhibition of lipid peroxidation of linoleic acid emulsion, respectively. Various antioxidant activities were compared with model antioxidants such as α-tocopherol, butylated hydroxyanisole and butylated hydroxytoulene.

Hedera helix C.B.Clarke Family: Araliaceae

The antioxidant activities of saponins α-hederin and hederasaponin-C from *Hedera helix* were investigated by Gülçin *et al*. (2004) using different antioxidant tests. α-hederin, hederasaponin-C exhibited a strong total antioxidant activity. At the concentration of 75µg/ml, these saponins showed 94, 86 per cent inhibition of lipid peroxidation of linoleic acid emulsion, respectively. Various antioxidant activities were compared with model antioxidants such as α-tocopherol, butylated hydroxyanisole and butylated hydroxytoulene.

Hedranthera batteri Family: Apocynaceae

Oke and Hamburger (2002) screened Nigerian medicinal plants for antioxidant activity. *Hedranthera batteri* exhibited low antioxidant activity.

Hedychium spicatum Buch.-Ham. ex D.Don Family: Zingiberaceae

Total phenolic compounds and antioxidant potential of *Hedychium spicatum* were evaluated by Rawat *et al*. (2011). Total phenolic compounds varied among populations from 4.70 mg GAE to 2.84 mg GAE/g dry weight. Three *in vitro* antioxidant assays, *i.e.*, ABTS, DPPH and FRAP assays, showed significant differences across populations. ABTS assay showed highest values of antioxidant potential ranging from 2.581 mM ascorbic acid equivalent per 100 g to 1.91 mM AAE per 100 g dry weight. All assays showed significant correlation with Total phenolic compounds. Total phenolic compounds showed a significant relationship to altitude.

Hedychium thyrsiforme Sm. Family: Zingiberaceae

Five flavonoids were isolated from the rhizome of *Hedychium thyrsiforme* and assayed by Mooi *et al*. (2003) for antioxidant and antitumor promoting activities. The antioxidant assays showed that 5,7,4'-trimethoxy-3-hydroxyflavone, 7,4-dimethoxy-3,5-dihydroxyflavone and 3,4'-dimethoxy-5,7-dihydroxyflavone had strong activities. Only two compounds 5,7,4'-trimethoxy-3-hydroxyflavone and 7,4-dimethoxy-3,5-dihydroxyflavone, were found to be strong DPPH free radical scavengers with fifty percent inhibition concentration (IC$_{50}$ values of 92 and 119 microM), respectively.

Hedyosmum angustifolium (Ruiz. et Pavon) Solms-Laub. Family: Chloranthaceae

The aerial parts of 17 Bolivian plants including *Hedyosmum angustifolium* were screened by Rosas-Romero and Saavedra (2005) to determine antioxidant activity. A methanol extract of each plant was prepared and portioned sequentially with hexane, chloroform, and ethyl acetate, leaving an aqueous solution. All extracts and their 5 fractions for a total of 102 samples, were evaluated using two techniques: an adaptation of the β-carotene bleaching technique using an emulsion of linoleic acid in water as the oxidizable substrate, and the DPPH free radical trapping technique. The results of the β-carotene bleaching technique were more discriminating and better related to the rancidity process under normal conditions; with this assay, 11 species provided at least one fraction with highly promising antioxidant activity. All species gave good results under the DPPH technique, and in most cases they performed better than BHA, which was used as a reference antioxidant (Rosas-Romero and Saavedra, 2005).

Hedyotis capitellata Wall. ex G.Don Family: Rubiaceae

Ahmad *et al.* (2005) investigated the antioxidant and radical-scavenging activities of methanolic extracts of seven *Hedyotis* species. The antioxidant activity was evaluated by the FTC and TBA methods while the radical scavenging activity was measured by the DPPH method. All tested extracts exhibited very strong antioxidant properties when compared to vitamin E (α-tocopherol) with per cent inhibition of 89-98 per cent in the FTC and 60-95 per cent in the TBA assays. In the DPPH method also the extract exhibited moderate radical scavenging activity. The results from the Griess assay showed that the tested extracts were weak inhibitors of NO synthase.

Hedyotis corymbosa (L.) Lam. Family: Rubiaceae

The methanolic extract of the aerial parts of *Hedyotis corymbosa* was screened by Sasikumar *et al.* (2010) for antioxidative activity using DPPH quenching assay, ABTS cation decoloration test, FRA scavenging capacity towards hydroxyl ion radicals and nitric oxide radical inhibition activity using established assay procedures. The plant yielded 210 gallic acid equivalent/100g phenolic content and 55 mg quercetin equivalent/100g flavonoid content. The extract exhibited high antiradical activity against DPPH, ABTS, nitric oxide and hydroxyl radicals with EC_{50} value of 82, 150, 130 and 170µg/ml, respectively. The FRP increased with increasing concentration of the sample. The antioxidant activity of the extract was comparable with that of the standard BHT. High correlation between total phenolic/flavonoid contents and scavenging potential of different reactive oxygen species indicated the polyphenols as the main antioxidants.

Hedyotis dichotoma Cav. Family: Rubiaceae

Ahmad *et al.* (2005) investigated the antioxidant and radical-scavenging activities of methanolic extracts of seven *Hedyotis* species. The antioxidant activity was evaluated by the FTC and TBA methods while the radical scavenging activity was measured by the DPPH method. All tested extracts exhibited very strong antioxidant properties when compared to vitamin E (α-tocopherol) with percent inhibition of 89-98 per cent in the FTC and 60-95 per cent in the TBA assays. In the DPPH method also *H. dichotoma* extract exhibited moderate radical scavenging activity. The results from the Griess assay showed that the tested extracts were weak inhibitors of NO synthase.

Hedyotis diffusa Willd. Family: Rubiaceae

Cai *et al.* (2004) evaluated the antioxidant activity and phenolic compounds in traditional Chinese medicinal plants associated with anticancer. The improved ABTS.+ method was used to systematically assess the total antioxidant capacity (Trolox equivalent antioxidant capacity, TEAC) of the medicinal extracts. The TEAC values and total phenolic content for methanolic extract of whole plant of *Hedyotis diffusa* were 113.8 µmol Trolox equivalent/100g dry weight (DW), and 0.90 g of gallic acid equivalent/100g DW respectively. A positive significant linear relationship between antioxidant activity and total phenolic content showed that phenolic compounds were the dominant antioxidant components in the tested herbs. Major types of phenolic compounds in *Hedyotis diffusa* include phenolic acids (*p*-coumaric acid), flavonols (kaempferol glycosides).

Hedyotis havilandi King Family: Rubiaceae

Ahmad *et al.* (2010) evaluated the antioxidant potential of 22 species of medicinal plants from Malaysian Rubiaceae including *Hedyotis havilandii*. FTC, TBA, TPC and the DPPH assays were employed. The tested extracts showed strong antioxidant potential with percent inhibition of 96.9 per cent in the FTC, and 95.2 per cent in TBA assays. The TPC of the extract was 12.26mg GAE/g PE. A good correlation was observed between total phenolic content and radical-scavenging activities.

Hedyotis herbacea L. Family: Rubiaceae

Ahmad *et al.* (2005) investigated the antioxidant and radical-scavenging activities of methanolic extracts of seven *Hedyotis* species. The antioxidant activity was evaluated by the FTC and TBA methods while the radical scavenging activity was measured by the DPPH method. All tested extracts exhibited very strong antioxidant properties when compared to vitamin E (α-tocopherol) with percent inhibition of 89-98 per cent in the FTC and 60-95 per cent in the TBA assays. In the DPPH method, *H. herbacea* exhibited the strongest radical scavenging activity with an IC_{50} value of 32µg/ml. The results from the Griess assay showed that the tested extracts were weak inhibitors of NO synthase.

Hedyotis philippinensis (Wild. ex Spreng.) Merr. ex C.B.Rob. Family: Rubiaceae

Ahmad *et al.* (2010) evaluated the antioxidant potential of 22 species of medicinal plants from Malaysian Rubiaceae including *Hedyotis philippinensis*. FTC, TBA, TPC and the DPPH assays were employed. The tested extracts showed strong antioxidant potential with percent inhibition of 95.9 per cent (leaves), 97.9 per cent (stems) in the FTC, and 94.7 per cent (leaves), 88.9 per cent (stems) in TBA assays. The TPC of the extract was 9.27 (leaves), 9.76 mg GAE/g PE and strong DPPH radical-scavenging activity with IC_{50} values of 114.19µg/ml (leaves) 130.78µg/ml (stems). A good correlation was observed between total phenolic content and radical-scavenging activities.

Hedyotis verticillata (L.) Lam. (Syn.: *Oldenlandia verticillata* L.) Family: Rubiaceae

Methanol exteracts of seven Malaysian medicinal plants, including *Hedyotis verticillata* were screened by Saha *et al.* (2004) for antioxidant and nitric oxide inhibitory activities. Antioxidant activity was measured by using FTC, TBA and DPPH free radical scavenging methods and Greiss assay was used for the measurement of nitric oxide inhibition in LPS and interferon-γ-treated RAW264.7 cells. All the extracts showed strong antioxidant activity comparable to or higher than that of α-tocopherol, BHT and quercetin in FTC and TBA methods. In the DPPH radical scavenging assay the extracts were weak. In the Greiss assay *Hedyotis verticillata* showed strong inhibitory activity on nitric oxide production in LPS and IFN-γ-induced RAW 264.7 cells.

Heinsis crinita (Afz.) G. Taylor Family: Rubiaceae

Odukoya *et al.* (2007) evaluated the antioxidant activity and total phenolic contents of Nigerian green leafy vegetables including *Heinsia crinita*.

Helianthus tuberosus L. Family: Asteraceae (Compositae)

English Name: Jerusalem artichoke

Souri *et al.* (2004) investigated the antioxidant activity of methanol extracts of Jerusalem artichoke, used as vegetable in Iranian diet against linoleic acid peroxidation. The extracts showed good antioxidant activity.

Aslan *et al.* (2010) studied the antidiabetic and antioxidant activities of the ethanol extracts of tubers of *Helianthus tuberosus*. TBARS and GSH livels in liver, kidney and hear tissues were measured by using spectrophotometric methods for antioxidant assay. Extracts induced significant alleviation on only kidney tissue TBARS levels (24.5 per cent). None of the extracts restored GSH levels in kidney, liver and heart tissues of diabetic rats.

Helichrysum arenarium Moench subsp. *armenium* Family: Asteraceae (Compositae)

The methanol extracts of 16 *Helichrysum* species were investigated by Albayrak *et al.* (2010) for their *in vitro* antioxidant, radical scavenging and antimicrobial activities. All the extracts showed strong antioxidant and radical scavenging activity. The total antioxidant capacity as ascorbic acid equivalent (AAE) in *Helichrysum arenarium* subsp. *armenium* was 157.29 mg/g dry extract in the phosphomolybdenum assay while in DPPH assay IC_{50} value was 16.61µg/ml. The total phenolic content of the extract was 89.02mg gallic acid equivalent (GAE)/g dry extract.

Helichrysum arenarium Moench subsp. *aucheri* Family: Asteraceae (Compositae)

Tepe *et al.* (2004) examined the *in vitro* antioxidant activities of methanol extracts of four *Helichrysum* species including *Helichrysum arenarium* subsp. *aucheri*. The extracts were screened for their possible antioxidant activity by two complementary test systems, namely DPPH free radical-scavenging and β-carotene/linoleic acid systems. In the first case, non-polar subfractions of the methanol extracts did not show any antioxidant activity.

The methanol extracts of 16 *Helichrysum* species were investigated by Albayrak *et al.* (2010) for their *in vitro* antioxidant, radical scavenging and antimicrobial activities. All the extracts showed strong antioxidant and radical scavenging activity. The total antioxidant capacity as ascorbic acid equivalent (AAE) in *Helichrysum arenarium* subsp. *aucheri* was 147.68 mg/g dry extract in the phosphomolybdenum assay while in DPPH assay IC_{50} value was 37.52µg/ml. The total phenolic content of the extract was 115.76mg gallic acid equivalent (GAE)/g dry extract.

Helichrysum arenarium Moench subsp. *rubicundum* Family: Asteraceae (Compositae)

Antioxidant and DPPH radical scavenging activities, reducing powers and the amount of total phenolic compounds of some medicinal Asteraceae species were studied by Özgen *et al.* (2004). Water extract of *Helichrysum arenarium* subsp. *rubicundum* showed good antioxidant activity with 80.6 per cent inhibition at 500µg/ml concentration. The DPPH radical scavenging activity was 89.0 per cent at 500µg/ml concentration and reducing power was 18.2 µg/ml Ascorbic acid equivalent. Total phenolic amount was 28.2µg/ml Gallic acid equivalent.

Helichrysum artvinense P.H.Davis et Kupicha Family: Asteraceae (Compositae)

The methanol extracts of 16 *Helichrysum* species were investigated by Albayrak *et al.* (2010) for their *in vitro* antioxidant, radical scavenging and antimicrobial activities. All the extracts showed strong antioxidant and radical scavenging activity. The total antioxidant capacity as ascorbic acid equivalent (AAE) in *Helichrysum artvinense* was 171.02 mg/g dry extract in the phosphomolybdenum assay while in DPPH assay IC_{50} value was 21.23µg/ml. The total phenolic content of the extract was 83.98mg gallic acid equivalent (GAE)/g dry extract.

Helichrysum chasmolycium P.H.Davis Family: Asteraceae (Compositae)

The methanol extracts of *Helichrysum* species were investigated by Albayrak *et al.* (2008) for their *in vitro* antioxidant and radical scavenging activities. All the extracts showed strong antioxidant and radical scavenging activity. The total antioxidant capacity as ascorbic acid equivalent (AAE) in *Helichrysum chasmolycium* was 147.88 mg/g dry extract in the phosphomolybdenum assay while in DPPH assay IC_{50} value was 25.33µg/ml.

Helichrysum chionophilum Boiss. et Balansa Family: Asteraceae (Compositae)

Tepe *et al.* (2004) examined the *in vitro* antioxidant activities of methanol extracts of four *Helichrysum* species including *Helichrysum chionophilum*. The extracts were screened for their possible antioxidant activity by two complementary test systems, namely DPPH free radical-scavenging and β-carotene/linoleic acid systems. In the first case, non-polar subfractions of the methanol extracts did not show any antioxidant activity, while the most active one was *H. chionophilum* (IC_{50} = 40.5µg/ml) among the polar subfractions.

The methanol extracts of 16 *Helichrysum* species were investigated by Albayrak *et al.* (2010) for their *in vitro* antioxidant, radical scavenging and antimicrobial activities. All the extracts showed strong antioxidant and radical scavenging activity. The total antioxidant capacity as ascorbic acid equivalent (AAE) in *Helichrysum chionophilum* was 140.43 mg/g dry extract in the phosphomolybdenum assay while in DPPH assay IC_{50} value was 53.10µg/ml. The total phenolic content of the extract was 106.97mg gallic acid equivalent (GAE)/g dry extract.

Helichrysum compactum Boiss. Family: Asteraceae (Compositae)

Extracts of the capitula of *Helichrysum compactum* showed antioxidant activity by inhibition of lipid peroxidation (Suzgec *et al.*, 2005).

The methanol extracts of 16 *Helichrysum* species were investigated by Albayrak *et al.* (2010) for their *in vitro* antioxidant, radical scavenging and antimicrobial activities. All the extracts showed strong antioxidant and radical scavenging activity. The total antioxidant capacity as ascorbic acid equivalent (AAE) in *Helichrysum compactum* was 165 mg/g dry extract in the phosphomolybdenum assay while in DPPH assay IC_{50} value was 27.32µg/ml. The total phenolic content of the extract was 79.59mg gallic acid equivalent (GAE)/g dry extract.

Helichrysum dasyanthum (Willd.) Sweet Family: Asteraceae (Compositae)

The methanol extracts of *Helichrysum* species were investigated by Lourens *et al.* (2004) for their radical scavenging activities. All the extracts showed strong radical scavenging activity. In *Helichrysum dasyanthum* in DPPH assay the IC_{50} value was 12.33µg/ml.

Helichrysum excisum (Thunb.) Less. Family: Asteraceae (Compositae)

The methanol extracts of *Helichrysum* species were investigated by Lourens *et al.* (2004) for their radical scavenging activities. All the extracts showed strong radical scavenging activity. In *Helichrysum excisum* in DPPH assay the IC_{50} value was 13.67µg/ml.

Helichrysum felinum Less. Family: Asteraceae (Compositae)

The methanol extracts of *Helichrysum* species were investigated by Lourens *et al.* (2004) for their radical scavenging activities. All the extracts showed strong radical scavenging activity. In *Helichrysum felinum* in DPPH assay the IC_{50} value was 26.71µg/ml.

Helichrysum goulandriorum Georgiadou Family: Asteraceae (Compositae)

The methanol extracts of 16 *Helichrysum* species were investigated by Albayrak *et al.* (2010) for their *in vitro* antioxidant, radical scavenging and antimicrobial activities. All the extracts showed strong antioxidant and radical scavenging activity. The total antioxidant capacity as ascorbic acid equivalent (AAE) in *Helichrysum goulandriorum* was 124.86 mg/g dry extract in the phosphomolybednum assay while in DPPH assay IC_{50} value was 23.92µg/ml. The total phenolic content of the extract was 114.41mg gallic acid equivalent (GAE)/g dry extract.

Helichrysum graveolens (M.Bieb.) Sweet Family: Asteraceae (Compositae)

The hypoglycemic, antihyperglycemic and antioxidant potentials of water and ethanol extracts of *Helichrysum graveolens* were evaluated by Aslan *et al.* (2007) by using *in vivo* method in normal and streptozotocin-induced diabetic rats. The antioxidant activity of these extracts was studied in liver, kidney and heart tissues. In order to determine antioxidant activity, tissue malondialdehyde and reduced glutathione levels were measured by using spectrophotometric methods. The experimental data obtained from water and methol extracts of capitulums confirmed the antioxidant activity.

The methanol extracts of 16 *Helichrysum* species were investigated by Albayrak *et al.* (2010) for their *in vitro* antioxidant, radical scavenging and antimicrobial activities. All the extracts showed strong antioxidant and radical scavenging activity. The total antioxidant capacity as ascorbic acid equivalent (AAE) in *Helichrysum graveolens* was 160.34 mg/g dry extract in the phosphomolybdenum assay while in DPPH assay IC_{50} value was 15.28µg/ml. The total phenolic content of the extract was 92.77mg gallic acid equivalent (GAE)/g dry extract.

Helichrysum heywoodianum P.H.Davis Family: Asteraceae (Compositae)

The methanol extracts of 16 *Helichrysum* species were investigated by Albayrak *et al.* (2010) for their *in vitro* antioxidant, radical scavenging and antimicrobial activities. All the extracts showed strong antioxidant and radical scavenging activity. The total antioxidant capacity as ascorbic acid equivalent (AAE) in *Helichrysum heywoodianum* was 191.97 mg/g dry extract in the phosphomolybdenum assay while in DPPH assay IC_{50} value was 22.23µg/ml. The total phenolic content of the extract was 93.85mg gallic acid equivalent (GAE)/g dry extract.

Helichrysum italicum G.Don Family: Asteraceae (Compositae)

The antioxidant properties of twenty medicinal herbs used in the traditional Mediterranean and Chinese medicine including *Helichrysum italicum* were studied by Schinella *et al.* (2002). Using free radical-generating systems *Helichrysum italicum* protected against enzymatic and non-enzymatic lipid peroxidation in model membranes and also showed scavenging property on the superoxide radical. All the extract were assayed at a concentration of 100µg/ml. Most of the extracts were weak scavengers of the hydroxyl radical.

The antioxidant activity of supercritical CO_2 extract of *H. italicum* dried flower heads derived from the commercial drug and from plants grown in different areas of north-east Italy with different culturing conditions was determined by Poli *et al.* (2003). The four kind of extracts were also tested for their ability to scavenge superoxide radicals. All extracts showed an antioxidant activity. The supercritical extracts obtained from commercial dried flower heads and from dried flower heads belonging to wild plants exhibited highest activity. These results established *H. italicum* supercritical extracts as important antioxidant solvent-free matrices in alimentary (*i.e.*, dietary, nutraceutical, flavouring and cosmetic fields).

Helichrysum kitianum Mill. Family: Asteraceae (Compositae)

The methanol extracts of 16 *Helichrysum* species were investigated by Albayrak *et al.* (2010) for their *in vitro* antioxidant, radical scavenging and antimicrobial activities. All the extracts showed strong antioxidant and radical scavenging activity. The total antioxidant capacity as ascorbic acid equivalent (AAE) in *Helichrysum kitianum* was 172.17 mg/g dry extract in the phosphomolybdenum assay while in DPPH assay IC_{50} value was 26.37µg/ml. The total phenolic content of the extract was 75.16mg gallic acid equivalent (GAE)/g dry extract.

Helichrysum longifolium DC. Family: Asteraceae (Compositae)

Aiyegoro and Okoh (2010) assessed the antioxidant potential and phytochemical constituents of crude aqueous extract of *Helichrysum longifolium* using tests involving inhibition of superoxide anions, DPPH, H_2O_2, NO and ABTS. Phytochemical analyses revealed the presence of tannins, flavonoids, steroids and saponins. The total phenolic content of the aqueous leaf extract was 0.4999 mg GAE/g of extract powder. The total flavonoid and proanthocyanidin contents of the plant were 0.705 and 0.005 mg GAE/g extract powder respectively. The percentage inhibition of lipid peroxide at the initial stage of oxidation showed antioxidant activity of 87 per cent compared to those of BHT (84.6 per cent) and gallic acid (96 per cent). Also, the percentage inhibition of malondialdehyde by the extract showed percentage inhibition of 78 per cent comparable to BHT (72.24 per cent) and Gallic acid (94.82 per cent).

Helichrysum noeanum Boiss. Family: Asteraceae (Compositae)

Tepe *et al.* (2004) examined the *in vitro* antioxidant activities of methanol extracts of four *Helichrysum* species including *Helichrysum noeanum*. The extracts were screened for their possible antioxidant activity by two complementary test systems, namely DPPH free radical-scavenging and β-carotene/linoleic acid systems. In the first case, non-polar subfractions of the methanol extracts did not show any antioxidant activity.

The methanol extracts of 16 *Helichrysum* species were investigated by Albayrak *et al.* (2010) for their *in vitro* antioxidant, radical scavenging and antimicrobial activities. All the extracts showed strong antioxidant and radical scavenging activity. The highest total antioxidant capacity as ascorbic acid equivalent (AAE) of 194.64 mg/g dry extract was obtained by *Helichrysum noeanum* in the phosphomolybdenum assay while in DPPH assay IC_{50} value was 18.46µg/ml. The total phenolic contents of the extract was 160.63mg gallic acid equivalent (GAE)/g dry extract.

Helichrysum orientale (L.) Gaertn. Family: Asteraceae (Compositae)

The methanol extracts of 16 *Helichrysum* species were investigated by Albayrak *et al.* (2010) for their *in vitro* antioxidant, radical scavenging and antimicrobial activities. All the extracts showed strong antioxidant and radical scavenging activity. The total antioxidant capacity as ascorbic acid equivalent (AAE) in *Helichrysum orientale* was 110.03 mg/g dry extract in the phosphomolybdenum assay while in DPPH assay IC_{50} value was 29.53µg/ml. The total phenolic content of the extract was 73.70mg gallic acid equivalent (GAE)/g dry extract.

Helichrysum pallasii (Spreng.) Ledeb. Family: Asteraceae (Compositae)

The methanol extracts of 16 *Helichrysum* species were investigated by Albayrak *et al.* (2010) for their *in vitro* antioxidant, radical scavenging and antimicrobial activities. All the extracts showed strong antioxidant and radical scavenging activity. The total antioxidant capacity as ascorbic acid equivalent (AAE) in *Helichrysum pallasii* was 118.99 mg/g dry extract in the phosphomolybdenum assay while in DPPH assay IC_{50} value was 26.23µg/ml. The total phenolic content of the extract was 94.13mg gallic acid equivalent (GAE)/g dry extract.

Helichrysum pamphylicum P.H.Davis et Kupicha Family: Asteraceae (Compositae)

The methanol extracts of *Helichrysum* species were investigated by Albayrak *et al.* (2008) for their *in vitro* antioxidant and radical scavenging activities. All the extracts showed strong antioxidant and radical scavenging activity. The total antioxidant capacity as ascorbic acid equivalent (AAE) in *Helichrysum pamphylicum* was 173.58 mg/g dry extract in the phosphomolybdenum assay while in DPPH assay IC_{50} value was 15.21μg/ml.

Helichrysum peshmenianum Erik. Family: Asteraceae (Compositae)

The methanol extracts of 16 *Helichrysum* species were investigated by Albayrak *et al.* (2010) for their *in vitro* antioxidant, radical scavenging and antimicrobial activities. All the extracts showed strong antioxidant and radical scavenging activity. The total antioxidant capacity as ascorbic acid equivalent (AAE) in *Helichrysum peshmenianum* was 125.47 mg/g dry extract in the phosphomolybdenum assay while in DPPH assay IC_{50} value was 35.55μg/ml. The total phenolic content of the extract was 66.74mg gallic acid equivalent (GAE)/g dry extract.

Helichrysum petiolare Hilliar et B.L.Burt. Family: Asteraceae (Compositae)

The methanol extracts of *Helichrysum* species were investigated by Lourens *et al.* (2004) for their radical scavenging activities. All the extracts showed strong radical scavenging activity. In *Helichrysum petiolare* in DPPH assay the IC_{50} value was 28.70μg/ml.

Helichrysum plicatum DC. subsp. *plicatum* Family: Asteraceae (Compositae)

Tepe *et al.* (2004) examined the *in vitro* antioxidant activities of methanol extracts of four *Helichrysum* species including *Helichrysum plicatum* subsp. *plicatum*. The extracts were screened for their possible antioxidant activity by two complementary test systems, namely DPPH free radical-scavenging and β-carotene/linoleic acid systems. In the first case, non-polar subfractions of the methanol extracts did not show any antioxidant activity.

Aslan *et al.* (2007) evaluated the hypoglycaemic and antioxidant potential of *Helichrysum plicatum* ssp. *plicatum* by using *in vivo* methods in normal and streptozotocin-induced-diabetic rats. The experimental data indicated that water and ethanol extracts of capitulums demonstrate significant antihyperglycaemic and antioxidant activity in streptozotocin-induced rats. In order to assess the role of polyphenolic components in the relevant activity, phenolic and flavonoid contents of each extract were also determined in terms of total phenols: 113.5 mg (GAE/g extract) and total flavonoids 50.5 mg (quercetin equivalent/1g extract) for ethanol extract, total phenols 75.9; flavonoids: 31.5 for water extract using Folin-Ciocalteu reagent.

The methanol extracts of 16 *Helichrysum* species were investigated by Albayrak *et al.* (2010) for their *in vitro* antioxidant, radical scavenging and antimicrobial activities. All the extracts showed strong antioxidant and radical scavenging activity. The total antioxidant capacity as ascorbic acid equivalent (AAE) in *Helichrysum plicatum* subsp. *plicatum* was 163.47 mg/g dry extract in the phosphomolybdenum assay while in DPPH assay IC_{50} value was 23.48μg/ml. The total phenolic content of the extract was 87.36mg gallic acid equivalent (GAE)/g dry extract.

Helichrysum plicatum DC. subsp. *polyphyllum* Family: Asteraceae (Compositae)

The methanol extracts of 16 *Helichrysum* species were investigated by Albayrak *et al.* (2010) for their *in vitro* antioxidant, radical scavenging and antimicrobial activities. All the extracts showed strong antioxidant and radical scavenging activity. The total antioxidant capacity as ascorbic acid equivalent (AAE) in *Helichrysum plicatum* subsp. *polyphyllum* was 152.64 mg/g dry extract in the phosphomolybdenum assay while in DPPH assay IC_{50} value was 13.23µg/ml. The total phenolic content of the extract was 154.64mg gallic acid equivalent (GAE)/g dry extract.

Helichrysum sanguineum Kostel. Family: Asteraceae (Compositae)

The methanol extracts of *Helichrysum* species were investigated by Albayrak *et al.* (2008) for their *in vitro* antioxidant and radical scavenging activities. All the extracts showed strong antioxidant and radical scavenging activity. The total antioxidant capacity as ascorbic acid equivalent (AAE) in *Helichrysum sanguineum* was 159.94 mg/g dry extract in the phosphomolybdenum assay while in DPPH assay IC_{50} value was 12.90µg/ml.

Helichrysum stoechas (L.) Moench. subsp. *barellieri* Family: Asteraceae (Compositae)

The methanol extracts of 16 *Helichrysum* species were investigated by Albayrak *et al.* (2010) for their *in vitro* antioxidant, radical scavenging and antimicrobial activities. All the extracts showed strong antioxidant and radical scavenging activity. The antioxidant capacity as ascorbic acid equivalent (AAE) was 188.26 mg/g dry extract. The highest IC_{50} value (7.95µg/ml) was observed for the extract of *Helichrysum stoechas* subsp. *barellieri* in the DPPH assay. The total phenolic contents of the extract was 94.16mg gallic acid equivalent (GAE)/g dry extract.

Helicia formosana Hemsl Family: Proteaceae

Chen *et al.* (2009) studied the antioxidant activity of leaves extract towards xanthine oxidase, lipoxygenase and tyrosinase activities.

Helicteres isora L. Family: Sterculiaceae

Kumar *et al.* (2008) investigated the effect of the aqueous extract of *Helicteres isora* bark on oxidative stress in the heart of rats during diabetes. An appreciable decrease in peroxidation products, TBARS, conjugated dienes and hydroperoxides was observed in the heart tissues of *Helicteres isora* (HI) treated diabetic rats. The decreased activities of key antioxidant enzymes such as superoxide dismutase, catalase, Glutathione peroxidase, glutathione S-transferase and glutathione in diabetic rats were brought back to near normal range upon HI treatment. Tolbutamide was used as the standard reference drug.

The antioxidant activity of the aqueous (hot) extract of *Helicteres isora* (AEHI) fruits was investigated in various *in vitro* models by Basniwal *et al.* (2009). The total polyphenolic content of the extract was found 7.04 per cent of AEHI, when compared to gallic acid and total flavonoid content was 2.4 mg/g of AEHI, when compared to rutin. Hydrogen peroxide radicals were inhibited at IC_{50} = 165 µg/ml, while ascorbic acid inhibited at 187.33µg/ml. AEHI inhibited the nitric oxide radical at IC_{50} = 820 µg/ml when it was compared with rutin as standard antioxidant with IC_{50} = 68.52µg/ml. Superoxide (SO) radical's inhibition was compared with quercetin and IC_{50} value was found more than 1000 µg/ml. Moreover, the results were observed in a concentration dependent manner. The study clearly indicated that the aqueous (hot) extract of *H. isora* has significant antioxidant activity.

Suthar *et al.* (2009) evaluated the antioxidant and antidiabetic activity of *Helicteres isora* fruits. The hot water extract of fruits was prepared and screened for its *in vitro* antioxidant activity using DPPH assay, β-carotene-lineolate model and microsomal lipid peroxidation or thiobarbituric acid reactive species assays and the IC_{50} values were calculated. The hot water extract showed maximum activity with IC_{50} value 25.12mug/ml for DPPH assay method, and low activity with IC_{50} value 740.64mug/ml for microsomal lipid peroxidation assay. In the β-carotene-lineolate model, the extract showed 45.63 per cent antioxidant activity.

Heliotropium glutinosum Phil. Family: Boraginaceae

From the resinous exudates of *Heliotropium glutinosum* a new aromatic geranyl derivative: 4-methoxy-3-[(2)-7'-methyl-3'- hydroxymethyl-2',6'-octadienyl] phenol (*1*) and three flavonoids: 5,3'-dihydroxy-7,4'-dimethoxyflavanone (2), 5,4'-dihydroxy-7-methoxyflavanone (3) and 4'-acetyl-5-hydroxy -7-methoxyflavanone (4) were isolated by Modak *et al.* (2007). Their antioxidant activities were evaluated using the bleaching of ABTS and DPPH derived cation radical methods and expressed in terms of FRE (fast reacting equivalents) and TRE (total reacting equivalents), where FRE is a good measure of the quick protection of a given compound against oxidants and TRE measures the degree of long-term protection of the antioxidant, or how effective it is against a strong oxidative stress.

Heliotropium indicum L. Family: Boraginaceae

Helindicine (*1*), a new pyrrolizidine alkaloid with unusual structural features, together with the known lycopsamine (2), were isolated by Souza *et al.* (2005) from the roots of *Heliotropium indicum*. Compounds *1* and 2 were assayed for antioxidant activity and showed moderate activity.

Heliotropium marifolium Retz. Family: Boraginaceae

Singh *et al.* (2009) evaluated the antioxidant activity of dichloromethane and methanolic extracts of 12 arid zone medicinal plants. *Heliotropium marifolium* showed appreciable antioxidant activity and per cent of inhibition of DPPH where RC_{50} (µg/ml) was 8.5 (Dichloromethane extract) and 6.5 (MeOH extract) per cent of inhibition of DPPH was 93.97 per cent in dichloromethane extract of *Heliotropium marifolium*. Key phytochemicals include heliotrine, europine and indicine.

Heliotropium sinuatum (Miers) I.M.Johnst. Family: Boraginaceae

From the resinous exudates of *Heliotroium sinuatum*, a new compound: 4-(3,5-dihydroxynonadecyl) phenol, together with eight previously described flavonoids, were isolated and their antioxidant activities were assessed by quenching measurements with ABTS and DPPH cation radicals (Modak *et al.*, 2003).

Heliotropium strigosum Willd. Family: Boraginaceae

Hussain *et al.* (2010) evaluated the antioxidant activity of *Heliotropium strigosum.* A crude extract and ethyl acetate, n-hexane, chloroform and aqueous fractions were obtained. The crude extract and fractions were screened for antioxidant properties. The plant showed excellent DPPH scavenging activity. Antioxidant activity was shown by ethyl acetate, n-hexane and aqueous fractions. Crude extract and chloroform fractions were lacking in DPPH scavenging activity.

Helixanthera parasitica Lour. Family: Loranthaceae

The aqueous extract of *Helixanthera parasitica* revealed a significant inhibitory effect in the cancer cell invasion, and showed antioxidant activity. However, anti-metastitic activity was not associated with the antioxidant activity of the aqueous extract (Lirdprapamongkol *et al.*, 2003).

Helleborus hercegovinus Martinis, *H. multifidus* Vis and *Helleborus odorus* Waldst. et Kit. Family: Ranunculaceae

Caker *et al.* (2011) evaluated possible antioxidative and antiproliferative activities of three *Helleborus* taxa. The dry leaves and roots of three *Helleborus* taxa were extracted with ethanol and water. A phytochemical evaluation of the selected extracts was performed using spectrophotometric methods and a 1,1-diphenyl-2-picrylhydrazyl (DPPH) radical scavenging activity assay was used for measuring the antioxidative activity of extracts. The phytochemical evaluation showed that the leaves contain high levels of total phenolic and flavonoid content. Results from the DPPH assay indicated that the activity of the ethanol and water extracts of the leaves was higher than that of positive control (thymol). Extracts from the roots of *H. odorus* also displayed higher antioxidant activity than the positive probe, while *H. mulifidus* and *H. hercegovinus* root extracts were less effective. A statistically significant correlation between total phenolic content and antioxidative properties indicates that these compounds contribute to the antioxidant activity.

Hemerocallis fulva L. Family: Liliaceae

English Name: Day lily.

Maeda *et al.* (2006) examined the phenolic contents of leaves of *Hemerocallis fulva* and its radical scavenging activities for DPPH radical. The polyphenol content was 48 mg-gallic acid equivalent/100 g) and DPPH scavenging activity was 164.3 µmol-Trolox eq./100 g).

Antioxidant properties of water and ethanol extracts prepared from the dried flowers were evaluated in terms of total antioxidant activity, reducing capacity, and metal chelating activity by Que *et al.* (2007). Extracts from daylily flowers exhibited strong antioxidant activity. Ethanol has more efficiency to extract antioxidants than water, and freeze-drying preserved higher activities than air-drying. Rutin, (+)-catechin, and gallic acid were identified in the extracts by HPLC, and were highly related to the antioxidant activities. The antioxidant activity was further evaluated by feeding mice with ethanol extract from freeze-dried daylily flowers for 60 days. The results demonstrated that the extract at dosage of 40-225 mg/100 g significantly increased the activity of SOD (superoxide dismutase) and reduced the lipid peroxidation in both blood and liver of rat.

The antioxidant properties of methanol extracts from daylily flowers during maturation were determined with antioxidant assays, including antioxidant activity, DPPH radical scavenging, superoxide anion radical scavenging and reducing power by Fu *et al.* (2009). Antioxidant compounds, such as phenolic compounds, ascorbic acid and β-carotene were also analysed. Significant variation in antioxidant properties and involved compounds was observed between different maturity stages of daylily flowers. The highest antioxidant activity was observed at stage III (flower opening) accompanying the highest content of ascorbic acid and phenolic compounds, while no significant difference of β-carotene contents was observed among the four maturity stages. Four individual phenolics, such as (+)-catechin, chlorogenic acid, rutin and quercetin were identified and quantified by HPLC. (+)-Catechin was the main phenolic compound identified in daylily flowers, accounting for about 74.11 per cent of total phenolics.

Liu *et al.* (2010) explored the antioxidant components and activities of daylily *Hemerocallis fulva* flowers among two growth areas, three flower ages and two processing methods. The results showed that the growth area, flower age and processing method all significantly influenced the functional components and antioxidant activities grown in mountainous areas of Taiwan. The total phenols, flavonoids and total chlorophylls of the methanol extracts of D3DF were 59.51, 70.76 and 5.67 per cent of the respective values of F3DF. The total phenols and anthocyanins of the methanol extracts of F1DF were 2.64 mg GAE/100 g dried basis (db) and 0.102 μmole/100 g db, which were significantly higher than the others. The total flavonoid contents of F1DF, F2DF, F3DF and D3DF were 20.83, 29.67, 31.65 and 22.30 mg QE/100 g db, respectively, with the F1DF level as the lowest. The reducing power showed that both the fresh and dried flowers were very weak. The amounts of most flavonoids in the flowers from Hualien were greater than those from Taidong.

Hemidesmus indicus (L.) R.Br. Family: Asclepiadaceae

Hemidesmus indicus is known for its use in the treatment for snakebites. An organic acid, isolated and purified from the root extract possessed viper venom inhibitory activity (Alam *et al.,* 1994). The compound designated HI-RVIF was isolated by solvent extraction, silica gel column chromatography and thin layer chromatography and was a homogenous in nature. The adjuvant effect of 2-hydroxy-4-methoxy benzoic acid from *H. indicus* in the generation of anti-venom antiserum was explored by Alam *et al.* (1998). The antiserum raised in presence of the compound showed the higher neutralization capacity (lethal and hemorrhage) when compared with the antiserum raised with venom alone. The pure compound potentiated the neutralization of the lethal action of venom by commercial equine polyvalent snake venom antiserum in experimental models. The same authors have also shown that it possessed potent anti-inflammatory, antipyretic and antioxidant properties (Alam *et al.,* 1998). The compound effectively neutralized inflammation induced by *Vipera ruselli* venom in male albino mice. It also neutralized free radical formation as estimated by TBARS and enhanced the activity of antioxidant enzyme superoxide dismutase. The anti-snake venom activity of this compound may at least be partly mediated through the above physiological process.

Prabhakan *et al.* (2000) reported protective effect of *Hemidesmus indicus* against rifampicin and isoniazid-induced hepatotoxicity in rats. They are of the opinion that the observed effects are probably due to a free radical scavenging activity of the coumarino lignoids hemidesmin-1 and hemidesmin-2 present in the extract.

Ravishankara *et al.* (2002) evaluated the antioxidant activity of methanolic extract of *Hemidesmus indicus* root bark in several *in vitro* and *ex vivo* models. The *in vitro* and *ex vivo* antioxidant potential of root bark was evaluated in different systems, *viz.* radical scavenging activity by DPPH reduction, superoxide radical scavenging activity in riboflavin/light/NBT system, nitric oxide (NO) radical scavenging activity in sodium nitroprusside/Greiss reagent system and inhibition of lipid peroxidation induced by iron-ADP-ascorbate in liver homogenate and phenylhydrazine induced haemolysis in erythrocyte membrane stabilization study. The extract was found to have different levels of antioxidant properties in the models tested. In scavenging DPPH and superoxide radicals, its activity was intense (EC_{50}=18.87 and 19.9 µg/ml respectively) while in scavenging NO radical, it was moderate. It also inhibited lipid peroxidation of liver homogenate (EC_{50}=43.8µg/ml) and the haemolysis induced by phenylhydrazide (EC_{50} = 9.74 µg/ml) confirming the membrane stabilizing activity.

The methanolic extract of *Hemidesmus indicus* roots was found to inhibit lipid peroxidation and scavenge hydroxyl and superoxide radicals *in vitro* (Mary *et al.*, 2003). The amount required for 50 per cent inhibition of lipid peroxide formation was 217.5 µg/ml. The concentrations needed to scavenge hydroxyl and superoxide radicals were 73.5 and 287.5 µg/ml, respectively. The intravenous administration of this extract (5mg/kg body weight) in rabbits delayed the plasma recalcification time and enhanced the release of lipoprotein lipase enzyme significantly.

Radioprotective effect of *Hemidesmus indicus* root extract on lipid peroxidation in rat liver microsomes and plasmid DNA was examined by Shetty *et al.* (2005). The root extract was found to protect microsomal membranes as evident from reduction in lipid peroxidation values.

Methanolic extracts of *Hemidesmus indicus* stem was evaluated for their antioxidant activity by ferric thiocyanate assay and compared with TBA method. The extract showed good antioxidant activity. 77.0 per cent decrease of DPPH standard solution was recorded for the extract at a concentration of 100µg/ml (Zahin *et al.*, 2009).

Surveswaran *et al.* (2010) surveyed antioxidant properties and total phenolic and flavonoid contents of 15 samples, representing 12 Indian medicinal plant species. Total antioxidant assay was performed using ABTS and FRAP methods. The roots of *Hemidesmus indicus* showed the highest OH radical scavenging activity. The highly significant and positive correlations between total antioxidant capacity parameters and total phenolic content indicated that the phenolic compounds contributed significantly to the antioxidant activity of the tested plant samples.

Murali *et al.* (2010) evaluated antioxidant activity of roots of *Hemidesmus indicus* var. *pubescens*. Methanol and aqueous extracts of the drug were evaluated for *in vitro* antioxidant activity against DPPH, ABTS, hydrogen peroxide, nitric oxide and superoxide radicals. Both extracts exhibited similar scavenging effects against ABTS, and superoxide radicals whereas DPPH and nitric oxide, the methanol extract was more effective. The study revealed that extracts of *H. indicus* var. *pubescens* roots possess antioxidant and free radical scavenging effects.

Henophyton deserti Coss. et Durieu. Family: Brassicaceae (Cruciferae)

Derbel *et al.* (2010) studied the chemical composition and biological potential of seed oil and leaf extracts of *Henophyton deserti*. Total flavonoids ranged between 45.66 and 181.2 mg QE g^{-1} and 2.03 and 38.95 mg QE g^{-1} dry weight (dw) in leaf and seed polar extracts, respectively. Rutin, Kaempferol 3-rutinoside, Diosmetin 7-O-Glucoside and Acacetin 7-O-glucoside flavonoids were tentatively identified in this plant. The profile of the seed fatty acids revealed oleic acid (27 per cent), linoleic acid (12 per cent) and linolenic acid (17 per cent). The highest antioxidant activity of 85.2 per cent and 67.5 per cent were obtained with methanol and ethyl acetate leaf extracts.

Heracleum nepalense D. Don Family: Apiaceae (Umbelliferae)

Dash *et al.* (2005) investigated the antioxidant activity of the methanol extract of *Heracleum nepalense* root. The methanol extract at 1000µg/ml and the ethyl acetate fraction at 50µg/ml exhibited significant antioxidant activity in ferrous sulphate induced lipid peroxidation, DPPH, hydroxyl radical and superoxide scavenging models.

Heracleum pastinacifolium C.Koch Family: Apiaceae (Umbelliferae)

Firuzi *et al.* (2010) analyzed the composition and biological activities of essential oils from *Heracleum pastinacifolium*. In the DPPH radical scavenging assay, *Heracleum pastinacifolium* oils showed highest activity with IC_{50} values of 7.3 mg/ml. Total phenolic content was 1.809 mg catechin equivalent/g essential oil. Antioxidant activity correlated well with the total phenolic content of the oils.

Heracleum persicum Desf. Family: Apiaceae (Umbelliferae)

Souri *et al.* (2004) isolated four furanocoumarins from the fruits of *Heracleum persicum*. The isolated constituents were tested by linolic acid peroxidation for their antioxidant activities and were found to be moderately active. Antioxidant activity of crude ethyl acetate extract was stronger than single isolated constituents (Souri *et al.*, 2004).

Antioxidant activities of ethanol extracts from seven Apiaceae fruits have been studied by Nickavar and Abolhasani (2009) by the DPPH radical scavenging test. All the studied extracts, including *Heracleum persicum*, showed antioxidant capability. For *Heracleum persicum* IC_{50} value of the DPPH scavenging activity was 294.00µg/ml and total flavonoid content was 22.23µg/mg. A positive correlation was found between the antioxidant potency and flavonoid content of the fractions.

Firuzi *et al.* (2010a) analyzed the composition and biological activities of essential oils from *Heracleum persicum*. In the DPPH radical scavenging assay, *Heracleum persicum* oils showed highest activity with IC_{50} values of 7.4 mg/ml. Total phenolic content was 1.353 mg catechin equivalent/g essential oil. Antioxidant activity correlated well with the total phenolic content of the oils.

Heracleum rechingeri Manden Family: Apiaceae (Umbelliferae)

Firuzi *et al.* (2010a) analyzed the composition and biological activities of essential oils from *Heracleum rechingeri*. In the DPPH radical scavenging assay, *Heracleum rechingeri* oils showed moderate antioxidant activity with IC_{50} values of 11.6 mg/ml. Total phenolic content was 0.399 mg catechin equivalent/g essential oil. Antioxidant activity correlated well with the total phenolic content of the oils.

Heracleum transcaucasicum Manden Family: Apiaceae (Umbelliferae)

Firuzi *et al.* (2010a) analyzed the composition and biological activities of essential oils from *Heracleum transcaucasicum*. In the DPPH radical scavenging assay, *Heracleum transcaucasicum* oils showed moderate antioxidant activity with IC_{50} values of 16.3 mg/ml.Total phenolic content was 0.390 mg catechin equivalent/g essential oil. Antioxidant activity correlated well with the total phenolic content of the oils.

Heritiera fomes Buch.-Ham. Family: Sterculiaceae

Wangensteen *et al.* (2009) evaluated the antioxidant activity of the EtOH extract of stem bark from *Heritiera fomes.* Trimeric, pentameric and hexameric procyanidins were identified in addition to highly polymeric material (average degree of polymerization 18-24). Bioactivity studies showed high DPPH radical scavenging and 15-lipoxygenase (15-LO) inhibiting activities of the bark extracts (EC_{50} = 19.4 and IC_{50} = 22 microg/ml, respectively) which could be ascribed to its high content of procyanidins. The procyanidins were also assayed as DPPH scavengers and 15-LO inhibitors, with EC_{50} and IC_{50} values in the range of 8-15 and 10-15 microg/ml, respectively.

Hernandia nymphaefolia (Presl.) Kubitzki Family: Hernandiaceae

Chen *et al.* (2001) reported that among the isolates of *Hernandia nymphaefolia* eight compounds showed effective antioxidant activities in scavenging the stable free radical, DPPH.

Heterosmilax erythrantha Baill. Family: Smilacaceae

Extracts of *Heterosmilax erythrantha* root showed remarkable free radical scavenging activity with IC_{50} value of 17.0µg/ml and significant inhibitory effects on lipid peroxidation with 88.0 per cent inhibition at the concentration of 50ng/ml (Thoung *et al.*, 2006).

Heterothalamus alienus (Spreeng.) Kuntze Family: Asteraceae (Compositae)

Borneo *et al.* (2008) evaluated the antioxidant activity of ethanolic extract of *Heterothalamus alienus*. The FRAP value was 826.2µmol of Fe(II)/g and DPPH radical scavenging activity was EC_{50} = 448.3.

Heterotheca inuloides Cass. Family: Asteraceae (Compositae)

Sesquiterpenoids, 7-hydroxy-3,4-dihydrocadalin and 7-hydroxycadelin and flavonoids, quercetin, kaempferol and their glycosides isolated from a Mexican medicinal plant *Heterotheca inuloides* was evaluated as an antioxidant by Haraguchi *et al.* (1996, 1997). Thes compounds showed potent scavenging activity on DPPH radical. Microsomal lipid peroxidation induced by Fe(II)-ADP/NADH or Fe(III)-ADP/NADH was inhibited by both terpenoids and flavonoids, though only flavonoids possessed superoxide anion scavenging activity in microsome. Flavonoids also scavenged enzymatically and non-enzymatically generated superoxide anion. On the other hand, mitochondrial lipid peroxidation induced by Fe(III)-ADP/NADH was inhibited only by sesquiterpenoids. Furthermore, these terpenes protected mitochondrial enzyme activity against oxidative stress. Kubo *et al.* (1996) also investigated the antioxidative properties of these sesquiterpenoids.

Hibiscus calyphyllus Cav. Family: Malvaceae

Odukoya *et al.* (2007) evaluated the antioxidant activity and total phenolic contents of Nigerian green leafy vegetables including *Hibiscus calyphyllus*.

Hibiscus aculeatus Roxb. Family: Malvaceae

Sunilson *et al.* (2008) reported gastroprotective and antioxidant activities of the roots of *Hibiscus aculeatus* in rats.

Hibiscus cannabinus L. Family: Malvaceae

Hibiscus cannabinus flowers were investigated by Mukherjee *et al.* (2010) for free-radical scavenging property *in vitro*, capacity to protect DNA from oxidative damage and inhibiting gelatinolytic activity of collagenase type I and II. For this purpose, flowers extracted with methanol and water and subjected to DPPH free radical scavenging assay, calculation of IC_{50} values, using vitamin C as a standard reducing power assay, DNA damage protection assay and inhibition of gelatinolytic activity of collagenase type I and II. The DPPH free radical scavenging activity raged from 440 to 700 µg/ml for different extracts vis-a-vis 39 µg/ml for standard vitamin C. A similar trend was visible in reducing power activity. Both the activities reflected a strong antioxidant potential of *Hibiscus cannabinus* flowers and in turn against stress. Further, both extracts at 100 µg/ml were efficient in protecting DNA against oxidative damage.

Hibiscus rosa-sinensis L. Family: Malvaceae

Antioxidant activity of alcoholic flower extract of *Hibiscus rosa-sinensis* was studied by Neeli *et al.* (2008) by DPPH reduction assay, scavenging of O_2, scavenging of H_2O_2 and NO scavenging. Alcoholic flower extract possessed significant antioxidant activity in all the models. In all the models at 250 µg/ml and 500 µg/ml showed significant activity at 50 and 100µg/ml.

Liu *et al.* (2008) evaluated the antioxidant activity of water, 50 per cent ethanol and 70 per cent acetone extracts of flowers of 15 plants of Hainan. 50 per cent ethanol extract from fresh flower of *Hibiscus rosa-sinensis* had best effect on eliminating O_2 and OH radicals.

Hibiscus sabdariffa L. Family: Malvaceae

English Name: Roselle.

The antioxidative activity of water extracts of three herbs, including the calyx of *Hibiscus sabdariffa* were investigsted by Osawa and Namiki (1981). The extracts showed marked antioxidative activity, not only in linoleic acid but also in liposome model systems, indicating that the three herbal water extracts may protect the cell from damage by peroxidation. The extracts possessed high contents of phenolic compounds and exhibited reducing power, revealing that these herbal extracts may contain reductones. The water extracts of three herbs also showed good hydrogen-donating abilities, indicating that they had effective activities as radical scavengers.

Suboh *et al.* (2004) investigated the antioxidant activity effects of various concentrations of methanol extracts of *Hibiscus sabdariffa* and six other medicinal plants. They reported that the extract significantly reduced 10mM hydrogen peroxide-induced lipid peroxidation caused by H_2O_2 in human erythrocytes.

The antioxidant actions of 80 per cent ethanolic extract of dried flowers of *Hibiscus sabdariffa* on lipid peroxidation (LPO), reduced glutathione (GSH), glutathione-s-transferase (GST), Catalase (CAT), superoxide dismutase (SOD) and vitamin C, were examined by Usoh *et al.* (2005) using a model of sodium arsenite induced oxidative stress in rats. Pretreatment with the extracts showed a significant dose-dependent increase in liver and decrease in whole blood activities of SOD and CAT, hence revealing the hepatoprotective and antioxidant effectiveness of the extracts. Furthermore, the extracts, evaluated (*in vitro*) by their capacity of quenching DPPH free radical, showed strong scavenging effects on DPPH free radical at concentration of 0.20mg/ml. The extracts at low and high concentrations showed no inhibitory effect on nitric oxide radical.

The antioxidant and free radical scavenging effects of two fractions of the ethanol extract (chloroform soluble fraction and ethyl acetate soluble fraction) obtained from its dried flowers were investigated (Farombi and Fakoya, 2005) and found that both the fractions scavenge hydrogen peroxide (79-94 per cent) at the dose of 500µg. Similarly the extracts showed inhibitory (70-80 per cent) effects on superoxide anions radicals at a dose of 100µg. The antioxidant activities of three varieties using liposome system have also been reported. Methanol and ethyl acetate extracts showed higher COX-1 enzyme inhibition than COX-2 inhibition (Christian *et al.*, 2006).

Hirunpanich *et al.* (2006) reported that the aqueous extracts from the dried calyx of *Hibiscus sabdariffa* possess both antioxidative effects against LDL oxidation and hypolipidemic effects *in vivo*.

Wong *et al.* (2006) investigated the antioxidant properties of 25 edible tropical plants including *Hibiscus sabdariffa* using DPPH and FRAP assays. Total phenolic content was also estimated. *Hibiscus sabdariffa* showed moderate antioxidant activity.

Kruawan and Kangsadalampai (2006) investigated the antioxidant activity and phenolic compound contents of water extract of *Hibiscus sabdariffa*. The extract exhibited strong scavenging activity against DPPH radicals (93.12 per cent) and their FRAP value was 2220 µmol/g. The content of phenolic compound in the extracts was determined using the Folin-Ciocalteu reagent and calculated as gallic acid equivalent (GAE). The herbal extract gave a high phenolic compound content (GAE 210.72 mg/g).

Hypolipidemic and antioxidant effects of ethanolic extract from dried calyx of *Hibiscus sabdariffa* (HSE) in alloxan-induced diabetic rats were evaluated by Farombi and Ige (2007). HSE at 200mg/kg attenuated the alloxan-induced decrease in the activities of SOD, CAT and the level of GSH by 36 per cent, 44 per cent and 64 per cent in the liver and by 20 per cent, 43 per cent and 85 per cent in the kidney of rats. HSE (200mg/kg) significantly decreased the alloxan-mediated increase in MDA and protein carbonyl levels in the liver by 44 per cent and 43 per cent and in kidneys by 45 per cent and 38 per cent respectively.

Antioxidant and antihyperlipidemic activities of the extracts of leaves and calyces of *Hibiscus sabdariffa* were investigated by Ochani and D'Mello (2009) by studying their *in vitro* inhibitory activity on lipid peroxidation and *in vivo* effects on cholesterol induced hyperlipidmia. Highest antioxidant activity was exhibited by ethanolic exract of calyces followed by ethanolic extract of leaves followed by aqueous extract of leaves. Results suggest that the ethanolic extract of calyces and leaves of *H. sabdariffa* containing polyphenols and flavonols possess significant antioxidant and antihyperlipidemic activities (Ochani and D'Mello, 2009)

The calyx of the roselle plant has long been recognised as a source of antioxidants. Mohd-Esa *et al.* (2010) evaluated the antioxidant activity, free radical scavenging and total phenolic content in other parts of the roselle plant. Roselle seed extracts were found to have the highest antioxidant activity and strongest radical-scavenging activity of all parts tested. Methanol extracts showed a positive correlation between phenolic content and antioxidant activity as measured by β-carotene bleaching assay and DPPH radical scavenging activity. The antioxidant efficacy of roselle seeds in a whole food systems was investigated by testing the effect on lipid oxidation in cooked beef patties stored at 4°C for 14 days. Results showed that patties treated with roselle seeds had reduced lipid oxidation compared to patties treated with BHT.

Antioxidant activities of eight leafy vegetables of Ghana including *Hibiscus sabdariffa* were assessed by Morrison and Twumasi (2010). The total antioxidant capacity (TAC) and Total Phenol content (TPC) in the methanol extracts (METE) and hydro-ethanol extracts (HETE) were measured. The TPC of the methanol extract of *Hibiscus sabdariffa* was 0.318 mg/ml TAE and TAE was 0.318 mg/ml Ascorbic acid equivalent (AAE) while TPC of hydro-ethanol extract was 0.463 mg/ml TAE and TAE was 0.445 mg/ml Ascorbic acid equivalent (AAE). Radical scavenging activity of methanol extract was EC_{50} = 0.1745mg/ml and HETE was 0.1212 mg/ml while Fe3+ reducing potential was EC_{50} = 0.1212 (METE) and 0.1212 (HETE).

Hibiscus tiliaceus L. Family: Malvaceae

Vadlapudi and Naidu (2009) evaluated the antioxidant potential of *Hibiscus tiliaceous*, a mangrove plant. The SOD was 0.82U/mg, Catalase was 0.42U/mg, Ascorbic acid was 290mg/100g, DPPH radical inhibition was 55 per cent and FRAP units were 650.

Hieracium cappadocicum Freyn Family: Asteraceae (Compositae)

Tepe *et al.* (2006) screened the antioxidative properties of the methanolic extracts of *Hieracium cappadocicum* by two complementary test systems, namely DPPH free radical-scavenging and β-carotene/linoleic acid. The extract exerted greater antioxidant activity with IC_{50} value of 30.0 µg/ml. In the β-carotene/linoleic acid test system, *H. cappadocicum* showed 55.1 per cent inhibition rate. Antioxidant activities of curcumin and ascorbic acid were also determined as positive controls in parallel experiments.

Hieracium pilosella L. Family: Asteraceae (Compositae)

The antioxidant activity of water, ethanol and methanol *Hieracium pilosella* extracts was reported by Stanojevi *et al.* (2009). The antioxidant activity of the investigated extracts slightly differs depending on the solvent used. The concentration of 0.30 mg/ml of water, ethanol and methanol extract was less effective in scavenging hydroxyl radicals (56.35, 58.73 and 54.35 per cent, respectively) in comparison with the DPPH radical scavenging activity (around 95 per cent for all extracts). The high contents of total phenolic compounds (239.59mg GAE/g of dry extract) and total flavonoids (79.13-82.18 mg RE/g of dry extract) indicated that these contribute to the antioxidative activity.

Himatanthus lancifolius (Muell.-Arg.) Woodson Family: Apocynaceae

The indole alkaloids obtained from the barks of *Himatanthus lancifolius* potently protected rats from experimentally induced gastric lesions and reduced gastric acid hypersecretion. An increase in the antioxidant activity as measured through glutathione S-transferase activity was observed in the hypersecretory but not in the ulcerative mode (Baggio *et al.*, 2005).

Hippomarathrum microcarpum Bieb. Family: Apiaceae (Umbelliferae)

Ozer *et al.* (2007) evaluated the *in vitro* antioxidant activity of the essential oil and methanol extract of *Hippomarathrum microcarpum*. The antioxidant activity of these extracts was assessed by the β-carotene bleaching test and the DPPH radical scavenging test. The inhibition of linoleic acid oxidation was very weak for both extracts tested. The inhibition percentages were found to be 22.9 and 33.5 per cent for methanol and essential oil, respectively, at the concentration of 2 g/l. The oil scavenged DPPH at higher concentrations (IC_{50} = 10.69 mg/ml), but the methanol extract exhibited no activity. The total phenolic content of the methanol extract was found to be 4.7 per cent.

Hippophae rhamnoides L. Family: Elaegnaceae

English Name: Seabuckthorn

Geetha *et al.* (2003) reported antioxidant activity of seabuckthorn on chromium induced oxidative stress in male albino rats. Different doses of the alcoholic leaf extract of seabuckthorn at a concentration of 100 and 250 mg/kg BW protected the animals from the chromium induced oxidative injury significantly.

Chauhan *et al.* (2007) screened seeds aqueous extract for antioxidant activities (reducing power, DPPH and liposome model system). The extract showed good antioxidant activity. Ercisli *et al.* (2007) investigated the chemical composition and antioxidant activity of berries of *Hippophae rhamnoides*. The total phenolic content of the berries ranged from 21.31 mg GAE/g dry weight basis to 55.38 mg GAE/g. The highest antioxidant activity was 93.54 per cent (similar to standard BHT at 200 mg/l) and the lowest was 80.38 per cent. There was no correlation between total phenolic content and the antioxidant activity.

Effects of an aqueous extract of seabuckthorn seed residues were examined on serum glucose, lipid profiles and antioxidant parameters in diabetic rats (Zhang *et al.*, 2010). The data showed that administration of the extract significantly lowered the serum glucose, triglyceride and nitric oxide levels in diabetic rats. More over, the extract treatment also increased serum superoxide dismutase activity and glutathione level markedly. These results showed that seabuckthorn has hypoglycemic, hypotriglyceridemic and antioxidant effects in streptozotocin-induced diabetic rats.

Holarrhena antidysenterica Wall. ex DC.Family: Apocynaceae

Methanolic extracts of *Holarrhena antidysenterica* bark was evaluated for their antioxidant activity by ferric thiocyanate assay and campared with TBA method. The extract showed moderate antioxidant activity. 20.06 per cent decrease of DPPH standard solution was recorded for the extract at a concentration of 100µg/ml (Zahin *et al.*, 2009).

Kalim *et al.* (2010) evaluated the antioxidant activity of Methanol (50 per cent) extract of *Holarrhena antidysenterica*. IC_{50} values for DPPH, ABTS, NO, OH, O_2 and ONOO were 98.84, 29.92, 29.23, 211.34, 83.49 and 880.51µg/ml. The extract contained significant amount of polyphenols.

Holoptelia integrifolia Roxb. Family:Ulmaceae

Reddy *et al.* (2007) evaluated the antioxidant and wound healing potentials of *Holoptelia integrifolia*. The antioxidant activity was evaluated by DPPH free radical scavenging activity using HPLC method. The IC_{50} values obtained from methanolic extracts of stem bark (TPC : 78.53 mg/g) and leaves (TPC 57.71 mg/g) were 37.66 and 53.36µg/ml. respectively.

Homalonema occulta (Lour.) Schott. Family: Araceae

Gan *et al.* (2010) evaluated the antioxidant activity of *Homalonema occulta* using the FRAP and TEAC assays, and its total phenolic content was measured by the Folin-Ciocalteu method. For FRAP assay the value was 32.30 µmol Fe(II)/g dry weight. For TEAC assay, the value was 35.57µmol Trolox/g dry weight. Total phenolic content was 3.40 mg gallic acid equivalent/g dry weight of plant material.

Hopea hainanensis Family: Dipterocarpaceae

Phytochemical investigation of the stem wood of *Hopea hainanensis* by Ge *et al.* (2009) lead to four new oligostilbenoids, hopeahainols C–F (**1–4**), and eight known ones (**5–12**). A resveratrol glucoside, eight resveratrol dimers, two trimers, and a tetramer were identified. All of the polyphenols were tested for their radical scavenging and total reducing capacities by measuring their capacity to scavenge the DPPH radical, anion superoxide radical, and also to induce the reduction of Mo (VI) to Mo (V). Most of the compounds exhibited potent antioxidant and radical scavenger capacity compared with the positive controls (resveratrol, ascorbic acid, and butylated hydroxyanisole (BHA)) at 0.4 mM concentration.

Hopea odorata Roxb. Family: Dipterocarpaceae

Hydromethanol extracts of 15 Bangladeshi medicinal plants, including *Hopea odorata*, were evaluated by Hasan *et al.* (2009) for antioxidant potential using DPPH radical scavenging assay. *Hopea odorata* leaves exhibited radical scavenging activity with an IC_{50} value of 33.03 µg/ml compared to the IC_{50} value of 5.15µg/ml as shown by the reference antioxidant ascorbic acid, in a dose dependent fashion.

Hopea ponga (Dennst.) Mabberly Family: Dipterocarpaceae

Seed wing of *Hopea ponga* were extracted by Sukesh *et al.* (2011) using different solvents of various polarities such as hexane, ethyl acetate, methanol and water and screened for phytochemicals such as alkaloids, flavonoids, phenols, tannins, saponins, and sterols. The antioxidant activity was assayed through *in-vitro* models such as DPPH, and reducing power.

Hordeum vulgare L. Family: Poaceae (Graminae)

English Name: Barley.

The antioxidative activity of water extracts of three herbs, including the roasted seed of *Hordeum vulgare* were investigsted by Osawa and Namiki (1981). The extracts showed marked antioxidative activity, not only in linoleic acid but also in liposome model systems, indicating that the three herbal water extracts may protect the cell from damage by peroxidation. The extracts possessed high contents of phenolic compounds and exhibited reducing power, revealing that these herbal extracts may contain reductones. The water extracts of three herbs also showed good hydrogen-donating abilities, indicating that they had effective activities as radical scavengers.

The antioxidative activity of phenolic extracts from barley was examined by autoxidation of methyl linoleate by Kahkonen *et al.* (1999). The processing seems to be important when evaluating the antioxidant activity of cereals as brans were more active than other products. This is obviously due to the localization of the phenolics in the grain: the outer layers of the grain (husk, pericarp, testa and aleurone cells) contain the greatest concentrations of total phenolic, whereas their concentrations are considerably lower in the endosperm layers.

Kim *et al.* (2007) investigated the relationship between phenolic compounds, anthocyanins content and antioxidant activity in colored barley germplasm. They reported that in colored barley, DPPH radical scavengning activity had a high positive correlation to the content of phenolic compounds and proanthocyanidins.

Two black barley (*Hordeum vulgare*) genotypes were fractionated in bran and flour fractions, examined, and compared for their free radical scavenging properties against TEAC, FRAP, TPC, phenolic acid composition, carotenoid composition and total anthocyanin content. Bran fractions had the highest antioxidant activities (1.9-2.3 mmol TEAC/100g) in both the grain genotypes and were 3-5-fold higher than the respective flour fractions (0.4-0.7 mmol TEAC/100g). *p*-coumaric acid was the predominant phenolic acid in bran fractions of barley genotypes. The highest contents of anthocyanins were found in the middlings of black barley genotypes (Siebenhandl *et al.*, 2007).

Korycinska *et al.* (2009) evaluated the antioxidant properties of rye bran alkylresorcinols (C15:0-C25:0) and extracts from whole-grain cereal products using their radical-scavenging activity on DPPH and the chemiluminescence method (CL). DPPH radical reduction varied from ~10 to ~60 per cent fro the alkylresorcinol homologues at concentrations from 5 to 300μM and was not dependent on the length of the alkyl side chain of the particular homologue. Differences in the EC_{50} values for the studied compounds were not statistically significant, the values varying from 157μM for homologue C23:0 to 195μM for homologue C15:0. Moreover, values of EC_{50} for all the alkylresorcinol homologues were significantly higher than those for Trolox and α-, δ-, γ-tocopherols, compounds with well-defined antioxidant activity and used as positive controls. CL inhibition was evaluated for all the tested alkylresorcinol homologues at concentrations of 5 and 10 μM and varied from ~27 to ~77 per cent. Similar to the DPPH method, the slight differences in CL inhibition suggest that the length of the alkyl side chain had no major impact on their antioxidant properties. The extracts from whole-grain products were added to the DPPH and CL reaction systems and their antioxidant activities were tested and compared with the total amount of alkylresorcinols evaluated in the extracts. DPPH radical and CL reduction for the whole-grain products varied from ~70 to ~43 per cent and from ~37 to ~91 per cent, respectively. A clear relationship between DPPH radical and CL reduction levels and the amount of total alkylresorcinols was obtained for whole-grain breakfast cereals, in which the reduction level decreased in the order rye>wheat>mixed>barley. Therefore, it may be considered that the antioxidant activity of alkylresorcinols could be of potential importance to the food industry, which is continuously searching for natural antioxidants for the production of food products during their processing and storage.

Qiangming *et al.* (2010) evaluated the antioxidant activities of malt extract from barley by various methods *in vitro* and *in vivo*. Scavenging effects on the hydroxyl and superoxide radicals and protection against reactive oxygen species induced lipid, protein and DNA damage were evaluated. The extract exhibited high antioxidant activities both *in vitro* and *in vivo*, evidenced by its ability to scavenge hydroxyl- and superoxide-radicals, high reducing power, and protection against biological macromolecular oxidative damage. Furthermore, malt extract prevented the decrease of antioxidant enzyme activities, decreased liver and brain malondialdehyde levels and carbonyl content, and improved total antioxidant capabioity in δ-galactose-treated mice. In conclusion these results demonstrated potential antioxidant activities and antiaging effect of malt.

Houttuynia cordata Thunb. Family: Saururaceae

Cai *et al.* (2004) evaluated the antioxidant activity and phenolic compounds in traditional Chinese medicinal plants associated with anticancer. The improved ABTS.+ method was used to systematically assess the total antioxidant capacity (Trolox equivalent antioxidant capacity, TEAC) of the medicinal extracts. The TEAC values and total phenolic content for methanolic extract of aerial parts of *Houttuynia cordata* were 359.3 μmol Trolox equivalent/100g dry weight (DW), and 2.19 g of gallic acid equivalent/100g DW respectively. A positive significant linear relationship between antioxidant activity and total phenolic content showed that phenolic compounds were the dominant antioxidant components in the tested herbs. Major types of phenolic compounds in *Houttuynia cordata* include flavonols (quercetin, quercitrin, rutin), phenolic acids (chlorogenic acid).

Chomnawang *et al.* (2007) reported that *Houttuynia cordata* had a moderate antioxidant activity. Antioxidant activity was determined by DPPH scavenging and NBT reduction assay.

Li *et al.* (2008) evaluated the antioxidant properties of 45 medicinal plants including *Houttuynia cordata* using FRAP and TEAC assays and the total phenolic contents of these plants were measured. *Houttuynia cordata* showed 84.52 µmol Fe(II)/g FRAP value, 53.17µmol Trolox/g of TEAC value and 12.50 mg GAE/g of phenolic content. A high correlation between antioxidant capacities and their phenolic contents indicated that phenolic compounds were a major contributor of antioxidant activity of these plants.

Chemical analysis of MeOH extract of the whole plant of *Houttuynia cordata* has resulted in the isolation of two new compounds, named as houttuynoside A and houttuynamide A, together with thirty-eight known compounds. Antioxidant activities of some isolated compounds were also evaluated. Among these compounds, quercitrin and quercetin-3-O-β-D-galactopyranoside showed radical scavenging property with IC_{50} values of 31 and 63 microM, respectively (Chou *et al.*, 2009).

Hovenia dulcis Thunb. Family: Rhamnaceae

The changes of the antioxidant activity of the vinegar of *Hovenia dulcis* during fermentation process were analyzed by Li *et al.* (2011). The antioxidant activity of *Hovenia dulcis* during fermentation process was determined in terms of phenol, flavoniods, scavenging activity of DPPH radicals, reducing capacity, total antioxidant ability. The results showed that the phenol content and scavenging activity of DPPH radicals of the *Hovenia dulcis* during fermentation process were increasing. Flavoniods content was decreasing during alcohol fermentation process and increasing during acetic acid fermentation process. Reducing capacity essentially remained the same. Total antioxidant ability was increasing during alcohol fermentation process, which was slightly reducing during acetic acid fermentation process.

Humulus lupulus L. Family: Cannabinaceae

English Name: Hop.

It is distributed in the North Temperate regions, largely cultivated. The antioxidative activity of phenolic extracts from hop was examined by autoxidation of methyl linoleate by Kahkonen *et al.* (1999). There was no antioxidant activity although the phenolics were 23.1 mg/g GAE.

Seven naturally derived components from hop plant (*Humulus lupulus*) extracts were tested for evaluation of biological activities affecting acne vulgaris by Yamaguchi *et al.* (2009). Antioxidant capacity was also evaluated with seven different methods based on different reactive oxygen species. Xanthohumol showed the highest activity in total oxygen radical absorbance capacity as well as singlet oxygen absorbance capacity.

Hunteria umbellata Family: Apocynaceae

Adejuwon *et al.* (2011) evaluated the anti-inflammatory and antioxidant activities of crude (HU) butanol (HU$_b$) and alkaloid (HU$_{Af}$) fractions of the water seed extract of *Hunteria umbellata*, in addition, to determining the phenolic content of HU and Hu$_b$. The fraction's antioxidant activities were evaluated using DPPH free radical, superoxide anions and nitric oxide scavenging methods, in addition, to determining the phenolic contents of the fractions using standard procedures. Results showed that HU$_b$ and HU$_{Af}$ at 0.2-0.8 mg/ml exhibited significant DPPH free radical, superoxide anion and nitric oxide scavenging activities with the most significant effect recorded for HU$_{Af}$. Their proanthocyanidin contents were estimated to be 38.90 and 53.67 mg/g of dry extract, respectively while their total flavonoid contents were estimated to be 0.50 and 11.78mg/g of dry extract, respectively. Also, their total phenolic contents were estimated to be 39.68 and 97.12 mg/g of dry extract.

Hybanthus enneaspermus (L.) F.Muell. Family: Violaceae

The hepatoprotective, curative and anti-oxidant properties of aqueous extract of *Hybanthus enneaspermus* used against CCl$_4$-induced liver damage in rats were investigated by Vuda *et al.* (2011). Liver damage was induced by CCl$_4$ (1ml/kg i.p.), and silymarin was used as a standard drug to compare hepatoprotective, curative and antioxidant effects of the extract. Rats were treated with aqueous extract of *H. enneaspermus* at a dose of either 200 or 400mg/kg after division into pre-treatment (once daily for 14 days before CCl$_4$ intoxication) and post-treatment (2, 6, 24 and 48h after CCl$_4$ intoxication) groups. Pre-treatment and post-treatment with aqueous extract of *H. enneaspermus* showed significant hepatoprotection by reducing the aspartate transaminase, alanine transaminase, and alkaline phosphatase enzymatic activities and total bilirubin levels which had been raised by CCl$_4$ administration. Pre- and post-treatment with aqueous extract significantly decreased hepatic lipid peroxidation as well as producing a corresponding increase in tissue total thiols. Post-treatment with aqueous extract improved ceruloplasmin levels.

Hydnocarpus annamensis (Gagnep.) Lescot et Sleumer Family: Flacourtiaceae

Two new phenolic glycosides 1 and 2 were isolated from the barks of *Hydnocarpus annamensis* along with 18 known compounds. Compounds 7, 12, 13 and 20 showed antioxidant capacities in the DPPH free-radical assay, with a scavenging effect of 60.9-69.4 per cent at 50µM, and 18.5-34.6 per cent at 10µM (Shi *et al.*, 2006).

Hydnocarpus wightiana Blume Family: Flacourtiaceae

Acetone extract of the seed hulls of *Hydnocarpus wightiana* possess strong free radicals (DPPH and ABTS) scavenging, α-glucosidase and moderate N-acetyl-β-D-glucosaminidase inhibitory activities (Reddy *et al.*, 2005). Further fractionation of the extract led to the isolation of hydnocarpin, luteolin and isohydnocarpin in substantial yields. All the compounds showed strong ABTS scavenging property. Only luteolin could display strong DPPH scavenging activity (Reddy *et al.*, 2005).

Hydnophytum formicarum Jack Family: Rubiaceae

Ahmad *et al.* (2010) evaluated the antioxidant potential of 22 species of medicinal plants from Malaysian Rubiaceae including *Hydnophytum formicarum*. FTC, TBA, TPC and the DPPH assays were employed. The tested extracts showed strong antioxidant potential with percent inhibition of 98.5 per cent in the FTC, and 92.0 per cent in TBA assays. The TPC of the extract was 120.63mg GAE/g PE and strong DPPH radical-scavenging activity with IC_{50} value of 22.40. A good correlation was observed between total phenolic content and radical-scavenging activities.

Hydrocotyle batrachium Hance Family: Apiaceae (Umbelliferae)

Huang *et al.* (2008) examined the antioxidant and antiproliferative activities of the ethanol and water extracts of four *Hydrocotyle* species from Taiwan including *Hydrocotyle batrachium*. The results showed that the water extracts of all the samples had higher antioxidant and antiproliferative activities than the ethanol extracts. All the tested extracts were weaker than the positive controls (BHT and GSH) in the antioxidant activity. The IC_{50} value of water extract of *H.batrachium* was 178.99 µg/ml while IC_{50} value of ethanol extract was 403.31µg/ml. They also found that the water extracts of all the samples had higher content of polyphenol compounds, but lower content of flavonoid compounds than the ethanol extracts. The ABTS radical scavenging assay, the TEAC values of the water extracts samples were in descending order: *H.nepalensis*> *H. setulosa* > *H. batrachium* > *H. sibthorpioides*.

Hydrocotyle nepalensis Hook. Family: Apiaceae (Umbelliferae)

Huang *et al.* (2008) examined the antioxidant and antiproliferative activities of the ethanol and water extracts of four *Hydrocotyle* species from Taiwan including *Hydrocotyle nepalensis*. The results showed that the water extracts of all the samples had higher antioxidant and antiproliferative activities than the ethanol extracts. All the tested extracts were weaker than the positive controls (BHT and GSH) in the antioxidant activity. The IC_{50} value of water extract of *H. nepalensis* was 84.20 µg/ml while IC_{50} value of ethanol extract was 314.51µg/ml. They also found that the water extracts of all the samples had higher content of polyphenol compounds, but lower content of flavonoid compounds than the ethanol extracts.

Hydrocotyle setulosa Hayata Family: Apiaceae (Umbelliferae)

Huang *et al.* (2008) examined the antioxidant and antiproliferative activities of the ethanol and water extracts of four *Hydrocotyle* species from Taiwan including *Hydrocotyle setulosa*. The results showed that the water extracts of all the samples had higher antioxidant and antiproliferative activities than the ethanol extracts. All the tested extracts were weaker than the positive controls (BHT and GSH) in the antioxidant activity. The IC_{50} value of water extract of *H. setulosa* was 117.01 µg/ml while IC_{50} value of ethanol extract was 197.87µg/ml. They also found that the water extracts of all the samples had higher content of polyphenol compounds, but lower content of flavonoid compounds than the ethanol extracts.

Hydrocotyle sibthorpioides Lam. Family: Apiaceae (Umbelliferae)

Huang *et al.* (2008) examined the antioxidant and antiproliferative activities of the ethanol and water extracts of four *Hydrocotyle* species from Taiwan including

Hydrocotyle sibthorpioides. The results showed that the water extracts of all the samples had higher antioxidant and antiproliferative activities than the ethanol extracts. All the tested extracts were weaker than the positive controls (BHT and GSH) in the antioxidant activity. The IC_{50} value of water extract of *H. sibthorpioides* was 375.96 µg/ml while IC_{50} value of ethanol extract was 919.47µg/ml. They also found that the water extracts of all the samples had higher content of polyphenol compounds, but lower content of flavonoid compounds than the ethanol extracts.

Hyeronima antioquiensis Cuatrec Family: Euphorbiaceae

Mosquera *et al.* (2007) studied the antioxidant activity of the extracts using DPPH free-radical scavenging method. The percentage of the antioxidant activity of the n-Hexane fraction was 20.181, Dichloromethane fraction was 45.75 and methanol fraction was 50.94.

Hyeronima macrocarpa Schltr. Family: Euphorbiaceae

Mosquera *et al.* (2007) studied the antioxidant activity of the extracts using DPPH free-radical scavenging method. The percentage of the antioxidant activity of the n-Hexane fraction was 19.05 Dichloromethane fraction was 52.643 and methanol fraction was 50.47.

Mosquera *et al.* (2009) evaluated the antioxidant activity of forty-six methanol plant extracts using DPPH radical scavenging assay. *Hyeronima macrocarpa* showed 37.1 per cent antioxidant activity.

Hygrophila auriculata (Schum.) Heine (Syn.: *Asteracantha longifolia* Nees) Family: Acanthaceae

Methanolic extracts of seeds showed hepatoprotective activity in rats with thioacetamide and paracetamol-induced liver damage and hepatocarcinogenesis (Singh and Handa, 1995; Ahmed *et al.*, 2001). Ahmed *et al.* (2001) showed that plant extract inhibits CCl_4 induced oxidative stress and proliferative response. The root extract was also studied for its *in vitro* antioxidant activity using ferric thiocyanate (FTC) and thiobarbituric acid (TBA) methods (Shanmugasundaram and Venkataraman, 2006).

The free radical scavenging potential of 50 per cent aqueous alcoholc extract of *Hygrophila auriculata* was studied by Vijayakumar *et al.* (2005, 2006) for DPPH scavenging activity, nitric oxide, hydroxyl radical and ferryl bipyridyl complex scavenging activity along with lipid peroxidation and total antioxidant activity using TBARS in rat liver homogenate. Integral antioxidative capacity was determined by photochemiluminescence assay. The extract showed good radical scavenging activity at various concentrations (200-1000 µg/ml) against DPPH (32.32-77.02 per cent) with moderate scavenging activity against nitric oxide (12.46-52.84 per cent), hydroxyl radical (11.69-55.26 per cent), ferryl bipyridyl complex (171.66-58.67 per cent) and lipid peroxidation (0l829-0.416 nmoles/mg protein).

Shanmugasundaram and Venkataraman (2006) studied hepatoprotective and *in vitro* antioxidant effects of root extract of *Hygrophila auriculata*. Ferric thiocyanate and TBA methods were followed for the assay. The extract exhibited significant hepatoprotective and antioxidant activities.

Hylocereus polyrhizus (F.A.C.Weber) Britton and Rose Family: Cactaceae

English Name: Pitahaya, Pitaya, Dragon fruit.

Vaillant *et al.* (2005) established that the pitahaya (*Hylocereus polyrhizus*) fruit exhibited high antiradical activities. Pitahya fruit has a low vitamin C content ranging from 116 to 171 µg/g of fresh pulp without seeds but it is rich in betacyanins (0.32 to 0.41 mg/g) and phenolic compounds (5.6 to 6.8 µmol equivalent GAC/g. It has a high antioxidant ORAC value of 8.8 to 11.3µmol equivalent Trolox/g.

Pigment identification and antioxidant properties of red dragon fruit (*Hylocereus polyrhizus*) were investigated by Rebecca *et al.* (2009, 2010). Total phenolic compounds were 86.10mg/050g of dried dragon fruit extract expressed GAE. The reducing power assay further confirmed the antioxidant activity present in dragon fruit where the reducing capability increased from 0.18 to 2.37 with the increase of dry weight sample from 0.03 to 0.5 g. The vanillin-HCl assay which measures the amount of condensed tannin showed that the dried dragon fruit sample had an equivalent of 2.30 mg catechin/g. The DPPH radical scavenging activity determination showed that the effective concentration (EC_{50}) for dragon fruit was 2.90mM vitamin C equivalents/g dried extract. The total polyphenol assay which expresses gallic acid as equivalent showed that there was 86.1 mg/g of total polyphenolic compound in 0.5 g dragon fruit dry extract and the reducing power increased from 0.178 to 2.365 in 0.03 to 0.50 g of sample (Rebecca *et al.*, 2009).

Kim *et al.* (2011) investigated the total polyphenol and flavonoid content, antioxidant activity against various free radicals. The total polyphenol and flavonoid contents of flesh and peel of white and red pitayas, collected from Jeju island, Korea. Peel extracts of both the red and white pitayas showed higher DPPH and alkyl radical-scavenging activity than did the corresponding flesh extracts. There was a direct correlation between the phenolic content and antioxidant effect.

Hymenocardia acida Tul. Family: Hymenocardiaceae

Sofidiya *et al.* (2006) evaluated the DPPH radical scavenging, total antioxidant activities, reducing power and contents of phenolic compounds in methanolic leaf extracts of *Hymenocardia acida*. The extract showed a dose-dependent effect both on the free radical scavenging DPPH and also on Fe^{3+} reducing power. *Hymenocardia acida* possess very high radical scavenging activity in both assays. Total antioxidant activities correlated with total phenols. A linear positive relationship existed between the reducing power and total phenolics.

Sofidiya *et al.* (2009) and Ogbunugafor *et al.* (2010) investigated the antioxidant activity of aqueous and methanolic extracts of leaves from *Hymenocardia acida*. The inhibition values of the extracts and quercetin were found to be very close, with no significant differences at a concentration of 0.05 mg/ml in their ability to inhibit DPPH. The total proanthocyanidins for both water and methanol extracts were 20.2 and 30.6 (catechin equivalent) while the total phenol contents were 20.0 and 35.6 mg/ml (tannic acid equivalent), respectively. Positive correlations R2=0.85, R2 =

0.94, R2 = 0.97 for DPPH, reducing power and ABTS. Linear regression analysis also produced a high correlation coefficient with total proanthocyanidins (DPPH, Rs = 0.69; ABTS, R2=0.94). Results also showed elevated activity of SOD while CAT activity was depressed in the treated rats. Effects on enzyme activities suggest that the extracts induced SOD activity while inhibiting CAT activity, indicating an altered oxidative status in the animals.

Hymenocrater longiflorus Benth. Family: Lamiaceae (Labiatae)

The antioxidant activity of *Hymenocrater longiflorus* was evaluated by Ahmadi *et al.* (2010). The essential oils were able to inhibit the bleaching of β-carotene. The percentage of inhibition was even close (64.4 per cent) to those found for the non-polar sub-fraction (chloroformic) (69.1 per cent) which presented the best activity. The polar fraction rich in phenolic compounds had the greatest activity.

Hyoscyamus niger L. Family: Solanaceae

Souri *et al.* (2004) evaluated the antioxidative activity of *Hyoscyamus niger* by linoleic acid peroxidation test using 1,3-diethyl-2-thiobarbituric acid as the reagent. Methanolic extract showed 91.19 per cent inhibition at 40μg concentration with IC_{50} value of 1.64μg.

Cai *et al.* (2004) evaluated the antioxidant activity and phenolic compounds in traditional Chinese medicinal plants associated with anticancer. The improved ABTS.+ method was used to systematically assess the total antioxidant capacity (Trolox equivalent antioxidant capacity, TEAC) of the medicinal extracts. The TEAC values and total phenolic content for methanolic extract of seed of *Hyoscyamus niger* were 153.3 μmol Trolox equivalent/100g dry weight (DW), and 0.87 g of gallic acid equivalent/100g DW respectively. A positive significant linear relationship between antioxidant activity and total phenolic content showed that phenolic compounds were the dominant antioxidant components in the tested herbs.

Hypericum androsaemum L. Family: Hypericaceae

Valent o *et al.* (2002c) investigated the scavenging activity of *Hypericum androsaemum* infusion against superoxide radical, hydroxyl radical and hypochlorous acid. The tested infusion mainly exhibited a potent scavenginag effect on superoxide radicals (although a non competitive inhibitory effect on xanthine oxidase was also observed). The infusion also acted as a moderate scavenger of hydroxyl radicals and hypochlorous acid. A phytochemical study of the infusin was also undertaken and nine phenolic compounds were identified.

Souri *et al.* (2004) evaluated the antioxidative activity of *Hypericum androsaemum* by linoleic acid peroxidation test using 1,3-diethyl-2-thiobarbituric acid as the reagent. Methanolic extract showed 93.37 per cent inhibition at 40μg concentration with IC_{50} value of 2.60μg.

Hypericum ascyron L. Family: Hypericaceae

Chon *et al.* (2009) evaluated the antioxidant activities and total phenolics level and cytotoxicity of young sprouts of some traditional Korean salad plants. Methanol extract of *Hypericum ascyron* at 63 mg/kg exhibited the highest dose-dependent DPPH radical scavenging activity by 91.2 per cent and highest amount of phenolics 197.1mg/kg.

Hypericum calycinum L. Family: Hypericaceae

From the MeOH extract of *Hypericum calycinum*, two caffeoylquinic acid derivatives, butyl chlorogenate (1), and chlorogenic acid (2), seven flavonoids, quercetin (3), quercitrin (4), hyperoside (5), isoquercitrin (6), miquelian (7), rutin (8) and 13,118-biapigenin (9) and two flavonols, (+)-catechin (10) and (-)-epicatechin (11) were isolated. Free radical scavenging activities of the isolated compounds were determined in *in vitro*, DPPH and nitric oxide (NO) scavenging models. The compounds showed strong DPPH and moderate NO scavenging activities in a concentration dependent manner. (+)-catechin and (-)-epicatechin were found to be the most active compounds with IC_{50} values of 4.16 and 4.67 µM for DPPH and 190 and 170 µM for NO scavenging activities, respectively (Kirmizibekmez *et al.*, 2009).

Hypericum densiflorum Pursch. Family: Hypericaceae

Three acylphloroglucinol derivatives, isolated from the hexane and acetone extracts of aerial parts of *Hypericum densiflorum* showed antioxidant activity in lipid peroxidation assay (Henry *et al.*, 2009).

Hypericum empetrifolium Willd. Family: Hypericaceae

Meral and Konyalioglu (2003) investigated the antioxidant effects of total extracts by phosphomolybdenum spectrophotometric method. The results clearly demonstrated that the extract has antioxidant activity. They also studied the total flavonoid content by using aluminium-chloride method. They found a correlation between the polyphenol and flavonoid contents and the antioxidant activity.

Hypericum hyssopifolium L. Family: Hypericaceae

Isolation and characterization of antioxidant phenolic compounds from the aerial parts of *Hypericum hyssopifolium* subsp. *elongatum* var. *elongatum* by activity-guided fractionation was carried by Cakir *et al.* (2003). Dried methanol extract was dissolved in distilled water, and then fractioned by re-extracting with petroleum ether, chloroform, and ethyl acetate, subsequently. Antioxidant and DPPH radical-scavenging activities of these fractions were determined. None of these fractions showed antioxidant activity, in contrast water and ethyl acetate fractions acted as prooxidant. However, the ethyl acetate fraction exhibited the highest DPPH radical-scavenging activity and the amount of phenolic compound was highest, too. Therefore ethyl acetate fraction was subjected to further separation by chromatographic methods. Thus five flavonoids (13, 118-biapigenin, quercetin, quercetin-3-O-α-L-arabinofuranoside, quercetin-3-O-β-D-galactopyranoside, quercetin-3-O–β-D-galactopyranoside,-7-O- β-D-glucopyranoside) and a naphthodianthrone (hypericin) were isolated, and their structures were determined. All isolated compounds showed antioxidant and DPPH radical-scavenging activities. Although, 13,118-biapigenin and hypericin were able to show highest antioxidant activity, they had the lowest DPPH radical-scavenging activities.

Hypericum jovis Greuter Family: Hypericaceae

Two new and four known phloroglucinol derivatives were isolated by Athanasas *et al*. (2004) from the dichloromethane extract of the Greek endemic plant *Hypericum jovis*. Their antioxidant activity was evaluated *in vitro* with the DPPH assay and in cell cultures using the DCFH-DA assay. All six compounds demonstrated significant antioxidant activity, while two of them possessed activity at a cellular level comparable to Trolox, protecting against ROS.

Hypericum papuanum Ridley Family: Hypericaceae

Ialibinone and hyperguinone B isolated from *H. papuanum* showed a strong reduction of oxygen production polymorphonuclear cells (Heilmann *et al*., 2003). Ialibinone showed antioxidant activity for H_2O_2/horseradish and peroxidase system and superoxide scavenging activity in a cytochrome C assay.

Hypericum perforatum L. Family: Hypericaceae

English Name: St. John's Wort.

A study was carried out by Budinevi *et al*. (1995) on the antioxidative activity of *Hypericum perforatum*. Using the ethanolic extracts of plant material good results were obtained.

Meral and Konyalioglu (2003) investigated the antioxidant effects of total extracts by phosphomolybdenum spectrophotometric method. The results clearly demonstrated that the extract has antioxidant activity. They also studied the total flavonoid content by using aluminium-chloride method. They found a correlation between the polyphenol and flavonoid contents and the antioxidant activity. Ivanova *et al*. (2005) also reported high phenolics content and antioxidant properties in *Hypericum perforatum.*

The free radical scavenging and antioxidant activities of a standardized extract of *Hypericum perforatum* (SHP) were examined by Benedi *et al*. (2004) for inhibition of lipid peroxidation, for hydroxyl radical scavenging activity and interaction with DPPH. Concentrations between 1 and 50 microg/ml of SHP effectively inhibited lipid peroxidation of rat brain cortex mitochondria induced by Fe^{2+}/ascorbate or NADPH system. The results showed that SHP scavenged DPPH radical in a dose dependent manner and also presented inhibitory effects on the activity of xanthine oxidase. In contrast hydroxyl radical scavenging occurs at high doses.

The antioxidant potentials of a total ethanolic extract of *Hypericum perforatum* (TE) and fractions were evaluated and correlated with their phenolic contents (Silva *et al*., 2005). Kaempferol 3-rutinoside and rutin-acetyl were identified in the extract. The free radical-scavenging properties of the TE (EC_{50} = 21µgdwb/ml) and fractions were studied using DPPH. Fractions containing flavonoids and/or caffeoyl-quinic acids were found to be the main contributors to the free radical-scavenging activity of the extract. Lipid peroxidation, induced with ascorbate/Fe2+, was significantly reduced in the presence of the extract (EC_{50} = 26 µgdwb/ml) and fractions containing flavonoids and/or caffeoylquinic acids. The fraction containing flavonoid aglycones was found to be responsible for a major part of the TE protection against lipid peroxidation. Hypericins and hyperforins made no significant contributions to the antioxidant properties of TE (Silva *et al*., 2005).

Kirca and Arslan (2008) investigated the antioxidant capacity and total phenol contents of *Hypericum perforatum*. The antioxidant capacity of methanolic extracts prepared from various parts of plants was evaluated by both trolox equivalent antioxidant capacity (TEAC) and DPPH assays, while the total phenolics were determined using the Folin-Ciocalteu method. *Hypericum perforatum* showed higher antioxidant activities in both TEAC and DPPH method.

The aerial parts of *Hypericum perforatum* were studied to assess their composition in basic categories of bioactive compounds and their antioxidant activity, employing accepted assays. The results reveal that floral buds, during the developmental stage, bear the highest concentrations of hyperforin followed by the flowers of the plant and comparable concentrations of hypericin and pseudohypericin. Among the various plant fractions, shoots and branches show significant antioxidant activity, a fact which may be accounted for by the high pehnolic content. The measurement of DPPH antioxidant activity of extracts delineated the pivotal contribution of hyperforin, adhyperforin and their analogues to this antioxidant property. The role of phenolic compounds on the total DPPH scavenging activity was minor except chlorogenic acid and its derivatives. No encouraging results were obtained as regards the inhibition of the oxidative damage of proteins. All the extracts promote the damage of proteins exhibiting a functional pro-oxidant role.

Zou *et al.* (2010) reported protective effects of a flavonod-rich extract of *Hypericum perforatum* L. against hydrogen peroxide-induced apoptosis in PC12 cells.

Spiridon *et al.* (2011) evaluated the total phenolic content and antioxidant activity of plants used in Romanian herbal medicine. *Hypericum perforatum* extracts showed good antioxidant activity.

Hypericum styphelioides Rich. Family: Hypericaceae

Two new antioxidative compounds have been isolated by Gamiotea-Turro *et al.* (2004) from the leaves of *Hypericum styphelioides.*

Hypericum triquetrifolium Turra Family: Hypericaceae

Meral and Konyalioglu (2003) investigated the antioxidant effects of total extracts by phosphomolybdenum spectrophotometric method. The results clearly demonstrated that the extract has antioxidant activity. They also studied the total flavonoid content by using aluminium-chloride method. They found a correlation between the polyphenol and flavonoid contents and the antioxidant activity.

The antioxidant activity of the methanol extract of aerial part and of flavonoids isolated therein was evaluated. The IC_{50} resulted between 0.062 and 1 mg/ml (Conforti *et al.*, 2002).

Hyphaene thebaica Mart. Family: Arecaceae

English Name: Doum palm.

Using Trolox assay, Cook *et al.* (1998) estimated the antioxidant activity of 17 wild edible plants of Niger Republic used for food and traditional medicine. They observed that *Hyphaene thebaica* possessed strong antioxidant activity.

A hot water extract of the fruit of *Hyphaene thebaica* was examined for its antioxidant activity (Hsu *et al.*, 2006). Hydrogen donating activity was 2.85 mmol ascorbic acid equivalent, Fe^{2+}-chelating activity was 1.78 mmol EDTA equivalent, hydroxyl radical scavenging activity was 192 mmol GAE, inhibition of substrate site-specific hydroxyl radical formation was 3.36 mmol GAE, superoxide radical scavenging activity was 1.78 mmol GAE, and reducing power was 3.93 mmol ascorbic acid equivalent.These values were of the same magnitude as antioxidant activity in black tea except for Fe^{2+}-chelating activity which was about 14 times more potent. The total phenolic content of the Doum palm was low, but the extract exhibited potent antioxidant activity in terms of GAE.

The antioxidant activity of the aqueous ethanolic extract of Doum leaves was studied by Eldahshan *et al.* (2008). Data showed that the extract scavenged superoxide anion radicals (IC_{50} = 1602 µg/ml) in a dose dependent manner using xanthine/hypoxanthine oxidase assay. Four major flavonoidal compounds were identified: Quercetin glucoside, Kaempferol rhamnoglucoside, Dimethoxyquercetin rhamnoglucoside.

Hypoxis hemerocallidea Fisch. et C.E.Mey. Family: Hypoxidaceae

English Name: African potato.

The norlignan glycoside hypoxoside, a major phytoconstituent of African potato (AP), its aglycon rooperol, and an aqueous preparation of lyophilized AP corms were screened for *in vitro* antioxidant activity using the DPPH and FRAP tests. Inhibition of quinolinic acid (QA) induced lipid peroxidation in rat liver tissue was studied *in vitro* using the TBA assay. Superoxide free radical scavenging activity was determined by the nitroblue tetrazolium assay. While rooperol and AP extracts reduced QA-induced lipid peroxidation in rat liver homogenates and significantly scavenged the superoxide anion at pharmacological doses, in comparison, hypoxoside was virtually devoid of activity. Since hypoxoside is converted to rooperol *in vivo* following administration of AP, the results indicate that the hypoxoside component in AP could have value as an antioxidant prodrug (Nair *et al.*, 2007).

Katere and Eloff (2008) investigated the chemical composition and antioxidant activity of fresh and dried aerial and dried underground parts of *Hypoxis hemerocallidea*, using DPPH assay. Components in the acetone extracts of both leaves and corms showed good antioxidant activity.

Hyptis fasciculata Benth. Family: Lamiaceae (Labiatae)

Silva *et al.* (2005) evaluated the antioxidant activity of ethanol and butanol extracts from aerial parts of fruit of *Hyptis fasciculata*.They evaluated the capacity of the extracts to inhibit the reduction of free radical DPPH and to protect *Saccharomyces cerevisae* cells against the lethal oxidative stress caused by tert-butylhydroperoxide (TBH). Ethanol and butanol extract of aerial parts of *Hyptis fasciculata* were able to increase the tolerance of *Saccharomyces cerevisae* to TBH and showed to be active as DPPH radical scavengers, thus indicating that this plant extracts could be considered as potential sources of antioxidants.

Silva *et al.* (2009) reported antioxidant activity of n-butanol fraction of aerial parts of *Hyptis fasciculata* and isoquercetin, a flavonoid identified in this species. DPPH radical scavenging method was followed.

Hyptis fruticosa Salzm. ex Benth. Family: Lamiaceae (Labiatae)

The ethanol extract (EE) of *Hyptis fruticosa* leaves and its n-Hexane, Chloroform, EtOAc and MeOH/H$_2$O partitions were evaluated by Andrade *et al.* (2010) for antioxidant activity. The EE and EtOAc partition present highest antioxidant potential (IC$_{50}$ = 35.00 and 36.67µg/ml DPPH, respectively), similar to the reference compound (IC$_{50}$ = 16.67µg/ml).

Hyptis suaveolens (L.) Poit. Family: Lamiaceae (Labiatae)

Narayanaswamy and Balakrishnan (2011) evaluated the antioxidant activity of 13 medicinal plants including leaf of *Hyptis suaveolens*. The aqueous extract showed highest 98 per cent inhibition of DPPH radical while ethanolic extract showed 90 per cent inhibition of DPPH radical.

Hyptis tetracephala Bordignon Family: Lamiaceae (Labiatae)

Mensor *et al.* (2001) evaluated the antioxidant activity of ethanol extracts of *Hyptis tetracephala* and partitions by DPPH radical scavenging assay. The extract exhibited significant free radical scavenging properties. Among the partitions the more polar ones (ethyl acetate and n-butanol) are those that generally have higher antioxidant activity.

Hyssopus officinalis L. Family: Lamiaceae (Labiatae)

English Name: Hyssop.

Two natural antioxidant preparations, namely Rosmol (liquid) and Rosmol-P (powder) were produced by Lugasi *et al.* (2006) by extraction of a mixture of medicinal plants belonging to the Lamiaceae family, such as *Rosmarinus officinalis, Prunella vulgaris, Hyssopus officinalis* and *Melissa officinalis*. The main active compound of the extract was supposed to be a phenolic (caffeic) acid derivative. The total polyphenol content of the preparation is very high, 8.72g/l for Rosmol and 93.7 g/kg for Rosmol-P. The products acted as primary and secondary antioxidants chelating transitional metal ions and inhibiting the autoxidation of linoleic acid. Rosmol and Romol-P scavenged free radicals formed during Fenton type reaction measured by chemiluminometry and also exhibited strong antioxidant property in Randox TAS measurement. The antioxidant activity of the products was unchanged after six months storage.

Kizil *et al.* (2010) evaluated the antioxidant activity of the essential oil of *Hyssopus officinalis* using the DPPH radical-scavenging method. The antioxidant activity of the essential oil was lower compared to BHT and ascorbic acid.

Icacina tricantha Oliv. Family: Icacinaceae

Sofidiya *et al.* (2006) evaluated the DPPH radical scavenging, total antioxidant activities, reducing power and contents of phenolic compounds in methanolic leaf extracts of *Icacina tricantha*. The extract showed a dose-dependent effect both on the free radical scavenging DPPH and also on Fe^{3+} reducing power. Total antioxidant activities correlated with total phenols. A linear positive relationship existed between the reducing power and total phenolics.

Ilex cornuta Lindl. Family: Aquifoliaceae

Zhu *et al.* (2009) investigated the major phenolic compounds, individual and total phenolics contents, and *in vitro* antioxidant properties (ABTS, DPPH, FRAP and OH assays) of Kudingcha genotypes from *Ilex cornuta*. Major phenolics in Kudingcha genotypes from *Ilex cornuta* were mono- and dicaffeoylquinic acids. All Kudingcha genotypes of *Ilex* exhibited significantly stronger antioxidant capacities. Within six *Ilex* genotypes, great variation existed in their composition of individual phenolic compounds and their antioxidant properties.

Ilex kudingcha C.J.Tseng. Family: Aquifoliaceae

Zhu *et al.* (2009) investigated the major phenolic compounds, individual and total phenolics contents, and *in vitro* antioxidant properties (ABTS, DPPH, FRAP and OH assays) of Kudingcha genotypes from *Ilex kudingcha*. Major phenolics in Kudingcha genotypes from *Ilex kudingcha* were mono- and dicaffeoylquinic acids. All Kudingcha genotypes of *Ilex* exhibited significantly stronger antioxidant capacities. Within six *Ilex* genotypes, great variation existed in their composition of individual phenolic compounds and their antioxidant properties.

Ilex paraguariensis A. St. Hil. Family: Aquifoliaceae

English Name: Mate.

Vander Jagt *et al.* (2002) evaluated the total antioxidant content of widely used medicinal plants of New Mexico. The plant *Ilex paraguriensis* contained the highest amount of antioxidant.

Paglioso *et al.* (2010) determined the methylxanthines, phenolic compounds and antioxidant activity of mate bark (residual biomass) and compare with those of mate leaves. The high antioxidant activity of mate bark and its high concentration of total polyphenols were apparent in both aqueous and methanolic extracts, the values of which were greater than those detected in the leaves. Of the phenolic acids identified, the levels of chlorogenic acid and 4,5-dicaffeoylquinic acid in the samples were significantly higher in the methanolic bark extract. With regard to methylxanthines, considerable concentrations were detected in the samples.

Ilex pubescens Hook. et Arn. Family: Aquifoliaceae

Gan *et al.* (2010) evaluated the antioxidant activities of 40 medicinal plants associated with prevention and treatment of cardiovascular and cerebrovascular diseases using FRAP and TEAC assays and their total phenolic contents were measured by the Folin-Ciocalteu methods. The antioxidant activity of *Ilex pubescens* was 44.41μmol Fe(II)/g in FRAP assay, 45.67μmol Trolox/gm in TEAC assay and Total phenolic content was 4.70 mg GAE/g.

Illicium anisatum L. Family: Illiciaceae

The essential oil of air-dried *Illicium anisatum* was analysed for chemical composition and antioxidant activity by Kim *et al*. (2009). The essential oil exhibited moderate DPPH scavenging activity.

Impatiens balsamina Family: Balsaminaceae

Gan *et al*. (2010) evaluated the antioxidant activities of 40 medicinal plants associated with prevention and treatment of cardiovascular and cerebrovascular diseases using FRAP and TEAC assays and their total phenolic contents were measured by the Folin-Ciocalteu methods. The antioxidant activity of *Impatiens balsamina* was 121.95μmol Fe(II)/g in FRAP assay, 107.60μmol Trolox/g TEAC assay and Total phenolic content was 8.42 mg GAE/g.

Impatiens bicolor L. Royle Family: Balsaminaceae

Antioxidant activity of different extracts of *Impatiens bicolor* were evaluated by Shahwar *et al*. (2010) through DPPH radical scavenging activity, phosphomolybdate and ferric thiocyanate methods. The DPPH radical scavenging activity of ethyl acetate extract was 92.2 per cent, total antioxidant activity was 0.495 and inhibition of lipid peroxidation was 68.3 per cent while total phenols was 492.2 mg GAE/g of crude extract.

Impatiens capensis Meerb. Family: Balsaminaceae

English Name: Spotted touch me not.

Borchardt *et al*. (2008) evaluated the antioxidant activity of methanol extracts of seeds of *Impatiens capensis* using DPPH antioxidant assay.The antioxidant value reported in μM Trolox/100g (TE) from DPPH radical scavenging activity of crude seeds extract was 70,940.

Impatiens edgeworthii Hook.f Family: Balsaminaceae

Antioxidant activity of different extracts of *Impatiens edgeworthii* were evaluated by Shahwar *et al*. (2010) through DPPH radical scavenging activity, phosphomolybdate and ferric thiocyanate methods. The DPPH radical scavenging activity of ethyl acetate extract was 71.0 per cent, total antioxidant activity was 0.473 and inhibition of lipid peroxidation was 72.3 per cent while total phenols was 429.4 mg GAE/g of crude extract.

Imperata cylindrica (L.) Beauv. Family: Poaceae (Graminae)

Cai *et al*. (2004) evaluated the antioxidant activity and phenolic compounds in traditional Chinese medicinal plants associated with anticancer. The improved ABTS.+ method was used to systematically assess the total antioxidant capacity (Trolox equivalent antioxidant capacity, TEAC) of the medicinal extracts. The TEAC values and total phenolic content for methanolic extract of root of *Imperata cylindrica* var. *major* were 98.2 μmol Trolox equivalent/100g dry weight (DW), and 0.65 g of gallic acid equivalent/100g DW respectively. A positive significant linear relationship between antioxidant activity and total phenolic content showed that phenolic compounds were the dominant antioxidant components in the tested herbs.

Gan *et al.* (2010) evaluated the antioxidant activities of 40 medicinal plants associated with prevention and treatment of cardiovascular and cerebrovascular diseases using FRAP and TEAC assays and their total phenolic contents were measured by the Folin-Ciocalteu methods. The antioxidant activity of *Imperata cylindrica* was 87.67µmol Fe(II)/g in FRAP assay, 43.23µmol Trolox/g and Total phenolic content was 4.88 mg GAE/g.

Indigofera coulutea (Burm.) Merril Family: Fabaceae (Papilionaceae)

Aqueous acetone extracts prepared from *Indigofera coulutea* were investigated for their phytochemical composition and their antioxidant activities. The total phenolic and flavonoid content of extracts were assessed by Folin-Ciocalteu and AlCl$_3$ methods, respectively. The extracts were evaluated for their antioxidant potentials using FRAP, DPPH and ABTS assays. Flavonoids, saponins, quinines, sterols/triterpenes and tannins were present in this species. Gallic acid, caffeic acid, rutin and myricetin were identified. *Indigofera coulutea* having highest phenolic content was also found to possess the best antioxidant activities. The results indicated a good correlation between antioxidant activities and total phenolic content (Bakasso *et al.*, 2008).

Indigofera macrocalyx Guill. et Perr. Family: Fabaceae (Papilionaceae)

Aqueous acetone extracts prepared from *Indigofera macrocalyx* were investigated for their phytochemical composition and their antioxidant activities. The total phenolic and flavonoid content of extracts were assessed by Folin-Ciocalteu and AlCl$_3$ methods, respectively. The extracts were evaluated for their antioxidant potentials using FRAP, DPPH and ABTS assays. Flavonoids, saponins, quinines, sterols/triterpenes and tannins were present in this species. Galangin and myricetin were identified in the extract. *Indigofera macrocalyx* having highest phenolic content was also found to possess the best antioxidant activities. The results indicated a good correlation between antioxidant activities and total phenolic content (Bakasso *et al.*, 2008).

Indigofera nigritana (Hook.f.) Merril Family: Fabaceae (Papilionaceae)

Aqueous acetone extracts prepared from *Indigofera nigritana* were investigated for their phytochemical composition and their antioxidant activities. The total phenolic and flavonoid content of extracts were assessed by Folin-Ciocalteu and AlCl$_3$ methods, respectively. The extracts were evaluated for their antioxidant potentials using FRAP, DPPH and ABTS assays. Flavonoids, saponins, quinines, sterols/triterpenes and tannins were present in this species. *Indigofera nigritana* having highest phenolic content was also found to possess the best antioxidant activities. The results indicated a good correlation between antioxidant activities and total phenolic content (Bakasso *et al.*, 2008).

Indigofera pulchra Willd. Family: Fabaceae (Papilionaceae)

Aqueous acetone extracts prepared from *Indigofera pulchra* were investigated for their phytochemical composition and their antioxidant activities. The total phenolic and flavonoid content of extracts were assessed by Folin-Ciocalteu and AlCl$_3$ methods, respectively. The extracts were evaluated for their antioxidant potentials using FRAP,

DPPH and ABTS assays. Flavonoids, saponins, sterols/triterpenes and tannins were present in this species (Bakasso *et al.*, 2008).

Indigofera tinctoria L. Family: Fabaceae (Papilionaceae)

Aqueous acetone extracts prepared from *Indigofera tinctoria* were investigated for their phytochemical composition and their antioxidant activities. The total phenolic and flavonoid content of extracts were assessed by Folin-Ciocalteu and $AlCl_3$ methods, respectively. The extracts were evaluated for their antioxidant potentials using FRAP, DPPH and ABTS assays. Flavonoids, saponins, quinines, sterols/triterpenes and tannins were present in this species. Gallic acid, quercitrin and myricetin were identified from the extract. *Indigofera tinctoria* having highest phenolic content was also found to possess the best antioxidant activities. The results indicated a good correlation between antioxidant activities and total phenolic content (Bakasso *et al.*, 2008).

Indigofera trifoliata L. Family: Fabaceae (Papilionaceae)

Ethanolic extract of aerial parts of *Indigofera trifoliata* was studied by Dhenge and Itankar (2008) for its *in vitro* antioxidant activity using DPPH and nitric oxide radical scavenging assay methods. The results of the study indicated that the ethanolic extract showed significant free radical scavenging activity.

Indigofera truxillensis Kunth Family: Fabaceae (Papilionaceae)

Antioxidant activity of the alkaloid indigo (2mg/kg, p.o) obtained from the leaves of *Indigofera truxillensis* were examined by Farias-Silva *et al.* (2007). The results suggested that the gastroprotective mechanisms of indigo include non-enzymatic antioxidant effects and the inhibition of polymorphonuclear neutrophil infiltration.

Inula aucherana DC. Family: Asteraceae (Compositae)

Crude extract from *Inula aucherana* was screened by Sengul *et al.* (2009) for its *in vitro* antioxidant and antimicrobial properties. Total phenolic content of extract of this plant was also determined. β-carotene bleaching assay and Folin-Ciocalteu reagent were used to determine total antioxidant activity and total phenols of plant extract. Antioxidant activity was 80.10 per cent and the total phenolic content was 6.57 mgGAE/g DW).

Inula britanica L. Family: Asteraceae (Compositae)

Cai *et al.* (2004, 2006) evaluated the antioxidant activity and phenolic compounds in traditional Chinese medicinal plants associated with anticancer. The improved ABTS.+ method was used to systematically assess the total antioxidant capacity (Trolox equivalent antioxidant capacity, TEAC) of the medicinal extracts. The TEAC values and total phenolic content for methanolic extract of inflorescence of *Inula britanica* were 356.3 µmol Trolox equivalent/100g dry weight (DW), and 2.37 g of gallic acid equivalent/100g DW respectively. A positive significant linear relationship between antioxidant activity and total phenolic content showed that phenolic compounds were the dominant antioxidant components in the tested herbs. Major types of phenolic compounds in *Inula britanica* include flavonols (quercetin and its glycosides), phenolic acids (caffeic acid, chlorogenic acid).

Antioxidant capacities of 56 selected Chinese medicinal plants, including *Inula britannica* were evaluated by Song *et al*. (2010) using the TEAC and FRAP assays, and their total phenolic content was measured by the Folin-Ciocaleteu method. *Inula britannica* showed moderate antioxidant activity with TEAC value of 96.12μmol Trolox/g, FRAP value of 142.31μmol Fe^{2+}/g and phenolic content was 12.83mg GAE/g.

Inula crithmoides L. Family: Asteraceae (Compositae)

The antioxidant activity of the essential oil of *Inula crithmoides* was evaluated by DPPH test and lipoxygenase assay by Giamperi *et al*. (2010). The essential oil exerted a good antioxidant activity in the protection of lipid peroxidation when compared with known antioxidants.

Inula helenium L. Family: Asteraceae (Compositae)

English Name: Elecampe.

Wojcikowski *et al*. (2007) evaluated the antioxidant activity of aerial parts of *Inula helenium*. Oxygen radical absorbance capacity of the ethanol extract, methanol extract and aqueous extracts were 12.42, 22.70 and 25.17μmol Trolox equivalent per g of dried starting material, respectively.

Spiridon *et al*. (2011) evaluated the total phenolic content and antioxidant activity of plants used in Romanian herbal medicine. Total phenolic concentration was determined using Folin-Ciocalteu phenol reagent method, while total flavonoids were measured using the aluminium chloride colorimetric method. *Inula helenium* extracts showed good antioxidant activity.

Inula racemosa Hook. f. Family: Asteraceae (Compositae)

Haseena Banu *et al*. (2010) evaluated hyperlipidemic and antioxidant activities of *Inula racemosa* roots. The results obtained in this study proved the efficacy of the hydroalcoholic extract of *Inula racemosa* has both hypolipidemic and antioxidant activities.

Inula viscosa (L.) Ait. Family: Asteraceae (Compositae)

The antioxidant properties of twenty medicinal herbs used in the traditional Mediterranean and Chinese medicine including *Inula viscosa* were studied by Schinella *et al*. (2002). Using free radical-generating systems *Inula viscosa* protected against enzymatic and non-enzymatic lipid peroxidation in model membranes and also showed scavenging property on the superoxide radical. All the extracts were assayed at a concentration of 100μg/ml. Most of the extracts were weak scavengers of the hydroxyl radical.

Al-Mustafa and Al-Thunibat (2008) evaluated the antioxidant activity of Jordanian medicinal plant *Inula viscosa* shoot. The level of antioxidant activity was determined by DPPH and ABTS assay in relation to total phenolic contents of the medicinally used part. They concluded that *Inula viscosa* shoot possess low antioxidant capacity (DPPH-TEAC < 20mg/g).

Ipomoea aquatica Forsk. Family: Convolvulaceae

English Name: Water convolvulus. Japanese: Ensai.

Huang *et al.* (2005) examined the possible antioxidant and antiproliferative activities of 95 per cent ethanol or water extract from *Ipomoea aquatica*. Ethanol extract of stem demonstrated a positive effect in DPPH staining when it was diluted to 6.25 mg dry matter/ml while all other fractions showed no effect at the same dilution. This fraction also had the highest content of total phenolic compounds, as well as the highest reducing power and FTC activity. Ethanol extract of leaf had the highest amount of flavonoids. Using DPPH calorimetric method, it was found that ethanol extract of stem had the highest radical-scavenging activity, followed by ethanol extract of leaf.

Maeda *et al.* (2006) examined the phenolic contents of leaf and stalk of *Ipomoea aquatica* and their radical scavenging activities by using DPPH assay. High contents of phenolic compounds and strong DPPH scavenging activities were observed in extracts from this species. They also showed radical scavenging activities for the superoxide radical and tert-butylhydroperoxyl radical (t-BuOO·).

Ipomoea batatas (L.) Lam. Family: Convolvulaceae

English Name: Sweet potato; Japanese name: Yaeyamakazura.

Trypsin Inhibitors (TIs), root storage proteins, purified from *Ipomoa batatas* roots (33kDa) had scavenging activity against DPPH radical (Hou *et al.*, 2001). There was positive correlation between scavenging effects against DPPH (2 to 22 per cent) and amounts of 33kDa TI (1.92 to 46 pmol). The scavenging activities of 33 kDa TI against DPPH were calculated from linear regression to be about one-third of those of glutathione between 5 and 80 pmol.

Huang *et al.* (2004) examined the possible antioxidant activity of different extracts from sweet potato. In the DPPH staining, ethanol extract of vein had the high radical-scavenging activity when it was diluted to 6.25 mg dry matter/ml. Among all the extracts, the highest amount of total phenolic and flavonoid compounds was found in the ethanol extract of vein. In the DPPH calorimetric method, it was found that ethanol extract of leaf had the highest radical-scavenging activity followed by water extract of vein. In the reducing power activity assay, it was found that the water extract of leaf had the highest reducing power activity, followed by ethanol extract of vein. Like phenolic compounds, the highest FTC activity was found in the ethanol extract of vein.

Leaves and stalk are used as vegetable in Okinawa Prefecture in Japan. Maeda *et al.* (2006) examined the phenolic contents of leaf of *Ipomoea batatas* and their radical scavenging activities by using DPPH assay. High contents of phenolic compounds and strong DPPH scavenging activities were observed in extracts from this species. They also showed radical scavenging activities for the superoxide radical and tert-butylhydroperoxyl radical (t-BuOO·).

Wong *et al.* (2006) investigated the antioxidant properties of 25 edible tropical plants including *Ipomoea batatas* using DPPH and FRAP assays. Total phenolic content was also estimated. *Ipomoea batatas* showed very high antioxidant activity. A strong correlation between TEAC values obtained for the DPPH assay and those for FRAP assay implied that compounds in the extracts were capable of scavenging DPPH radical and reducing ferric ions. A satisfactory correlation of TPC with TEAC suggested that polyphenols in the extracts were partly responsible for the antioxidant activities while its correlation with CCA was poor, indicating that polyphenols might not be the main cupric ion chelators.

Phytochemical flavonols, carotenoids and the antioxidant capacity of leaves of *Ipomoea batatas* was evaluated by Lako *et al.* (2006). *Ipomoea batatas* leaves have the highest Total antioxidant capacity (650 mg/100g) and are rich in Total polyphenol content (TPP) (270mg/100g), quercetin (90mg/100g) and β-carotene (13mg/100g).

The total phenol contents (TPC) of leaf extracts from 116 varieties of sweet potato cultivated in China were determined by Xu *et al.* (2010) by Folin-Ciocalteu method, in addition, the crude extract(CE) of Pushu 53 leaves, with relatively higher TPC and its fractions of chloroform, ethyl acetate, n-butanol and water, were prepared and subjected to the determination of TPC, evaluation of antioxidant activities *in vitro* by assays of DPPH, ABTS and FRAP, and analysis of phenolic constituents by HPLC and HPLC-mass spectrometry. Their results showed that the extracts demonstrated potential antioxidant activity, and that a satisfactory correlation between TPC and antioxidant activity was observed. In addition, the main bioactive compounds for the antioxidant activity of the extract were polyphenols, especially derivatives of caffecylquinic acid (CQA), such as 5-CQA, 3,4-diCQA, 3,5-diCQA and 4,5-diCQA. Their contents (8.95 per cent) were 0.41, 3.41, 4.09 and 1.04 per cent, respectively, accounting for 79.6 per cent of TPC in CE (11.24 per cent) (Xu *et al.,* 2010).

Antioxidant activities of eight leafy vegetables of Ghana including *Ipomoea batatas* were assessed by Morrison and Twumasi (2010). The total antioxidant capacity (TAC) and Total Phenol content (TPC) in the methanol extracts (METE) and hydro-ethanol extracts (HETE) were measured. The TPC of the methanol extract of *Ipomoea batatas* was 0.455 mg/ml TAE and TAE was 0.412 mg/ml Ascorbic acid equivalent (AAE) while TPC of hydro-ethanol extract was 0.639 mg/ml TAE and TAE was 0.621 mg/ml Ascorbic acid equivalent (AAE). Radical scavenging activity of methanol extract was EC_{50} = 0.1746mg/ml and HETE was 0.1307 mg/ml while Fe^{3+} reducing potential was EC_{50} = 0.1307 (METE) and 0.1307 (HETE).

Adefegha and Oboh (2011) reported that cooking decreases the vitamin C contents in the leafy vegetable *Ipomoea batatas*, while it increased the phenolic content and antioxidant activities.

Ipomoea obscura (L.) Ker-Gawl. Family: Convolvulaceae

Srinivasan *et al.* (2007) reported antioxidant activity of *Ipomoea obscura*.

Ipomoea quamoclit L. Family: Convolvulaceae

Hydromethanol extracts of 15 Bangladeshi medicinal plants, including *Ipomoea quamoclit*, were evaluated by Hasan *et al.* (2009) for antioxidant potential using DPPH radical scavenging assay. *Ipomoea quamoclit* aerial parts exhibited radical scavenging activity with an IC_{50} value of 25.96 µg/ml compared to the IC_{50} value of 5.15µg/ml as shown by the reference antioxidant ascorbic acid, in a dose dependent fashion.

Ipomoea reniformis (Roxb.) Choisy Family: Convolvulaceae

In vitro antioxidant activities of *Ipomoea reniformis* herb methanolic extract was carried out by Sanja *et al.* (2009). The antioxidant property was evaluated using DPPH, superoxide, hydroxyl and nitric oxide scavenging activity in *in vitro* and by reducing power. The results indicated that the extract inhibited DPPH, superoxide hydroxyl and nitric oxide scavenging activity at IC_{50} value 5.0µg/ml and 149.9 µg/ml respectively against corresponding standards; ascorbic acid (4.3 µg/ml), curcumin (4.9 µg/ml), vitamin E (30.4 µg/ml) and rutin (145.7 µg/ml). In addition the extract had effectively reducing power. The results obtained in this study indicate that the methanolic extract can be a potential source of natural antioxidant agents due to presence of phenolic and flavonoids in the extract.

Iris versicolor L. Family: Iridaceae

Wojcikowski *et al.* (2007) evaluated the antioxidant activity of aerial parts of *Iris versicolor*. Oxygen radical absorbance capacity of the ethyl extract, methanol extract and aqueous extracts were 11.53, 112.10 and 145.02µmol Trolox equivalent per g of dried starting material, respectively.

Irvingia gabonensis (Aubry-Lecomte ex O' Rorke) Baill. Family: Ixonanthaceae

English Name: African mango

Agbor *et al.* (2005) evaluated the antioxidant capacity of 14 herbs from Cameroon. Methanol and HCl in methanol extracts were analyzed using two different antioxidant assay methods (Folin-Ciocalteu method and FRAP method). The 1.2M HCl in methanol extract had significantly higher antioxidant capacity than the methanolic extract. The FRAP antioxidant values were significantly higher than the Folin antioxidant values. *Irvingia gabonensis* tops the FRAP free antioxidant list.

Irvingea grandifolia (Engl.) Engl. Family: Ixonanthaceae

The effects of aqueous extract of *Irvingea grandifolia* bark on the oxidative status of normal rabbits were monitored by Omonkhua and Onoagbe (2008). They seemed protective against lipid peroxidation.

Iryanthera juruensis Warb. Family: Myristichaceae

Santos *et al.* (2009) evaluated the antioxidant activity of extracts from nine plant species belonging to the Brazilian flora acting on the free radicals DPPH and TEMPOL (4-hydroxy-2,2,6,6-tetramethyl-1-piperidinyloxy-1-oxyl) by electron paramagnetic resonance (EPR), and acting on the hydroxyl radical by the spin trapping technique generated by a Fenton reaction. Results showed that the extracts of *Iryanthera juruensis* display the strongest antioxidant activities with DPPH scavenging IC_{50} value of 0.028 mg/ml (leaf) and 0.41 mg/ml (seed) while Hydroxyl scavenging IC_{50} value was 1.676 (leaf), 1.252 mg/ml (seed).

Iryanthera lancifolia Ducke Family: Myristichaceae

Lock *et al.* (2005) evaluated the *in vitro* antioxidant activity of 40 Peruvian plants using DPPH assay. *Iryanthera lanciifolia* showed highest antioxidant activity with EC_{50} value of 14.08µg/ml.

Iryanthera sagotiana (Benth.) Warb. Family: Myristicaceae

Davino *et al.* (1998) investigated the antioxidant activity by measuring malondialdehyde (MDA) production through thiobarbituric acid (TBA) reaction and chemiluminescence emission. Thes results suggest that the antioxidant activity of the hexane and ethanolic extracts is due to the presence of the flavonoid glycosides, since the ethyl acetate fraction which contains large amount of these compounds showed the greatest activity.

Isatis indigotica Fort. Family: Brassicaceae (Cruciferae)

Cai *et al.* (2004) evaluated the antioxidant activity and phenolic compounds in traditional Chinese medicinal plants associated with anticancer. The improved ABTS.+ method was used to systematically assess the total antioxidant capacity (Trolox equivalent antioxidant capacity, TEAC) of the medicinal extracts. The TEAC values and total phenolic content for methanolic extract of root of *Isatis indigotica* were 63.3 µmol Trolox equivalent/100g dry weight (DW), and 0.45 g of gallic acid equivalent/100g DW respectively. A positive significant linear relationship between antioxidant activity and total phenolic content showed that phenolic compounds were the dominant antioxidant components in the tested herbs. Major types of phenolic compounds in *Isatis indigotica* include anthraquinones.

Isatis tinctoria L. Family: Brassicaceae (Cruciferae)

Li *et al.* (2008) evaluated the antioxidant properties of 45 medicinal plants including *Isatis tinctoria* using FRAP and TEAC assays and the total phenolic contents of these plants were measured. *Isatis tinctoria* showed low antioxidant activity with 12.21 µmol Fe(II)/g FRAP value, 5.81µmol Trolox/g of TEAC value and 4.18 mg GAE/g of Phenolic content.

Ixeridium gracile (DC.) Shin Family: Asteraceae (Compositae)

Phytochemical investigation of *Ixeridium gracile* by Ma *et al.* (2007) lead in the isolation and identification of twelve flavonoids and two coumarins. The free radical-scavenging potential of different extract fractions as well as of the pure compounds towards the DPPH radical were evaluated and are discussed in terms of structure-activity relationship. The flavonoids were found to be the major constituents contributing to the free-radicaal scavenging activity of *I.gracile*, but the high concentration of coumarin additionally contributed to the observed activity.

Ixeris dentata (Thunb.) Nakai Family: Asteraceae (Compositae)

Heo *et al.* (2009) investigated the total phenolic content, antioxidant activity and cytotoxicity of methanol extracts from aerial parts of 11 Korean medicinal salad plants. Methanol extract of aerial parts showed highest phenolic content (16.4mg/100g) and DPPH radical scavenging activity at 50 microg/ml was 86.4 per cent. In conclusion, the studied salad plants have a high phenolics content and high antioxidant activity. These plants dose dependently increased DPPH free radical scavenging activity. The total phenolics level was highly correlated with free radical scavenging activity.

Ixora chinensis Lam. Family: **Rubiaceae**

Liu *et al.* (2008) evaluated the antioxidant activity of water, 50 per cent ethanol and 70 per cent acetone extracts of flowers of 15 plants of Hainan. All extracts of *Ixora chinensis* showed some antioxidant effects.

Ixora coccinea L. Family: **Rubiaceae**

Antioxidant activity of the methanol extract of *Ixora coccinea* was determined by Saha *et al.* (2008) by DPPH free radical scavenging assay, reducing power and total antioxidant capacity using phosphomolybdenum method. Preliminary phytochemical screening revealed that the extract of the flower possesses flavonoids, steroids and tannin materials. The extract showed significant activities in all antioxidant assays compared to the standard antioxidant in a dose dependent manner and remarkable activities to scavenge reactive oxygen species (ROS) may be attributed to the high amount of hydrophilic phenolics. In DPPH radical scavenging assay the IC_{50} value of the extract was found to be 100.53μg/ml while ascorbic acid had IC_{50} value 58.92μg/ml.

Ixora williamsii Sandwith Family: **Rubiaceae**

Liu *et al.* (2008) evaluated the antioxidant activity of water, 50 per cent ethanol and 70 per cent acetone extracts of flowers of 15 plants of Hainan. All extracts of *Ixora williamsii* showed some antioxidant effects. Water extract of fresh flowers of *Ixora williamsii* had better effects on eliminating OH.

Jacaranda micrantha Cham. Family: **Bignoniaceae**

Menezes *et al.* (2004) investigated the antioxidant activity of hydroalcoholic extracts of bark of *Jacaranda micrantha* by DPPH assay and phosphomolybdenum method. The relative activity in phosphomolybdenum complex was 0.38 while scavenging activity against DPPH in IC_{50} (μg/ml) was 46.29.

Jacaratia mexicana A.DC. Family: **Bignoniaceae**

Ruiz-Terán *et al.* (2008) evaluated the antioxidant activity of aerial parts of *Jacaratia mexicana* using DPPH and β-carotene-linoleic acid bleaching assay. The extract displayed antioxidant activity comparable to the commercial antioxidant BHA. For the methanolic extract analysed, a clear relation between the total phenolic content of the extracts and their antioxidant activity was found.

Jasminum fruticans L. Family: **Oleaceae**

Serteser *et al.* (2009) studied the antioxidant activity of 50 per cent aqueous methanol extracts by various antioxidant assays, including free radical scavenging, hydrogen peroxide scavenging and metal (Fe^{2+}) chelating activities. The methanolic extract examined in the assay showed the strongest activities. The DPPH radical scavenging effects of leaves extract was 0.525 while fruit extracts showed 0.533. Fe^{2+}

chelating activity (per cent) of leaves extract was 31.24 while fruit extract showed 33.54 and Hydrogen peroxide inhibition activity of leaves extract was 37.02 per cent while fruit extracts showed 35.32 per cent.

Jasminum grandiflorum L. Family: Oleaceae

Umamaheswari *et al*. (2007) reported antiulcer and *in vitro* antioxidant activities of *Jasminum grandiflorum*.

Jasonia montana Vahl. Family: Asteraceae (Compositae)

Hussain (2011) investigated the antioxidant activities of the ethanolic and aqueous extracts of aerial parts of *Jasonia montana* in streptozotocin-induced diabetic rats. The treatment of diabetic rats with aerial parts extracts resulted in a significant increase in reduced glutathione, superoxide dismutase, catalase, glutathione peroxidase and glutathione-S-transferase in the liver and kidney of diabetic rats.

Jatropha curcas L. Family: Euphorbiaceae

Getwichit *et al*. (2008) investigated the physical properties, chemical compositions and antioxidant activity of *Jatropha* oil detoxified by bentonite (JOB). The detoxification of *Jatropha* oil by bentonite reduced the viscosity value from 162.87 to 130.44 cp, and increased the time for the induction of autooxidation from 7.88 to 9.66 h. JOB contained saturated fatty acids, such as palmitic acid (13.9$ per cent), stearic acid (6.81 per cent) and arachidic acid (0.22 per cent). The main unsaturated fatty acids in JOB were oleic acid (45.78 per cent, linoleic acid (32.28 per cent), palmitoleic acid (0.77 per cent) and α-linoleic acid (0.20 per cent). The vitamin E content of JOB was 54.61mg/100g. JOB at 1.0, 5.0, 10.0, 15.0 and 20.0 mg/ml exhibited antioxidant activities of 37.067, 63.175, 72.320, 72.898 and 77.592 per cent, respectively. The IC_{50} value of JOB was 2.2950 mg/ml. JOB was characterized by less viscosity, greater lightness, paler yellow colour, longer time for the induction of autooxidation and greater autooxidation efficiency than non-detoxified *Jatropha* oil.

The methanolic fraction of leaves of *Jatropha curcas* was evaluated by Balaji *et al*. (2009) against hepatocellular carcinoma induced by Aflatoxin B1. Treatment with extract of *Jatropha curcas* decreased the levels of elevated serum enzymes, lipid levels, bilirubin and increase in the protein and uric acid levels. The levels of lipid peroxides and activity of enzymatic antioxidants superoxide dismutase (SOD), catalase (CAT), glutathione peroxidase (GPx), glutathione-S-transferase (GST) and glutathione reductase (GR) were determined in liver homogenates. Marked increase in lipid peroxide levels and concomitant decrease in enzymatic antioxidant levels were observed in carcinoma induced rats, while methanolic extract of *Jatropha curcas* treatment reversed conditions to near normal levels. Liver histopathology showed that the extract reduced the incidence of liver lesions, lymphocytic infiltrations and hepatic necrosis induced by AFB1 in rats. These results suggest that methanolic fraction of leaves of *Jatropha curcas* could prevent liver against the AFB1-induced oxidative damage in rats, which may be due to its capability to induce the *in vivo* antioxidant system.

El Diwani *et al*. (2009) studied the antioxidant activity of ethanol and petroleum ether extracts from roots, stem, leaves and nodes of *Jatropha curcas* by DPPH free radical method. The study revealed that the roots and stems were the strongest in all tests followed by leaves and nodes. Roots crude extract showed $IC_{50} = 0.521$ mg/ml. The presence of phenolic compounds and tannins were screened in Folin-Ciocalteau assay, HPLC and X-rays diffraction.

Antioxidant and chelating activity of *Jatropha curcas* protein hydrolysates was investigated by Gallegos-Tintore *et al.* (2010). Free radical scavenging activity was measured by DPPH assay and reducing power was determined by measuring the formation of Prussian blue at 700 nm. Reducing power of *J. curcas* protein hydrolysates increased after 20 min with increasing hydrolysis time and up to 50 min. Iron chelating activity increases with hydrolysis time from 30 min and beyond. Highest chelating activity was reached at 50 min hydrolysis and then decreased at 60 min of hydrolysis. Copper chelating activity was higher at initial times of the protein hydrolysis was around 60 per cent similar to the one observed for iron chelation.

Sawant *et al.* (2010) investigated the *in vitro* antioxidant activity of hydroalcoholic extract of leaves of *Jatropha curcas* by using DPPH radical scavenging activity, nitric oxide radical scavenging activity, hydroxyl radical scavenging activity, reducing power method and hydrogen peroxide radical scavenging activity. Significant results were obtained in the estimated parameters, thus concluded that the extract of leaves of *Jatropha curcas* possess potential source of antioxidants of natural origin.

A comprehensive study on the phytochemical contents and biological activities of the methanolic extract from different parts of *Jatropha curcas* was conducted by Oskoueian *et al.* (2011). The extracts of different plant parts contained various levels of phenolics, flavonoids and saponins. Letex and leaf extracts showed the highest antioxidant activity.

Narayanaswamy and Balakrishnan (2011) evaluated the antioxidant activity of 13 medicinal plants including fruit of *Jatropha curcas*. The aqueous extract showed 80 per cent inhibition of DPPH radical while ethanolic extract showed 40 per cent inhibition of DPPH radical.

Jatropha gaumeri Greenm. Family: Euphorbiaceae

The methanolic extract of leaves of *Jatropha gaumeri* showed antioxidant activity (Can-Ake *et al.*, 2004). The antioxidant assays were inhibition of bleaching of β-carotene and reduction of DPPH. The bioassay-guided purification of the crude leaf extract allowed the identification of β-sitosterol and the triterpenes α-amyrin, β-amyrin and taraxasterol, as the metabolites responsible for the antioxidant activity.

Jatropha gossypifolia L. Family: Euphorbiaceae

Povichit *et al.* (2010) investigated phenolic content and free radical scavenging effect of *Jatropha gossypifolia* by DPPH and ABTS assays, anti-lipid peroxidation activity by TBARS and for antiglycation activity. The results revealed that the total phenolic content showed good correlation with free radical scavenging by ABTS and anti-lipid peroxidation by TBARS, but showed no correlation with antiglycation. The extract of *Jatropha gossypifolia* demonstrated a moderate antioxidant effect and also showed a moderate antiglycation effect. The IC_{50} were 1.45mg/ml for the DPPH method, TEAC value of 0.42 mg/Trolox/mg sample for the ABTS method, IC_{50} of 0.42 mg/ml for the TBARS method and IC_{50} of 8.40 for the antiglycation method.

Antioxidant activity of different extracts of *Jatropha gossypifolia* were evaluated by Shahwar *et al.* (2010) through DPPH radical scavenging activity, phosphomolybdate and ferric thiocyanate methods. The antioxidant activity was weak. The DPPH radical scavenging activity of ethyl acetate extract was 15.6 per cent, total antioxidant activity was 0.311 and inhibition of lipid peroxidation was 55.2 per cent while total phenols was 296.0 mg GAE/g of crude extract.

Narayanaswamy and Balakrishnan (2011) evaluated the antioxidant activity of 13 medicinal plants including leaf of *Jatropha gossypifolia*. Aqueous extract showed 70 per cent inhibition of DPPH radical while ethanolic extract showed 90 per cent inhibition of DPPH radical.

Jatropha mollissima (Pohl.) Baill. Family: Euphorbiaceae

De Melo *et al.* (2010) evaluated the antioxidant capacity and tannin content in *Jatropha mollissima*. IC_{50} value of DPPH radical scavenging was 54.09µg/ml while tannin content was 0.08 mg/g.

Jatropha multifida L. Family: Euphorbiaceae

Narayanaswamy and Balakrishnan (2011) evaluated the antioxidant activity of 13 medicinal plants including fruit of *Jatropha multifida*. The aqueous and ethanolic extracts showed 90 per cent inhibition of DPPH radical.

Jatropha panduraefolia Andrews Family: Euphorbiaceae

Antioxidant activity of *Jatropha panduraefolia* and four other species have been investigated by Adib *et al.* (2010) by DPPH assay. *J. panduraefolia* was found to have a moderate antioxidant activity. Methanolic extract of leaves showed free radical scavenging activity with an IC_{50} value of 160µg/ml, Hexane soluble fraction of methanolic extract was with IC_{50} value of 165µg/ml, Carbon tetrachloride soluble fraction of methanolic extract was with IC_{50} value of 104µg/ml and chloroform soluble fraction of methanolic extract had IC_{50} value of 91µg/ml.

Juglans regia L. Family: Juglandaceae

English Name: Walnut.

Cai *et al.* (2004) evaluated the antioxidant activity and phenolic compounds in traditional Chinese medicinal plants associated with anticancer. The improved ABTS.+ method was used to systematically assess the total antioxidant capacity (Trolox equivalent antioxidant capacity, TEAC) of the medicinal extracts. The TEAC values and total phenolic content for methanolic extract of seed of *Juglans regia* were 279.1 µmol Trolox equivalent/100g dry weight (DW), and 0.74 g of gallic acid equivalent/100g DW respectively. A positive significant linear relationship between antioxidant activity and total phenolic content showed that phenolic compounds were the dominant antioxidant components in the tested herbs. Major types of phenolic compounds in *Juglans regia* include tannins.

Different cultivars of walnut grown in Portugal, were investigated in what concerns phenolic compounds and antioxidant properties. Antioxidant activity was assessed by the reducing power assay, the scavenging effect on DPPH radicals and β-carotene linoleate model system. In a general way, all of the studied walnut leaves cultivars presented high antioxidant activity EC_{50} values lower than 1mg/ml (Pareira *et al.*, 2007).

Total phenols, antioxidant potential and antimicrobial activity of walnut green husks were studied by Oliveira *et al.* (2008). Total phenols content was determined by colorimetric assay and their amount ranged from 32.61 mg/g of GAE to 74.08 mg/g of GAE. The antioxidant capacity of aqueous extracts was assessed through reducing power assay, scavenging effects on DPPH radicals and β-carotene lineolate model system. A concentration dependent antioxidant capacity was verified in reducing power and DPPH assays with EC_{50} values lower than 1 mg/ml for all the tested extracts.

Miliauskas *et al.* (2009) investigated the radical scavenging activity of acetone, ethyl acetate and methanol extracts of leaves and stems at growing stage using DPPH and ABTS assays. They found that Methanol extracts were most effective DPPH radical scavengers. Ethyl acetate and acetone extracts were considerably less effective radical scavengers. The amount of flavonoid in the extract was found to be 13.8 mg/g.

Juncus effusus L. Family: Juncaceae

Gan *et al.* (2010) evaluated the antioxidant activity of *Juncus effusus* using the FRAP and TEAC assays, and its total phenolic content was measured by the Folin-Ciocalteu method. For FRAP assay the value was 56.69μmol Fe(II)/g dry weight. For TEAC assay, the value was 34.77μmol Trolox/g dry weight. Total phenolic content was 3.00 mg gallic acid equivalent/g dry weight of plant material.

Jussiaea repens L. Family: Onagaraceae

Extracts of *Jussiaea repens* aerial part/stem showed remarkable free radical scavenging activity with IC_{50} value of 7.7μg/ml and significant inhibitory effects on lipid peroxidation with 88.8 per cent inhibition at the concentration of 50ng/ml (Thoung *et al.*, 2006).

Justicia flava (Forssk.) Vahl Family: Acanthaceae

Odukoya *et al.* (2005) evaluated the antioxidant activity of Nigerian dietary spices including *Justicia flava*. The antioxidant activity (expressed as per cent inhibition of oxidation) was 49.62 per cent, reducing power was 9.77 per cent and Total phenolic content was 107.88 mg/100g Tannic acid Equivalent. Antioxidant activity correlated significantly and positively with total phenolics while there was no linear correlation between total antioxidant activity and reducing power neither between reducing power and total phenolic content.

Kadsura longipedunculata Finet et Gagnepain Family: Schisandraceae

Mulyaningsih *et al.* (2010) determined the chemical composition and antioxidant activity of the essential oil from stem bark of *Kadsura longipedunculata*. The essential oil showed radical scavenging activity with IC_{50} value of 3.06 mg/ml. Camphene and borneol from the essential did not show antioxidant activity.

Kaempferia galanga L. Family: Zingiberaceae

Chanwitheesuk *et al.* (2005) investigated the antioxidant activity of rhizome of *Kaempferia galanga* using a β-carotene bleaching method. The contents of plant chemicals such as vitamin C, Vitamin E, carotenoids, tannin and total phenolics, were also determined. Methanolic extract showed antioxidant activity. *Kaempferia galanga* has antioxidant index of 3.15, Vitamin C 5.37 mg per cent, Vitamin E 0.0035mg per cent, Total carotenes 1.91 mg per cent, Total xanthophylls 1.59mg per cent, Tannins 4.48mg per cent and Total phenolics of 26.4mg per cent.

Kalanchoe pinnata (Lam.) Pers. (Syn.: *Bryophyllum pinnatum* (Lam.) Oken) Family: Crassulaceae

Harika *et al.* (2007) evaluated the aqueous extract of *Kalanchoe pinnata* for its protective effects on gentamicin-induced nephrotoxicity in rats. *In vitro* studies revealed that the aqueous leaf extract possesses significant antioxidant as well as oxidative radical scavenging activities. The EC_{50} value of 116.25µg/ml for DPPH free radical scavenging activity, 90µg/ml nitric oxide scavenging activity and 125µg/ml anti lipid peroxidation activity.

Kelussia odoratissima Mozaff. Family: Apiaceae (Umbelliferae)

The antioxidant activity of the methanolic extract of *Kelussia odoratissima* was evaluated by Ahmadi *et al.* (2007) using β-carotene bleaching assay, reducing power, thiocyanate, accelerated oxidation of sunflower oil, DPPH radical scavenging. In the DPPH and reducing power models the antioxidant activity of the plant extract was generally found to be less effective than that of ascorbic acid, but it was comparable to and/or greater than the activities of α-tocopherol and BHT. The methanolic extract inhibited the oxidation of sunflower oil at 60°C more efficiently than did BHT.

Khaya senegalensis (Desr.) A.Juss. Family: Meliaceae

Karou *et al.* (2005) evaluated the polyphenol content and antioxidant activity of *Khaya senegalensis* by ABTS assay. Polyphenols in the lyophilized extract of leaves was 34.91 per cent, while in bark 47.19 per cent. The antioxidant activity of leaves was 1.50µmol Trolox/µg in the Phosphomolybdenum assay and 15.47 µmol Trolox/µg in ABTS assay, while bark showed 2.21µmol Trolox/µg in the Phosphomolybdenum assay and 21.97 µmol Trolox/µg in ABTS assay. The total phenolic compounds were highly correlated with the antioxidant activities.

The chloroform and methanolic extracts of 124 Egyptian plant species belonging to 56 families were investigated by Moussa *et al.* (2011) and compared their antioxidant activity by DPPH scavenging assay. Among the 124 plant species tested 18 exhibited extremely high antiradical activity (>80 per cent inhibition). The IC_{50} value of DPPH radical scavenging of *Khaya senegalensis* was 29.61µg/ml, while total phenolic and flavonoid contents were 149.36 mg Tannic acid equivalent/g extract and 215.06 mg Rutin equivalent/g extract, respectively. Correlation coefficient between DPPH radical scavenging activity and the total phenolic and flavonoid contents suggest that phenolics and flavonoids in the extracts were partly responsible for the antiradical activities.

Kigelia africana (Lam.) Benth. (Syn.: *Kigelia pinnata* (Jacq.) DC) Family: Bignoniaceae

An ethanol extract of *Kigelia africana* has been shown to possess some antioxidant activity (Olaleye and Rocha, 2007, 2008).

Kochia scoparia (L.) Schrad. Family: Chenopodiaceae

Gan *et al.* (2010) evaluated the antioxidant activity of *Kochia scoparia* using the FRAP and TEAC assays, and its total phenolic content was measured by the Folin-Ciocalteu method. For FRAP assay the value was 103.22 µmol Fe(II)/g dry weight. For TEAC assay, the value was 68.83µmol Trolox/g dry weight. Total phenolic content was 8.63 mg gallic acid equivalent/g dry weight of plant material.

Koelreuteria henryi Dummer Family: Sapindaceae

Lee *et al.* (2006) investigated the antioxidant activity of ethanol extract of leaves using DPPH, hydroxyl and superoxide radical scavenging and reducing power activities as well as the induction of heme oxygenase-1 (HO-1). The IC_{50} values against DPPH, radical were 3.6µg/ml, OH radical 0.36µg, superoxide radical 19.8µg/ml and total phenolics was 3627.9 mg of GAE/g.

Chen *et al.* (2009) studied the antioxidant activity of leaves extract towards xanthine oxidase, lipoxygenase and tyrosinase activities. The inhibitory effect of acetone extract on xanthine oxidase activity was more than 50 per cent, tyrosine activity was almost 70 per cent and lipoxygenase activity was 100 per cent.

Krameria triandra Ruiz et Pavon Family: Polygonaceae

Carini *et al.* (2002) evaluated the antioxidant/photoprotective potential of a standardized *Krameria triandra* root extract. The extract was significantly more active (IC_{50} =0.28µg/ml) than standard antioxidant α-tocopherol (IC_{50} = 6.37µg/ml).

Lactuca sativa L. Family: Asteraceae (Compositae)

English Name: Lettuce.

Souri *et al.* (2004) investigated the antioxidant activity of methanol extracts of lettuce, used as vegetable in Iranian diet against linoleic acid peroxidation. The extracts showed good antioxidant activity.

Cai *et al.* (2004) evaluated the antioxidant activity and phenolic compounds in traditional Chinese medicinal plants and vegetables associated with anticancer. The improved ABTS.+ method was used to systematically assess the total antioxidant capacity (TEAC) of the medicinal extracts. The TEAC values and total phenolic content for methanolic extract of Chinese lettuce were 128.4 µmol Trolox equivalent/100g dry weight (DW), and 0.78 g of gallic acid equivalent/100g DW respectively.

Souri *et al.* (2008) evaluated the antioxidant activity against linoleic acid peroxidation using 1,3-diethyl-2-thiobarbituric acid as reagent. Antioxidant activity (IC_{50}) against peroxidation of linoleic acid (2mg/ml) was 14.28 and phenolic content was 168.56 mg/100g dry weight. The results of this study showed that there is no significant correlation between antioxidant activity and phenolic content of the studied plant materials and phenolic content could not be a good indicator of antioxidant capacity.

Lactuca sativa L. var. *capitata* L. Family: Asteraceae (Compositae)

English Name: Head Lettuce; Japanese: Retasu.

Maeda *et al.* (2006) examined the phenolic contents of leaf and its radical scavenging activities for DPPH radical. The polyphenol content was 27.6 mg-gallic acid equivalent/100 g) and DPPH scavenging activity was 81.9 µmol-Trolox eq./100 g).

Lactuca scariola L. Family: Asteraceae (Compositae)

The antioxidant activity of *Lactuca scariola* was investigated by Kim (2001) by measuring the radical scavenging effect on DPPH radical. The methanolic extract of the aerial parts of *Lactuca scariola* showed strong radical scavenging activity. The EtOAc soluble fraction exhibited a stronger activity than the other. Quercetin-3-O-β-D-glucopyranoside, luteolin-7-O-β-D-glucopyranoside, luteolin, quercetin and kaempferol, together with 1 β-, 13-dihydrolactucin were isolated from the EtOAc soluble fraction as active ingredients.

Ladambergia macrocarpa Family: Rubiaceae

Mosquera *et al.* (2007) studied the antioxidant activity of the extracts of *Ladambergia macrocarpa* using DPPH free-radical scavenging method. The percentage of the antioxidant activity of the n-Hexane fraction was 13.832, Dichloromethane fraction was 54.023 and methanol fraction was 45.54.

Lagerstroemia indica L. Family: Lythraceae

Fu *et al.* (2010) evaluated the antioxidant capacity and total phenolic contents of 56 wild fruits from South China using FRAP and TEAC assays. A high correlation between antioxidant capacity and phenolic compounds could be the main contributor to the measured antioxidant activity. The results showed that fruits of *Lagerstroemia indica* has possessed the highest antioxidant capacity and total phenolic contents.

The chloroform and methanolic extracts of 124 Egyptian plant species belonging to 56 families were investigated by Moussa *et al.* (2011) and compared their antioxidant activity by DPPH scavenging assay. Among the 124 plant species tested 18 exhibited extremely high antiradical activity (>80 per cent inhibition). The IC_{50} value of DPPH radical scavenging of *Lagerstroemia indica* was 23.74µg/ml, while total phenolic and flavonoid contents were 107.06 mg Tannic acid equivalent/g extract and 223.16 mg Rutin equivalent/g extract, respectively. Correlation coefficient between DPPH radical scavenging activity and the total phenolic and flavonoid contents suggest that phenolics and flavonoids in the extracts were partly responsible for the antiradical activities.

Lagerstroemia speciosa L. Family: Lythraceae

English Name: Banaba.

Unno *et al.* (1998) evaluated the antioxidant activity of the water extract of leaves of *Lagerstroemia speciosa*. Banaba extract showed strong antioxidative activity in linoleic acid autooxidation system. Banaba extract had a potent radical scavenging action on DPPH radicals and superoxide radicals generated by hypoxanthine/xanthine oxidase system. *In vitro* lipid peroxidation of rat liver homogenate induced by tert-butyl hydroperoxide (BHP) was inhibited by the addition of banaba extract in a dose-dependent manner.

The *in vitro* antioxidant activity of the successive extracts (ethyl acetate, ethanol, methanol and water) of the leaves of *Lagerstroemia speciosa* were studied by Priya *et al.* (2008) by examining their superoxide, hydroxyl ion scavenging and by measuring lipid peroxidation. The ethyl acetate and ethanol extracts were found to possess greater antioxidant property than the methanol and water extracts.

The hydroalcoholic extract of leaves of *Lagerstroemia speciosa* was studied for antioxidant activity on different *in vitro* models namely DPPH assay, Hydrogen peroxide and nitric oxide radical scavenging method and superoxide radical scavenging by alkaline DMSO method. The extract showed dose-dependent free radical scavenging property in the tested models. *Lagerstroemia* showed IC_{50} value of 4.75µg/ml for DPPH method, which was comparable to that of ascorbic acid (IC_{50} = 2.75 µg/ml) and rutin (7.89µg/ml). For hydrogen peroxide method, IC_{50} value was found to be 28.00µg/ml, which compares favourable with ascorbic acid (IC_{50} = 187.33 µg/ml) and rutin (35.26 µg/ml). In Nitric oxide model IC_{50} value was found to be 750 which is very low when compared to rutin (Anil *et al.*, 2010).

Fu *et al.* (2010) evaluated the antioxidant capacity and total phenolic contents of 56 wild fruits from South China using FRAP and TEAC assays. A high correlation between antioxidant capacity and phenolic compounds could be the main contributor to the measured antioxidant activity. The results showed that fruits of *Lagerstroemia speciosa* possessed the highest antioxidant capacity and total phenolic contents.

Laggera pterodonta (DC.) Benth. Family: Asteraceae (Compositae)

Hepatoprotective and antioxidative effects of total phenolics from *Laggera pterodonta* on chemical-induced injury in primary cultured neonatal rat hepatocytes were investigated by Wu *et al.* (2007). DPPH and superoxide radicals scavenging activities of Total phenolics from *Laggera pterodonta* (TPLP) were determined. TPLP afforded much stronger protection than the reference drug silibinin.

Lamium album L. Family: Lamiaceae (Labiatae)

English Name: Dead nettle.

Matkowski and Piotrowska (2006) studied the antioxidative effects of methanolic extract of *Lamium album* with the use of three *in vitro* assays (DPPH assay, phosphomolybdenum method and lipid peroxidation). The herb showed strong antioxidant activity.

Lamium maculatum L. Family: Lamiaceae (Labiatae)

Matkowski *et al.* (2008) evaluated the antioxidant activity of aerial parts of *Lamium maculatum* using DPPH, OH radical scavenging assays, reducing power assay and inhibition of deoxyribose degradation. Crude MeOH extract was stronger than any fraction. EC_{50} value of DPPH radical scavenging was 105.40µg/ml for MeOH extract, 106.54µg/ml for Dichloromethane extract, 91.24µg/ml for BuOH extract and 70.59µg/ml for ethyl acetate extract.

Lamium purpureum L. Family: Lamiaceae (Labiatae)

Matkowski and Piotrowska (2006) studied the antioxidative effects of methanolic extract of *Lamium purpureum* with the use of three *in vitro* assays (DPPH assay, phosphomolybdenum method and lipid peroxidation). The herb showed strong antioxidant activity.

Landolfia owariensis P.Beauv. Family: Apocynaceae

Oke and Hamburger (2002) studied the antioxidant activity of ethanol extract of leaf using DPPH assay. The extract showed moderate antioxidant activity.

Lannea acida A. Rch. Family: Anacardiaceae

The total phenolic and flavonoid contents of hydroalcoholic extract (70 per cent v/v ethanol/distilled water) from the bark of *Lannea acida* and two other species of *Lannea* were determined by the method of Folin Ciocalteu and $AlCl_3$ by spectrophotometry (Outtara *et al.*, 2010). Antioxidant activity was determined by the method of DPPH and compared with quercetin. *L. acida* exhibited the highest total phenolic contents (40.55 g GAE/100g) which correlated with better antioxidant activity IC_{50} = 345.72µg/ml). The total flavonoids 8.70 g QE/100g were recorded in the extract.

Lannea microcarpa Engl. Family: Anacardiaceae

Polyphenol content and antioxidant activity of fourteen wild edible fruits from Burkina Faso were investigated by Lamien-Meda *et al.* (2008). The data obtained show that the total phenolic and total flavonoid levels were significantly higher in the acetone than in the methanol extracts. *Lannea microcarpa* fruit had higher flavonoid content and higher antioxidant activity.

The total phenolic and flavonoid contents of hydroalcoholic extract (70 per cent v/v ethanol/distilled water) from the bark of *Lannea microcarpa* and two other species of *Lannea* were determined by the method of Folin Ciocalteu and $AlCl_3$ by spectrophotometry (Outtara *et al.*, 2010). Antioxidant activity was determined by the method of DPPH and compared with quercetin. *L. microcarpa* exhibited the higher total phenolic contents (40.07 g GAE/100g) which correlated with better antioxidant activity (IC_{50} = 450.33µg/ml). The total flavonoids of 6.45 g QE/100g were recorded in the extract.

Lannea velutina A. Rich Family: Anacardiaceae

Diallo *et al.* (2001) demonstrated with DPPH assay that *Lannea velutina* possessed potent antioxidant activity.

The total phenolic and flavonoid contents of hydroalcoholic extract (70 per cent v/v ethanol/distilled water) from the bark of *Lannea velutina* and two other species of *Lannea* were determined by the method of Folin Ciocalteu and $AlCl_3$ by spectrophotometry (Outtara *et al.*, 2010). Antioxidant activity was determined by the method of DPPH and compared with quercetin. *L. velutina* exhibited the high total phenolic contents (38.04 g GAE/100g) which correlated with better antioxidant activity IC_{50} = 478.68µg/ml). Furthermore, the highest content of total flavonoids (11.02 g QE/100g) were recorded with *L. velutina*.

Lantana camara L. Family: Verbenaceae

Mensor *et al.* (2001) evaluated the antioxidant activity of ethanol extracts of *Lantana camara* and partitions by DPPH radical scavenging assay. The extract exhibited significant free radical scavenging properties. Among the partitions the more polar ones (ethyl acetate and n-butanol) are those that generally have higher antioxidant activity.

Basu and Hazra (2006) evaluated the nitric oxide scavenging activity of extracts of *Lantana camara* with different solvents. Notable Nitric oxide scavenging activity was exhibited *in vitro* by some extracts (IC^{50}< 0.2mg/ml). The extracts showed marked inhibition (60-80 per cent) *ex vivo* at a dose of 80µg/ml without appreciable cytotoxic effect on the cultured macrophages.

Lantana magnibracteata Tronc. Family: Verbenaceae

The aerial parts of 17 Bolivian plants including *Lantana camara*, *Lantana magnibracteata* and *Lantana trifolia* were screened by Rosas-Romero and Saavedra (2005) to determine antioxidant activity. A methanol extract of each plant was prepared and portioned sequentially with hexane, chloroform, and ethyl acetate, leaving an aqueous solution. All extracts and their 5 fractions for a total of 102 samples, were evaluated using two techniques: an adaptation of the β-carotene bleaching technique using an emulsion of linoleic acid in water as the oxidizable substrate, and the DPPH free radical trapping technique. The results of the β-carotene bleaching technique were more discriminating and better related to the rancidity process under normal conditions; with this assay, 11 species provided at least one fraction with highly promising antioxidant activity. All species gave good results under the DPPH technique, and in most cases they performed better than BHA, which was used as a reference antioxidant (Rosas-Romero and Saavedra, 2005).

Lantana trifolia L. Family: Verbenaceae

Mensor *et al.* (2001) evaluated the antioxidant activity of ethanol extracts of *Lantana trifolia* and partions by DPPH radical scavenging assay. The extract exhibited significant free radical scavenging properties. Among the partions the more polar ones (ethyl acetate and n-butanol) are those that generally have higher antioxidant activity.

Laportea aestuans L. Family: Urticaceae

Antioxidant activities of eight leafy vegetables of Ghana including *Laportea aestuans* were assessed by Morrison and Twumasi (2010). The total antioxidant capacity (TAC) and Total Phenol content (TPC) in the methanol extracts (METE) and hydro-ethanol extracts (HETE) were measured. The TPC of the methanol extract of *Laportea aestuans* was 0.407 mg/ml TAE and TAE was 0.406 mg/ml Ascorbic acid equivalent (AAE) while TPC of hydro-ethanol extract was 0.602 mg/ml TAE and

TAE was 0.517 mg/ml Ascorbic acid equivalent (AAE). Radical scavenging activity of methanol extract was EC_{50} = 0.2326mg/ml and HETE was 0.1586 mg/ml while Fe^{3+} reducing potential was EC_{50} = 0.1586 (METE) and 0.1586 (HETE).

Larrea tridentata Sesse et Moq. ex DC.) Coville Family: Zygophyllaceae

Three lignans together with 10 known compounds were isolated from the leaves of *Larrea tridentata* by Abou-Gazar *et al.* (2004). Their antioxidant activities against intracellular reactive oxygen species were evaluated in HL-60 cells.

Lasianthera africana P.Beauv. Family: Icacinaceae

Odukoya *et al.* (2007) evaluated the antioxidant activity and total phenolic contents of Nigerian green leafy vegetables including *Lasianthera africana*.

Lasianthus cyanocarpus Jack Family: Rubiaceae

Ahmad *et al.* (2010) evaluated the antioxidant potential of 22 species of medicinal plants from Malaysian Rubiaceae including *Lasianthus cyanocarpus*. FTC, TBA, TPC and the DPPH assays were employed. The tested extracts showed strong antioxidant potential with per cent inhibition of 96.3 per cent in the FTC, and 87.7 per cent in TBA assays. The TPC of the extract was 8.61mg GAE/g PE. A good correlation was observed between total phenolic content and radical-scavenging activities.

Lasianthus maingayi Hook.f. Family: Rubiaceae

Ahmad *et al.* (2010) evaluated the antioxidant potential of 22 species of medicinal plants from Malaysian Rubiaceae including *Lasianthus maingayi*. FTC, TBA, TPC and the DPPH assays were employed. The tested extracts showed strong antioxidant potential with percent inhibition of 96.1 per cent in the FTC, and 91.8 per cent in TBA assays. The TPC of the extract was 14.03mg GAE/g PE. A good correlation was observed between total phenolic content and radical-scavenging activities.

Lasianthus oblongus King et Gamble Family: Rubiaceae

Methanol exteracts of seven Malaysian medicinal plants, including *Lasianthus oblongus* were screened by Saha *et al.* (2004) for antioxidant and nitric oxide inhibitory activities. Antioxidant activity was measured by using FTC, TBA and DPPH free radical scavenging methods and Greiss assay was used for the measurement of nitric oxide inhibition in LPS and interferon-γ-treated RAW264.7 cells. All the extracts showed strong antioxidant activity comparable to or higher than that of α-tocopherol, BHT and quercetin in FTC and TBA methods. In the DPPH radical scavenging assay the extracts were weak. In the Greiss assay *Lasianthus oblongus* showed strong inhibitory activity on nitric oxide production in LPS and IFN-γ-induced RAW 264.7 cells.

Lasianthus pilosus Wight Family: Rubiaceae

Ahmad *et al.* (2010) evaluated the antioxidant potential of 22 species of medicinal plants from Malaysian Rubiaceae including *Lasianthus pilosus*. FTC, TBA, TPC and the DPPH assays were employed. The tested extracts showed strong antioxidant potential with percent inhibition of 96.8 per cent in the FTC, and 88.3 per cent in TBA assays. The TPC of the extract was 8.55mg GAE/g PE. A good correlation was observed between total phenolic content and radical-scavenging activities.

Lathyrus binatus Pancic Family: Fabaceae (Papilionaceae)

Godevac *et al.* (2007) evaluated the antioxidant activity of nine Fabaceae members including *Lathyrus binatus* using DPPH radical scavenging capacity, TEAC values by ABTS radical cation and inhibition of liposome peroxidation. The plant exhibited strong antioxidant capacity in all the tested methods.

Launaea pinnatifida Cass. Family: Asteraceae (Compositae)

The antioxidant activities of petroleum ether, chloroform, ethanol and water extracts of the leaves of *Launaea pinnatifida* were determined by Nagalapur and Paramjyothi (2010) by ferric reducing power, free radical scavenging activity by DPPH and hydroxyl radical scavenging activity. The total amount of phenols and flavonoids were estimated to be 179.46 mg/g as gallic acid equivalents and 87.46 mg/g as quercetin equivalents, respectively in ethanolic extract. The ethanolic extract exhibited the significant activity against DPPH free radical compared to the other extracts. Ethanol and water extracts gave similar and significant high levels of hydroxyl radical scavenging activities when compared with the standard. Chloroform and petroleum ether extracts showed almost similar results.

Laurocerasus officinalis Roem. Family: Rosaceae

The antioxidant activity of the fruit was investigated by Kolayali *et al.* (2003) using TLC plate and ferric thiocyanate methods. Its antioxidative character was also tested utilizing hydroxyl, DPPH, and superoxide radical scavenging activity measurements, using BHT, vitamin C and Trolox as references. Its antioxidant and radical scavenging activities were comparable to or higher than those of the reference antioxidants.

Laurus nobilis L. Family: Lauraceae

English Name: Laurel

Demo *et al.* (1998) reported antioxidant α-tocopherol from the Hexane extract of *Laurus nobilis*. Antioxidative activity of *Laurus nobilis* leaves, bark and fruit methanolic extracts (crude and defatted) were studied by Simic *et al.* (2003) on the level of lipid peroxidation in liposomes, induced by Fe^{2+}/ascorbate system and measured spectrophotometrically by TBA-test. The most significant inhibition of lipid peroxidation was obtained with methanolic extracts of laurel bark (70.6 per cent of inhibition was obtained with 1.0 mg of crude extract).

For laurel, antioxidant activity has been found for methanolic and ethanolic extracts (Kang *et al.*, 2002). The ethanol extract of *Laurus nobilis* leaves was screened for antioxidant activity by Pabuccuoglu *et al.* (2003). The ethanol extract showed antioxidant activity based on scavenging of ABTS radical cation as the percentage inhibition of absorbance at 734 nm and this activity was concentration-dependent.

Mariutti *et al.* (2008) evaluated the antioxidant activity of garlic employing ABTS and DPPH radical assay. DPPH EC_{50} value was 3.9g/kg while DPPH-TEAC value was 53mM/g. ABTS-TEAC value was 57mM/g.

Hydrodistillated extracts from laurel were assessed for their total phenol content, and antioxidant (iron(III) reduction, inhibition of linoleic acid peroxidation, iron (II) chelation, DPPH radical-scavenging and inhibition of hydroxyl radical-mediated 2-deoxy-D-ribose degradation, site and nonsite-specific) activities by Hinneburg *et al.* (2005). The extracts from Laurel possessed the highest antioxidant activities except iron chelation. In linoleic acid peroxidation assay, 1 g of laurel extract was effective as 21 of trolox. Thus laurel extract is promising alternative to synthetic substances as food ingredients with antioxidant activity.

The chloroform and methanolic extracts of 124 Egyptian plant species belonging to 56 families were investigated by Moussa *et al.* (2011) and compared their antioxidant activity by DPPH scavenging assay. Among the 124 plant species tested 18 exhibited extremely high antiradical activity (>80 per cent inhibition). The IC_{50} value of DPPH radical scavenging of *Laurus nobilis* was 25.30µg/ml, while total phenolic and flavonoid contents were 88.76 mg Tannic acid equivalent/g extract and 205.16 mg Rutin equivalent/g extract, respectively. Correlation coefficient between DPPH radical scavenging activity and the total phenolic and flavonoid contents suggest that phenolics and flavonoids in the extracts were partly responsible for the antiradical activities.

Lavandula angustifolia Mill. (Syn: *Lavandula vera* DC.) Family: Lamiaceae (Labiatae)

English Name: Lavender

Lavender is an important source of a thoroughly studied essential oil, while antioxidant properties of this plant are much less documented. Some reports on the antioxidant properties of this plant are somewhat contradictory, most likely due to the differences in the assessment methodology. For instance, Dapkevicius *et al.* (1998) did not detect antioxidant activity of various plant extracts in the model linoleic acid-β-carotene system, while Hohman *et al.* (1999) reported, that aqueous methanolic extracts of lavender were effective in lipid peroxidation media. Miliuskas *et al.* (2004) reported weak antioxidant activity with the extracts of *Lavendula angustifolia* in DPPH radical scavenging assay and ABTS radical cation decolourisation assay.

Yang *et al.* (2010) reported antioxidant activity of the essential oil from *Lavandula angustifolia*. Lavender oil was most effective for inhibiting linoleic acid peroxidation after 10 days.

Spiridon *et al.* (2011) evaluated the total phenolic content and antioxidant activity of plants used in Romanian herbal medicine. Total phenolic concentration was determined using Folin-Ciocalteu phenol reagent method, while total flavonoids were measured using the aluminium chloride colorimetric method. *Lavandula angustifolia* extracts showed good antioxidant activity.

Lavandula x intermedia Emeric ex Loiseleur Family: Lamiaceae (Labiatae)

English Name: Lavandin.

The phenolic content of Lavandin waste obtained after the distillation of essential oils was investigated. The antioxidant activity of different fractions as well as their phenolic content was evaluated by different methods (Torras-Claveria *et al.*, 2007).

Lavandula officinalis Chaix ex Vill. Family: Lamiaceae (Labiatae)

The ability of *Lavandula officinalis* extract to act as a free radical scavenger or hydrogen donor was revealed by DPPH radical-scavenging activity assay (Bouayed *et al.*, 2007). A positive correlation between total phenolic or flavonoid contents and antioxidant activity was found.

Lawsonia inermis L. (syn.: *Lawsonia alba* L.) Family: Lythraceae

English Name: Henna.

Ahmed *et al.* (2000) evaluated the efficacy of *Lawsonia inermis* (syn: *Lawsonia alba*) in the alleviation of carbon tetrachloride –induced oxidative stress.

Modulator effect of 80 per cent ethanol extract of leaves of henna on drug metabolizing phase I and Phase II enzymes, antioxidant enzymes, lipid peroxidation in the liver of Swiss Albino mice was studied by Dasgupta *et al.* (2003). The hepatic glutathione S-transferase and DT-diaphorase specific activites were elevated above basal level by *L. inermis* extract treatment. With reference to antioxidant enzyme the investigated doses were effective in increasing the hepatic glutathione reductase (GR), superoxide dismutase (SOD) and catalase activities significantly elevated in liver. Among the extrahepatic organs examined (forstomach, kidney and lung) glutathione S-transferase and DT-diaphorase level were increased in a dose dependent manner (Dasgupta *et al.*, 2003). Chloroform extract of leaves of *Lawsonia inermis* had shown the highest activity (87.6 per cent) followed by α-tocopherol (62.5 per cent) by using FTC method and based on TBA method significant activity (55.7 per cent) compared to α-tocopherol (44.4 per cent) (Endrini *et al.*, 2007). Total phenolic compound was 2.56 and 1.45 mg tannic per mg of Henna dry matter as extracted with methanol and water respectively (Prakash *et al.*, 2007). Lawsone (2-hydroxy-1,4-naphthaquinone) is the main ingredient of *L. inermis*. During oxidation of 100µM phenanthridine by guinea pigs aldehyde oxidase formation of superoxide anion and hydrogen peroxide at 6-10 per cent and 85-90 per cent respectively. Lawsone inhibits the production of superoxide anion and substrate oxidation more potently than hydrogen peroxide, the IC_{50} value of Lawsone with phenanthridine oxidation by aldehyde oxidase was 9.3µM, which in excess of 15 fold of maximal plasma concentrations of Lawsone, indicating a high degree of safety margin (Omar, 2005).

p-Coumaric acid, 2-methoxy-3-methyl-1,4-naphthaquinone, apiin, lawsone, apigenin, luteolin and cosmosiin, isolated form the methanolic extract of henna leaves showed antioxidant activity comparable to that of ascorbic acid (Mikhaeil *et al.*, 2004).

Phenolic compounds and antioxidant activity of Henna leaves extracts was investigated by Hosein and Zinab (2007). Two solvent (water and methanol) were used to prepare extract of Henna leaves. Water extract in comparison with the methanolic one had been more efficient. BHA and BHT at 200 ppm and methanolic extract at 800 ppm and 1400 ppm had equal TBA and peroxide value in soybean oil. Also the antioxidant activity of water and methanolic extracts was determined by using the rancimat method on refined soybean oil and compared with the induction period of synthetic antioxidants (BHA, BHT, TBHQ). It was shown that extraction method has significant effect on phenolic compound and antioxidant activity of Henna extract.

Hexavalent chromium Cr (VI) is a very strong oxidant which consequently causes high cytotoxicity through oxidative stress. Prevention of Cr (VI)-induced cellular damage has been sought in the study of Guha *et al.* (2009) in aqueous and methanolic extracts of *Lawsonia inermis*. The extracts showed significant potential in scavenging free radicals (DPPH) and ABTS and Fe(3+), and in inhibiting lipid peroxidation.

Total phenol, antioxidant and free radical scavenging activities of some medicinal plants, including *Lawsonia inermis* fruits were studied by Prakash *et al.* (2007). Fruits of *Lawsonia inermis* had comparatively lower phenols (75.8 mg/g) but exhibited good ARP (14.4) and reducing power (0.6ASE/ml).

Ledum groenlandicum Retzius Family: Ericaceae

English Name: Labrador tea.

Methanolic extracts of Labrador tea *Ledum groenlandicum* showed a strong antioxidant activity using the ORAC method and a cell based-assay. Moreover, the twig and leaf extracts showed significant anti-inflammatory activity, inhibiting NO release, respectively, by 28 and 17 per cent at 25µg/ml in LPS-stimulated RAW 264.7 macrophages (Dufour *et al.*, 2007).

Leea asiatica (L.) Ridsdale Family: Leeaceae

Dalal *et al.* (210) investigated the antioxidant activity of methanolic extract of *Leea asiatica* using DPPH radical, superoxide anion radical, nitric oxide radical and hydroxyl radical scavenging assay. The free radical scavenging activity of methanolic extract of leaves of *Leea asiatica* increased in a concentration dependent manner.

Leea guineense G. Don Family: Leeaceae

From leaves of *Leea guineense* three hydrophilic flavonoids were isolated and identified as quercetin-3'-sulphate-3-0-alpha-L-rhamnopyranoside, quercetin-3,3'-disulphate and a new flavonoid sulphate, quercetin-3,3',4'-trisulphate, together with kaempferol, quercetin, quercitrin, mearnsitrin, gallic acid and ethyl gallate. Their antioxidant effect on free radical scavenging was evaluated in the DPPH assay (Op de Beck *et al.*, 2003).

Leea indica (Burm. f.) Merr. Family: Leeaceae

Methanol exteracts of seven Malaysian medicinal plants, including *Leea indica* were screened by Saha *et al.* (2004) for antioxidant and nitric oxide inhibitory activities. Antioxidant activity was measured by using FTC, TBA and DPPH free radical scavenging methods and Greiss assay was used for the measurement of nitric oxide inhibition in LPS and interferon-γ-treated RAW264.7 cells. All the extracts showed strong antioxidant activity comparable to or higher than that of α-tocopherol, BHT and quercetin in FTC and TBA methods. The extract of *Leea indica* showed strong DPPH free radical scavenging activity comparable with quercetin, BHT and Vit C.

Lens culinaris Medik. Family: Fabaceae (Papilionaceae)

Amarowicz *et al.* (2010) evaluated the free radical-scavenging capacity, antioxidant activity, and phenolic composition of green lentil. Phenolic compounds present in the preparations showed antioxidant and radical scavenging properties as revealed by a β-carotene-liniolate model system, the total antioxidant activity (TAA) method, the DPPH scavenging activity assay and a reducing power assay. Data from these tests showed the greatest efficacies coming from the tannins. Catechin and epicatechin glucosides, procyanidin dimers, quercetin diglycoside, and *trans-p-*coumaric acid were the dominant phenolics in green lentils.

Lens esculenta Moench. Family: Fabaceae (Papilionaceae)

English Name: Lentil.

Sreeramulu *et al.* (2009) evaluated the antioxidant activity of lentil by DPPH, scavenging assay, FRAP assay and reducing power. Phenolic content of the lentil seed was 112.91 mg/100g. DPPH radical scavenging activity expressed as trolox equivalent, FRAP (μmol/g) and reducing power (mg/g) were 0.44, 68 and 1.03, respectively.

Leontice smirnowii Trautv. Family: Leonticaceae

Gülçin *et al.* (2006) screened antiradical and antioxidant activity of monodesmoids and crude extract from *Leontice smirnowii* tuber. Experiment revealed that MLS and CELS have an antioxidant effect concentration-dependently. Total antioxidant activity was performed according to FTC method. At the 30μg/ml concentration, the inhibition effects of MLS and CELS on peroxidation of linoleic acid emulsion were found to be 95.3 and 95.6 per cent, respectively. On the other hand, percentage inhibition of BHA, BHT, α-tocopherol and trolox were found to be 98.2, 98.5, 84.0 and 87.9 per cent inhibition of peroxidation of linoletic acid emulsion, respectively, at the same concentration. In addition MLS and CELS had effective DPPH radical scavenging, superoxide anion radical scavenging, hydrogen peroxide scavenging, reducing power and metal chelating activity.

Leonurus cardiaca L. Family: Lamiaceae (Labiatae)

English Name: Motherwort.

Matkowski and Piotrowska (2006) studied the antioxidative effects of methanolic extract with the use of three *in vitro* assays (DPPH assay, phosphomolybdenum method and lipid peroxidation). The extract showed strong antioxidant activity.

The aqueous extracts of 13 oriental medicinal plants were examined for their reducing power, scavenging ability towards superoxide and hydroxyl radicals and their inhibitory effect on lipid peroxidation. All extracts tested were found to be highly active on scavenging of superoxide radicals (Nam and Kang, 2004).

Matkowski *et al.* (2008) evaluated the antioxidant activity of aerial parts of *Leonurus cardiaca* using DPPH, OH radical scavenging assays, reducing power assay and inhibition of deoxyribose degradation. EC_{50} value of DPPH radical scavenging was 27.27μg/ml for MeOH extract, 39.14μg/ml for Dichloromethane extract, 4.45μg/ml for BuOH extract and 12359μg/ml for ethyl acetate extract.

Leonurus heterophyllus Sweet Family: Lamiaceae (Labiatae)

Gan *et al.* (2010) evaluated the antioxidant activities of 40 medicinal plants. The antioxidant activity of *Leonurus heterophyllus* was very low: 14.26μmol Fe(II)/g in

FRAP assay, 28.35μmol Trolox/g TEAC assay and Total phenolic content was 3.03 mg GAE/g.

Leopoldia comosa (L.) Parl (Syn.: *Muscari comosum*) Family: Liliaceae

A total of 27 extracts from non-cultivated and weedy vegetables traditionally consumed by ethnic Albanians in southern Italy were tested for their free radical scavenging activity (FRSA) in the DPPH screening assay, for their *in vitro* non-enzymatic inhibition of bovine brain lipid peroxidation and for their inhibition of xanthine oxidase. In both antioxidant assays strong activity was shown for *Leopoldia comosa* bulbs. The activity was comparable to quercetin and *Rhodiola rosea* extract (Pieroni *et al.*, 2002).

Lepechinia graveolens (Reg.) Epling. Family: Lamiaceae (Labiatae)

A bioguided separation of *Lepechinia graveolens* for antioxidant activity was carried out by Parejo *et al.* (2004). The radical scavenging activity of each fraction as well as that of the isolated compounds, was tested using three different methods. The major isolated antioxidant compounds were identified as luteolin-7-O-glucononide, rosmarinic acid and rosmarinic acid methyl ester. The major phenolic compound was found to be rosmarinic acid.

Lepechinia meyenii (Walp.) Epling Family: Lamiaceae (Labiatae)

Lock *et al.* (2005) evaluated the *in vitro* antioxidant activity of 40 Peruvian plants using DPPH assay. *Lepeceinia meyenii* showed highest antioxidant activity with EC_{50} value of 16.65μg/ml.

Lepidagathis cristata Willd. Family: Acanthaceae

Alcoholic extract of *Lepidagathis cristata* showed significant free radical scavenging activity *in vitro* (Achliya *et al.*, 2003). The phytochemical screening revealed presence of flavonoids and glycosides.

Lepidagathis trinervis Nees Family: Acanthaceae

Singh *et al.* (2009) evaluated the antioxidant activity of dichloromethane and methanolic extracts of 12 arid zone medicinal plants. *Lepidagathis trinervis* showed appreciable antioxidant activity and per cent of inhibition of DPPH where RC_{50} (μg/ml) was 20 (Dichloromethane extract) and 8 (MeOH extract). per cent of inhibition of DPPH was highest 96.17 per cent) in methanol extract of *Lepidagathis trinervis*.

Lepidium apetalum Willd. Family: Brassicaceae (Cruciferae)

Antioxidant capacities of 56 selected Chinese medicinal plants, including *Lepidium apetalum* were evaluated by Song *et al.* (2010) using the TEAC and FRAP assays, and their total phenic content was measured by the Folin-Ciocalteu method. *Lepidium apetalum* showed moderate antioxidant activity with TEAC value of 47.23μmol Trolox/g, FRAP value of 34.64μmol Fe^{2+}/g and phenolic content was 5.91mg GAE/g.

Lepidium perfoliatum L. Family: Brassicaceae (Cruciferae)

English Name: Alyssum

Souri *et al.* (2008) evaluated the antioxidant activity against linoleic acid peroxidation using 1,3-diethyl-2-thiobarbituric acid as reagent. Antioxidant activity (IC_{50}) against peroxidation of linoleic acid (2mg/ml) was 78.54 and phenolic content was 276.19 mg/100g dry weight. The results of this study showed that there is no significant correlation between antioxidant activity and phenolic content of the studied plant materials and phenolic content could not be a good indicator of antioxidant capacity.

Lepidium sativum L. Family: Brassicaceae (Cruciferae)

English Name: Garden-cress.

Souri *et al.* (2004) investigated the antioxidant activity of methanol extracts of garden-cress, used as vegetable in Iranian diet against linoleic acid peroxidation. The extracts showed good antioxidant activity. Garden-cress had exceptionally high antioxidant activity even higher than that of quercetin.

Leptadenia hastata (Pers.) Decne Family: Asclepiadaceae

Using Trolox assay, Cook *et al.* (1998) estimated the antioxidant activity of 17 wild edible plants of Niger Republic used for food and traditional medicine. They observed that *Leptadenia hastata* possessed strong antioxidant activity.

Lespedeza capitata Michx. Family: Fabaceae

Borchardt *et al.* (2008) evaluated the antioxidant activity of methanol extracts of seeds of *Lespedeza capitata* using DPPH antioxidant assay.The antioxidant value reported in µM Trolox/100g (TE) from DPPH radical scavenging activity of crude seeds extract was 43,945.

Leucaena leucocephala de Wit Family: Mimosaceae

English Name: Snbabul, White popinac.

Chanwitheesuk *et al.* (2005) investigated the antioxidant activity of *Leucaena leucocephala* using a β-carotene bleaching method. The contents of plant chemicals such as vitamin C, Vitamin E, carotenoids, tannin and total phenolics, were also determined. Methanolic extract showed antioxidant activity. *Leucaena leucocephala* has antioxidant index of 9.37, Vitamin C 48.5 mg per cent, Vitamin E 0.0202mg per cent, Total carotenes 0.64 mg per cent, Total xanthophylls 3.18mg per cent, Tannins 60.6mg per cent and Total phenolics of 405mg per cent.

The phenolic compounds of five southern Thai indigenous vegetables including white popinac (*Leucaena leucocephala*) leaf were extracted using different solvents. The extracts were analyzed for total phenolic content using the Folin-Ciocalteu procedure, free radical scavenging capacity by using ABTS and DPPH methods, and reducing capacity by using FRAP assay. The acetone extract possessed the highest total phenolic content followed by methanolic and distilled water extracts, respectively. The acetone extract showed higher free radical scavenging capacity and reducing capacity than those of their methanolic and water extracts counterparts respectively (Panpipat *et al.*, 2010).

Leucas aspera (Willd.) Link Family: Lamiaceae (Labiatae)

The ethanolic extract of *Leucas aspera* root was subjected to DPPH free radical scavenging assay for screening antioxidant activity (Rahman *et al.*, 2007). The extract showed a significant free radical scavenging activity with an IC_{50} of 8µg/ml.

The crude methanol extract of *Leucas aspera* leaves showed strong DPPH and superoxide radical-scavenging activities compared to other polarity-based extracted fractions (Meghashri *et al.*, 2010). The activity-guided repeated fractionation of the methanol extract yielded a compound that exhibited strong antioxidant activity. Based on various physicochemical and spectroscopic analysis the bioactive compound isolated was elucidated as 5,7-dihydroxy-2-[14-methoxy-15-propyl phenyl]-4H-chromen-4-one (leucasin). Radical-scavenging potential and strong inhibition of lipid peroxidation in a liposome model were observed at a leucasin concentration of 40 ppm. These results demonstrated the antioxidant potency of leucasin which could be the basis for its alleged health-promoting potential.

Leucas cephalotes (Roth) Spreng. Family: Lamiaceae (Labiatae)

Mathur *et al.* (2010) carried out phytochemical investigation and antioxidant activity of *Leucas cephalotes*. Traditional solvent extraction (TSE) using four solvents – methanol, aqueous, hexane and petroleum ether were utilized to determine the content of antioxidants. DPPH assay and superoxide anion radical scavenging activity were used to determine the antioxidant capacity. The most antioxidant capacity was achieved using methanol as the solvent followed by aqueous extracts. The antioxidant capacity with in the specific extract of plant was found to be correlated with the total phenolic content. The radical scavenging activity by DPPH radical scavenging method (IC_{50}) was 72.57µg/ml (hexane extract), 75.10 µg/ml (petroleum ether extract), 39.2 (aqueous) and 47.15 (methanol extract).

Leucas indica L. Family: Lamiaceae (Labiatae)

A new phenylethanoid glycoside along with five known phenyl ethanoid glycosides were isolated from the aerial parts of *Leucas indica*. Compounds 1-6 exhibited significant antioxidant activity in DPPH radical assay method. These compounds were also found to be moderate inhibitors of xanthine oxidase enzyme (Mostafa *et al.*, 2007).

Levisticum officinale Koch Family: Apiaceae (Umbelliferae)

Wojcikowski *et al.* (2007) evaluated the antioxidant activity of aerial parts of *Levisticum officinale*. Oxygen radical absorbance capacity of the ethyl extract, methanol extract and aqueous extracts were 7.69, 50.17 and 21.06µmol Trolox equivalent per g of dried starting material, respectively.

Licania arborea Seem. Family: Chrysobalanaceae

Ruiz-Terán *et al.* (2008) evaluated the antioxidant activity of aerial parts of *Licania arborea* using DPPH and β-carotene-linoleic acid bleaching assay. The extract displayed antioxidant activity comparable to the commercial antioxidant BHA. For the methanolic extract analysed, a clear relation between the total phenolic content of the extracts and their antioxidant activity was found.

Licania licaniaeflora Blake Family: Chrysobalanaceae

Braca *et al.* (2002) screened the antioxidant activity of flavonoids from *Licania licaniaeflora*. All the isolated compounds exhibited DPPH radical scavenging activity: quercetin derivatives showed the strongest action, while the flavanone 8-hydroxy-naringenin and kaempferol 3-O-alpha-rhamnoside had the lowest.

Licaria martiniana (Mez.) Kostern. Family: Lauraceae

Alcantara *et al.* (2010) reported antioxidant activity of ethe essential oil of *Licaria martiniana* from Brazil. Authors followed DPPH method coupled to TLC. The IC_{50} values were generally >1000µg/ml, substantially superior to that of quercetin, the reference used by the authors.

Ligaria cuneifolia (R. et P.) Tiegh. Family: Loranthaceae

Borneo *et al.* (2009) analyzed the total phenols content (Folin-Ciocalteau assay) and antioxidant capacity (ferric reducing/antioxidant power – FRAP) of 41 plants from Cordoba (Argentina). *Ligaria cuneifolia* exhibited the highest value of total phenols (100.2 mg GAE/g) and antioxidant capacity (1862 µmol of Fe(II)/g). A significant linear correlation was found between phenols content and antioxidant capacity.

Ligularia fischeri Turcz. Family: Asteraceae (Compositae)

Choi *et al.* (2007) evaluated the antioxidant activity of *Ligularia fischeri* leaves. Ethanol extract displayed activity against DPPH, H_2O_2 radicals significantly, decreased the levels of malondialdehyde measured by Ferric thiocyanate and TBA. Like antioxidant activity, the reducing power of the extract was also good.

Shang *et al.* (2010) detected 4 major antioxidant compounds from *Ligularia fischeri* leaves using an on-line HPLC-ABTS screening system, which can determine the antioxidant activity based on a decrease in absorbance of 734nm after postcolumn reaction of HPLC-separated antioxidants with the ABTS. The active compounds include 5-O-caffeoylquinic acid (5-CQA), 3,4-di-O-caffeoylquinic acid (3,4DCQA), 3,5-di-O-caffeoylquinic acid (3,5-DCQA), and 4,5-di-O-caffeoylquinic acid (4,5-DCQA). These 4 isomers comprised over 10 per cent of dried leaves with 3,5-DCQA being the most abundant compound. The radical scavenging activity of each isomer was also evaluated simultaneously through online HPLC-ABTS method, which showed 94 per cent antioxidant acitivity of the ethanol extract derived from caffeoylquinic acids. Among these isomers, 3,4-DCQA contained the most strong antioxidant activity while 3,5-DCQA accounted for the highest radical scavenging capacity due to having the highest content.

Ligusticum chuanxiong Hort Family: Apiaceae (Umbelliferae)

Jeong *et al.* (2009a) reported antioxidant activity of the essential oil of *Ligusticum chuanxiong* using DPPH and ABTS scavenging assay. Twenty constituents in the essential oil were identified and they showed good antioxidant properties, in that IC_{50} value in DPPH and ABTS showed 1.58 and 1.58µg/ml.

Gan *et al.* (2010) evaluated the antioxidant activities of 40 medicinal plants. The antioxidant activity of *Ligusticum chuanxiong* was 137.53µmol Fe(II)/g in FRAP assay, 72.20µmol Trolox/g TEAC assay and Total phenolic content was 5.51 mg GAE/g.

Ligusticum sinense Oliv. Family: Apiaceae (Umbelliferae)

Antioxidant capacities of 56 selected Chinese medicinal plants, including *Ligusticum sinense* were evaluated by Song *et al.* (2010) using the TEAC and FRAP assays, and their total phenolic content was measured by the Folin-Ciocalteu method. *Ligusticum sinense* showed low antioxidant activity with TEAC value of 87.80µmol Trolox/g, FRAP value of89.84µmol Fe^{2+}/g and phenolic content was 11.99mg GAE/g.

Ligustrum delavayanum Hariot Family: Oleaceae

The free radical scavenging activity of the water infusions, different organic solvent extracts and some constituents from leaves was assessed with the aid of DPPH radical by Nagy *et al.* (2006). Among the samples screened, water infusions had the strongest free radical scavenging capacity. From the tested compounds scavenging active flavonoid aglycones were present in the most active chloroform fractions from both leaves samples.

Ligustrum japonicum Thunb. Family: Oleaceae

Katsube *et al.* (2004) screened aqueous ethanol extract of *Ligustrum japonicum* for antioxidant activity using DPPH radical scavenging activity and LDL oxidation induced by copper ion. Total phenolic content was also measured for comparisons with antioxidant activity in LDL. The study revealed high levels of LDL antioxidant activity in *Ligustrum japonicum*.

Ligustrum lucidum Ait. Family: Oleaceae

Cai *et al.* (2004) evaluated the antioxidant activity and phenolic compounds in traditional Chinese medicinal plants associated with anticancer. The improved ABTS.+ method was used to systematically assess the total antioxidant capacity (Trolox equivalent antioxidant capacity, TEAC) of the medicinal extracts. The TEAC values and total phenolic content for methanolic extract of fruit of *Ligustrum lucidum* were 234.4 µmol Trolox equivalent/100g dry weight (DW), and 1.76 g of gallic acid equivalent/100g DW respectively. A positive significant linear relationship between antioxidant activity and total phenolic content showed that phenolic compounds were the dominant antioxidant components in the tested herbs. Major types of phenolic compounds in *Ligustrum lucidum* include phenolic (secoiridoid) glycosides (nuzhenide, oleuropein).

Oleanolic acid from *Ligustrum lucidum* had hypoglycemic, hypolipidemic and antioxidant efficacy in the diabetic rats (Gao *et al.*, 2009).

Ligustrum robustum (Roxb.) Blume Family: Oleaceae

Ligustrum is used for tea preparation in China. Phytochemical investigation of the ethanol extract of the leaves of *Ligustrum robustum* monitored by a bioassay involving the hemolysis of red blood cells induced by 2,2'-azo-bis(2-amidinopropane)dihydrochloride, led to the isolation of three new glycosides, ligurobustosides M, N and O, along with 10 known ones. Seven of the glycosides showed stronger antioxidant effects than the standard, trolox (He *et al.*, 2003).

Zhu *et al.* (2009) investigated the major phenolic compounds, individual and total phenolics contents, and *in vitro* antioxidant properties (ABTS, DPPH, FRAP and OH assays) of Kudingcha genotypes from *Ligustrum robustum*. Major phenolics in Kudingcha genotypes from *Ligustrum robustum* were phenylethanoid and monoterpenoid glycosides. Among the Kudingcha genotypes *Ligustrum* genotypes exhibited lesser antioxidant capacity than *Ilex* genotypes.

Ligustrum sinense Lour. Family: Oleaceae

Three active glycosides that afford protection to red blood cell membrane to resist hemolysis induced by a peroxyl radical initiator, 2,2-azo-bis-(2-amidinopropane) dihydrochloride were isolated from the MeOH extract by Ouyang *et al.* (2003).

Ligustrum vulgare L. Family: Oleaceae

The free radical scavenging activity of the water infusions, different organic solvent extracts and some constituents from leaves was assessed with the aid of DPPH radical by Nagy *et al.* (2006). Among the samples screened, water infusions had the strongest free radical scavenging capacity. From the tested compounds scavenging active flavonoid aglycones were present in the most active chloroform fractions from both leaves samples.

Limnophila geofrayi Family: Scrophulariaceae

The chloroform extract of aerial part showed antimycobacterial and antioxidant activities (Suksamaran *et al.*, 2003). Isothymusin isolated from the aerial part exhibited antioxidant activity against the radical scavenging ability of DPPH.

Limonium bicolor (Bunge) Kuntze Family: Plumbaginaceae

In a study on the preparation and antioxidant capacity of *Limonium bicolor* Zhen *et al.* (2009) reported that compared with the traditional extraction method by hot water the ultrasonic-microwave synergistic extraction can shorten the extraction time and enhance the extraction efficiency. The polysaccharides of *Limonium bicolor* had the obvious antioxidant activity.

Extraction process of flavonoids in *Limonium bicolor* and its antioxidant function were studied by Li *et al.* (2010). The antioxidant capacity of flavonoids in *Limonium bicolor* was estimated with the test of hydroxyl radical inhibition and lipid antioxidation. The test of antioxidant capacity showed that radical OH was eliminated significantly. Lipid rancidification was inhibited and the role of these effect increased with the increase of flavonoid concentration.

Limonium wrightii O.Kuntze Family: Plumbaginaceae

Free radical scavenging action of *Limonium wrightii* was examined by Aniya *et al.* (2002) *in vitro* and *in vivo* by using electron spin resonance spectrometer and chemiluminescence analyzer. A water extract of *L.wrightii* showed a strong scavenging action for the DPPH or superoxide anion and moderate for hydroxyl radical. The extract also depressed production of reactive oxygen species from polymorphonuclear leukocytes stimulated by phorbor-12-myristate acetate and inhibited lipid peroxidation in rat liver microsomes. When the extract was given intraperitoneally to mice prior to Carbon tetrachloride (CCl_4) treatment, CCl_4-induced liver toxicity, as seen by an elevation of serum aspartate aminotransferase and alanine aminotransferase activities was significantly reduced. Gallic acid was identified as the active component of *L. wrightii* with a strong free radical scavenging action. Results of Aniya *et al.* (2002) demonstrate the free radical scavenging action of *L. wrightii* and that gallic acid contributes to these actions.

Linaria pyramidata Boiss. Family: Scrophulariaceae

Souri *et al.* (2004) evaluated the antioxidative activity of *Linaria pyramidata* by linoleic acid peroxidation test using 1,3-diethyl-2-thiobarbituric acid as the reagent. Methanolic extract showed 79.31 per cent inhibition at 40µg concentration with IC_{50} value of 5.57µg.

Lindelofia stylosa (Kar. et Kir.) Brand. Family: Boraginaceae

Seven phenyl propanoids isolated from the aerial parts of *Lindelofia stylosa* were studied for their antioxidant properties by Choudhary *et al.* (2005).

Lindera pulcherrima (Nees) Benth. ex Hook.f. Family: Lauraceae

Joshi *et al.* (2010) evaluated the antioxidant activity of seven Himalayan Lauraceae species. The essential oil of *Lindera pulcherrima* was able to inhibit linoleic acid oxidation. Furanodienone and curzerenone constituted the major components of the oil.

Lindera strychnifolia (Bl.) Fern.-Vill. Family: Lauraceae

Methanol aqueous extract of *Lindera strychnifolia* was screened by Kim *et al.* (1997) for free radical scavenging activity using the DPPH free radical generating system. The extract showed free radical scavenging activity.

Lindernia anagallis (N.L. Burman) Pennell Family: Scrophulariaceae

Shyur *et al.* (2005) examined antioxidant activities of 26 medicinal herbal extracts that have been popularly used as folk medicines in Taiwan. The results of scavenging DPPH radical activity show that, among the 26 tested medicinal plants, *Lindernia anagallis* exhibited strong activities and its IC_{50} value for DPPH radical was 36µg/ml.

Linum usitatissimum L. Family: Linaceae

English Name: Linseed, Flax.

The antioxidative activity of phenolic extracts from seeds of flax was examined by autoxidation of methyl linoleate by Kahkonen *et al.* (1999). Phenolics are practically nil (0.8mg/g GAE). However antioxidant activity was 35 per cent at 500 ppm level and 18 per cent at 5000ppm level.

Souri *et al.* (2008) evaluated the antioxidant activity of seed against linoleic acid peroxidation using 1,3-diethyl-2-thiobarbituric acid as reagent. Antioxidant activity (IC_{50}) against peroxidation of linoleic acid (2mg/ml) was 53.52 and phenolic content was 21.76 mg/100g dry weight. The results of this study showed that there is no significant correlation between antioxidant activity and phenolic content of the studied plant materials and phenolic content could not be a good indicator of antioxidant capacity.

Zanwar *et al.* (2010) evaluated the *in vitro* antioxidant activity of ethanolic extract of *Linum usitatissimum*. The results indicated significant dose-dependent inhibition against DPPH radical, reducing power, superoxide anion radical scavenging, hydroxyl scavenging, metal chelating and hydrogen peroxide scavenging by the ethanolic extract.

Lippia adoensis Hochst. ex Walp. var. *koseret* Family: Verbenaceae

Mikre *et al.* (2007) reported relatively weak antioxidant activity of the essential oil from the leaves of *Lippia adoensis* var. *koseret.*

Lippia alba (Mill.) N.E.Br. ex Britton et P. Wilson Family: Lamiaceae (Labiatae)

In-vitro antioxidant activity of methanolic leaves and flowers extract of *Lippia alba* was determined by Ara and Nur (2009) by DPPH free radical scavenging assay. The Reducing power of extracts was also determined. Ascorbic acid was used as standard and positive control for both the analysis. All the analysis was made with the use of UV-Visible Spectrophotometer. The methanolic leaves and flowers extracts of *Lippia alba* had shown very significant DPPH radical scavenging activity compared to standard antioxidant. The DPPH radical scavenging activity of the extract was increased with the increasing concentration. In DPPH free radical scavanging assay IC_{50} value of leaves and flowers extracts of *Lippia alba* was found to be 34.4 µg/ml. The results concluded that the extracts have a potential source of antioxidants of natural origin.

Lippia boliviana Rusby Family: Verbenaceae

The aerial parts of 17 Bolivian plants including *Lippia boliviana* were screened by Rosas-Romero and Saavedra (2005) to determine antioxidant activity. A methanol extract of each plant was prepared and portioned sequentially with hexane, chloroform, and ethyl acetate, leaving an aqueous solution. All extracts and their 5 fractions for a total of 102 samples, were evaluated using two techniques: an adaptation of the β-carotene bleaching technique using an emulsion of linoleic acid in water as the oxidizable substrate, and the DPPH free radical trapping technique. The results of the β-carotene bleaching technique were more discriminating and better related to the rancidity process under normal conditions; with this assay, 11 species provided at least one fraction with highly promising antioxidant activity. All species gave good results under the DPPH technique, and in most cases they performed better than BHA, which was used as a reference antioxidant (Rosas-Romero and Saavedra, 2005).

Lippia citriodora Kunth Family: Verbenaceae

Valent o *et al.* (2002a) examined the superoxide radical, hydroxyl radical and hypochlorous acid scavenging activities of *Lippia ctriodora* infusion. The results demonstrated that this infusion has a potent superoxide radical scavenging activity and a moderate scavenging activity of hydroxyl radical and hypochlorous acid.

Lippia dulcis Trevir. Family: Verbenaceae

Cirsimaritin, eupatorin, 5,3'-dihydroxy-6,7,4',5'-tetramethoxyflavone isolated from the aerial parts of *Lippia dulcis* exhibited almost the same antioxidant activity as that of α-tocopherol, and decaffeoylverbascoside, acetoside and icoacetoside were identified as stronger antioxidants than α-tocopherol using the ferric thiocyanate method (Ono *et al.*, 2005).

Lippia graveolens Kunth. Family: Verbenaceae

Three Mexican *Lippia graveolens* oils with different chemical compositions as well as their microcapsules were evaluated in terms of antiradical activities. Sánchez-Arana *et al.* (2010) concluded that microencapsulation increased the anti-radical activity from fourfold to eightfold.

Lippia multiflora Mold. Family: Verbenaceae

Volatile constituents and antioxidant activity of essential oils from *Lippia multiflora* growing in Gabon were investigated by Agnaniet *et al.* (2005). An interesting antioxidant activity was found for the second chemotype.

The antioxidant and antiradical activities of the essential oil of the leaves of *Lippia multiflora* was found to be low comparatively to that of BHT (Avlessi *et al.*, 2005).

Lippia sidoides Cham. Family: Verbenaceae

Monteiro *et al.* (2007) reported topical anti-inflammatory, gastroprotective and antioxidant effects of the essential oil of *Lippia sidoides* leaves.

Liquidamber formosana Hance Family: Altingiaceae

Antioxidant activity of *Liquidambar formosana* extracts was assessed by using DPPH, reducing and Phosphomolybdenum assays (Wang *et al.*, 2009). All extracts exhibited excellent antioxidant activities and were superior to BHT. There was a good correlation between antioxidant activities and TP/TF content.

Gan *et al.* (2010) evaluated the antioxidant activity of *Liquidamber formosana* using the FRAP and TEAC assays, and its total phenolic content was measured by the Folin-Ciocalteu method. For FRAP assay the value was 118.44 μmol Fe(II)/g dry weight. For TEAC assay, the value was 81.88μmol Trolox/g dry weight. Total phenolic content was 5.58 mg gallic acid equivalent/g dry weight of plant material.

Liriope spicata L. Family: Liliaceae

DPPH scavenging activities of 80 per cent methanolic leaf extracts of three cultivars (small leaf SL, bigh leaf BL; thin leaf TL) of Mai-Men-Dong (*Liriope spicata*) were analyzed by spectrophotometry. The concentrations required for 50 per cent inhibition (IC_{50}) of DPPH radicals were 81.08, 96.97 and 53.78 mg/ml respectively. The ethyl acetate-soluble fraction exhibited the highest DPPH scavenging activity. The IC_{50} of ethyl acetate-soluble fractions of SL, BL and TL for DPPH radical scavenging activity were 41.55, 24.55 and 53.33mg/ml respectively. A positive correlation between DPPH radical scavenging and phenolic contents has been observed. The IC_{50} of hot water extracts of Sl, Bl and TL for DPPH radical scavenging activities were 378.97, 171.12 and 95.84 mg/ml respectively. All three hot water extracts can effectively scavenge hydroxyl radical using ESR spectrometry. The IC_{50} against hydroxyl radical were 80.8, 69.7 and 116 mg/ml respectively for SL, BL and TL cultivars (Hou *et al.*, 2004).

Litchi chinensis (Gaertn.) Sonn. Family: Sapindaceae

English Name: Lyohee.

Alfredo (2007) reported high antioxidant activity of Lichi fruits.

The main oligomeric procyanidins from litchi pericarp (*Litchi chinensis*) were isolated and identified. The effects of oligomeric procyanidins, A2 and epicatechin from litchi pericarp on free-radical scavenging were determined by using a chemiluminiscent method. The result showed that all of them had a strong scavenging effect on OH and the IC_{50} values were 2.60 μg/ml, 1.75μg/ml and 1.65μg/ml for oligomeric procyanidins, A2 and the trimeric procyanidins respectively. The antioxidant activities of A-type dimeric and trimeric procyanidins seemed to be related to the number of hydroxyls in their molecular structures (Liu *et al.*, 2007).

The experiments were performed by Sun *et al.* (2010) to extract and purify substrates for polyphenol oxidase (PPO) from pericarp tissue of postharvest litchi fruit. Two purified PPO substrates were identified as (-)-epicatechin and procyanidin A2. The results showed that (-)-epicatechin exhibited stronger antioxidant capability than procyanidin A2, in terms of reducing power and scavenging activities of DPPH radical, hydroxyl radical and superoxide radical. Furthermore, (-)-epicatechin content in pericarp tissue tended to decrease with increasing skin browning index of litchi fruit during storage at 25°C. Thus, these two compounds can be used as potential antioxidants in litchi waste and the fresh pericarp tissue of litchi fruit exhibited a better utilization value.

The pericarps of 10 litchi cultivars from two production seasons were studied by Wang *et al.* (2011) for their phenolic concentrations and composition and activities of antioxidation. The total phenolic concentrations of dried litchi pericarp ranged from 51 to 102g/kg. Diffeent cultivars varied considerably in phenolic concentration and composition. The extracts of litchi pericarps exhibited considerably high FRAP and strong activities of DPPH scavenging, lipid peroxidation inhibiting and oxidative DNA damage protection. The crude ethanol extracts of pericarp displayed significantly higher FRAP and stronger DPPH scavenging capacity than BHT and catechin. Among the tested cultivars concentrations of major phenolics had significant linear correlations with FRAP and DPPH scavenging activities.

Lithocarpus pachyphyllus (Kurz.) Rehd. Family: Fagaceae

In vitro antioxidant activities of three sweet dihydrochalcone glucosides from the leaves of *Lithocarpus pachyphyllus*, trilobatin 2"-acetate (1), phloridzin (2) and trilobatin (3), were investigated by Yang *et al.* (2004). The IC_{50} values for compound 1-3 of lipid peroxidation in rat liver homogenate were 261, 28, 88μM, respectively. Compounds 1-3 increased SOD activity with EC_{50} values of 575, 167, 128 μM, and GSH-Px activity with EC_{50} values of 717, 347, 129 μM, respectively, and showed only weak DPPH radical scavenging activity.

Lithospermum erythrorhizon Sieb. et Zucc. Family: Boraginaceae

Seven compounds, deoxyshikonin (1), β,β-dimethylacrylshikonin (2), isobutyl-shikonin (3), shikonin (4), 5,8-dihydroxy-2(1-methoxy-4-methyl-3-pentenyl)-1,4-naphthalenedione (5), β-sitosterol (6) and a mixture of two caffeic acid esters (7) were isolated by Han *et al.* (2008) from *Lithospermum erythrorhizon* and identified by spectroscopic methods. The antioxidant activities of the seven compounds were compared and evaluated through Rancimat method, reducing power and radical scavenging activity. Results showed that, except compound 6, another 6 compounds all exhibited obvious antioxidant activities against four different methods. Compounds 4 and 7 exerted much more potent antioxidant effects on retarding the lard oxidation than that of BHT and both were found to exhibit strong reducing power. In addition compounds 1-5 all exerted very good radical scavenging activities toward ABTS but showed moderate inhibition of DPPH, while compound 7 presented as a powerful radical scavenger against both ABTS and DPPH.

Cai *et al.* (2004, 2006) evaluated the antioxidant activity and phenolic compounds in traditional Chinese medicinal plants associated with anticancer. The improved ABTS.+method was used to systematically assess the total antioxidant capacity (Trolox equivalent antioxidant capacity, TEAC) of the medicinal extracts. The TEAC values and total phenolic content for methanolic extract of *Lithospermum erythrorhizon* were 413.9 µmol Trolox equivalent/100g dry weight (DW), and 0.92 g of gallic acid equivalent/100g DW respectively. A positive significant linear relationship between antioxidant activity and total phenolic content showed that phenolic compounds were the dominant antioxidant components in the tested herbs. Major types of phenolic compounds in *Lithospermum erythrorhizon* include naphthoquinones (shikonin, acetyl shikonin, alkannin).

Li *et al.* (2008) evaluated the antioxidant properties of 45 medicinal plants including *Lithospermum erythrorhizon* using FRAP and TEAC assays and the total phenolic contents of these plants were measured. *Lithospermum erythrorhizon* showed 29.59 µmol Fe(II)/g FRAP value, 24.83µmol Trolox/g of TEAC value and 7.80 mg GAE/g of phenolic content.

Litsea cubeba Pers. Family: Lauraceae

There is antioxidant serum in the market by the name 'Juice Beauty' which is a polyherbal preparation and one of the phytoconstituents is *Litsea cubeba* (http://www.skin-one.com/juice-beauty-antioxidant-serum.html).

The antioxidant activity of *Litsea cubeba* was studied by Hwang *et al.* (2005) in terms of three different assay systems: DPPH assay, peroxidase/guaiacol assay and TBA test. The methanol and its fractions showed remarkable antioxidant activity in comparison with α-tocopherol and ascorbic acid.

Litsea glutinosa (Lour.) C.B.Robinson Family: Lauraceae

Kshirsagar and Upadhyay (2009) investigated the radical scavenging activity of stem and twig of *Litsea glutinosa*. Both methanol extracts and sequential extracts of stem showed high antioxidant activity. Successive ethyl acetate extracts have shown equal antioxidant activity as that of successive methanol (Kshirsagar and Upadhyay, 2009).

Lobelia chinensis Lour. Family: Campanulaceae

Cai *et al.* (2004) evaluated the antioxidant activity and phenolic compounds in traditional Chinese medicinal plants associated with anticancer. The improved ABTS.+method was used to systematically assess the total antioxidant capacity (Trolox equivalent antioxidant capacity, TEAC) of the medicinal extracts. The TEAC values and total phenolic content for methanolic extract of whole plant of *Lobelia chinensis* were 57.5 µmol Trolox equivalent/100g dry weight (DW), and 0.39 g of gallic acid equivalent/100g DW respectively. A positive significant linear relationship between antioxidant activity and total phenolic content showed that phenolic compounds were the dominant antioxidant components in the tested herbs. Major types of phenolic compounds in *Lobelia chinensis* include flavonol glycosides.

Li *et al.* (2008) evaluated the antioxidant properties of 45 medicinal plants including *Lobelia chinensis* using FRAP and TEAC assays and the total phenolic contents of these plants were measured. *Lobelia chinensis* showed low antioxidant activity with 16.47 µmol Fe(II)/g FRAP value, 11.42µmol Trolox/g of TEAC value and 4.72 mg GAE/g of phenolic content.

Loiseleuria procumbens (L.) Desv. Family: Ericaceae

A stilbene and dihydrochalcones isolated from the whole plant of *Loiseleuria procumbens* showed scavenging properties towards DPPH radical and antioxidant properties in test with lysozyme (Cuendet *et al.*, 2000).

Lomatogonium carinthiacum (Wulfen) Reichb. Family: Gentianaceae

Myagmar and Aniya (2000) evaluated the free radical scavenging action of some medicinal herbs growing in Mongolia. The aqueous extract of nine herbs including *Lomatogonium carinthiacum* were used. The free radical scavenging action was determined *in vitro* and *ex vivo* by using ESR spectrometer and chemiluminiscence analyzer. The results showed that *Lomatogonium carinthiacum* extract possess strong scavenging action of DPPH, superoxide and hydroxyl radicals. *Lomatogonium carinthiacum* also depressed reactive oxygen production from polymorphonuclear leukocytes stimulated by phorbol-12-myristate *ex vivo*.

Lonicera japonica Thunb. Family: Caprifoliaceae

English Name: Flos Lonicera.

Cai *et al.* (2004, 2006) evaluated the antioxidant activity and phenolic compounds in traditional Chinese medicinal plants associated with anticancer. The improved ABTS.+ method was used to systematically assess the total antioxidant capacity (Trolox equivalent antioxidant capacity, TEAC) of the medicinal extracts. The TEAC values and total phenolic content for methanolic extract of floral bud of *Lonicera japonica* were 589.1 µmol Trolox equivalent/100g dry weight (DW), and 3.63 g of gallic acid equivalent/100g DW respectively. A positive significant linear relationship between antioxidant activity and total phenolic content showed that phenolic compounds were the dominant antioxidant components in the tested herbs. Major types of phenolic compounds in *Lonicera japonica* include phenolic acids (chlorogenic acid), flavones (luteolin, luteolin-7-glucosdie).

Choi *et al.* (2007) evaluated the antioxidant effects of *Lonicera japonica* via DPPH radical, total ROS, hydroxyl radical and peroxynitrite assays. Among the methanolic extract and the dichloromethane, ethyl acetate, n-butanol and water fractions, the EtOAc fraction of *Lonicera japonica* exhibited marked scavenging/inhibitory activities, as follows: IC_{50} values of 4.37, 27.58, 0.47 and 12.13µg/ml in the DPPH, total ROS, ONOO- and OH assays, respectively. Luteolin, caffeic acid, protocatechuic acid, isorhamnetin-3-O-β-D-glucopyranoside, quercetin-3-O-β-D-glucopyranoside and luteolin 7-O-β-D-glucopyranoside evidenced marked scavenging activities, with IC_{50} values of 2.08-11.76 µg/ml for DPPH radicals and 1.47 µM for ONOO-.

Li *et al.* (2008) evaluated the antioxidant properties of 45 medicinal plants including *Lonicera japonica* using FRAP and TEAC assays and the total phenolic contents of these plants were measured. *Lonicera japonica* flower showed 261.05 µmol Fe(II)/g FRAP value, 121.97µmol Trolox/g of TEAC value and 27.36 mg GAE/g of Phenolic content. A high correlation between antioxidant capacities and their phenolic contents indicated that phenolic compounds were a major contributor of antioxidant activity of these plants.

Lophatherum gracile Brongn. Family: Poaceae (Graminae)

Li *et al.* (2008) evaluated the antioxidant properties of 45 medicinal plants including *Lophatherum gracile* using FRAP and TEAC assays and the total phenolic contents of these plants were measured. *Lophatherum gracile* showed 67.90 µmol Fe(II)/g FRAP value, 33.37µmol Trolox/g of TEAC value and 12.11 mg GAE/g of Phenolic content.

Loranthus parasiticus (L.) Merr. Family: Loranthaceae

Gan *et al.* (2010) evaluated the antioxidant activity of *Loranthus parasiticus* using the FRAP and TEAC assays, and its total phenolic content was measured by the Folin-Ciocalteu method. For FRAP assay the value was 580.02 µmol Fe(II)/g dry weight. For TEAC assay, the value was 457.00µmol Trolox/g dry weight. Total phenolic content was 29.67 mg gallic acid equivalent/g dry weight of plant material.

Ludwigia octovalvis (Jacq.) P.H. Raven Family: Onagraceae

Shyur *et al.* (2005) examined antioxidant activities of 26 medicinal herbal extracts that have been popularly used as folk medicines in Taiwan. The results of scavenging DPPH radical activity show that, among the 26 tested medicinal plants, *Ludwigia octovalvis* exhibited strong activities and its IC_{50} value for DPPH radical was 4.6µg/ml. Superoxide anion scavenging activity was $IC_{50} = 26$µg/ml. It was also observed that at 1 mg/ml *Ludwigia octovalvis* exhibited significant protection on φ x174 supercoiled DNA against strand cleavage induced by UV irradiated H_2O_2 with a superior compatible effect to that of catechin.

Luffa acutangula (L.) Roxb. Family: Cucurbitaceae

English Name: Angular loofah, Ridge gourd.

Ansari *et al.* (2005) evaluated the antioxidant activity of five vegetables traditionally consumed by South-Asian migrants in Bradford, Yorkshire, UK. Extracts from *Luffa acutangula* showed a significant difference in the FRSA between the extract obtained by using cold maceration and that prepared by boiling the plant in the solvent under reflux, suggesting the chemical composition of the plant changed during heating process, leading to an increase in the amount of antioxidant components.

A study was carried out by Raghu *et al.* (2010) to determine the antioxidant activity of aqueous extracts of ten selected common vegetables. For *Luffa acutangula* the IC_{50} value for DPPH scavenging was 1.70mg/ml while standard was 2.45µg/ml. Reducing power of the extract was moderate.

Luffa cylindrica M.J.Roem. Family: Cucurbitaceae

English Name: Smooth gourd, Loofah; Japanese: Hechima.

Maeda *et al.* (2006) examined the phenolic contents of Loofah and its radical scavenging activities for DPPH radical and found them to be very low. The polyphenol content was 7.9 mg-gallic acid equivalent/100 g and DPPH scavenging activity was 49.7 µmol-Trolox eq./100 g.

Luffa echinata Roxb. Family: Cucurbitaceae

Kumar *et al.* (200) investigated antioxidant activity of alcoholic (50 per cent) extract of the plant *Luffa echinta* using inhibition of lipid peroxidaation, hydroxyl radical scavenging activity and interaction with DPPH. It was found that the test extract exhibited a considerable inhibition of lipid peroxidation and possessed hydroxyl scavenging activity. Evaluation of antiradical scavenging activity showed significant interaction with DPPH.

Lumnitzera racemosa Willd. Family: Combretaceae

Vadlapudi and Naidu (2009) evaluated the antioxidant potential of *Lumintzera racemosa*, a mangrove plant. The SOD was 1.39U/mg, Catalase was 0.18U/mg, Ascorbic acid was 160mg/100g, DPPH radical inhibition was 79 per cent and FRAP units were 672.

Lupinus angustifolius L. Family: Fabaceae (Papilionaceae)

English Name: Blue lupin.

The antioxidative activity of phenolic extracts from seeds of blue lupin was examined by autoxidation of methyl linoleate by Kahkonen *et al.* (1999). Both phenolics and antioxidant activity was very low. Phenolics were 4.7 mg/g GAE while the antioxidant activity was 10 per cent at 500 ppm level.

Lupinus mutabilis Sweet Family: Fabaceae (Papilionaceae)

Penarrieeta *et al.* (2005) measured the Total Antioxidant Capacity (TAC) in some Andean foods. Eight Andean foods, including *Lupinus mutabilis* were analyzed by two methods ABTS and FRAP to assess TAC. TAC value of *Lupinus mutabilis* was good in both water-soluble fraction and water-insoluble fraction.

Lycianthes acutifolia (R. et P.) Bitter Family: Solanaceae

Mosquera *et al*. (2009) evaluated the antioxidant activity of forty-six methanol plant extracts using DPPH radical scavenging assay. The plant extracts that showed greatest antioxidant activity were *Lycianthes acutifolia* with 37.7 per cent of antioxidant activity.

Lycianthes radiata (Sendt.) Bitter Family: Solanaceae

Mosquera *et al*. (2009) evaluated the antioxidant activity of forty-six methanol plant extracts using DPPH radical scavenging assay. The plant extracts that showed greatest antioxidant activity were *Lycianthes radiata* with 41.5 per cent of antioxidant activity.

Lycium barbarum L. Family: Solanaceae

Ren *et al*. (1995) reported antioxidant activity of *Lycium barbarum* fruit.

Cai *et al*. (2004, 2006) evaluated the antioxidant activity and phenolic compounds in traditional Chinese medicinal plants associated with anticancer. The improved ABTS.+ method was used to systematically assess the total antioxidant capacity (Trolox equivalent antioxidant capacity, TEAC) of the medicinal extracts. The TEAC values and total phenolic content for methanolic extract of root bark and fruit of *Lycium barbarum* were 188.8 and 490.8 µmol Trolox equivalent/100g dry weight (DW) respectively, and 0.93 and 2.58 g of gallic acid equivalent/100g DW respectively. A positive significant linear relationship between antioxidant activity and total phenolic content showed that phenolic compounds were the dominant antioxidant components in the tested herbs. Major types of phenolic compounds in root bark of *Lycium barbarum* include phenolic acids (cinnamic acid) while fruit contained coumarins (scopoletin).

Total antioxidant capacity of *Lycium barbarum* fruit water decoction, crude polysaccharide extracts and purified polysaccharide fractions were assessed by Luo *et al*. (2004) using trolox equivalent antioxidant capacity (TEAC) and oxygen radical absorbance capacity (ORAC) assay. Total antioxidant capacity assay showed that all three *Lyceum barbarum* extracts/fractions possessed antioxidant activity. However, water and methanolic fruit extracts and crude polysaccharide extracts were identified to be rich in antioxidants (*e.g*., carotenoids, riboflavin, ascorbic acid, thiamine, nicotinic acid).

Wu *et al*. (2005) evaluated the antioxidant activities of *Lycium barbarum*. The results showed that aqueous extracts of these drugs exhibited antioxidant activities in a concentration-dependent manner. The extract displayed an inhibitory effect on $FeCl_2$-ascorbic acid and lipid peroxidation in rat liver homogenate *in vitro*. The extract exhibited the IC_{50} value 0.77-2.55µg[soI]ml) in all model systems tested in this study.

Antioxidant activity of polysaccharides extracted from *Lycium barbarum* was evaluated by six established *in vitro* methods. The polysaccharides showed notable inhibitory activity in the β-carotene-linoleate model system in a concentration-dependent manner. Furthermore, it exhibited a moderate concentration-dependent inhibition of DPPH-radical. The multiple antioxidant activity of the polysaccharides was evident as it showed significant reducing power, superoxide scavenging ability, inhibition of mice erythrocyte hemolysis mediated by peroxyl free radicals and also ferrous ion chelating potency (Li *et al.*, 2007).

Lycium chinense Miller Family: Solanaceae

Wong *et al.* (2006) investigated the antioxidant properties of 25 edible tropical plants including *Lycium chinense* using DPPH and FRAP assays. Total phenolic content was also estimated. *Lycium chinense* showed moderate antioxidant activity.

Li *et al.* (2008) evaluated the antioxidant properties of 45 medicinal plants including *Lycimum chinense* using FRAP and TEAC assays and the total phenolic contents of these plants were measured. *Lycium chinense* showed 37.40 µmol Fe(II)/g FRAP value, 25.18µmol Trolox/g of TEAC value and 6.22 mg GAE/g of Phenolic content.

Zhang *et al.* (2010) investigated the cytoprotective effect of the fruits of *Lycium chinense* against oxidative stress-induced hepatotoxicity. Lycium extract scavenged the DPPH free radicals, intracellular ROS, hydroxyl radicals, and superoxide. Lycium extract recovered activities of CAT, SOD and GSH-Px decreased by H_2O_2. Lycium extract decreased DNA damage, lipid peroxidation and protein carbonyl values increased by H_2O_2 exposure.

Lycopersicon esculentum Mill. Family: Solanaceae

English Name: Tomato.

The total antioxidant activity of tomato was measured by Wang *et al.* (1996) using automated oxygen radical absorbance capacity (ORAC) assay. On the basis of the wet weight of the fruits (edible portion), tomato had moderate ORAC activity (1.89micromoles of Trolox equivalents per gram). On the basis of dry weight of the fruits tomato had the highest ORAC activity (37.8µmoles/gTrolox equivalent). The contribution of the fruit pulp fraction (extracted with acetone) to the total ORAC activity of a fruit was usually less than 10 per cent.

The antioxidative activity of phenolic extracts from fruits of tomato was examined by autoxidation of methyl linoleate by Kahkonen *et al.* (1999). The total phenolics were 21mg/g GAE while the antioxidant activity was 52 per cent at 500 ppm level.

Agostini *et al.* (2004) carried out studies to determine the antioxidant capacity of the flavonoids in tomatoes. The results indicated that the antioxidant capacity of the flavonoids of tomatoes was of 0.278 equivalent µM Trolox.

Cai *et al.* (2004) evaluated the antioxidant activity and phenolic compounds in traditional Chinese medicinal plants and vegetables associated with anticancer. The improved ABTS.+ method was used to systematically assess the total antioxidant capacity (TEAC) of the medicinal extracts. The TEAC values and total phenolic content for methanolic extract of tomato were 149.4 µmol Trolox equivalent/100g dry weight (DW), and 0.42 g of gallic acid equivalent/100g DW respectively.

Kim *et al.* (2004) examined the effects of intake of lycopene or tomato extract, a rich source of lycopene, on acute liver injury caused by the oxidant CCl_4. Feeding with tomato extract (10 per cent tomato powder) but not with lycopene beadlets), partially inhibited CCl_4-induced hepatic injury based on the serum activities of sorbitol dehydrogenase and aspartate aminotransferase. No effect was seen for either lycopene or tomato extract on serum beta-glucuronidase activity, a marker of lysosomal injury. It was observed that tomato extract, but not lycopene, partially protected against acute liver injury due to chemically-induced oxidant stress (Kim *et al.*, 2004).

Huang *et al.* (2004a) evaluated the antioxidant activities of various fruits and vegetables produced in Taiwan. Tomato seeds showed the highest antioxidant activity. At the level of 1 g fresh sample, low-density lipoprotein peroxidation was inhibited by at least 90 per cent by tomato meat. The total phenolic content was significantly correlated with antioxidant activities measured by TBA and Iodometric assays.

Tomato is a versatile vegetable that is consumed fresh as well as in the form of processed products. More recently, there has been renewed attention given to the antioxidant content of tomatoes because many epidemiological studies suggested that regular consumption of fruits and vegetables, including tomatoes, can play an important role in preventing cancer and cardiovascular problems. Toor and Savage (2005) studied the major antioxidant activity in different fractions (skin, seeds and pulp) of three tomato cultivars. It was found that the skin fraction of all cultivars had significantly higher levels of total phenolics, total flavonoids, lycopene, ascorbic acid and antioxidant activity compared to their pulp and seed fractions. The amount of antioxidants in each fraction was calculated on the basis of their actual fresh weights in whole tomato and it was found that the skin and seeds of the three cultivars on average contributed 53 per cent of the total phenolics, 52 per cent to the total flavonoids, 48 per cent to the total lycopene, 43 per cent to the total ascorbic acid and 52 per cent to the total antioxidant activity present in tomatoes. These results show that removal of skin and seeds of tomato during home cooking and processing results in a significant loss of all the major antioxidants. Therefore, it is important to consume tomatoes along with their skin and seeds, in order to attain maximum health benefits. This study suggests that the skin and seed fractions of tomato are very rich source of antioxidant compounds and the incorporation of skin and seeds fraction during home consumption or processing could lead to about a 40-53 per cent increase in the amount of all the major antioxidants in the final product. Therefore, removal of these fractions during home cooking or processing results in a loss of their potential health benefits.

Odriozola-Serrano *et al.* (2008) assessed the feasibility of using modified atmospheric packaging (5kPa O_2 + 5 kPa CO_2) to maintain the antioxidant properties of fresh-cut tomatoes during shelf-life through storage at different temperatures (5, 10, 15 and 20°C). Health-related compounds, antioxidant capacity, microbiological counts, physicochemical parameters and in-package atmosphere of tomato slices were determined. Initial lycopene, vitamin C and phenolic contents and physicochemical parameters of tomato slices were well maintained for 14 days at 5°C. Lycopene and total phenolic contents were enhanced over time in tomato slices stored at 15 and 20°C. However, this increase in antioxidant compounds of fresh-cut tomatoes during storage may be associated with excessive amounts of CO_2 in the packages due to microbial growth. Although keeping tomatoe slices at temperatures above 10°C increased their antioxidant content, the shelf life of the product was reduced by up 4 days. A storage temperature of 5°C is appropriate for maintaining the microbiological shelf-life of fresh-cut tomatoes for up to 14 days and also allows the antioxidant properties of tomato slices to be retained over this period, thus reducing wounding stress and deteriorative changes.

Hsu *et al.* (2008) investigated the lipid-lowering effects and antioxidant mechanisms of tomato paste. MDA and diene conjugation assays indicated the potent antioxidant activity of the tomato paste. The increased activities of superoxide dismutase (SOD), catalase (CAT) and glutathione peroxidase (GSH-Px), further supported the antioxidant effects of tomato paste. Two dimension-gel electrophoresis (2-DE) analysis revealed that carbonic anhydrase III and adenylate kinase 2 may be two important regulators involved in antilipid and antioxidant effects of tomato paste (Hsu *et al.*, 2008).

Fourteen commercial cultivars of tomato were analyzed by Walia *et al.* (2010) for their antioxidant composition. There was significant difference in lycopene and phenolic contents between red and yellow cultivars. Red cultivars had higher lycopene content (2.735 to 6.552 mg/100g) than yellow cultivars (0.769 to 1.238 mg/100g). Mean total polyphenolic content and total antioxidant activity in red cultivars was also higher than those in yellow cultivars. Overall cherry tomatoes had highest phenolic content and appeared to be a promising cultivar in terms of their health promoting effects (Walia *et al.*, 2010).

A study was carried out by Raghu *et al.* (2010) to determine the antioxidant activity of aqueous extracts of ten selected common vegetables. For *Lycopersicon esculentum* the IC_{50} value for DPPH scavenging was 1.64mg/ml while standard was 2.45µg/ml. Reducing power of the extract was moderate.

Lycopus lucidus Hand.-Maz. Family: Asteraceae (Compositae)

Cai *et al.* (2004) evaluated the antioxidant activity and phenolic compounds in traditional Chinese medicinal plants associated with anticancer. The improved ABTS.+ method was used to systematically assess the total antioxidant capacity (Trolox equivalent antioxidant capacity, TEAC) of the medicinal extracts. The TEAC values and total phenolic content for methanolic extract of aerial parts of *Lycopus lucidus* were 226.7 µmol Trolox equivalent/100g dry weight (DW), and 0.93 g of gallic acid equivalent/100g DW respectively. A positive significant linear relationship between antioxidant activity and total phenolic content showed that phenolic compounds were the dominant antioxidant components in the tested herbs. Major types of phenolic compounds in *Lycopus lucidus* include flavones (acciin, linarin), phenolic acids (chlorogenic acid).

Three phenolic compounds, rosmarinic acid, methyl rosmarinate, ethyl rosmarinate and two flavonoids, luteolin, luteolin-7-O-β-D-glucuronide methyl ester were isolated from the aerial parts of *Lycopus lucidus* by Woo and Piao (2004). These compounds exhibited potent antioxidative activity on the NBT superoxide scavenging assay.

The aqueous extracts of 13 oriental medicinal plants were examined for their reducing power, scavenging ability towards superoxide and hydroxyl radicals and their inhibitory effect on lipid peroxidation. The reducing power of Herba Lycopi was more apparent than with other plants (Nam and Kang, 2004). All extracts tested were found to be highly active on scavenging of superoxide radicals.

Slusarczyk *et al.* (2008) evaluated the antioxidant properties of different extracts from *Lycopus lucidus* (crude methanolic extract, dichloromethane, diethyl ether and ethyl actate subfractions) and to correlate their antioxidant potential to the composition of polyphenols. The main antioxidant compound from the methanol extract was confirmed to be rosmarinic acid. Other identified phenolic acids that are likely to contribute to the antioxidant potential were: ferulic, caffeic, chlorogenic, *p*-hydroxybenzoic acid, *p*-coumaric acid and protocatechuic acid. The most active ethyl acetate extract contained highest level of rosmarinic acid, but the luteolin glycoside were also the major determinants of the antioxidant activity.

Gan *et al.* (2010) evaluated the antioxidant activities of 40 medicinal plants. The antioxidant activity of *Lycopus lucidus* was 138.69µmol Fe(II)/g in FRAP assay, 120.98µmol Trolox/g TEAC assay and Total phenolic content was 7.87 mg GAE/g.

Lysimachia christinae Hance Family: Primulaceae

Gan *et al.* (2010) evaluated the antioxidant activity of *Lysimachia christinae* using the FRAP and TEAC assays, and its total phenolic content was measured by the Folin-Ciocalteu method. For FRAP assay the value was 88.85 µmol Fe(II)/g dry weight. For TEAC assay, the value was 65.30µmol Trolox/g dry weight. Total phenolic content was 6.99 mg gallic acid equivalent/g dry weight of plant material.

Lysimachia vulgaris L. Family: Primulaceae

English Name: Yellow loosestrife.

The antioxidative activity of phenolic extracts from fruits of *Lysimachia vulgaris* was examined by autoxidation of methyl linoleate by Kahkonen *et al.* (1999). The total phenolics were 13.5 mg/g GAE while the antioxidant activity was 44 per cent at 500 ppm level and 65 per cent at 5000 ppm level.

Lythrum alatum Pursh. Family: Lythraceae

Borchardt *et al.* (2008) evaluated the antioxidant activity of methanol extracts of seeds of *Lythrum alatum* using DPPH antioxidant assay.The antioxidant value reported in µM Trolox/100g (TE) from DPPH radical scavenging activity of crude seeds extract was 206,154.

Lythrum salicaria L. Family: Lythraceae

English Name: Purple loosestrife; Turkish: Tibbi hevhulma.

Tunalier *et al.* (2007) reported antioxidant and anti-inflammatory activities and composition of *Lythrum salicaria*. Free radical scavenging activity (DPPH assay), iron (III) reductive activity, capacity of the inhibition of linoleic acid peroxidation and MDA formation were used for antioxidant activity assay. Total phenolics, flavonoids and flavonols were determined.

Borchardt *et al.* (2008) evaluated the antioxidant activity of methanol extracts of seeds of *Lythrum salicaria* using DPPH antioxidant assay.The antioxidant value reported in µM Trolox/100g (TE) from DPPH radical scavenging activity of crude seeds extract was 261,384.

Lee *et al.* (2010) evaluated the antioxidant activity of organic solvent fractions from *Lythrum salicaria* root. All the fractions showed effective antioxidant activities on DPPH radical and $CuSO_4$-induced oxidation of human LDL and E fraction showed the highest inhibitory effect (98.1 per cent at 50µg/ml) on linoleic acid autoxidation, which was more effective than α-tocopherol (82.4 per cent).

Mabea montana Muell.-Arg. Family: Euphorbiaceae

Mosquera *et al.* (2007) studied the antioxidant activity of the extracts using DPPH free-radical scavenging method. The percentage of the antioxidant activity of the n-Hexane fraction was 39.683, Dichloromethane fraction was 45.06 and methanol fraction was 50.47.

Macadamia integrifolia Maiden et Betche Family: Proteaceae

Wall (2010) investigated functional lipid characteristics, oxidative stability, and antioxidant activity of madamia nut (*Macadamia integrifolia*) cultivars. Cultivars that had the greatest oxidative stability also had high total lipid-soluble antioxidant capacity. Tocopherols were not detected in most macadamia nut samples, but macadamia kernels contained significant amounts of tocotrienols and squalene for all cultivars tested.

Machaerium villosum Vogel Family: Fabaceae (Leguminosae)

Santos *et al.* (2009) evaluated the antioxidant activity of extracts from nine plant species belonging to the Brazilian flora acting on the free radicals DPPH and TEMPOL. Results showed that the extracts of *Machaerium villosum* display moderate antioxidant activities with DPPH scavenging IC_{50} value of 0.717 mg/ml while Hydroxyl scavenging IC_{50} value was 10.720 mg/ml.

Machilus zuihoensis Hayata Family: Lauraceae

Hou *et al.* (2003) evaluated the antioxidant activity of 70 per cent aqueous acetone extract of *Machilus zuihoensis* by various assays including DPPH, hydroxyl radicals and reducing power assay. The extract exhibited stronger activity against DPPH radicals (7.9 µg/ml) and inhibited formation of OH generated in the Fenton reaction system (0.8µg/ml). The total phenolics was 51.7 mg of GAE/g).

Maclura pomifera (Raf.) Schneid. Family: Moraceae

The major constituents of fruits are the prenylated isoflavones, osajin and pomiferin. The antioxidant activities showed the significant difference in antioxidant profile of highly active pomiferin comparable to used reference compounds, while osajin showed only low activities (Tsao *et al.*, 2003).

Pomiferin and osajin have been isolated from the total acetonic extract of *Maclura pomifera*. Effect of total acetonic extract, pomiferin and osajin on the autoxidation of purified sunflower triacylglycerol were studied by Hamed and Hussain (2005). Pomiferin showed a high antioxidant activity whereas total acetonic extract showed moderate and osajin revealed low activity.

Maclura tinctoria L. Family: Moraceae

Four chalcone glycosides, including three new natural products, and three flavanones were isolated from the methanol extract of stem bark of *Maclura tinctoria* (Cioffi *et al.*, 2003). The antioxidant activity of all the isolated compounds was determined by measuring free radical scavenging activity using two different assays, namely, the TEAC assay and the coupled oxidation of β-carotene and linoleic acid (autooxidation assay). These results showed that compound 3'-(3-methyl-2-butenyl)-4-O- β-D-glucopyranosyl 4,2'-dihydroxychalcone was the most active in both antioxidant assays (Cioffi *et al.*, 2003).

Macropanax dispermus (Bl.) Kuntze Family: Araliaceae

Chanwitheesuk *et al.* (2005) investigated the antioxidant activity of *Macropanax dispermus* using a β-carotene bleaching method. The contents of plant chemicals such as vitamin C, Vitamin E, carotenoids, tannin and total phenolics, were also determined. Methanolic extract showed antioxidant activity. *Macropanax dispermus* has antioxidant index of 1.27, Vitamin C 31.0 mg per cent, Vitamin E 0.0064mg per cent, Total carotenes 3.89 mg per cent, Total xanthophylls 1.06mg per cent, Tannins 37.4mg per cent and Total phenolics of 651 mg per cent.

Macrotyloma uniflorum (Lam.) Verdc. Family: Fabaceae (Papilionaceae)

Total phenolics and the antioxidative properties of two varieties of horsegram were studied by Siddhuraju and Manian (2007). The black seeds contained relatively high levels of total phenolics and tannins than the brown seeds. The extracts were subjected to assess their potential antioxidant activities using systems such as DPPH, ABTS, FRAP, linoleic acid emulsion, O_2., and.OH. The superoxide anion radical-scavenging activity was found to be higher in 70 per cent acetone extract of both raw and dry heated seeds of the respective varieties at the concentration of 600µg in the reaction mixture. The DPPH radical and ABTS cation radical-scavenging activities were well proved and related with the ferric-reducing/antioxidant capacity of the extracts. In general, all extracts exhibited good antioxidant activity (53.3-73.1 per cent) against the linoleic acid emulsion system but were significantly lower than the synthetic antioxidant, BHA (93.3 per cent).

Madhuca indica Gmel. Family: Sapotaceae

Madhucosides A and B isolated from the bark of *Madhuca indica* showed significant inhibitory effects on both superoxide release from polymorphonuclear cells in a NBT reduction assay and hypochlorous acid generation from neutrophils assessed in a lumunol-enhanced chemiluminiscence assay (Pawar and Bhutani, 2004).

Kaushik *et al.* (2010) evaluated the antioxidant activities of alcoholic extract of bark of *Madhuca indica*. The extract was screened for free radical scavenging effects at various concentrations (100, 300 and 500µg/ml) by superoxide free radical scavenging activity and DPPH free radical scavenging method. All these antioxidant activities were concentration dependent.

Maesa lanceolata Forssk. Family: Myrsinaceae

The isomeric acylated benzoquinones with shorter alkyl substitutes isolated from *Maesa lanceolata* showed most prominent antioxidant and antiproliferative effect on HL-60 cell line (IC_{50} values 6.2 and 2.2 µg/ml) (Muhammad *et al.*, 2003).

Magnolia liliflora Desr. Family: Magnoliaceae

Antioxidant capacities of 56 selected Chinese medicinal plants, including *Magnolia liliflora* were evaluated by Song *et al.* (2010) using the TEAC and FRAP assays, and their total phenolic content was measured by the Folin-Ciocalteu method. *Magnolia liliflora* showed moderate antioxidant activity with TEAC value of 49.19µmol Trolox/g, FRAP value of 118.53µmol Fe^{2+}/g and phenolic content was 10.98mg GAE/g.

Magnolia officinalis Rehd. et Wils. Family: Magnoliaceae

Cai *et al.* (2004, 2006) evaluated the antioxidant activity and phenolic compounds in traditional Chinese medicinal plants associated with anticancer. The improved ABTS.+ method was used to systematically assess the total antioxidant capacity (Trolox equivalent antioxidant capacity, TEAC) of the medicinal extracts. The TEAC values and total phenolic content for methanolic extract of bark of *Magnolia officinalis* were 727.5 µmol Trolox equivalent/100g dry weight (DW), and 3.15 g of gallic acid equivalent/100g DW respectively. A positive significant linear relationship between antioxidant activity and total phenolic content showed that phenolic compounds were the dominant antioxidant components in the tested herbs. Major types of phenolic compounds in *Magnolia officinalis* include neolignans (magnolol, ospmagnolol), tannins.

Gan *et al.* (2010) evaluated the antioxidant activity of *Magnolia officinalis* using the FRAP and TEAC assays, and its total phenolic content was measured by the Folin-Ciocalteu method. For FRAP assay the value was 257.45 µmol Fe(II)/g dry weight. For TEAC assay, the value was 188.70µmol Trolox/g dry weight. Total phenolic content was 9.68 mg gallic acid equivalent/g dry weight of plant material.

Magonia glabrata St. Hill. Family: Sapindaceae

The ethanolic extract of the fruit bark yielded shikimic acid, scopoletin, sitosterol glycoside and 2-O-methyl-L-inositol (Lemos *et al.*, 2006). Extract showed moderate activity as a radical scavenger.

Mahonia aquifolium (Pursh.) Nutt. Family: Berberidaceae

English Name: Oregon grape.

The effects of the extract of the bark of *Mahonia aquifolium* and its main constituents (berberine, berbamine, oxyacanthine) on lipoxygenase, lipid peroxidation in phospholipids liposomes induced by 2,2'-azo-(bis-2-amidinopropane), deoxygenase degradation, and their reactives against the free radical DPPH have been studied. The extract of *M. aquifolium* inhibited 5-LO with an IC_{50} value of 50µM, whereas no appreciable effects were observed by its constituent alkaloids. Reactivity against DPPH increased in the following order: berberine<*M.aquifolium*<Oxyacanthine<

berbamine. Pro-oxidant effects of *M. aquifolium* or its constituents can be excluded, since deoxyribose degradation was not influenced as determined by the release of malondialdehyde. The most prominent feature of *M. aquifolium* was its efficacy in inhibition of lipid peroxidation (IC_{50} = 5µM) which was not mediated by the alkaloids berberine, berbamine, and oxyacanthine.

Four protoberberine alkaloids, berberine, oxyberberine, jatrorrhizine, columbamine, and two aporphine alkaloids, magnoflorine and corytuberine, isolated from *Mahonia aquifoliium*, were tested for lipoxygenase inhibition. Oxyberberine, corytuberine and columbamine were the most potent lipoxygenase inhibitors tested, whereas berberine and magnoflorine exhibited only low potencies. A strong linear correlation between lipoxygenase inhibition and lipid antioxidant properties of these compounds was found (Misik *et al.*, 1995). Six bisbenzylisoquinoline (BBIQ) alkaloids, oxyacanthine, armoline, baluchistine, berbamine, obamegine, aquifoline, isolated from *Mahonia aquifolium*, were tested for lipoxygenase inhibition. Berbamine and oxyacanthine were the most potent lipoxygenase inhibitors, whereas aromoline and baluchistine exhibited only very low potencies. Oxyacanthine and berbamine were also among the most active compounds to inhibit lipid peroxidation. Between the results of lipoxygenase inhibition and the lipid peroxidation a linear correlation was found.

Kardosova and Machova (2006) investigated the ability of polysaccharides, isolated from stems, to inhibit peroxidation of soyabean lecithin liposomes by OH radicals. All polysaccharides exhibited antioxidant activity.

The antioxidant activities of three alkaloids isolated from *Mahonia aquifolium* – berberine, jatrorrhizine and magnoflorine- were studied for their antioxidant activity by Rackova *et al.* (2004) by DPPH radical scavenging assay. Both alkaloids bearing free phenolic groups – jatrorrhizine and magnoflorine-showed better activities in both systems used than berberine not bearing any readily abstractable hydrogen on its skeleton. The former two showed antiperoxidative efficiency in DOPC liposomal membrane comparable to that of an effective scavenger of peroxyl radicals stobadine- and higher than that of Trolox. Rackova *et al.* (2007) evaluated the DPPH radical scavenging activity of roots and stem bark of *Mahonia aquifolium* to elucidate the rate of possible lipid-derived radical scavenging in the mechanism of the enzyme inhibition. In conclusion, their results indicate that although the direct scavenging of free radicals can not be ruled out in the mechanism of lipoxygnase inhibition by the *Mahonia aquifolium* extract and its two representative fractions containing BBIQ and protoberberine alkaloids.

Mahonia leschenaultii (Wall. ex Wt. et Arn.) Takeda Family: Berberidaceeae

Karuppusamy *et al.* (2011) evaluated the anthocyanins, ascorbic acid, total phenolics and flavonoids and antioxidant activity of methanol extracts of fruits of *Mahonia leschenaultii*. The anthocyanins were 8.58 CGE/100g, Ascorbic acid 69.9 AAE/100g, Total phenolics 86.8GAE/100g, Total flavonoids 95.5QE/100g and antioxidant activity (DPPH) 361.2µg/ml.

Majorana hortensis Moench. Family: Lamiaceae (Labiatae)

English Name: Sweet Marjoram.

Kosar *et al.* (2003) prepared water extracts from steam distilled essential oil-extracted *Majorana hortensis.* The HPLC-DPPH on-line method was applied to the qualitative and quantitative analysis of this plant extract. There was a strong correlation between the scavenging (negative) peak area and the concentration of the radical scavenging reference substances used. The radical scavenging compounds within the extracts were determined as benzoic acid and hydroxycinnamic acid derivatives, flavonoids and diterpenoids according to their retention time and UV spectral data. Rosmarinic acid and carnosic acid were identified as the dominant radical scavengers in these extracts by this method.

The antioxidant property of the supercritical fluid extraction (SFE) was markedly higher than that of BHA and hydrodistillation extract, which might be due to the presence of higher contents of gamma-terpinene, terpinolene and thymol (El-Ghorab *et al.,* 2004).

Water-soluble extract from *Majorana hortensis* was screened by Dorman *et al.* (2004) for antioxidant properties in a battery of six *in vitro* assays. The extract demonstrated varying degrees of efficacy in each screen. The extract contained Folin-Ciocalteu reagent-reactive substances, which was confirmed by the presence of polar phenolic analytes (*i.e.,* hydroxybenzoates, hydroxycinnamates, and flavonoids).

Mariutti *et al.* (2008) evaluated the antioxidant activity of sweet marjoram employing ABTS and DPPH radical assay. DPPH EC_{50} value was 17.84g/kg while DPPH-TEAC value was 14.8mM/g. ABTS-TEAC value was 25mM/g.

Majorana syriaca L. Family: Lamiaceae (Labiatae)

Al-Bandak *et al.* (2009) evaluated the antioxidant activity of ethyl acetate extract of *Majorana syriaca* in yellowfin tuna. The extract exhibited significant antioxidant activity in Yellowfin tuna.

Mallotus japonicus (Thunb.) Muell.-Arg. Family: Euphorbiaceae

Katsube *et al.* (2004) screened aqueous ethanol extract of *Mallotus japonicus* for antioxidant activity using DPPH radical scavenging activity and LDL oxidation induced by copper ion. Total phenolic content was also measured for comparisons with antioxidant activity in LDL. The study revealed high levels of LDL antioxidant activity in *Mallotus japonicus.*

Mallotus oppositifolium Mull. Family: Euphorbiaceae

Methanol extract of leaves of Nigerian plant, *Mallotus oppositifolium* harvested from western Nigeria has been shown to possess antioxidant and ani-inflammatory activities in β-carotene linoleate model system and the carrageenan-induced rat paw oedema animal model (Farombi *et al.,* 2001). Thin layer chromatographic analysis of this extract revealed the presence of four phenolic spots, two of which were flavonoids.

Mallotus repandus (Willd.) Muell.-Arg. Family: Euphorbiaceae

The active oxygen species scavenging potencies of *Mallotus repandus* extracts were evaluated by Lin *et al.* (1995) by the electron spin resonance (ESR) spin-trapping technique. The ethyl acetate fraction of *Mallotus repandus* (stem) showed the greatest superoxide radical scavenger activity and the n-Hexane fraction of *Mallotus repandus* (stem as well as root) the greatest hydroxyl radical scavenger activity.

Malpighia emarginata DC. Family: Malpighiaceae

English Name: Acerola.

Acerola is a wild plant from Central America. This fruit is well known as an excellent food source of vitamin C, and it also contains phytochemicals such as carotenoids and polyphenols. Mezadri *et al.* (2008) evaluated the antioxidant capacity of hydrophilic extracts of acerola pulps and juices by ABTS, ORAC and DPPH methods. Antioxidant activity values obtained for acerola juice were higher than those reported for other fruit juices particularly rich in polyphenols such as strawberry, grape and apple juices among others. Vitamin C, total phenol index (TPI), total anthocyanins and polyphenolic compounds by HPLC, as main factors responsible for antioxidant activity were determined. Contents in total ascorbic acid ranged from 6.32 to 9.20g/kg of pulp and 9.44 to 17.97g/l of juice. Five different polyphenolic compounds were identified in the samples by means of HPLC and diode-array detection: chlorogenic acid, (-)epigallocatechin gallate, (-)-epicatechin, procyanidin B1 and ruitin, being the two last predominant. By means of solid phase extraction three soluble polyphenolic fractions (phenolic acids, anthocyanins and flavonoids were separated from the different sample extracts and their respective antioxidant activities calculated. Among them, phenolic acids were the main contributors to the antioxidant activity.

Malpighia glabra L. Family: Malpighiaceae

English Name: Acerola.

Schmourlo *et al.* (2007) evaluated the antioxidant activity of aqueous extract of fruits of *Malpighia glabra* by DPPH radical scavenging method. The extract showed DPPH radical scavenging activity above 90 per cent at $100\mu g/ml$. The antioxidant activity may be due to its contents of Vitamin C and anthocyanins.

Malus docmeri Chev. Family: Rosaceae

Hou *et al.* (2003) evaluated the antioxidant activity of 70 per cent aqueous acetone extract of *Malus docmeri* by various assays including DPPH, hydroxyl radicals and reducing power assay. The extract exhibited stronger activity against DPPH radicals with IC_{50} value of 23.1 $\mu g/ml$ and inhibited formation of OH generated in the Fenton reaction system with IC_{50} value of 0.7 $\mu g/ml$). The amount of phenolics was 2.2 mg of GAE/g.

Malus pumila Mill. Family: Rosaceae

English Name: Apple.

The total antioxidant activity of apple was measured by Wang *et al.* (1996) using automated oxygen radical absorbance capacity (ORAC) assay. On the basis of the wet weight of the fruits (edible portion), apple had moderate ORAC activity (2.18micromoles of Trolox equivalents per gram). On the basis of dry weight of the fruits apple had the moderate ORAC activity (13.2μmoles/gTrolox equivalent). The contribution of the fruit pulp fraction (extracted with acetone) to the total ORAC activity of a fruit was usually less than 10 per cent.

The antioxidative activity of phenolic extracts from fruits of apple was examined by autoxidation of methyl linoleate by Kahkonen *et al.* (1999). Both phenolics and antioxidant activity of the flowers were low. The total phenol contents of the two apple varieties studied were almost similar (11.9 and 12.1 mg/g GAE). However, apples exserted strong antioxidant activities.

Eberhardt *et al.* (2000) reported antioxidant activity of fresh apples.

The relationship between phenolic composition and free radical scavenging activity of apple peel and pulp was investigated in fruit produced according to both organic and integrated agricultural methods (Chinnici *et al.*, 2004). In peels, the antioxidant activity was found to be flavonols, flavanols and procyanidins, which accounted for about 90 per cent of the total calculated activity whereas in pulps, the (Total Antioxidant Capacity (TAC) was primarily derived from flavanols (monomers and polymers) together with hydroxycinnamates (Chinnici *et al.*, 2004).

Agostini *et al.* (2004) carried out studies to determine the antioxidant capacity of the flavonoids in red apples with and without skin. The results indicated that the antioxidant capacity of the flavonoids of red apples with and without skin was of 0.259 equivalent μM Trolox.

Cai *et al.* (2004) evaluated the antioxidant activity and phenolic compounds in traditional Chinese medicinal plants and fruits associated with anticancer. The improved ABTS.+ method was used to systematically assess the total antioxidant capacity (TEAC) of the medicinal extracts. The TEAC values and total phenolic content for methanolic extract of Fuji apple were 92.7 μmol Trolox equivalent/100g dry weight (DW), and 0.48 g of gallic acid equivalent/100g DW respectively. Washington red apple showed 133.27 μmol Trolox equivalent/100g dry weight (DW), and 0.75 g of gallic acid equivalent/100g DW respectively.

Malus sylvestris Mill. subsp. *orientalis* (Uglitzk.) Browicz. var. *orientalis* Family: Rosaceae

Serteser *et al.* (2009) studied the antioxidant activity of 50 per cent aqueous methanol extracts by various antioxidant assays, including free radical scavenging, hydrogen peroxide scavenging and metal (Fe^{2+}) chelating activities. The DPPH radical scavenging effect of fruit extract of *Malus sylvestris* subsp. *orientalis* var. *orientalis* was 0.732. Fe^{2+} chelating activity (per cent) of fruit extract is 45.32 and Hydrogen peroxide inhibition activity of fruit extract was 61.32 per cent.

Malva neglecta Wallr. Family: Malvaceae

Antioxidant and radical scavenging activities, reducing powers and the amount of total phenolic compounds of aqueous and methanolic extracts of *Malva neglecta* were studied by Mavi *et al.* (2004). The extract showed moderate antioxidant activity.

Malva sylvestris L. Family: Malvaceae

Antioxidant capacity of the aqueous extract of *Malva sylvestris* was measured by DellaGreca *et al*. (2009) by its ability to scavenge the DPPH and superoxide anion radicals and to induce the formation of a molybdenum complex. Analysis of the extract led to the isolation of eleven compounds: 4-hydroxybenzoic acid, 4-methoxybenzoic acid, 4-hydroxy-3-methoxybenzoic acid, 4-hydroxycinnamic acid, ferulic acid, methyl 2-hydroxydihydrocinnamate, scopoletin, N-feruloyl tyramine, a sesquiterpene, (3R, 7E)-3-hydroxy-5,7-megastigmadien-9-one and (10E, 15Z)-9,12,13-trihydroxyoctadeca-10,15-dienoic acid. The antioxidant activities of all these compounds are reported.

Malva verticillata L. Family: Malvaceae

Gan *et al*. (2010) evaluated the antioxidant activity of *Malva verticillata* using the FRAP and TEAC assays, and its total phenolic content was measured by the Folin-Ciocalteu method. For FRAP assay the value was 30.67 µmol Fe(II)/g dry weight. For TEAC assay, the value was 20.32µmol Trolox/g dry weight. Total phenolic content was 2.18 mg gallic acid equivalent/g dry weight of plant material.

Mammea africana Sabine Family: Clusiaceae (Guttiferae)

Nguelefack-Mbuyo *et al*. (2010) evaluated the *in vitro* antioxidant activity of the aqueous and methylene chloride/methanol extracts of *Mammea africana* and isolated the active principles. DPPH, Nitric oxide production and β-carotene bleaching tests were followed for the antioxidant assay. The aqueous and methylene chloride/methanol extracts exhibited a concentration dependent antiradical activity on DPPH with respective EC_{50} of 2.00 and 5.24µg/ml. The purification of the extract yielded two coumarins: 4-phenylcoumarins and 4-n-propylcoumarins that exhibited potent radical scavenging activity with EC_{50} of 7.61 and 22.92µg/ml, respectively. Plant extracts, fractions and compounds tested substantially increase NO production at high concentrations and exerted a protective effect against the oxidation of β-carotene at all the concentrations used; with a 400 per cent inhibition obtained with methylene chloride fraction at the concentration of 300µg/ml.

Mammea americana L. Family: Clusiaceae (Guttiferae)

Antioxidant-guided fractionation of *Mammea americana* seeds resulted in the identification of three new isoprenylated coumarins, mammea B/BAhydroxycyclo F, mammea E/BC and mammea E/BD together with twelve known isoprenylated coumarins, and two flavonols. The fifteen isoprenylated coumarins were screened for their antioxidant activity in the DPPH free-radical assay. New coumarins displayed high antioxidant activity in the DPPH free-radical assay (IC_{50} = 53.0, 46.1 and 45.1 µM) (Nguelefack-Mbuyo *et al*., 2010).

Mammea siamensis T. Anders. Family: Clusiaceaee (Guttiferae)

Leelapornpisid *et al*. (2007) reported antioxidant activity of the absolutes (obtained by solvent extraction) of *Mammea siamensis* (IC_{50} = 0.3271 mg/ml).

Mangifera indica L. Family: Anacardiaceae

English Name: Mango.

A stem bark extract of *Mangifera indica* was tested *in vitro* for its antioxidant activity using commonly accepted assays. It showed a powerful scavenging activity against hydroxyl radicals and hypochlorous acid and acted as an iron chelator. The extract also showed a significant inhibitory effect on the peroxidation of rat-brain phospholipids and inhibited DNA damage by bleomycin or copper-phenanthroline systems (Martinez *et al.*, 2000).

Scartezzini and Speroni (2000) are of the opinion that this plant contains antioxidant principles, that can explain and justify their use in traditional medicine in the past as well as the present.

Maisuthisakul and Pasuk (2008) carried out studies on antioxidant activities on eleven cultivars of Thai mango seed kernels using DPPH assay and phenolic compounds of the seed kernel extracts. The results showed that all the mango seed kernels showed antioxidant activities comparable to the α-tocopherol. The phenolic acids of mango seed kernels were in the form of free phenolic acid (42-56 per cent) more than esterified phenolic acid (10-19 per cent) and insoluble bound phenolic acid (15-20 per cent). The antioxidant activities in the mango seed kernel extracts did not correlate with the yields of extracts or yields of seed kernels from mango seeds, however the activities related with phenolic, flavonoid and phenolic acid contents.

Ayoola *et al.* (2008) carried phytochemical screening and antioxidant activities of the ethanolic extract of stem bark of *Mangifera indica*. The crude extract was six times less potent (than vitamin C) with a maximum inhibition of 83.8 per cent at 5mg/ml. They are of the opinion that the antioxidant activity of this plant may contribute to their claimed antimalarial activity.

The antioxidant and antiproliferative properties of flesh and peel of mango were investigated by Kim *et al.* (2010). The mango peel extract exhibited stronger free radical scavenging ability on DPPH and alkyl radicals than mango flesh extract, in a dose dependent manner. According to Kim *et al.* (2010) mango peel, a major by-product obtained during the processing of mango product exhibited good antioxidant activity and may serve as a potential source of phenolics with anticancer activity.

Niwano *et al.* (2011) summarized their research for herbal extracts with potent antioxidant activity obtained from a large scale screening based on superoxide radical. Experiments with the Fenton reaction and photolysis of H_2O_2 induced by UV irradiation demonstrated that extract of *Mangifera indica* (Kernel) have potent ability to directly scavenge OH radicals. Furthermore, the scavenging activities against O_2 and OH of extracts of *Mangifera indica* (kernel) proved to be heat resistent.

Manihot esculenta Crantz. Family: Euphorbiaceae

English Name: Cassava

Wong *et al.* (2006) investigated the antioxidant properties of 25 edible tropical plants including *Manihot esculenta* using DPPH and FRAP assays. Total phenolic content was also estimated. *Manihot esculenta* showed moderate antioxidant activity.

Patnibul *et al.* (2008) screened 18 vegetables for antioxidant activity using silica gel Thin Layer chromatography followed by spraying with DPPH. *Manihot esculenta* exhibited strong antioxidant activity.

Ndidi and Akeem (2011) investigated the antioxidant enzyme activity during postharvest deterioration of cassava root tubers.

Suresh *et al.* (2011) investigated the total phenol contents, flavonoids, anthocyanins of cassava and its antioxidant activities using leaf stalks. In the case of total phenol content acetone extract was found to have maximum phenol content followed with acidified methanol and methanol. In the case of anthocyanin content, acidified methanol extract gave maximum yield followed with methanol and acetone extracts. DPPH radical scavenging was 68 per cent (1 per cent methanol HCl extract), hydroxyl radical scavenging activity was 93.3 per cent (acetone extract), superoxide radical scavenging activity was 56.8 per cent (methanol extract), metal chelating activity was 80.1 per cent (methanol extract) and reducing power was 0.535 per cent (methanol extract).

Flours of 10 cassava varieties were screened by Eleazu *et al.* (2011) for their nutritional composition and antioxidant activities. All the varieties were found to possess antioxidant activities as evaluated by phenolic composition of the methanolic extracts of the flours and reducing power tests.

Manihot utilissima Pahl Family: Euphorbiaceae

Salawu *et al.* (2011) evaluated the cellular antioxidant activities of green leafy vegetables including *Manihot utilissima* against peroxyl radical-induced oxidation in HepG2 cells. HPLC/DAD/MS investigation of the ethanolic extract of the vegetal matter revealed the presence of 4 phenolic compounds. At a concentration of 1 mg/ml the quenching of peroxyl radical in HepG2 cells revealed that *Manihot utilissima* has low antioxidant activity. Prooxidant activities were observed at lower concentrations (0.1, 0.25 and 0.5mg/ml).

Manilkara zapota (L.) P. van Royen Family: Sapotaceae

English Name: Sapodilla

Thirty-eight types of fruits commonly consumed in Singapore were systematically analyzed by Isabelle *et al.* (2010) for their hydrophilic oxygen radical absorbance capacity (H-ORAC), total phenolic content (TPC), ascorbic acid (AA) and various lipophilic antioxidants. Among all the fruits tested sapodilla had the highest H-ORAC and TPC per gram fresh weight.

Mansonia gagei Drumm. Family: Sterculiaceae

Eleven compounds isolated from the heartwood of *Monsonia gagei* were tested for radical scavenging properties. Ansonone N was the only isolated product to show radical scavenging properties (Tiew *et al.*, 2003).

Marrubium astracanicum Jacq. Family: Lamiaceae (Labiatae)

Firuzi *et al.* (2010) evaluated the antioxidant activity and total phenolic content of 24 Lamiaceae species growing in Iran. *Marrubium astracanicum* showed low antioxidant activity with 16.4μM/g DW FRAP value and IC_{50} = 618.3μg/ml of DPPH radical scavenging activity. It also showed low phenolic content of 6.2 mg CE/g.

Marrubium globosum Montbr. et Auch. Benth. subsp. *globosum* Family: Lamiaceae (Labiatae)

Sarikurkcu *et al.* (2008) examined the chemical composition and *in vitro* antioxidant activity of the essential oil and sub-fractions of the methanol extract. Antioxidant activities of the samples were determined by three different test systems namely DPPH, β-carotene/linoleic acid and reducing power assay. In DPPH system, the weakest radical scavenging activity was exhibited by the essential oil (1203.38μg/ ml). Antioxidant activity of the polar sub-fraction of methanol extract was superior to the all samples tested with an EC_{50} value of 157.26μg/ml). In the second case, the inhibition capacity (per cent) of the polar sub-fraction of the methanol extract (97.39 per cent) was found the strongest one, which is almost equal to the inhibition capacity of positive control BHT(97.44 per cent). In the case of reducing power assay, a similar activity pattern was observed as given in the first two systems. Polar sub-fraction was the strongest radical reducer when compared with the non-polar one, with an EC_{50} value of 625.63μg/ml. The amount of the total phenolics was highest in polar sub-fraction 25.60μg/ml). A positive correlation was observed between the antioxidant activity potential and total phenolic level of the extracts.

Marrubium vulgare Mill. Family: Lamiaceae (Labiatae)

English Name: Horehound.

Vander Jagt *et al.* (2002) evaluated the total antioxidant content of widely used medicinal plants of New Mexico. The plant *Marrubium vulgare* contained the higher amount of antioxidant.

Souri *et al.* (2004) evaluated the antioxidative activity of *Marrubium vulgare* by linoleic acid peroxidation test using 1,3-diethyl-2-thiobarbituric acid as the reagent. Methanolic extract showed 89.51 per cent inhibition at 40μg concentration with IC_{50} value of 5.61μg.

Matkowski and Piotrowska (2006) studied the antioxidative effects of methanolic extract with the use of three *in vitro* assays (DPPH assay, phosphomolybdenum method and lipid peroxidation). The herbal extract showed strong antioxidant activity. Matkowski *et al.* (2008) evaluated the antioxidant activity of aerial parts of *Marrubium vulgare* using DPPH, OH radical scavenging assays, reducing power assay and inhibition of deoxyribose degradation. EC_{50} value of DPPH radical scavenging was 36.69μg/ml for MeOH extract, 35.51μg/ml for Dichloromethane extract, 23.22μg/ml for BuOH extract and 11.67μg/ml for ethyl acetate extract.

Firuzi *et al.* (2010) evaluated the antioxidant activity and total phenolic content of 24 Lamiaceae species growing in Iran. *Marrubium vulgare* showed good antioxidant activity with 20.3μM/g DW FRAP value and IC_{50} = 554.1μg/ml of DPPH radical scavenging activity. It showed phenolic content of 4.6 mg CE/g.

Marsdenia glabra Costantin Family: Asclepiadaceae

Chanwitheesuk *et al.* (2005) investigated the antioxidant activity of *Marsdenia glabra* using a β-carotene bleaching method. The contents of plant chemicals such as vitamin C, Vitamin E, carotenoids, tannin and total phenolics, were also determined. Methanolic extract showed antioxidant activity. *Marsdenia glabra* has antioxidant index of 2.06, Vitamin C 22.7 mg per cent, Vitamin E 0.0018mg per cent, Total carotenes 8.92 mg per cent, Total xanthophylls 7.42mg per cent, Tannins 4.47mg per cent and Total phenolics of 51.5 mg per cent.

Matricaria aurea Schultz. Family: Asteraceae (Compositae)

Al-Mustafa and Al-Thunibat (2008) evaluated the antioxidant activity of Jordanian medicinal plant *Matricaria aurea* shoot. The level of antioxidant activity was determined by DPPH and ABTS assay in relation to total phenolic contents of the medicinally used part. They concluded that *Matricaria aurea* shoot possess low antioxidant capacity (DPPH-TEAC < 20mg/g).

Matricaria chamomilla L. Family: Asteraceae (Compositae)

Bajpai *et al.* (2005) investigated the phenolic contents and antioxidant activity of some food and medicinal plants including *Matricaria chamomilla*. The leaves and roots were found to have moderate phenolic contents and antioxidant activity. The TPC of leaves and roots of *Matricaria chamomilla* were 15.4 and 8.4 mg/g GAE, respectively. The antioxidant activity of *Matricaria chamomilla*, as assayed by auto-oxidation of β-carotene and linoleic acid and expressed as the per cent of inhibition relative to the control were, 76.9 per cent and 43.7 per cent for leaves and roots, respectively.

Matricaria matricarioides (Less.) Porter Family: Asteraceae (Compositae)

English Name: Pineapple weed.

The antioxidative activity of phenolic extracts from herb of Pineapple weed was examined by autoxidation of methyl linoleate by Kahkonen *et al.* (1999). Both phenolics and antioxidant activity of the flowers were low. The phenolics were 4.2 mg/g GAE while antioxidant activity was 19 per cent at 500 ppm level and 21 per cent at 5000 ppm level.

Matricaria recutita L. Family: Asteraceae (Compositae)

English Name: German chamomile.

The use of German chamomile teas and medicinal preparations has a long tradition in various countries. Although chamomile contains a great number of polyphenolic compounds (Hurrell *et al.*, 1999), it was reported that antioxidative properties of its extracts in rapeseed oil were not distinct (Lionis *et al.*, 1998). Miliuskas *et al.* (2004) investigated the radical scavenging activity using DPPH and ABTS assays. The extracts of German chamomile showed weak radical scavenging activity.

The antioxidative activity of phenolic extracts from fowers of Chamomile was examined by autoxidation of methyl linoleate by Kahkonen *et al.* (1999). Both phenolics and antioxidant activity of the flowers were low. The phenolics were 9.1 mg/g GAE while antioxidant activity was 16 per cent at 500 ppm level.

Maytenus aquifolium Mart. Family: Celastraceae

Ethyl alcohol extract obtained from leaves and root bark revealed the presence of five compounds exhibiting antioxidant properties toward β-carotene (Corsino *et al.*, 2003). The isolates were investigated for their redox properties by cyclic voltametry and for their radical scavenging abilities through spectrophotometric assay on the reduction of DPPH (Corsino *et al.*, 2003).

Maytenus ilicifolia Mart. Family: Celastraceae

Vellosa *et al.* (2006) evaluated the crude ethanolic extract as a potential antioxidant source using an assay based on the bleaching of the radical ABTS+ and by HOCl scavenger capacity. Trolox and uric acid were used as positive controls. The results indicated the root bark as a great source of antioxidants based on its potential as scavenger of radicals.

Maytenus imbricata Mart. Family: Celastraceae

The free radical scavenging activity using DPPH, the reducer power and the total phenolic concentration of extracts and compounds isolated from leaves, branches and roots of *Maytenus imbricata* were evaluated by Silva *et al.* (2009). Some extracts, a mixture of phenolic compounds and epicatechin showed higher reducing power and free radical scavenging activities in comparison with the standard BHA and gallic acid used in the assays. The ethyl acetate extract from leaves showed higher total phenolic content and also higher reducing power and DPPH radical scavenging than the other extracts. These facts indicate that there are some relations between phenolic concentration in the extract and the antioxidant activity and reducing power.

Maytenus krukovii A.C.Sm. (=*Maytenus chuchuhuasha* Raymond-Hamet et Colas) Family: Celastraceae

The hydroalcoholic extract of *Maytenus krukovii* bark was investigated for its antioxidant and radical scavenging activity by Bruni *et al.* (2006). The extract showed radical scavenging capacity comparable to that of most synthetic and natural reference compounds. In particular, in the DPPH test and in the β-carotene bleaching test the crude extract produced a dose-dependent inhibition.

Medicago sativa L. Family: Fabaceae (Papilionaceae)

Rana *et al.* (2010) studied the *in vitro* antioxidant activity of alcoholic extract of *Medicago sativa* using different models. It's antioxidant activity was estimated by IC_{50} value and the values were 100.38 µg/ml (DPPH radical scavenging), 12.33 µg/ml (ABTS radical scavenging), 115.79µg/ml (Iron chelating activity) and 49.06µg/ml (lipid peroxidation), 21.77µg/ml (nitric oxide scavenging) and 15.91µg/ml (alkaline DMSO). In all the testing a significant correlation existed between concentrations of the extract and percentage inhibition of free radical metal chelation or inhibition of lipid peroxidaation.

Bora and Sharma (2011) investigated the neuroprotective and antioxidant activity of methanol extract of *Medicago sativa* on Ischemia and reperfusion-induced cerebral injury in mice. The extract directly scavenged free radicals generated against a stable radical DPPH and O_2^- and also inhibited XD/XO conversion and resultant O_2^- production. Pretreatment with the extract markedly reduced cerebral infarct size XO, O_2^- and TBARS levels significantly restored GSH, SOD and T-SH levels.

Melaleuca alternifolia Cheel. Family: Myrtaceae

English Name: Australian Tea Tree.

Antioxidant activity of Australian tea tree Oil (TTO) was determined by Kim *et al.* (2004) using two different assays. In the DPPH assay, 10 µl/ml crude TTO in methanol had approximately 80 per cent free radical scavenging activity, and in the hexane/hexanoic acid assay, 200 µl/ml crude TTO exhibited 60 per cent inhibitory activity against the oxidation of hexanol to hexanoic acid over 30 days. These results were equivalent to the antioxidant activities of 30 µM BHT in both tests at the same experimental conditions, *i.e.*, α-terpinene, α-terpinolene, and γ-terpinene, in the crude TTO were separated and identified chromatographically. Their antioxidant activities decreased in the following order in both assays: α-terpinene, α-terpinolene, and γ-terpinene.

Melaleuca leucadendron L. Family: Myrtaceae

Fu *et al.* (2010) evaluated the antioxidant capacity and total phenolic contents of 56 wild fruits from South China using FRAP and TEAC assays. A high correlation between antioxidant capacity and phenolic compounds could be the main contributor to the measured antioxidant activity. The results showed that fruits of *Melaleuca leucadendron* possessed the highest antioxidant capacity and total phenolic contents.

Melaleuca leucandra L. Family: Myrtaceae

The antioxidant activity of essential oils from leaf and fruit of *Melaleuca leucandra* was determined by Pino *et al.* (2010) by three different *in vitro* assays, DPPH radical, TBARS and ABTS radical cation, and significant activities were evidenced for all of them.

Melampyrum barbatum L. Family: Scrophulariaceae

Stajner *et al.* (2009) examined the antioxidant and free radical scavenging activities of red and yellow forms of *Melampyrum barbatum*. They reported the results concerning flower and leaf antioxidant enzyme activities (superoxide dismutase, catalase, guaiacol peroxidase and glutathione peroxidase), reduced glutathione quantity, flavonoids, photosynthetic pigments and soluble protein contents and quantities of malonyldialdehyde, (*)OH and O(2)(*-) radicals; total antioxidant capacity was determined by FRAP method and scavenger activity by DPPH method. Lipofuscin 'plant age pigments' were also determined. According to their results, flowers of the red form exhibited the highest antioxidant capacity.

Melastoma decemfidum Roxb. ex Jack Family: Melastomataceae

Sarju *et al.* (2011) investigated the content of kaempferol and naringenin and antioxidant activity of methanol extracts from the leaves of *Melastoma decemfidum* plants. The antioxidant assay was carried out by the DPPH radical-scavenging. The combination of pure compound of naringenin and kaempferol had strongest antioxidant activity.

Melastoma malabathricum L. Family: Melastomataaceae

Deny *et al.* (2007) evaluated the antioxidant activity of the extracts and the isolated compounds from the flowers of *Melastomaa malabathricum*. The ethyl acetate extract yielded three compounds – naringenin, kaempferol and kaempferol-3-O-D-glucoside and methanol extract gave kaempferol-3-O-(2″,6″-di-O-*p*-trans-coumaroyl) glucoside and kaempferol-3-O-D-glucoside. The crude extracts and isolated compound showed antioxidant activity.

Zakaria (2007) evaluated the free radical scavenging properties of *Melastoma malabathricum* using DPPH and superoxide anion radical scavenging assays. The extract was found to show remarkable antioxidant activity in both assays. The DPPH rdical scavenging activity was 98.30 per cent and superoxide scavenging activity was 96.80 per cent. Phytochemicals screening of this plant demonstrated the presence of flavonols, triterpenes, tannins, saponins and steroids.

Melastoma sanguineum Sims. Family: Melastomataceae

Fu *et al.* (2010) evaluated the antioxidant capacity and total phenolic contents of 56 wild fruits from South China using FRAP and TEAC assays. A high correlation between antioxidant capacity and phenolic compounds could be the main contributor to the measured antioxidant activity. The results showed that fruits of *Melastoma sanguineum* possessed the highest antioxidant capacity and total phenolic contents.

Melicope ptelefolia Champ. ex Benth. Family: Rutaceae

Abbas *et al.* (2006) evaluated the antioxidant activity of *Melicope ptelefolia* using DPPH assay and nitric oxide inhibition using Griess assay. Methanolic extract showed DPPH radical scavenging and nitric oxide inhibition in murine peritoneal macrophages.

Melilotus officinalis (L.) Medik. ex Desr. Family: Fabaceae (Papilionaceae)

Miliauskas *et al.* (2004) investigated the radical scavenging activity of acetone, ethyl acetate and methanol extracts of leaves, stems and flowers at initial blossom stage using DPPH and ABTS assays. They found that methanol extracts were most effective DPPH radical scavengers. Ethyl acetate and acetone extracts were considerably less effective radical scavengers. High content of phenolics in the extracts correlates with its antiradical activity, confirming that phenolic compounds are likely to contribute to the radical scavenging activity of this plant extracts.

Pourmorad *et al.* (2006) carried out a systematic record of the relative antioxidant activity in selected Iranian medicinal plant species. The total phenol content of *Melilotus officinalis*, measured by Folin Ciocalteu method was 22.3 mg/g GAE. Flavonoid contents in terms of quercetin equivalent was78.3 mg/g. DPPH radical scavenging effect of the extracts was detemined spectrophotometrically. The highest radical scavenging effect was observed in *Melilotus officinalis* with IC_{50} = 0.018 mg/ml.The potency of radical scavenging effect of *M.officinalis* extract was about 4 times greater than synthetic antioxidant BHT. The greater amount of phenolic compounds lead to more potent radical scavenging effect as shown by *M. officinalis* extract.

Melissa officinalis L. Family: Lamiaceae (Labiatae)

English Name: Lemon balm.

The antioxidant effects of aqueous methanolic extracts from *Melissa officinalis* were investigated in enzyme-dependent and enzyme-independent lipid peroxidation systems. The extract caused a considerable concentration-dependent inhibition of lipid peroxidation. Phenolic components present in the plant extract were evaluated for antioxidant activity and were found effective in both tests (Hohmann *et al.*, 1999).

de Souza *et al.* (2004) reported antioxidant activity of the essential oil of *Melissa offcinalis* as evidenced by reduction of DPPH radical.

Mimica-Dukic *et al.* (2004) reported antioxidant activity of the essential oil isolated from *Melissa officinalis*. The examined essential oil exhibited very strong RSC reducing the DPPH radical formation (IC_{50}-7.58µg/ml) and OH radical generation (IC_{50}-1.74µg/ ml) in a dose-dependent manner. The most powerful scavenging compounds were monoterpene aldehydes and ketones (neral/geranial citronellol, isomethone, and menthone) and mono- and sesquiterpene hydrocarbons (E-caryophyllene). Very strong inhibition of LP, particularly in the $Fe^{2+}H_2O$, system of induction (94.59 per cent for 2.13µg/ml), was observed in both cases, also in a dose-dependent manner.

Two natural antioxidant preparations, namely Rosmol (liquid) and Rosmol-P (powder) were produced by Lugasi *et al.* (2006) by extraction of a mixture of medicinal plants belonging to the Lamiaceae family, such as *Rosmarinus officinalis, Prunella vulgaris, Hyssopus officinalis* and *Melissa officinalis*. The main active compound of the extract was supposed to be a phenolic (caffeic) acid derivative. The total polyphenol content of the preparation was very high, 8.72g/l for Rosmol and 93.7 g/kg for Rosmol-P. The products acted as primary and secondary antioxiants chelatng transitional metal ions and inhibiting the autoxidation of linoleic acid. Rosmol and Romol-P scavenged free radicals formed during Fenton type reaction measured by chemiluminometry and also exhibited strong antioxidant property in Randox TAS measurement. The antioxidant activity of the products was unchanged after six months storage.

Ivanova *et al.* (2005) also reported high phenolics content and antioxidant properties in *Melissa officinalis*.

The ability of *Melissa officinalis* extract to act as a free radical scavenger or hydrogen donor was revealed by DPPH radical-scavenging activity assay (Bouayed *et al.*, 2007). A positive correlation between total phenolic or flavonoid contents and antioxidant activity was found.

Spiridon *et al.* (2011) evaluated the total phenolic content and antioxidant activity of plants used in Romanian herbal medicine. *Melissa officinalis* extracts showed good antioxidant activity.

Melodorum fruticosum L. Family: Annonaceae

Pripdeevech and Chukeatirote (2010) investigated the chemical composition, antifungal and antioxidant activities of essential oil and various extracts of *Melodorum fruticosum* flowers. The dichloromethane extracts were evaluated to be superior to all extracts tested with an IC_{50} value of 87.6µg/ml whereas other extracts showed their IC_{50} values ranging from 100.13 to 194.50µg/ml

Mentha aquatica L. Family: Lamiaceae (Labiatae)

Serteser *et al.* (2009) studied the antioxidant activity of 50 per cent aqueous methanol extracts by various antioxidant assays, including free radical scavenging, hydrogen peroxide scavenging and metal (Fe^{2+}) chelating activities. The DPPH radical scavenging effects of leaves extract was 0.612. Fe^{2+} chelating activity (per cent) of leaves extracts was 38.96 and Hydrogen peroxide inhibition activity of leaves extracts was 46.32.

Ebrahimzadeh *et al.* (2010) examined the antioxidant and antihemolytic activities of *Mentha aquatica* aerial part. The extract showed moderate antioxidant activity in some models. In DPPH radical scavenging model, extract showed potent activity (IC_{50} was 46.05µg/ml). The extract showed moderate nitric oxide scavenging activity between 0.2 and 0.8mg/ml. The extract exhibited weak antioxidant activity in peroxidation inhibition through linoleic acid emulsion system.

Mentha arvensis L. Family: Lamiaceae (Labiatae)

Chanwitheesuk *et al.* (2005) investigated the antioxidant activity of *Mentha arvensis* using a β-carotene bleaching method. The contents of plant chemicals such as vitamin C, Vitamin E, carotenoids, tannin and total phenolics, were also determined. Methanolic extract showed antioxidant activity. *Mentha arvensis* has antioxidant index of 10.9, Vitamin C 12.8 mg per cent, Vitamin E 0.0294mg per cent, Total carotenes 4.48 mg per cent, Total xanthophylls 26.5mg per cent, Tannins 21.0mg per cent and Total phenolics of 70.0 mg per cent.

Wong *et al.* (2006) investigated the antioxidant properties of 25 edible tropical plants including *Mentha arvensis* using DPPH and FRAP assays. Total phenolic content was also estimated. *Mentha arvensis* showed highest antioxidant activity. A strong correlation between TEAC values obtained for the DPPH assay and those for FRAP assay implied that compounds in the extracts were capable of scavenging DPPH radical and reducing ferric ions. A satisfactory correlation of TPC with TEAC suggested that polyphenols in the extracts were partly responsible for the antioxidant activities while its correlation with CCA was poor, indicating that polyphenols might not be the main cupric ion chelators.

Zakaria *et al.* (2008) studied the antioxidant properties of the methanol extract of leaves and stems of *Mentha arvensis* by DPPH radical scavenging activity. The extracts showed antioxidant activity comparable to butylated hydroxytoulene (BHT).

Mentha cordifolia Opiz. ex Fresen Family: Lamiaceae (Labiatae)

Chanwitheesuk *et al.* (2005) investigated the antioxidant activity of *Mentha cordifolia* using a β-carotene bleaching method. The contents of plant chemicals such as vitamin C, Vitamin E, carotenoids, tannin and total phenolics, were also determined. Methanolic extract showed antioxidant activity. *Mentha cordifolia* has antioxidant index of 7.45 Vitamin C 11.3 mg per cent, Vitamin E 0.0016mg per cent, Total carotenes 2.58 mg per cent, Total xanthophylls 4.24g per cent, Tannins 73.7mg per cent and Total phenolics of 280.0 mg per cent.

Mentha haplocalyx Briq. Family: Lamiaceae (Labiatae)

Antioxidant capacities of 56 selected Chinese medicinal plants, including *Mentha haplocalyx* were evaluated by Song *et al.* (2010) using the TEAC and FRAP assays, and their total phenic content was measured by the Folin-Ciocaleteu method. *Mentha haplocalyx* showed moderate antioxidant activity with TEAC value of 87.80µmol Trolox/g, FRAP value of 175.06µmol Fe^{2+}/g and phenolic content was 12.08mg GAE/g.

Mentha longifolia L. Family: Lamiaceae (Labiatae)

English Name: Mint.

Mkaddem *et al.* (2009) investigated chemical composition and antimicrobial and antioxidant activities of *Mentha longifolia* essential oil. The antioxidant activity by ABTS assay showed IC_{50} values of 476.3. The DPPH assays have resulted in a moderate IC_{50} (>8000mg/l).

Spiridon *et al.* (2011) evaluated the total phenolic content and antioxidant activity of plants used in Romanian herbal medicine. Total phenolic concentration was determined using Folin-Ciocalteu phenol reagent method, while total flavonoids were measured using the aluminium chloride colorimetric method. *Mentha longifolia* extracts showed good antioxidant activity.

Mentha longifolia (L.) Huds. subsp. *longifolia* Family: Lamiaceae

English Name: Horsemint.

Souri *et al.* (2004) investigated the antioxidant activity of methanol extracts of Horsemint, used as vegetable in Iranian diet against linoleic acid peroxidation. The extracts showed good antioxidant activity.

Serteser *et al.* (2009) studied the antioxidant activity of 50 per cent aqueous methanol extracts by various antioxidant assays, including free radical scavenging, hydrogen peroxide scavenging and metal (Fe^{2+}) chelating activities. The methanolic extract examined in the DPPH assay showed the strongest activities. The DPPH radical scavenging effects of leaves extract was 0.596. Fe^{2+} chelating activity (per cent) of leaves extracts was 37.23 and Hydrogen peroxide inhibition activity of leaves extracts was 44.29.

Mentha longifolia (L.) Huds. subsp. *typhoides* (Briq.) Harley var. *typhoides* Family: Lamiaceae (Labiatae)

English Name: Horsemint

Serteser *et al.* (2009) studied the antioxidant activity of 50 per cent aqueous methanol extracts by various antioxidant assays, including free radical scavenging, hydrogen peroxide scavenging and metal (Fe^{2+}) chelating activities. The methanolic extract examined in the DPPH assay showed the strongest activities. The DPPH radical scavenging effects of leaves extract was 0.620. Fe^{2+} chelating activity (per cent) of leaves extracts was 38.23 and Hydrogen peroxide inhibition activity of leaves extracts was 47.32.

The aerial parts of *Mentha longifolia* ssp. *typhoides* var. *typhoides* were screened by Gursoy *et al.* (2009) for their possible antioxidant and antimicrobial activities in addition to their phenolic contents. Antioxidant activity was employed by two complementary test systems: DPPH free radical scavenging and β-carotene/linoleic acid. The amount of total phenolics was 93.47 μg/mg.

Bhatti *et al.* (2010) investigated total phenolics and antioxidant activity of *Mentha longifolia*. Horsemint showed the total phenolic contents range from 2.95 GAE g/100 g, total flavonoids 2.00 CE g/100g, percentage of inhibition of peroxidation was 93.1-97 and low IC_{50} values 12.46 – 22.09 μg/ml) for DPPH free radical scavenging. The antioxidant activity of horsemint extracts showed significant differences with different solvent extraction.

Ebrahimzadeh *et al.* (2010) investigated the antioxidant and antihemolytic activities of *Mentha longifolia* aerial part employing seven *in vitro* assay systems. The extract showed moderate antioxidant activity in some models. IC_{50} for DPPH radical-scavenging activity was 12.6μg/ml. The extract showed potent nitric oxide-scavenging activity between 100 and 800μg/ml. The extracts showed good Fe^{+3} chelating ability, IC_{50} were 766.6μg/ml. The extract exhibited low antioxidant activity in the linoleic acid model but were capable of scavenging hydrogen peroxide in a concentration dependent manner.

Mentha piperita L. Family: Lamiaceae (Labiatae)

English Name: Pepper mint.

The essential oil of *Mentha piperita* was screened by Yadegarinia *et al.* (2006) for its possible antioxidant activity by two complementary test systems, namely DPPH free radical scavenging and β-carotene/linoleic acid systems. The essential oil exhibited antioxidant activity.

Limonene and menthol in the leaves of the plant have been found to be anticancerous (Mishra *et al.*, 2007). Mathur *et al.* (2010) carried out phytochemical investigation and antioxidant activity of *Mentha piperita*. Traditional solvent extraction (TSE) using four solvents – methanol, aqueous, hexane and petroleum ether were utilized to determine the content of antioxidants. DPPH assay and superoxide anion radical scavenging activity were used to determine the antioxidant capacity. The most antioxidant capacity was achieved using methanol as the solvent followed by aqueous extracts. The antioxidant capacity with in the specific extract of plant was found to be correlated with the total phenolic content. The radical scavenging activity by DPPH radical scavenging method (IC_{50}) was 42.57μg/ml (hexane extract), 45.10 μg/ml (petroleum ether extract), 32.10 (aqueous) and 36.15 (methanol extract).

Yang *et al.* (2010) reported antioxidant activity of the essential oil from *Mentha piperita*. Radical scavenging activity against ABTS radical was highest in peppermint oil. Ebrahimzadesh *et al.* (2010) evaluated antioxidant activity of aerial parts of *Mentha piperita.* IC_{50} for DPPH radical-scavenging activity was 129.3µg/ml. The extract showed potent nitric oxide scavenging activity between 0.1 and 1.6 mg/ml. The extract showed good Fe^{2+} chelating ability, IC_{50} was 698.3µg/ml. The extract also exhibited low antioxidant activity in the linoleic acid model but were capable of scavenging hydrogen peroxide in a concentration dependent manner.

Mentha pulegium L. Family: Lamiaceae (Labiatae)

English Name: Pennyoryal.

Souri *et al.* (2004) evaluated the antioxidative activity of *Mentha pulegium* by linoleic acid peroxidation test using 1,3-diethyl-2-thiobarbituric acid as the reagent. Methanolic extract showed 81.44 per cent inhibition at 40µg concentration with IC_{50} value of 0.57µg, which is approximately in the range of α-tocopherol.

Serteser *et al.* (2009) studied the antioxidant activity of 50 per cent aqueous methanol extracts by various antioxidant assays, including free radical scavenging, hydrogen peroxide scavenging and metal (Fe^{2+}) chelating activities. The methanolic extract examined in the DPPH assay showed the strongest activities. The DPPH radical scavenging effects of leaves extract was 0.659. Fe^{2+} chelating activity (per cent) of leaves extracts was 40.22 and Hydrogen peroxide inhibition activity of leaves extracts was 59.22

Kamkar *et al.* (2010) evaluated the antioxidative activities of the essential oil, menthol and water extracts of Iranian pennyroyal in vegetable oil during storage. Antioxidant capacity of the essential oil and extract were determined using DPPH and β-carotene linoleic acid methods. Results showed that DPPH and β-carotene linoleic acid assay findings on the *M. pulegium* extracts were comparable to those found on β-hydroxy toluene. Water extract was more potent than the methanol extract. Essential oil did not show considerable antioxidative effect.

Mentha spicata L. subsp. *spicata* Family: Lamiaceae (Labiatae)

English Name: Spirmint.

Souri *et al.* (2004) investigated the antioxidant activity of methanol extracts of spirmint, used as vegetable in Iranian diet against linoleic acid peroxidation. The extracts showed good antioxidant activity.

Serteser *et al.* (2009) studied the antioxidant activity of 50 per cent aqueous methanol extracts by various antioxidant assays, including free radical scavenging, hydrogen peroxide scavenging and metal (Fe^{2+}) chelating activities. The methanolic extract examined in the DPPH assay showed the strongest activities. The DPPH radical scavenging effects of leaves extract was 0.604. Fe^{2+} chelating activity (per cent) of leaves extracts was 39.64 and Hydrogen peroxide inhibition activity of leaves extracts was 54.32

Kiselova *et al.* (2006) studied the antioxidant activity of *Mentha spicata*. Antioxidant activity was measurd by ABTS cation radical decoloration assay and the total polyphenol content was measured according to the Folin-Ciocalteu method. *Mentha spicata* extract exhibited higher antioxidant activity than mate. A positive correlation between antioxidant activity and polyphenol content was found, suggesting that the antioxidant capacity of the aqueous plant extract is due to a greater extent to their polyphenols.

Extracts of *Mentha spicata* were investigated by Hosseinimehr *et al.* (2007) for their total flavonoids, phenol contents and their radical scavenging activity using DPPH assays. Quercetin and butylated hydroxytoluene were used as standard reference with well documented antioxidant activity. Methanolic extract sample showed free radical scavenging activity. A correlation between radical scavenging capacitiy of the extract with total phenolic compounds content was observed.

Souri *et al.* (2008) investigated the antioxidant activity and radical scavenging activity of methanolic extract of leaves. The antioxidant activity of this plant extract against linoleic acid peroxidation expressed as IC_{50} was 4.16ng/ml and radical scavenging activity against DPPH expressed as IC_{50} was 1.14 µg/ml.

Nikolova and Dzhurmanski (2009) evaluated the free radical scavenging activity of total methanol extract of *Mentha spicata*. The extract exhibited strong antioxidant activity and its IC_{50} value for DPPH radical was 20.19 µg/ml.

The antioxidant activity of *Mentha spicata* aerial part was investigated by Ebrahimzadeh *et al.* (2010) employing eight *in vitro* assay system. The extract showed moderate antioxidant activity in some models. IC_{50} for DPPH radical-scavenging activity was 105.8µg/ml. The extract showed potent nitric oxide-scavenging activity between 0.1-1.6 mg/ml. The extracts showed weak Fe^{2+} chelating ability. IC_{50} were 757.4µg/ml. The extract also exhibited low antioxidant activity in the linoleic acid model but were capable of scavenging hydrogen peroxide in a concentration dependent manner.

Mentha spicata L. subsp. *tomentosa* Family: Lamiaceae (Labiatae)

Serteser *et al.* (2009) studied the antioxidant activity of 50 per cent aqueous methanol extracts by various antioxidant assays, including free radical scavenging, hydrogen peroxide scavenging and metal (Fe^{2+}) chelating activities. The methanolic extract examined in the assay showed the strongest activities. The DPPH radical scavenging effects of leaves extract was 0.583. Fe^{2+} chelating activity (per cent) of leaves extracts was 34.42 and Hydrogen peroxide inhibition activity of leaves extracts was 56.31.

Mentha viridis L. Family: Lamiaceae (Labiatae)

Arumugam *et al.* (2004) studied the *in vitro* screening of antioxidant activity in different fractions of *Mentha viridis*. The ABTS/HRP/H_2O_2 decoloration method was used for evaluation of antioxidant activity. Total antioxidant activity was estimated in different solvent fractions (hexane, chloroform, ethylacetate and water of ethanolic extract of dried leaf powder of *Mentha viridis*. Total antioxidant activity was higher for ethyl acetate fraction (83 per cent) followed by aqueous (75 per cent), chloroform (51 per cent) and hexane fractions (47 per cent). Relative antioxidant activity of ethyl acetate and water fractions were comparable to the known standards but less in hexane and chloroform fractions. The results indicate that polyphenolics were responsible for enhanced antioxidant activity than the leaf pigments (Arumugam *et al.*, 2004).

Mkaddem *et al.* (2009) investigated chemical composition and antimicrobial and antioxidant activities of *Mentha viridis* essential oil. The antioxidant activity by ABTS assay showed IC_{50} values of 195.1. The DPPH assays have resulted in a moderate IC_{50} (>3476.3mg/l).

Mesembryanthemum crystallinum L. Family: Aizoaceae

Shoot extracts from edible halophytic *Mesembryanthemum* species were evaluated by Hanen *et al.* (2009) for their effects against O(2)(-), DPPH and ABTS radicals. The reducing power, chelating ability, inhibition of lipid peroxidation and of β-carotene bleaching were also evaluated. Moreover, the total phenolic, flavonoid and condensed tannin contents were evaluated. *In vitro* biotests showed a significant difference in the antioxidant capacities of the species studied. *M. crystallinum* showed the best activity in iron-chelating test with the lowest EC_{50} value (2.13mg/ml). Phenolic content was 1.4 mg GAE g DW.

Mesembryanthemum edule L. Family: Aizoaceae

Shoot extracts from edible halophytic *Mesembryanthemum* species were evaluated by Hanen *et al.* (2009) for their effects against O(2)(-), DPPH and ABTS radicals. The reducing power, chelating ability, inhibition of lipid peroxidation and of β-carotene bleaching were also evaluated. Moreover, the total phenolic, flavonoid and condensed tannin contents were evaluated. *In vitro* biotests showed a significant difference in the antioxidant capacities of the species studied. *Mesembryanthemum edule* was found to exhibit the higher antioxidant activity. *M. edule* exhibited high phenolic levels, especially in total phenols (70mg of GAE/g dry weight). Flavonoids Phloretin, quercetin and avicularin were most abundant (1, 0.84 and 0.66 mg/g DW, respectively).

Mesembryanthemum nodiflorum L. Family: Aizoaceae

Shoot extracts from edible halophytic *Mesembryanthemum* species were evaluated by Hanen *et al.* (2009) for their effects against O(2)(-), DPPH and ABTS radicals. The reducing power, chelating ability, inhibition of lipid peroxidation and of β-carotene bleaching were also evaluated. Moreover, the total phenolic, flavonoid and condensed tannin contents were evaluated. *In vitro* biotests showed a significant difference in the antioxidant capacities of the species studied. Phenolic content was 1.7 mg GAE g DW.

Mesona procumbens Hemsl. Family: Lamiaceae (Labiatae)

Chinese Name: Hsian-tsao

Lai *et al.* (2001) evaluated the antioxidant activity of crude hsian-tsao leaf gum extracted by sodium bicarbonate solutions and precipitated by 70 per cent ethanol. The antioxidative activities, including the radical-scavenging effects, Fe(2+)-chelating ability, and reducing power as well as the inhibition of FeSO(4)-H(2)O(2)-induced malondialdehyde formation in rat tissue homogenate were studied *in vitro*. It was found that the antioxidative effect provided by hsian-tsao leaf gum was strongly concentration dependent. In general, the antioxidative activity increased with increasing gum concentration, to a certain extent, and then leveled off with further increase in gum concentration. From comparison of the IC(50) values for different antioxidative reactions it seemed that hsian-tsao leaf gum was more effective in scavenging superoxide radicals than chelating Fe(2+) or scavenging DPPH radicals. As compared to the commercial antioxidants, hsian-tsao leaf gum showed less scavenging effect on the DPPH radical and reducing power but better superoxide radical-scavenging effect and Fe(2+)-chelating ability than α-tocopherol and BHT.

Mezoneuron benthamianum Baill. Family: Caesalpiniaceae

Antioxidant and free radical activities of *Mezoneuron benthamianum* using DPPH spectrophotometric and TBA lipid peroxidation assays were conducted by Dickson *et al.* (2006).The petroleum spirit and chloroform extracts showed strong free radical scavenging activity with IC_{50} values of 15.33 and 19.72 µg/ml, respectively.

Michelia alba DC. Family: Magnoliaceae

Leelapornpisid *et al.* (2007) reported antioxidant activity of the absolutes (obtained by solvent extraction) *Michelia alba* (IC_{50} = 0.7155 mg/ml). Michephyll A, isolated from the leaves of *Michelia alba* exhibited the antioxidant activity (Wang *et al.*, 2010).

Michelia champaca L. Family: Magnoliaceae

Hossain *et al.* (2009) investigated the antioxidant activity of the crude methanol extract of *Michelia champaca* leaf. The extract showed remarkable antioxidant activity in DPPH radical scavenging activity, nitric oxide scavenging activity and total antioxidant capacity assays. In both DPPH radical and NO scavenging assay, the extract exhibited strong antioxidant activity and the IC_{50} values in DPPH radical and NO scavenging assays were found to be 30.07µg/ml and 15.42µg/ml, respectively while the IC_{50} values of ascorbic acid were 12.5µg/ml and 4.07µg/ml, respectively. Total antioxidant activity of the extract increased in a dose dependent manner.

Hydromethanol extracts of 15 Bangladeshi medicinal plants, including *Michelia champaca*, were evaluated by Hasan *et al.* (2009) for antioxidant potential using DPPH radical scavenging assay. *Michelia champaca* leaves exhibited radical scavenging activity with an IC_{50} value of 22.43 µg/ml compared to the IC_{50} value of 5.15µg/ml as shown by the reference antioxidant ascorbic acid, in a dose dependent fashion.

Vivek Kumar *et al.* (2010) evaluated the antioxidant activity of various extracts of flowers of *Michelia champaca*. The antioxidant activity increased with increase in amount of extract (5-20 mg). DPPH free radical-scavenging activity of methanol, ethanol and aqueous extracts of *Michelia champaca* flowers, Gallic and ascorbic acid standards were found to be 80.56 per cent, 90.20 per cent, 81.32 per cent, 91.34 per cent and 93.64 per cent, respectively.

Miconia lemannii Cogn. Family: Melastomataceae

Mosquera *et al.* (2009) evaluated the antioxidant activity of forty-six methanol plant extracts using DPPH radical scavenging assay. The plant extracts that showed greatest antioxidant activity were *Miconia lehmannii* with 45.3 per cent of antioxidant activity.

Microdesmis keayana J. Leonard (Syn.: *Microdesmis puberula*) Family: Pandaceae

Microdesmis keayana is an African tropical plant whose roots are used in traditional medicine for erection impairment but the compounds responsible for its action were unknown. Two major alkaloids isolated from the roots were tested by Zamble *et al.* (2006, 2009) for vasorelaxing properties using isolated rat aertic rings precontracted by phenylephrine to confirm its traditional use. Influence of the alkaloids on the endothelial production of endothelial nitric oxide synthase was measured by quantitative polymerase chain reaction analysis. Scavenging activities were assessed versus DPPH and reactive oxygen species such as superoxide anion and hydrogen peroxide in cell-free and cellular systems. The results showed that keanidine B and Keaynine had significant vasorelaxing properties. This effect could be due to their strong antioxidant activity (Zamble *et al.*, 2009).

Microlabium candidum (Griseb.) H.Rob. Family: Asteraceae (Compositae)

Borneo *et al.* (2008) evaluated the antioxidant activity of ethanolic extract of *Microlabium candidum*. The FRAP value was 443.2µmol of Fe(II)/g and DPPH radical scavenging activity was EC_{50} = 198.0.

Micromeria graeca (L.) Benth. et Reichnb. Family: Lamiaceae (Labiatae)

Couladis *et al.* (2003) screened Greek aromatic plants from the Lamiaceae family for the antioxidant activity. Of the 21 plants tested they found ethanol extracts prepared from *Micromeria graeca*, as well as 9 other plants exhibited the same antioxidant activity as α-tocopherol in their ability to inhibit bleomycin-Fe(II) complex-induced arichidonic acid superoxidation to MDA.

Mikania buchtienii B.L.Rob. Family: Asteraceae (Compositae)

The aerial parts of 17 Bolivian plants including *Mikania buchtienii* were screened by Rosas-Romero and Saavedra (2005) to determine antioxidant activity. A methanol extract of each plant was prepared and portioned sequentially with hexane, chloroform, and ethyl acetate, leaving an aqueous solution. All extracts and their 5 fractions for a total of 102 samples, were evaluated using two techniques: an adaptation of the β-carotene bleaching technique using an emulsion of linoleic acid in water as the oxidizable substrate, and the DPPH free radical trapping technique. The results of the β-carotene bleaching technique were more discriminating and better related to the rancidity process under normal conditions; with this assay, 11 species provided at least one fraction with highly promising antioxidant activity. All species gave good results under the DPPH technique, and in most cases they performed better than BHA, which was used as a reference antioxidant (Rosas-Romero and Saavedra, 2005)

Mikania leiostachya Benth. Family: Asteraceae (Compositae)

Mosquera *et al.* (2007) studied the antioxidant activity of the extracts using DPPH free-radical scavenging method. The percentage of the antioxidant activity of the n-Hexane fraction was 35.374, Dichloromethane fraction was 37.471 and methanol fraction was 38.5.

Mikania psilostachya DC. Family: Asteraceae (Compositae)

Extracts and fractions from *Mikania psilostachya* among the nine Bolivian plants were attributed to the phenolic compounds present in this bioactive species (Parejo *et al.*, 2005).

Mikania scandens (L.) Willd. Family: Asteraceae (Compositae)

Hasan *et al.* (2009) evaluated the antioxidant potential of the hydromethanol extract of the leaves of *Mikania scandens* using DPPH radical scavenging and nitric oxide scavenging and reducing power assays. A dose dependent scavenging of DPPH radical scavenging and NO was observed with good reducing power with the extract. In DPPH radical scavenging assay, the IC_{50} value of the extract was 375.40µg/ml while the IC_{50} value for the reference ascorbic acid was 55.89µg/ml. The IC_{50} values of the extract and ascorbic acid were 220.43 and 125.10µg/ml, respectively in NO scavenging assay.

Millettia dielsiana Harms Family: Fabaceae (Papilionaceae)

Gan *et al.* (2010) evaluated the antioxidant activities of 40 medicinal plants associated with prevention and treatment of cardiovascular and cerebrovascular diseases. *Millettia dielsiana* showed the highest antioxidant activity 790.79µmol Fe(II)/g in FRAP assay, 615.79µmol Trolox/g TEAC assay and Total phenolic content was 41.93 mg GAE/g. A high correlation between antioxidant capacity and total phenolic content indicated that phenolic compounds could be the main contributor of the antioxidant activity of these plants.

Millettia pulchra Kurz. var. *microphylla* Dunn Family: Fabaceae (Papilionaceae)

Lee *et al.* (2006) investigated the antioxidant activity of ethanol extract of leaves using DPPH, hydroxyl and superoxide radical scavenging and reducing power activities as well as the induction of heme oxygenase-1 (HO-1). The IC_{50} values against DPPH, radical were 10.6µg/ml, OH radical 2.14µg, superoxide radical 265.6µg/ml and total phenolics was 29.4 mg of GAE/g.

Millingtonia hortensis L. Family: Bignoniaceae

Leelapornpisid *et al.* (2007) reported antioxidant activity of absolutes (obtained by solvent extraction) of *Millingtonia hortensis*.

Mimosa hamata Willd. Family: Mimosaceae

Jain *et al.* (2009) evaluated the antioxidant activity of roots, stems and leaves of *in vivo* and *in vitro* plants using DPPH method. The extract of *in vivo* roots was found to be good scavenger of DPPH radicals (RS_{50} 5µg/ml) comparable to standards.

Singh *et al.* (2009) evaluated the antioxidant activity of dichloromethane and methanolic extracts of 12 arid zone medicinal plants. *Mimosa hamata* showed appreciable antioxidant activity and per cent of inhibition of DPPH where RC_{50} (µg/ml) was 9 (Dichloromethane extract) and 6.5 (MeOH extract). Per cent of inhibition of DPPH was 88.32 in methanol extract of *Mimosa hamata*. Key phytochemicals include gallic acid, mimonoside A-C, 4-ethylgallic acid.

Mimosa pudica L. Family: Mimosaceeae

The ethanol extract of *Mimosa pudica* leaves was evaluated for its hepatoprotective and antioxidant activities against CCl_4-induced liver damage, in wistar albino rats. The substantially elevated levels of serum GOT, GPT, ALP and total bilirubin, due to CCl_4 treatment, were restored towards near normal by *Mimosa pudica*, in a dose (Muthukumaran *et al.*, 2011).

Zhang *et al.* (2011) estimated total flavonoid and total phenolic contents of ethanol extracts and their antioxidant activity by DPPH and FRAP assays. The results showed that leaf extracts contained the highest amount of total flavonoids and total phenolics. The sequence of antioxidant activity of the ethanol extracts was as follows: leaf>the whole plant>seed>stem; the sequence of the 5 flavonoid monomers was as follows: 5,7,3′,4′-tetrahydroxy-6-C-[β-D-apiose(1→4)]- β-D-glycopyranosyl flavone (1)> isorientin (2) > orientin (3) > isovitexin (4) > vitexin (5) and the antioxidant activity of the compound 1 was equivalent to the synthetic antioxidant trolox or a bit stronger than trolox, and significant correlations were found among the active ingredient contents and the results of the antioxidant activity.

Suneetha *et al.* (2011) evaluated the hepatoprotective and antioxidant activity of *Mimosa pudica*. The enzyme levels of catalase and reduced GSH were significantly increased and malondialdehyde content significantly decreased in the group treated with *Mimosa pudica* at a dose of 400 mg/kg.

Mimosa verrucosa Benth. Family: Mimosaceae

The *in vitro* antioxidant and free radical scavenging properties of bark extract of *Mimosa verrucosa* used as anti-inflammatory agent in the Brazilian state of Bahis, was studied using different bioassays (Desmarchelier *et al.*, 1999). The total reactive antioxidant potential (TRAP) of the aqueous and methanolic extracts was determined by monitoring the intensity of luminal-enhanced chemiluminescence (CL), using 2,2′-azo-bis(2-amidinopropane) as a peroxyl radical source.The extract studied was active in this method.

Mimusops elengi L. Family: Sapotaceae

Antioxidant potential of the methanol extract of the leaves of *Mimusops elengi* was evaluated by Saha *et al.* (2008) by using DPPH scavenging assay, reducing power and total antioxidant capacity. The extract showed significant activities in all antioxidant assays compared to the reference antioxidant ascorbic acid and in a dose dependent manner. In DPPH scavenging assay the IC_{50} value of the extract was found to be 43.26µg/ml while the IC_{50} value of the reference standard ascorbic acid was 58.9µg/ml. Total antioxidant activity was also found to increase in a dose dependent manner. Moreover, *M. elengi* extract showed stronger reducing power.

Gayathri and Suresh (2010) evaluated the antioxidant potential of *Mimusops elengi* in alloxan-induced diabetic mice. Aqueous extract of *Mimusops elengi* (MEAQE) exhibited reducing power as well as DPPH and OH radical scavenging activity *in vitro*. MEAQE scavenged the DPPH radical in concentration dependent manner. The IC_{50} of MEAQE and ascorbic acid were found to be 23.66 and 8 µg/ml respectively. MEAQE possesses activity of scavenging hydroxyl free radical in concentration dependent manner. The IC_{50} on MEAQE and ascorbic acid were found to be 42.38 and 20.05µg/ml respectively.

Minuartia lineata Bornm. Family: Caryophyllaceae

Souri *et al.* (2004) evaluated the antioxidative activity of *Minuartia lineata* by linoleic acid peroxidation test using 1,3-diethyl-2-thiobarbituric acid as the reagent. Methanolic extract showed 96.16 per cent inhibition at 40µg concentration with IC_{50} value of 3.34µg.

Misodendrum punctulatum Banks ex DC. Family: Myzodendraceae

Aerial hemiparasitic plant endemic to temperate forest of Chile and Argentina that parasitize *Nothofagus*. The peroxyl radical scavenging activity of a dry methanol extract of misoolen drum punctulatum was determined by means of luminal enhanced chemiluminiscence assay, allowing to calculate the total reactive antioxidant potential (TRAP) index equal to 239 + 26µm expressed on Trolox equivalents. The flavon-3-ol catechin (1) and the phenylbutanone derivative myzodendrone (2) were identified through assay-guided fractionation as active metabolites present in the extract, and their structures were elucidated by chemical spectroscopic analysis. Compounds 1 and 2 were highly effective as free radical scavengers (TRAP =1257µm and 1018µm, respectively) when compared to Trolox (TRAP =144µm), used as a standard. Catechin (1) also reduced the production of thiobarbituric acid reactive substances (TBARS) in rat liver homogenates with IC_{50} =26µg/ml, superior to that obtained for Trolox IC_{50}= 73µg/ml.

Mitragyna rotundifolia (Roxb.) O.Kuntze Family: Rubiaceae

Methanol extracts of twig of *Mitragyna rotundifolia* showed good activity in dose dependent study, *i.e.*, 50 per cent reduction of DPPH is achieved at the dose of 0.03mg/ml that is comparable to the standards (which are pure molecules), *viz.*, catechin and Trolox (Kshirsagar and Upadhyay, 2009). Moreover, their activities were better than that of curcumin for which 50 per cent reduction was achieved by 0.04 mg/ml.

Different solvent extracts of leaves and bark of *Mitragyna rotundifolia* were evaluated by Kang *et al.* (2009) by DPPH, ABTS, and FRAP assays, respectively for antioxidant properties. Polar solvent extracts (n-butanol and ethyl acetate) had relatively higher antioxidant activity than nonpolar solvent extracts (petroleum ether). The n-butanol extract also exhibited a higher phenolic and flavonoid content than the other solvent extracts did. Two phenolic and four flavonoid compounds were isolated from the ethyl acetate leaf extracts. Compound 5 showed highest antioxidant activity.

Mollugo nudicaulis Lam. Family: Molluginaceae

Singh *et al.* (2009) evaluated the antioxidant activity of dichloromethane and methanolic extracts of 12 arid zone medicinal plants. *Mollugo nudicaulis* showed appreciable antioxidant activity and per cent of inhibition of DPPH where RC_{50} (µg/ml) was 15.5 (Dichloromethane extract) and 6.5 (MeOH extract). per cent of inhibition of DPPH was 90.82 per cent in methanol extract of *Mollugo nudicaulis*. Key phytochemicals include cyanogenic glycosides, terpenes, saponins.

Mollugo pentaphylla L. Family: Molluginaceae

Maharana *et al.* (2010) evaluated *in vitro* antioxidant activity of aqueous extract of aerial parts of *Mollugo pentaphylla*. The extract showed the total antioxidant activity of 98.66 mg ascorbic acid equivalent/g as compared to 117.83 mg of the reference standard ascorbic acid. The reducing power of the extract was found to be significant and in a concentration dependent manner. The extract showed 96.5µg/ml for DPPH radical, 381.4µg/ml for superoxide radical, 432.7µg/ml for H_2O_2 radical and 247.5µg/ml for nitric oxide radical.

Momordica charantia L. Family: Cucurbitaceae

English Name: Bitter gourd; Japanese: Nigauri.

Scartezzini and Speroni (2000) are of the opinion that this herb contains antioxidant principles that can explain and justify their use in traditional medicine in the past as well as the present.

A significant decrease in malondialdehyde (MDA) levels and a significant increase in antioxidant activity were observed by Chandra *et al.* (2003) when *Momordica charantia* was administered to diabetic rats.

Sabu and Kuttan (2003) investigated the antioxidant activity of Indian herbal drugs, including *Momordica charantia*, in rats with alloxan-induced diabetes. The extract of *Momordica charantia* showed antidiabetic and antioxidant activity.

Naik *et al.* (2003) evaluated the antioxidant activity of aqueous extract of *Momordica charantia* for its potential as antioxidant. The antioxidant activity of these extracts was tested by studying the inhibition of radiation induced lipid peroxidation in liver microsomes at different doses in the rages of 100-600 Gy as estimated by TBARS.

Ansari *et al.* (2005) evaluated the antioxidant activity of five vegetables traditionally consumed by South-Asian migrants in Bradford, Yorkshire, UK. Extracts from Bitter gourd showed a significant difference in the FRSA between the extract obtained by using cold maceration and that prepared by boiling the plant in the solvent under reflux, suggesting the chemical composition of the plant changed during heating process, leading to an increase in the amount of antioxidant components.

Chanwitheesuk *et al.* (2005) investigated the antioxidant activity of *Momordica charantia* using a β-carotene bleaching method. The contents of plant chemicals such as vitamin C, Vitamin E, carotenoids, tannin and total phenolics, were also determined. Methanolic extract showed antioxidant activity. *Momordica charancita* has antioxidant index of 1.49, Vitamin C 13.8 mg per cent, Vitamin E 0.0024mg per cent, Total carotenes 1.31 mg per cent, Total xanthophylls 0.54mg per cent, Tannins 4.48mg per cent and Total phenolics of 37.0 mg per cent.

Bajpai *et al.* (2005) investigated the phenolic contents and antioxidant activity of some food and medicinal plants including *Momordica charantia*. The fruits, leaves, stem and seeds were found to have low phenolic contents and low antioxidant activity. The TPC of fruits, leaves, stem and seeds of *Momordica charantia* were 5.8, 15.9, 9.1 and

4.8 mg/g GAE, respectively. The antioxidant activity of *Momordica charantia*, as assayed by auto-oxidation of β-carotene and linoleic acid and expressed as the per cent of inhibition relative to the control were, 44.3 per cent, 48.1 per cent, 35.9 per cent and 20.1 per cent for fruits, leaves, stem and seeds respectively.

Maeda *et al.* (2006) examined the phenolic contents of bitter gourd and its radical scavenging activities for DPPH radical. The polyphenol content was 35.9 mg-gallic acid equivalent/100 g) and DPPH scavenging activity was 174.7 μmol-Trolox eq./ 100 g).

Liu *et al.* (2010) isolated antioxidant triterpenoids from the stems of *Momordica charantia*. These three new compounds displayed ABTS radical cation scavenging activity with IC_{50} values of 268.5, 352.1 and 458.9μM, respecively and an inhibitory effect of xanthine oxidase (XO) activity with IC_{50} values of 142.3, 36.8 and 124.9 μM, respectively.

Saeed *et al.* (2010) evlauted the antioxidant activity of *Momordica charantia*. The results obtained from DPPH assay and reduing power activity indicated that bitter gourd extracts exhibits potent antioxidant activity. Bitter gourd's flakes extract possess potent free radical scavenging activities (63.20 per cent) followed by seed (33.05 per cent), DPPH (per cent inhibition) at 2 mg/ml concentration.

Biswas *et al.* (2010) analyzed antioxidant activity of the aqueous extract of leaves of *Momordica charantia* in different systems of assay. Aqueous extract showed superoxide radical scavenging activity with IC_{50} = 478.29μg/ml; Hydroxyl radical scavenging IC_{50} = 713.45μg.ml; Lipid peroxidation IC_{50} = 1615.18μg/ml.

A study was carried out by Raghu *et al.* (2010) to determine the antioxidant activity of aqueous extracts of ten selected common vegetables. For *Momordica charantia* the IC_{50} value for DPPH scavenging was 1.84mg/ml while standard was 2.45μg/ml. Reducing power of the extract was moderate.

Momordica cochinchinensis Lour. Family: Cucurbitaceae

Gan *et al.* (2010) evaluated the antioxidant activity of 40 medicinal plants. *Momordica charantia* showed very low antioxidant activity of 6.16 μmol Fe(II)/g in FRAP assay and 6.15μmol in TEAC assay. Total phenolic content was also very low (0.87 mg GAE/g).

Momordica dioica Roxb. Family: Cucurbitaceae

The alcoholic extract of *Momordica dioica* roots significantly reduced CCl_4 induced hepatotoxicity in rats upon oral administration (200mg/kg), as judged from the serum enzyme levels. The alcoholic extract inhibited the formation of oxygen derived free radicals (ODFR) *in vitro* with 4000μg/ml ascorbic system (Shreedhara and Vaidya, 2006).

Jain *et al.* (2008) evaluated the antioxidant and hepatoprotective activity of ethanolic and aqueous extracts of *Momordica dioica* leaves. The *in vivo* antioxidant and free radical scavenging activities were screened which were positive for both ethanolic and aqueous extracts.

Momordica grosvenori Swingle Family: Cucurbitaceae

Song *et al.* (2007) evaluated the effect of *Momordica grosvenori* (MG) on oxidative stress pathways in renal mitochondria of normal and alloxan-induced diabetic mice and involvement of heme oxygenase-1. Low dose of MG attenuated diabetic nephropathy symptoms partially, inhibited lipid peroxidation, up-regulated HO-1 and Mn-SOD mRNA expression, and increased HO-1 activity.

Antioxidant capacities of 56 selected Chinese medicinal plants, including *Momordica grosvenori* were evaluated by Song *et al.* (2010) using the TEAC and FRAP assays, and their total phenolic content was measured by the Folin-Ciocaleteu method. *Lepidium apetalum* showed moderate antioxidant activity with TEAC value of 63.17µmol Trolox/g, FRAP value of 41.28µmol Fe^{2+}/g and phenolic content was 12.22mg GAE/g.

Monechma ciliatum (Jacq.) Milne-Redhead. Family: Acanthaceae

English Name: Black mahlab.

Mariod *et al.* (2010) investigated the antioxidant activities of phenolic rich fractions (PRFs) from crude methanolic extract (CME) and its fractions using ethyl acetate, hexane and water of black mahlab seed cake. The antioxidant activity determined by the DPPH method revealed that black mahlab PRFs had the highest antioxidant activity.

Moricandia arvesnis (L.) DC. Family: Brassicaceae (Cruciferae)

Caffeic acid derivative 7 and the quercetin triglycoside 2 isolated from the leaves of *Moricandia arvensis* proved to possess the most potent scavenging activity (Braham *et al.*, 2005). Skandrani *et al.* (2010) reported that petroleum ether and methanol leaf extracts from *Moricandia arvensis* promote antiproliferation of human cancer cells, induce apoptosis, and enhance antioxidant activity. Using lipid peroxidation inhibitory assay, the antioxidant capacity of *M. arvensis* extracts was evaluated by the ability of each extract to inhibit malondialdehyde formation. It was revealed that ethyl acetate extract and total aligomer flavonoids extracts are the most active in scavenging the hydroxyl radicals. These two extracts showed important free radical scavenging activity toward the DPPH radical whereas the chloroform extract exhibited the highest TEAC against ABTS+ radical (Skandrani *et al.*, 2007).

Morinda citrifolia L. Family: Rubiaceae

English Name: Noni.

Research by Zin *et al.* (2002) has revealed the potential of *Morinda citrifolia* as a source of various antioxidants in roots, fruits and leaves. Ramamoorthy and Bono (2007) studied antioxidant activity, total phenolic and flavonoid content of *Morinda citrifolia* fruit extracts from various extraction processes. The results of this study showed that the highest antioxidant activity, total phenolic content and total flavonoid content were exhibited by the extracts obtained by high pressure extraction with ethyl acetate as solvent. It has been reported that *M. citrifolia* fruit contains relatively larger quantity of non-polar antioxidant compounds. However, among the ethylacetate extracts spray dried extract exhibited highest antioxidant activity and total flavonoid content and vacuum dried, the highest total phenolic content. Ramamoorthy and

Bono (2007) are of the opinion that the polarity of the solvent had significant impact on the extraction of phytochemicals such as antioxidants, phenolics and flavonoids from the fruit matrices of *Morinda citrifolia*.

Root bark of *Morinda citrifolia* was successively extracted with hexane and methanol. From the hexane extract, β-sitosterol and fatty acid mixture was isolated and palmitic, steric, oleic and linolic acid mixture were identified by GLC. From the methanol extract anthraquinones were isolated by TLC/HPLC. Details of characterization of individual components and their antioxidant activities were discussed by Singh *et al.* (2004).

Compounds isolated from *Morinda citrifolia* were screened by Su *et al.* (2005) for antioxidant activity in terms of DPPH and ONOO bioassays. The neolignan, americanin A, was found to be a potent antioxidant in these assays.

Noni (*Morinda citrifolia*) juice has demonstrated antioxidant activity *in vitro* and *in vivo*. To test this activity in humans noni juice from Tahiti was evaluated by Wang *et al.* (2009) in a 30 day double blind placebo to controlled clinical trial with 285 current heavy smokers. The results suggest an antioxidant activity from noni juice in humans exposed to tobacco smoke.

Morinda lucida Benth. Family: Rubiaceae

Akinmoladun *et al.* (2010) assessed the phytochemical constituents and antioxidant and free radical scavenging activities of *Morinda lucida* using seven different antioxidant assay methods. The results suggest that the methanolic extracts possess significant antioxidant and radical scavenging activity that may be due to the phytochemical content of the plant.

Morinda officinalis How. Family: Rubiaceae

Soon and Tan (2002) evaluated the hypoglycemic and antioxidant activities of *Morinda officinalis* in streptozotocin-induceed diabetic rats. The 10-day oral administration of the extract reduced the fasting serum glucose, hepatic and renal TBARS level and significantly increased the hepatic SOD and CAT acivities as well as GSH levels.

Wu *et al.* (2006) evaluated the antioxidant activity of *Morinda officinalis*. It was shown that the extraction from *Morinda officinalis* could obviously scavenge superoxide anion and hydroxyl radicals.

Moringa oleifera Lam. (Syn.: *Moringa pterigosperma*) Family: Moringaceae

Water, aqueous methanol and aqueous ethanol extract of freeze-dried leaves of *Moringa oleifera* from different agroclimatic regions were examined by Sidduraju and Becker (2003) for radical scavenging capacities and antioxidant activities. All leaf extracts were capable of scavenging peroxyl and superoxyl radicals. Similar scavenging activities for different solvent extracts of each collection were found for the stable DPPH radical. Among the three different moringa samples, both methanol and ethanol extracts of Indian origins showed the highest antioxidant activities, 65.1 and 66.8 per cent, respectively, in the β-carotene-linoleic acid system. Nonetheless, increasing concentration of all the extracts had significantly increased reducing power, which may in part be responsible for their antioxidant activity. The major bioactive compounds of phenolics were found to be flavonoid groups such as quercetin and kaempferol.

Bajpai *et al*. (2005) investigated the phenolic contents and antioxidant activity of some food and medicinal plants including *Moringa oleifera*. The leaves were found to have moderate phenolic contents and antioxidant activity. The TPC of leaves of *Moringa oleifera* was 20.9 mg/g GAE, respectively. The antioxidant activity of *Moringa oleifera*, as assayed by auto-oxidation of β-carotene and linoleic acid and expressed as the per cent of inhibition relative to the control was, 79.6 per cent for leaves.

Phytochemical flavonols, carotenoids and the antioxidant capacity of leaves of *Moringa oleifera* was evaluated by Lako *et al*. (2006). *Moringa oleifera* leaves have moderate Total antioxidant capacity (260 mg/100g) and are rich in Total polyphenol content (TPP) (290mg/100g), quercetin (100mg/100g), kaempferol (34 mg/100g) and β-carotene (34mg/100g).

Wong *et al*. (2006) investigated the antioxidant properties of 25 edible tropical plants including *Moringa oleifera* (=*M.pterigosperma*) using DPPH and FRAP assays. Total phenolic content was also estimated. *Moringa oleifera* showed moderate antioxidant activity.

The antioxidant activity of *Moringa oleifera* was examined by Charde *et al*. (2009) by DPPH radical scavenging method. The different diluted extract showed antioxidant activity in range of 16.24 per cent to 32.78 per cent.

Moringa peregrina (Forssk.) Fiori Family: Moringaceae

Marwah *et al*. (2007) investigated the antioxidant activity of wound-healing herbs using *in vitro* DPPH and phosphomolybdenum assay methods. Out of the 19 plants screened, the aqueous ethanol extract of *Moringa peregrina* showed the best DPPH scavenging activities with per cent IP value 87.8 per cent, after 15 min of incubation at an effective test concentration of 50µg/ml. The Total Antioxidant Capacity as gallic acid equivalent of 814mg/g of ethanol extract was obtained in the phosphomolybdenum assay. Total phenolic content was 454 mg GAE/g of ethanol extract.

Morus alba L. Family: Moraceae

Aqueous extract of *Morus alba* inhibited $FeSO_4/H_2O_2$ induced lipid peroxidation in rat liver homogenate and protected against t-butyl hydroperoxide caused Ac2F cell damage (Cheong *et al*., 1998).

Nine flavonoids were isolated from the leaves of *Morus alba* (Kim *et al*., 1999). Quercetin-3-O-beta-D-glucopyranosyl-(1→6)-beta-D-glucopyranoside and quercetin exhibited significant radical scavenging effect on DPPH radical (Kim *et al*., 1999).

Chemical investigation of the EtOH extract of *Morus alba* as guided by free radical scavenging activity, furnished 5,7-dihydroxycoumarin 7-methyl ether (1), two prenylflavones, cudraflavone B (2) and cudraflavone C (3) and oxyresveratrol (4). Compound 1 and 4 showed superoxide scavenigng effects with the IC_{50} values of

19.1 and 3.81µM respectively. Compound 4 exhibited a DPPH free radical scavenging effect (IC_{50} = 23.4µM).

Chang *et al.* (2002) evaluated the antioxidant activity and free radical scavenging capacity of Korean medicinal plants using commonly accepted assays. They were extracted with dichloromethane, methanol or ethanol, respectively and selected for the best antioxidant results. Each sample under assay condition showed a dose dependent free radical scavenging effect of DPPH and a dose-dependent inhibitory effect of xanthine oxidase and lipid peroxidation. Among plant extracts the root bark of *Morus alba* showed stronger SC_{50} and ID_{50} values than other plant extracts.

Cai *et al.* (2004, 2006) evaluated the antioxidant activity and phenolic compounds in traditional Chinese medicinal plants associated with anticancer. The improved ABTS.+ method was used to systematically assess the total antioxidant capacity (Trolox equivalent antioxidant capacity, TEAC) of the medicinal extracts. The TEAC values and total phenolic content for methanolic extract of fruit of *Morus alba* were 132.6 µmol Trolox equivalent/100g dry weight (DW), and 1.96 g of gallic acid equivalent/100g DW respectively. A positive significant linear relationship between antioxidant activity and total phenolic content showed that phenolic compounds were the dominant antioxidant components in the tested herbs. Major types of phenolic compounds in *Morus alba* include flavonols (rutin), stilbenes (oxyresveratrol).

Extracts of *Morus alba* leaf showed remarkable free radical scavenging activity with IC_{50} value of 13.6µg/ml and significant inhibitory effects on lipid peroxidation with 78.9 per cent inhibition at the concentration of 50ng/ml (Thoung *et al.*, 2006).

Four known flavonoids were isolated from mulberry leaf and their structures were identified as kaempferol-3-O-beta-D-glucopyranoside (1), quercetin-3-O-beta-D-glucopyranoside (2), quercitrin(3) and morin-3-O-beta-D-glucopyranoside (4). All the compounds showed DPPH and ABTS radicals scavenging activity (Jiang *et al.*, 2008).

Serteser *et al.* (2009) studied the antioxidant activity of 50 per cent aqueous methanol extracts by various antioxidant assays, including free radical scavenging, hydrogen peroxide scavenging and metal (Fe^{2+}) chelating activities. The DPPH radical scavenging effect of fruit extract was 0.890. Fe^{2+} chelating activity (per cent) of fruit extract was 49.98 and Hydrogen peroxide inhibition activity of fruit extract was 66.43 per cent.

Combined far-infrared radiation with hot-air convection (FIR-HA) drying was used by Wanyo *et al.* (2010) for improving colour and antioxidant property of mulberry leaf. Their results have demonstrated that FIR-HA should be considered as a suitable drying method for mulberry tea with respect to preserving its antioxidant properties and phenolic compounds.

Antinephritic activity of morusin isolated from *Morus alba* was evaluated in mice with glomerular disease. ESR spectroscopy demonstrated that morusin increased the radical intensity of sodium ascorbate by about two times. Morusin showed weak scavenging activity against superoxide anion radical.

Gan *et al*. (2010) evaluated the antioxidant activity of *Morus alba* using the FRAP and TEAC assays, and its total phenolic content was measured by the Folin-Ciocalteu method. For FRAP assay the value was 46.96 µmol Fe(II)/g dry weight. For TEAC assay, the value was 33.03µmol Trolox/g dry weight. Total phenolic content was 1.96 mg gallic acid equivalent/g dry weight of plant material.

Povichit *et al*. (2010) investigated phenolic content and free radical scavenging effect of *Morus alba* by DPPH and ABTS assays, anti-lipid peroxidation activity by TBARS and for antiglycation activity. The results revealed that the total phenolic content showed good correlation with free radical scavenging by ABTS and anti-lipid peroxidation by TBARS, but showed no correlation with antiglycation. The extract of *Morus alba* demonstrated a moderate antioxidant effect and also showed a moderate antiglycation effect. The IC_{50} were 0.85mg/ml for the DPPH method, TEAC value of 0.10 mg/Trolox/mg sample for the ABTS method, IC_{50} of 2.60 mg/ml for the TBARS method and IC_{50} of 0.62 for the antiglycation method.

Vadlamudi and Naidu (2010) evaluated the antioxidant activity of ethanolic extracts of *Morus alba* leaves. SOD activity was 1.5U/mg, Catalase activity was 1.2U/mg while vitamin C content was 50 mg/100g.

Antioxidant capacities of 56 selected Chinese medicinal plants, including *Morus alba* were evaluated by Song *et al*. (2010) using the TEAC and FRAP assays, and their total phenolic content was measured by the Folin-Ciocaleteu method. Bark of root of *Morus alba* showed moderate antioxidant activity with TEAC value of 67.22µmol Trolox/g, FRAP value of 21.67µmol Fe^{2+}/g and phenolic content was 5.34mg GAE/g while leaf showed 74.19µmol Trolox/g of TEAC, 65.79 µmol Fe^{2+}/g of FRAP and 10.94 mg of phenolic content.

Morus bombycis Koidzumi Family: Moraceae

Jin *et al*. (2005) investigated the antioxidant activity and liver protective effect of *Morus bombycis*. Aqueous extract had higher superoxide scavenging activity than other types of extracts. The aqueous extract at a dose of 100mg/kg showed significant hepatoprotective activity when compared with that of a standard agent. The water extract recovered the CCl_4-induced liver injury and showed antioxidant effects in assays of FeCl(2)-ascorbic acid-induced lipid peroxidation in rats.

Jin *et al*. (2006) investigated hepatoprotective activity and antioxidant effect of the 2,5-dihydroxy-4,3'-bi(beta-D-glucopyranosyloxy)-trans-stilbene that purified from *Morus bombycis* roots against CCl_4-induced damage in roots. The compound displayed dose-dependent superoxide radical scavenging activity (IC_{50} =430.2µg/ml).

Morus macroura Miq. Family: Moraceae

Fractionation of the ethanolic extract of the stem bark of *Morus macroura* resulted in the isolation of four new Diels-Alder type adducts, named guangsangons K-N, together with two known compounds, muberofuran G and K (Dai *et al*., 2004). The isolated new compounds showed antioxidant activity *in vitro*, with the inhibitory rates of MDA being from 91.8 to 100.0 per cent at concentrations of 10-5 mol/l.

Morus mongolica (Bureau) Schneid. Family: Moraceae

Mongolicin B, mongolicin F and kuwanon O isolated from the stem and root bark of *Morus mongolica* were active as antioxidant inhibition of liver microsomal lipid peroxidation induced by Fe^{2+} Cys system with inhibitory ratios of 83.6 per cent, 86 per cent and 63 per cent at a concentration of 10-5 mol/l (Kang *et al.*, 2006).

Morus nigra L. Family: Moraceae

English Name: Black mulberry.

The antioxidant effect of three different extracts of *Morus nigra* fruit (fruit juice, hydroalcoholic and polyphenolic) on haemoglobin glycosylation, peroxidative damage to human erythrocytes, liver hepatocytes of rats and human low-density lipoprotein (LDL) was studied by Naderi *et al.* (2004). The results showed that all three extracts inhibited haemoglobin glycosylation induced by glucose to differing degrees, haemolysis of human erythrocytes induced by hydrogen peroxide was also inhibited. The production of malondialdehyde peroxidative damage to plasma membranes of isolated rat hepatocytes induced by tert-butyl hydroperoxide (tBH) was also studied. Inhibition of lipid peroxidation of LDL induced by copper (II) ion was achieved during this study (Naderi *et al.*, 2004). The results suggest that *Morus nigra* fruit has a protective action against peroxidative damage to biomembranes and biomolecules.

Serteser *et al.* (2009) studied the antioxidant activity of 50 per cent aqueous methanol extracts by various antioxidant assays, including free radical scavenging, hydrogen peroxide scavenging and metal (Fe^{2+}) chelating activities. The DPPH radical scavenging effect of fruit extract was 0.825. Fe^{2+} chelating activity (per cent) of fruit extract is 49.98 and Hydrogen peroxide inhibition activity of fruit extract was 66.21 per cent.

Al-Mustafa and Al-Thunibat (2008) evaluated the antioxidant activity of Jordanian medicinal plant *Morus nigra* leaves. The level of antioxidant activity was determined by DPPH and ABTS assay in relation to total phenolic contents of the medicinally used part. They concluded that *Morus nigra* leaves have moderate antioxidant capacity (DPPH-TEAC range 20-80mg/g).

Özgen *et al.* (2009) investigated the phytochemical and antioxidant properties of *Morus nigra* fruits. The average Total Phenol Contents of fruits are 2737µg GAE/g fw. *Morus nigra* had the richest amount of anthocyanin with an average of 571µg cy-3-glu/g fw. They found that FRAP, TEAC, TPC and TMA were significantly correlated with each other. Fructose (5.27g/100ml) and glucose (5.81 g/100ml) were determined to be major sugars. *M. nigra* displayed a higher TA (2.05g/100ml) with citric acid as the major acid.

Antioxidant potential and radical scavenging activity of the various organic and aqueous extracts of leaves, stem and fruit of *Morus nigra* were investigated by Iqbal *et al.* (2010) using ABTS decolourization assay, FRAP assay, DPPH assay, total phenolic content (TPC) assay and total antioxidant activity. Using TPC assay the amount of total phenolics ranged from 1.3048 to 5.8287, 0.413 to 2.9394 and 0.5346 to 6.2266 mg/l for different fractions of leaves, stem and fruit of *Morus nigra* respectively.

Good correlation was found between TPCs and TEAC values determined by ABTS+ decolouration assay and FRAP assay. Employing total antioxidant assay using linoleic acid emulsion sysm, methanol, 1-butanol, aqueous and ethyl acetate fractions of all the parts showed strong peroxyl radicals scavenging activity.

Morus rubra L. Family: Moraceae

English Name: Red mulberry.

Özgen *et al.* (2009) investigated the phytochemical and antioxidant properties of *Morus rubra* fruits. The average total phenol contents of fruits were 1603µg GAE/g fw. They found that FRAP, TEAC, TPC and TMA were significantly correlated with each other. Fructose (5.27g/100ml) and glucose (5.81 g/100ml) were determined to be major sugars. *M. rubra* displayed good TA (0.78g/100ml) with citric acid as the major acid.

Mosla chinensis Maxim. Family: Lamiaceae (Labiatae)

Cao *et al.* (2009) analyzed the essential oil composition, antimicrobial and antioxidant properties of *Mosla chinensis*. The essential oil showed significantly higher antioxidant activity than that of the methanol extract.

Mouriri pusa Gardner Family: Melastomataceae

Andreo *et al.* (2006) reported in *Mouriri pusa* and several flavonoids and tannins which were attributed the antioxidant activity.

Mucuna pruriens (L.) DC. Family: Fabaceae (Papilionaceae)

The antioxidant activities of a methanolic extract of mucuna beans (*Mucuna pruriens* var. *utilis*) and several non-protein amino/imino acids, namely L-3,4-dihydroxyphenylalanine (L-dopa), L-3-carboxy-6,7-dihydroxy-1,2,3,4-tetrahydroisoquinoline (compound I), (")-1-methyl-3-carboxy-6,7-dihydroxy-1,2,3,4-tetrahydroisoquinoline (compound II) and 5-hydroxytryptophan (5-HTP), were evaluated. By virtue of their hydrogen-donating ability, all the tested compounds and the mucuna seed extract showed excellent reducing power, with the highest values being recorded for L-dopa in a dose-dependent manner. Similarly, as compared with synthetic antioxidants (BHT and BHA) and quercetin, all the tested compounds and the seed extract were found to be more potent in free radical-scavenging activity against DPPH radicals. Hydroxyl radicals (OH$^{•}$) and superoxide anion radicals (O$_2^{•-}$) were effectively scavenged by the tested compounds, with the exception that no scavenging activity of 5-HTP was observed on (O$_2^{•-}$) up to a concentration of 2 mg ml^{-1}, as was also the case for BHA. Among the tested non-protein amino/imino acids and seed extract the highest peroxidation-inhibiting activity (95 per cent) was recorded for 5-HTP. On the other hand, in the linoleic acid/β-carotene-bleaching system, L-dopa, compound I and compound II acted as pro-oxidants, whereas the seed extract showed only weak antioxidant activity as in the linoleic acid emulsion system (Sidduraju and Becker, 2003).

Rajeswar *et al.* (2005) reported that methanol extract of *Mucuna pruriens* seeds at 10-320µg/ml concentration exhibited strong superoxide anion radical scavenging activity which was equal to quercetin standard.

Parkinson's disease is a neurodegenerative disorder for which no neurorestorative therapeutic treatment is currently available. Oxidative stress plays an important role in the pathophysiology of Parkinson's disease. The ancient Indian medical system traditionally uses *Mucuna pruriens* to treat Parkinson's disease. The antioxidant activity of *Mucuna pruriens* was demonstrated by its ability to scavenge DPPH radicals, ABTS radicals and reactive oxygen species. *Mucuna pruriens* significantly inhibited the oxidation of lipids and deoxyribose sugar. *Mucuna pruriens* exhibited divalent iron chelating activity and did not show genotoxic/mutagenic effect on the plasmid DNA. These results suggest that the neuroprotective and neurorestorative effect of *Mucuna pruriens* may be related to its antioxidant activity independent of the symptomatic effect. In addition, the drug appears to be therapeutically safe in the treatment of patients with Parkinson's disease (Dhanasekaran *et al.* (2008).

Mukia maderaspatana (L.) M.Roem. Family: Cucurbitaceae

Petrus *et al.* (2011) evaluated the antioxidant activity of fresh leaves of *Mukia maderaspatana*. The leaves extract exhibited potent *in vitro* antioxidant/radical scavenging (ABTS and FRAP assays) and metal (Ferrous ion) chelating activities and also inhibited β-carotene bleaching. Saponin has been isolated and determined as the major phenolic antioxidant. The amount of phenolics (292.392 mg GAE), flavonoid (247.079 QE), carotenoids (0.812 mg), vitamins C (17.046 mg) and E (0.194 mg) and saponin (220.800mg) present and the antioxidant capacities (301.926 mg vitamin C equivalent antioxidant capacity) are reported for 100g of fresh leaves.

Muntingia calabura L. Family: Elaeocarpaceae

Zakaria (2007) evaluated the free radical scavenging properties of *Muntingia calabura* using DPPH and superoxide anion radical scavenging assays. The extract was found to show remarkable antioxidant activity in both assays. The DPPH radical scavenging activity was 94.80 per cent and superoxide scavenging activity was 83.70 per cent. Phytochemicals screening of this plant demonstrated the presence of flavonols, triterpenes, tannins, saponins and steroids.

Murraya koenigii (L.) Spreng Family: Rutaceae

English Name: Curry leaf.

Khan *et al.* (1996, 1997) studied the antioxidant effects of *Murraya koenigii* leaves fed with high fat diet. The study revealed that this species alter the peroxidation (TBA reactive substances) level to a beneficial extent. Concentration of malondialdehyde showed a significant decrease, while hydroperoxides and conjugated dienes were significantly increased in liver and heart of both the experimental groups. SOD and Catalase activity was found to be increased in liver and heart and kidney were lowered in rats administered extract of the plants of this species. Glutathione reductase, glutathione peroxidase and glutathione S-transferase activity showed a sharp increase in the experimental groups compared with controls.

Sabu and Kuttan (2003) investigated the antioxidant activity of Indian herbal drugs, including *Murraya koenigii*, in rats with alloxan-induced diabetes. The extract of *Murraya koenigii* showed antidiabetic and antioxidant activity.

Wong *et al.* (2006) investigated the antioxidant properties of 25 edible tropical plants including *Murraya koenigii* using DPPH and FRAP assays. Total phenolic content was also estimated. *Murraya koenigii* showed good antioxidant activity.

The *in vitro* antioxidant properties of different extracts (water, alcohol, alcohol-water, hexane or chloroform extract) of curry leaves were evaluated by Ningappa *et al.* (2008). The alcohol:water (1:1) extract of curry leaves showed the highest antiodixant and free radical scavenging activity. It inhibited membrane lipid peroxidation by 76 per cent, at 50µg/ml, scavenged 93 per cent of superoxides at 200µg/3ml and scavenged approximately 90 per cent of hydroxyl and DPPH radicals at 4-5-fold lower concentrations compared to the other tested extracts.

Huda-Faujan *et al.* (2007, 2009) reported antioxidative activities of aqueous and methanolic extracts of *Murraya koenigii* and four other plants. The analysis carried out was total phenolic content, ferric reducing power, ferric thiocyanate (FTC) and TBA tests. The phenolic content was 24.62 mg (aqueous extract) and 38.60 mg (methanolic extract) TAE/100 g fresh weight and antioxidant activity of 67.67 (aqueous extract) and 70.60 per cent (methanolic extract) using FTC method. Increasing the concentration of the extracts resulted in increased ferric reducing antioxidant power. Total phenolic content has positive correlation with antioxidant capacity.

Mathur *et al.* (2010) carried out phytochemical investigation and antioxidant activity of *Murraya koenigii*. Traditional solvent extraction (TSE) using four solvents – methanol, aqueous, hexane and petroleum ether were utilized to determine the content of antioxidants. DPPH assay and superoxide anion radical scavenging activity were used to determine the antioxidant capacity. The most antioxidant capacity was achieved using methanol as the solvent followed by aqueous extracts. The antioxidant capacity with in the specific extract of plant was found to be correlated with the total phenolic content. The radical scavenging activity by DPPH radical scavenging method (IC_{50}) was 50.07µg/ml (hexane extract), 52.18 µg/ml (petroleum ether extract), 43.10 (aqueous) and 46.00 (methanol extract).

Musa paradisiaca L. Family: Musaceae

English Name: Banana.

The total antioxidant activity of banana was measured by Wang *et al.* (1996) using automated oxygen radical absorbance capacity (ORAC) assay. On the basis of the wet weight of the fruits (edible portion), banana had moderate ORAC activity (2.21micromoles of Trolox equivalents per gram). On the basis of dry weight of the fruits banana had the highest ORAC activity (9.0µmoles/gTrolox equivalent). The contribution of the fruit pulp fraction (extracted with acetone) to the total ORAC activity of a fruit was usually less than 10 per cent.

The antioxidant activity of flavonoids from banana (*Musa paradisiaca*) was studied in rats fed normal as well as high fat diets (Vijayakumar *et al.*, 2008). Concentrations of peroxidation products namely malondialdehyde, hydroperoxides and conjugated dienes were significantly decreased whereas the activities of catalase and superoxide dismutase were enhanced significantly. Concentrations of glutathione were also elevated in the treated animals.

The antioxidant capacity and phenol content of local cutivar of banana (known as *pisang mas*) (*Musa paradisiaca*) was studied by Alothman *et al.* (2009). Three solvent systems were used (methanol, ethanol and acetone) at three different concentrations (50, 70 and 90 per cent) and with 100 per cent distilled water. The antioxidant capacity of the fruit extracts was evaluated using a ferric reducing/antioxidant power assay and the free radical-scavenging capacity was evaluated using 2,2-diphenyl-1-picrylhydrazyl radical-scavenging assays. The polyphenol content was 24.4 to 72.2 Gallic Acid Equivalent (GAE)/100g. High phenol content was significantly correlated with high antioxidant capacity.

Faller and Fialho (2010) evaluated the polyphenol content and antioxidant capacity in organic and conventional plant foods. The results showed that organic fruits tend to have higher hydrolysable polyphenol contents than conventional ones. Fruit peels also showed higher concentration of polyphenols than pulp, reaching, for bananas twice the amount found in pulps, which reflected in higher antioxidant capacity.

Mutisia friesiana Cabrera Family: Asteraceae (Compositae)

The water-soluble extract of *Mutisia friesiana* showed radical scavenging activity in the DPPH discoloration assay (Viturro *et al.*, 1999). Bioassay-guided isolation led to caffeic acid derivatives and flavonoids as the main active compounds. After hydrolysis, caffeic acid and quercetin proved to be the bioactive principles.

Myracrodendruon urundeuva M.Allemao Family: Anacardiaceae

Tannin-enriched fraction isolated from the stem bark of *Myracrodendruon urundeuva* showed mitochondrial protection and antioxidant activity in neuronal cells (Nobre Junior *et al.*, 2007).

Myrianthus arboreus P.Beauv. Fanily: Urticaceae

Odukoya *et al.* (2007) evaluated the antioxidant activity and total phenolic contents of Nigerian green leafy vegetables including *Myrianthus arboreus*.

Biapa *et al.* (2007) investigated the phytochemical and antioxidant properties of four medicinal plants used in Cameroon, including *Myrianthus arboreus*. Four extracts (methanol, hydro-ethanolic, aqueous and hydrolysed) from each of the plants were prepared and analysed. Folin Ciocalteu, FRAP and DPPH assays were used to determine the antioxidant potential. The highest antioxidant activites were observed in the hydrolysed extracts of each plant, while the aqueous extract showed the least activity irrespective of the method used.

Myrica gale L. Family: Myricaceae

Antioxidant and radical scavenging effects of a diethyl ether extract of the fruit exudates of *Myrica gale* and of C-methylated dihydrochalcones isolated from it were studied by Mathiesen *et al.* (1995). The main antioxidant of the extract myriagalone B (MyB), inhibited lipid peroxidation in hepatocytes with an IC_{50} value of 23µM, whereas in mitochondria the value was 5.2µM. The fruit extract itself inhibited peroxidation in hepatocytes with an IC_{50} value of 7.0µM calculated according to its MyB content, and in mitochondria with and IC_{50} of 1.7µM. Other myricagalones were considerably less active as antioxidants Both MyB and the fruit extract caused scavenging of the DPPH radical with IC_{50} value of 32µM and 14µM, respectively. Peroxidation in linoleic acid catalyzed by soybean 15-lipoxygenase was inhibited by MyB (IC_{50} + 23µM calculated as MyB; corresponding to an extract concentration of 71µg/ml. However, the extract content of myricagalone A, itself a fairly potent inhibitor of 15-lipoxygenase, may contribute significantly to the latter effect.

Myrica rubra Sieb. et Zucc. Family: Myricaceae

English Name: Bayberry

Characterization of anthocyanins and flavonols and radical scavenging activity assays of extracts of four Chinese bayberry (*Myrica rubra*) varieties with different fruit colors were carried out by Bao *et al.* (2005). Both DPPH and ABTS cation assays indicated that the black varieties (Biji and Hunan) demonstrated much higher radical scavenging activities than the pink (Fenhong) and yellow (Shuijing) varieties, which may be attributed to much higher levels of anthocyanins, flavonoids and total phenolics in the black varieties. Biji and Hunan had 6.49 and 6.52 mM Trolox equivalent antioxidant capacity (TEAC) per 100 g of fresh weight, whereas the pink (Fenhong) and yellow (Shuijing) bayberries had 1.32 and 1.31 mM TEAC/100g.

Myriostachya wightiana (Nees ex Steud.) Hook.f. Family: Poaceae (Graminae)

Vadlapudi and Naidu (2009) evaluated the antioxidant potential of *Myriostachya wightiana*, a mangrove plant. The SOD was 1.04U/mg, Catalase was 0.47U/mg, Ascorbic acid was 300mg/100g, DPPH radical inhibition was 38.2 per cent and FRAP units were 770.

Myristica argentea Warb. Family: Myristicaceae

Calliste *et al.* (2010) isolated a new dilignan, argenteane, from nutmeg mace. On the basis of experimental measurements of the lipid peroxidation inhibition, argenteane appeared to be an antioxidant as powerful as vitamin E. The density functional theory (DFT) calculations of the O-H bond dissociation enthalpies (BDEs) correlated with the capacity to scavnge free radicals.

Myristica fragrans Houtt. Family: Myristichaceae

English Name: Nutmeg.

The fruit of nutmeg beneficial against arthritis, muscular aches and pains, rheumatism, improves circulation. It protects nervous system and provides immunity against bacterial infection. Camphene, eugenol, γ-terpinene, isoeugenol, lauric acid, myrcene, palmitic acid and terpene-4-ol in nutmeg possess antioxidative properties (Mishra *et al.*, 2007).

Souri *et al.* (2008) evaluated the antioxidant activity of fruit against linoleic acid peroxidation using 1,3-diethyl-2-thiobarbituric acid as reagent. Antioxidant activity (IC_{50}) against peroxidation of linoleic acid (2mg/ml) was 7.29 and phenolic content was 543.7 mg/100g dry weight. The results of this study showed that there is no significant correlation between antioxidant activity and phenolic content of the studied plant materials and phenolic content could not be a good indicator of antioxidant capacity.

Mariutti *et al.* (2008) evaluated the antioxidant activity of nutmeg employing ABTS and DPPH radical assay. DPPH EC_{50} value was 9.6g/kg while DPPH-TEAC value was 27.0mM/g. ABTS-TEAC value was 77mM/g.

Two new phenolic compounds were isolated from the fruits of *Myristica fragrans* and their antioxidant activities were studied using DPPH assay. These compounds showed significant antioxidant activity.

Aqueous extract of seed of *Myristica fragrans* evaluated for its phytochemical constituents, antinutrients and antioxidant properties. The results showed that akaloids, saponins, anthraquinones, cardiac glycosides, flavonoids and phlobatannins were present while tannins were absent in the aqueous extract. The phytate content was 564.11 mg/100g while antioxidant indices of 100mg/100g, 44 per cent and 0.6 were obtained for the ascorbic acid value, radical scavenging activity and reducing power respectively (Olaleye *et al.*, 2006).

Six diarylbutane lignans 1-5 and one aryltetralin lignan were isolated from the methanol extract of *Myristica fragrans* seeds. The compounds were evaluated for LDL-oxidant activity to identify the most potent LDL-antioxidant 3 with IC_{50} value of 2.6µM in TBARS assay. In a macrophage –mediated LDL oxidation, the TBARS formation was inhibited by compound 3 (Kwon *et al.*, 2008).

Antioxidants in aqueous extract of *Myristica fragrans* suppressed mitosis and cyclophosphamide induced chromosomal aberration in *Allium cepa* (Akinboro *et al.*, 2011).

Myristica malabarica Lamk. Family: Myristicaceae

The DPPH assay of the ether, methanol and aqueous extracts of the spice *Myristica malabarica* revealed the methanol extract to possess the best antioxidant activity (Patro *et al.*, 2005). 2-acylresorcinol and four diarylnonaoids, isolated from the methanol extract, of which the diarylnonaoid, malabaricone C, showed the maximum DPPH scavenging activity. Malabaricone C could prevent both Fe(II) and 2,2'-azobis(2-amidinopropane) dihydrochloride-induced lipid peroxidation (LPO) of rat liver mitochondria more efficiently than curcumin. The anti LPO activity of malabaricone C was attributed to its better radical scavenging and Fe(II) chelation capacities. The radioprotective activity of malabaricone C was found to correlate with its OH radical scavenging property.

Myrothamnus flabellifolia Welw. Family: Myrothamnaceae

The antioxidant status of South African resurrection plant, *Myrothamnus flabellifolia,* a short woody shrub, is reported to correlate with its revival (Kranner *et al.*, 2002).

Myrtus communis L. Family: Myrtaceae

English Name: Myrtle.

Demo *et al.* (1998) reported antioxidant α-tocopherol from the Hexane extract of *Myrtus communis*. The aqueous extract, the total flavonoids oligomer fraction, and the ethyl acetate and methanol extracts of *Myrtus communis* showed an important free-radical scavenging activity towards the DPPH radical (Hayder *et al.*, 2004).

Souri *et al.* (2004) evaluated the antioxidative activity of *Myrtus communis* by linoleic acid peroxidation test using 1,3-diethyl-2-thiobarbituric acid as the reagent. Methanolic extract showed 90.57 per cent inhibition at 40μg concentration with IC_{50} value of 2.40μg, which was approximately in the range of α-tocopherol.

Wannes *et al.* (2010) evaluated the antioxidant activity of essential oils of different parts of *Myrtus communis* var. *italica* leaf, stem and flower. Leaf and flower oils had the best antioxidant activities, nevertheless inferior to those of BHT and BHA. The authors attributed such weak activities to the low level of phenolic compounds (eugenol in flowers) or even their absence (stem and leaf), because the extracts also studied in this present work presented higher activities.

Mimica-Dukic *et al.* (2010) evaluated the DPPH radical scavenging activity of *Myrtus communis* essential oil. Two samples of the oils evaluated exhibited moderate DPPH scavenging activity, with IC_{50} values of 6.24 mg/ml and 5.99mg/ml.

The chloroform and methanolic extracts of 124 Egyptian plant species belonging to 56 families were investigated by Moussa *et al.* (2011) and compared their antioxidant activity by DPPH scavenging assay. Among the 124 plant species tested 18 exhibited extremely high antiradical activity (>80 per cent inhibition). The IC_{50} value of DPPH radical scavenging of *Myrtus communis* was 21.53μg/ml, while total phenolic and flavonoid contents were 126.96 mg Tannic acid equivalent/g extract and 241.56 mg Rutin equivalent/g extract, respectively. Correlation coefficient between DPPH radical scavenging activity and the total phenolic and flavonoid contents suggest that phenolics and flavonoids in the extracts were partly responsible for the antiradical activities.

Nardostachys jatamansi DC. Family: Valerianaceae

The role of antioxidant properties of hydroethanolic extract of rhizome of *Nardostachys jatamansi* in alleviation of the symptoms of the chronic fatigue syndrome was investigated by Lyle *et al.* (2009a,b). Biochemical analysis showed that chronic fatigue syndrome significantly increased lipid peroxidation, nitrite and superoxide dismutase levels and decreased catalase level in rat brain. Administration of *Nardostachys jatamansi* extract (NJE 200 and 500 mg/kg) tended to normalize both augmented lipid peroxidation nitrite, superoxide dismutase activities and catalase level significantly. *In vitro* antioxidant activity of NJE was studied by measuring the free radical scavenging activity. NJE showed potent antioxidant activity and significantly reversed the stress-induced elevation of LPO and NO levels and decreases the catalase activity in the brain.

Nasturtium aquatica Family: Brassicaceae (Cruciferae)

In a screening of South African indigenous food plants for antioxidant activity by testing for inhibition of lipid peroxidation *Nasturtium aquatica* had high activity (Lindsey *et al.*, 2006). The activity in the aqueous extract was lost by boiling.

Nasturtinum officinale R.Br. Family: Brassicaceae (Cruciferae)

English Name: Watercress; Japanese: Kureson.

Maeda *et al.* (2006) examined the phenolic contents of leaves of watercress and its radical scavenging activities for DPPH radical. The polyphenol content was 83.9

mg-gallic acid equivalent/100 g) and DPPH scavenging activity was 424.9 µmol-Trolox eq./100 g).

Wojcikowski *et al.* (2007) evaluated the antioxidant activity of aerial parts of *Nasturtium officinale*. Oxygen radical absorbance capacity of the methanol extract and aqueous extracts were 214.43 and 457.98µmol Trolox equivalent per g of dried starting material, respectively.

Extracts of watercress were evaluated by Ozen (2009) for total antioxidant activity by ferric thiocyanate method, total reducing power by potassium ferricyanide reduction method, DPPH radical scavenging activities, superoxide anion radical scavenging activites *in vitro* and lipid peroxidation *in vivo*. The ethanolic extract was found as the most active in total antioxidant activity, reducing power, DPPH radicals and superoxide anion radicals scavenging activities. Administration of the ethanol extract to rats decreased lipid peroxidation in liver, brain and kidney. These results lead to the conclusion that *Nasturtium officinale* extracts show relevant antioxidant activity by means of reducing cellular lipid peroxidation and increasing antioxidant activity, reducing power, free radical and superoxide anion radical scavenging activities. In addition, total phenolic compounds in the aqueous and ethanolic extract were determined as pyrocatechol.

Nauclea latifolia Sm. Family: Rubiaceae

Lagnika *et al.* (2011) evaluated the antioxidant activity of *Nauclea latifolia* using DPPH radical scavenging aassay. The extract showed good radical scavenging activity with IC_{50} value of 1.56µg/ml while the standard L-Ascorbic acid had IC_{50} value of 1.1µg/ml.

Nelumbo nucifera Gaertn. Family: Nymphaeaceae

The antioxidant activity of *Nelumbo nucifera* stamens was evaluated by Jung *et al.* (2003) for their potential to scavenge stable DPPH free radicals, inhibit total reactive oxygen species generation, in kidney homogenates using 2',7'-dichlorodihydrofluorescein diacetate (DCHF-DA), and scavenge authentic peroxynitrites(ONOO-). A MeOH extract of the stamens showed strong antioxidant activity in the ONOO- system, and marginal activity in the DPPH and total ROS systems, so were therefore fractionated with several organic solvents, such as dichloromethane, ethyl acetate and n-butanol. The EtOAc soluble fraction, which exhibited strong antioxidant activity in all the model systems tested, was further purified by repeated silica gel Sephadex LH-20 column chromatographies. Seven known flavonoids along with β-sitosterol glucopyranoside (8) were isolated. Compound 1 possessed good activities in all the model systems tested. Compounds 2 to 7 showed scavenging activities in the DPPH and ONOO-tests, while compounds 3 and 4 were only active in the ONOO-test. Conversely, compound 8 showed no activities in any of the model systems tested (Jung *et al.*, 2003).

Antioxidant actity of hydro-alcoholic extract of *Nelumbo nucifera* seeds was studied using *in vitro* and *in vivo* models (Rai *et al.*, 2006). Total phenolic content in the extract was found to be 7.61 per cent (w/w). The extract exhibited strong free radical scavenging activity as evidenced by low IC_{50} values in both DPPH (6.12 µg/ml) and nitric oxide (84.86µg/ml) methods. The values were found to be less than those of rutin, the standard used.

The ethanolic leaf extract of *Nelumbo nucifera* was exposed to γ-irradiation, and its antioxidant activities, total polyphenols and colour characteristics were studied to discern its potential ability as a food or cosmetic materials. The results demonstrated that the radical scavenging activities and total polyphenols of the γ-irradiated leaf extract of *Nelumbo nucifera* were not observed to be significantly different.

An ethanolic extract of the leaves of *Nelumbo nucifera* was studied by Huang *et al.* (2010) for its *in vitro* and *in vivo* antioxidant activity.The results showed the antioxidant activity of lotus leaf extract (LLE) at 100mg/kg. The main flavonoids and phenolic compounds of LLE were: Catechin glycoside and five flavonoid glycoside derivatives: mircitrin-3-3-O-glucoside, hyperin, isoquercitrin, quercetin-3-O-rhamnoside and astragalin.

Neolitsea sericea (Blume) Koidz. var. *aurata* (Hayata) Hatusima Family: Magnoliaceae

Lee *et al.* (2006) investigated the antioxidant activity of ethanol extract of leaves using DPPH, hydroxyl and superoxide radical scavenging and reducing power activities as well as the induction of heme oxygenase-1 (HO-1). The IC_{50} values against DPPH, radical were 1296µg/ml, OH radical 0.24µg, superoxide radical 70.4µg/ml and total phenolics was 20.8 mg of GAE/g.

Nepeta cataria L. Family: Lamiaceae (Labiatae)

English Name: Catnip.

Essential oil and methanol extract of *Nepeta cataria* were investigated by Adiguzel *et al.* (2009) for their antioxidant activity. In the DPPH assay, the extract showed slight antioxidant activity where as the essential oil remained inactive. In the latter case, both the extract and the essential oil exerted weak activity having inhibition ratios of linoliec acid oxidation at 16.4 per cent and 27 per cent respectively. The weak antioxidative nature of the extract could be attributed to the low phenolic content, estimated as gallic acid equivalent at 22.6µg/ml or 2.26 per cent.

Nepeta flavida Hub.-Mor. Family: Lamiaceae (Labiatae)

Tepe *et al.* (2007) examined the chemical composition and *in vitro* antioxidant activity of the essential oil and various extracts of *Nepeta flavida*. The samples were subjected to a screening for their possible antioxidant activities by using DPPH and β-carotene-linoleic acid assays. In DPPH assay the IC_{50} value of the essential oil was determined to be 42.8µg/ml. Among the extracts the strongest activity was exhibited by the polar sub-fraction of the methanol extract with IC_{50} value of 63.2 µg/ml. In the and β-carotene-linoleic acid system, the essential oil exhibited 86.3 per cent inhibition against linoleic acid oxidation.

Nepeta foliosa Moris Family: Lamiaceae (Labiatae)

The chemical constituents of the essential oil of *Nepeta foliosa* were characterized and their antioxidant activity was evaluated by Giamperi *et al.* (2009) by two *in vitro* systems.

Nepeta glomerulosa Boiss. subsp. *glomerulosa* Family: Lamiaceae (Labiatae)

Souri *et al*. (2004) evaluated the antioxidative activity of *Nepeta glomerulosa* subsp. *glomerulosa* by linoleic acid peroxidation test using 1,3-diethyl-2-thiobarbituric acid as the reagent. Methanolic extract showed 94.78 per cent inhibition at 40µg concentration with IC_{50} value of 4.40µg.

Nepeta ispahanica Boiss. Family: Lamiaceae (Labiatae)

Salehi *et al*. (2007) investigated the antibacterial and antioxidant properties of the essential oil and various extracts of *Nepeta ispahanica* from Iran. The free radical scavenging activity of n-BuOH subfraction of the methanol extract (IC_{50} = 37µg/ml) was superior to all other extracts, while the oil was the least effective.

Nepeta meyeri Benth. Family: Lamiaceae (Labiatae)

Cigremis *et al*. (2010) screened the antioxidant and antimicrobial activities of acetone extract of *Nepeta meyeri*. The antioxidant properties of the extract were investigarted by using various methods established *in vitro* systems such as DPPH, nitric oxide (NO) radical scavenging activity. Acetone extract of the plant examined exhibited significant concentration dependent inhibition of DPPH and NO radical. Furthermore, *Nepeta meyeri* showed very high reducing power. In DPPH radical and NO* scavenging assays the IC_{50} value of extract was 672.2µg/ml and 165.32µg/ml, respectively. The amounts of total phenolic compounds were also determined and 12.86 µg pyrocatechol equivalents of phenols were detected in the extract.

Nepeta nuda L. ssp. *nuda* Family: Lamiaceae (Labiatae)

Antioxidant activity of *Nepeta nuda* ssp. *nuda* essential oil rich in nepetalactones from Greece was studied by Gkinis *et al*. (2010). Neutralization of stable DPPH radical ranged from 10.83 per cent (2.50µg/ml) to 58.64 per cent (50.00µg/ml) for verticillaster oil and from 6.25 per cent (2.50µg/ml) to 57.79 per cent (50.00µg/ml) for leaf oil. The essential oil from verticillasters had significant effects on lipid peroxidation (in the range of 41.18-59.23 per cent), compared to tert-butylated hydroxytoulene (37.04 per cent). In contrast, the essential oil from leaves exhibited prooxidant activity at the highest concentration applied.

Nephelium lappaceum L. Family: Sapindaceae

English Name: Rambutan.

The rind of rambutan, which is normally discarded, was found to contain extremely high antioxidant activity when assessed using several methods (Palanisamy *et al*., 2008). Although having a yield of only 18 per cent, the ethanolic rambtan rind extract had a total phenolic content of 762 mg GAE/g extract, which was comparable to that of a commercial preparation of grape seed extract.

Ether, methanolic and aqueous extracts of lyophilized *Nephelium lappaceum* peels and seeds were evaluated by Thitilertdecha *et al*. (2008) for phenolic constituents, antioxidant and antibacterial activities. High amounts of phenolic compounds were found in the peel extracts and the highest content was in the methanolic fraction (542.2mg/g dry extract). Several potential antioxidant activities, including reducing power, β-carotene bleaching, linoleic peroxidation and free radical scavenging activity, were evaluated. The peel extracts exhibited higher antioxidant activity than the seed extracts in all methods determined. The methanolic fraction was found to be the most active antioxidant as shown by their 50 per cent DPPH inhibition concentration, 4.94µg/ml. The results indicated this fraction exhibited greater DPPH radical scavenging activity than BHT and ascorbic acid.

Experiments on the isolation and identification of the active constituents from the fruit peel if *Nephelium lappaceum* were conducted by Thitlertdecha *et al.* (2010) and on their antioxidant acitivity using lipid peroxidation inhibition assay. The methanolic extract of *N. lappaceum* fruit peels exhibited strong antioxidant properties. Ellagic acid, corilagin and geraniin were isolated from the fruit peel. These compounds accounted for 69.3 per cent of the methanolic extract, with geraniin (56.8 per cent) as the major component, and exhibited much greater antioxidant activities than BHT in both lipid peroxidation (77-186 fold) and DPPH (42-87 fold) assays.

Nephelium longan (Lour.) Hook. (Syn.: *Dimocarpus longan* Lour.) Family: Sapindaceae

The petroleum ether, chloroform and ethyl acetate fractions of ethanol extract of leaf and stem from the plant *Nephelium longan* was subjected to antioxidant, antibacterial and cytotoxic activity by Ripa *et al.* (2010). All the fractions showed potent antioxidant activity, of which the ethyl acetate and chloroform fraction of leaf demonstrated the strongest antioxidant activity with IC_{50} value of 44.28 and 44.31 µg/ml, respectively.

Neptunia oleracea Lour. Family: Mimosaceae

Chanwitheesuk *et al.* (2005) investigated the antioxidant activity of *Neptunia oleracea* using a β-carotene bleaching method. The contents of plant chemicals such as vitamin C, Vitamin E, carotenoids, tannin and total phenolics, were also determined. Methanolic extract showed antioxidant activity. *Neptunia oleracea* has antioxidant index of 1.16, Vitamin C 12.9 mg per cent, Vitamin E 0.0066mg per cent, Total carotenes 3.18 mg per cent, Total xanthophylls 1.06mg per cent, Tannins 21mg per cent and Total phenolics of 104mg per cent.

Nerium oleander L. Family: Apocynaceae

Souri *et al.* (2004) evaluated the antioxidative activity of *Nerium oleander* by linoleic acid peroxidation test using 1,3-diethyl-2-thiobarbituric acid as the reagent. Methanolic extract showed 92.56 per cent inhibition at 40µg concentration with IC_{50} value of 1.51µg.

Newbouldia laevis Seem. Family: Bignoniaceae

Akinmoladun *et al.* (2010) assessed the phytochemical constituents and antioxidant and free radical scavenging activities of *Newbouldia laevis* using seven different assays. Significant antioxidant and radical scavenging activity was observed that may be due to the phytochemical content of the plant.

Oloyede *et al.* (2010) evaluated the antioxidant activity of the hexane, ethyl acetate, butanol and aqueous extracts of the aerial parts of *Newbouldia laevis*. Antioxidant activity determined by ferric thiocyanate method showed that the species provided atleast one fraction with highly promising antioxidant activity. Hexane, ethyl acetate, butanol and water extract at various concentrations (50, 100, 200 and 500 µg/ml) showed antioxidant activities in a concentration dependent manner, the butanol extract however at the concentration of 500µg/ml showed 70-78 per cent antioxidant activity which is very close to that of 500µg/ml of α-tocopherol (82 per cent), the reference compound.

Nicandra physaloides (L.) Gaertn. Family: Solanaceae

In a study with 32 methanolic extracts, from Brazilian Caatinga plants, through DPPH radical scavenging assay showed that the species *Nicandra physaloides* displayed good antioxidant activity with IC_{50} = 4.2 mg/l (David *et al.*, 2007).

Nigella sativa L. Family: Ranunculaceae

English Name: Black cumin.

Kanter *et al.* (2003, 2005) evaluated the hepatoprotective effect of *Nigella sativa* on lipid peroxidation, antioxidant enzyme systems and liver enzymes in carbon tetrachloride-treated rats. The CCl_4 treatment for 60 d increased the lipid peroxidation and liver enzymes, and also decreased the antioxidant enzyme levels. *Nigella sativa* treatment for 60d decreased the elevated lipid peroxidation and liver enzyme levels and also increased the reduced antioxidant enzyme levels.

Suboh *et al.* (2004) investigated the antioxidant activity effects of various concentrations of methanol extracts of *Nigella sativa* and six other medicinal plants. They reported that the extract significantly reduced 10mM hydrogen peroxide-induced protein degradation, lipid peroxidation, loss of deformability and increased osmotic fragility caused by H_2O_2 in human erythrocytes.

Souri *et al.* (2008) evaluated the antioxidant activity of seed against linoleic acid peroxidation using 1,3-diethyl-2-thiobarbituric acid as reagent. Antioxidant activity (IC_{50}) against peroxidation of linoleic acid (2mg/ml) was 146.84 and phenolic content was 122.67 mg/100g dry weight. The results of this study showed that there is no significant correlation between antioxidant activity and phenolic content of the studied plant materials and phenolic content could not be a good indicator of antioxidant capacity.

Nigella volatile oil contains a considerable phenolic content (2.69 per cent); mostly carvacrol and thymohydroquinone. The oil exhibited antioxidant and radical scavenging activities on DPPH at various concentrations compared with the known synthetic antioxdants such as butylhydroxyanisol (BHA), that might be due to the considerable concentration of phenolic contents (El-Ghorab, 2003). The antioxidant activity of hexane extract of seeds was described by Mehta *et al.* (2009). The antioxidant activity of methanolic extract of *Nigella sativa* seeds was investigated by Okeola *et al.* (2010). The extract showed strong antioxidant property.

Nonea pulla (L.) DC. Family: Boraginaceae

Myagmar and Aniya (2000) evaluated the free radical scavenging action of some medicinal herbs growing in Mongolia. The aqueous extract of nine herbs including

Nonea pulla were used. The free radical scavenging action was determined *in vitro* and *ex vivo* by using ESR spectrometer and chemiluminiscence analyzer. The results showed that *Nonea pulla* extract possess strong scavenging action of DPPH, superoxide and hydroxyl radicals. *Nonea poulla* also depressed reactive oxygen production from polymorphonuclear leukocytes stimulated by phorbol-12-myristate *ex vivo*.

Nothapodytes nimmoniana (J.Graham) Mabberly Family: Icacinaceae

Namdeo *et al.* (2010) investigated the phenolic content and antioxidant activity of methanolic extract of different parts of *Nothapodytes nimmoniana*. Among the various plant parts phenolic content ranged from 281.0 to 450.0 mg GAE/100g DW). Fruit showed maximum antioxidant activity with an IC_{50} value of 0.177mg/ml for DPPH radical, 0.177mg/ml for H_2O_2 radical, 0.167mg/ml for superoxide radical and 0.175 mg/ml for nitric oxide radical. Fruit showed maximum anti-lipid peroxidation effect (0.362 mg/ml) with high reducing potential 3.65.

Seasonal discrepancy in the antioxidant potential from bark of *Nothapodytes nimmoniana* was evaluated by Pai *et al.* (2010) using two different methods (DPPH and FRAP) and were related to total phenolic as well as flavonoid content. Higher yield of phenolic and flavonoid were observed during summer (May), similarly the antioxidant activities also were high.

Nothopanax scutellarius (Burm.f.) Merr. Family: Araliaceae

Extracts from 11 vegetables of Indonesian origin were screened for flavonoid content, total phenolics, and antioxidant activity by Andarwulan *et al.* (2010). Flavonoid content in *Nothopanax scutellarius* was 5.43mg/100g fresh weight.

Notopterygium incisum Ting Family: Apiaceae (Umbelliferae)

Antioxidant capacities of 56 selected Chinese medicinal plants, including *Notopterygium incisum* were evaluated by Song *et al.* (2010) using the TEAC and FRAP assays, and their total phenolic content was measured by the Folin-Ciocaleteu method. Bark of root of *Notopterygium incisum* showed moderate antioxidant activity with TEAC value of 62.94μmol Trolox/g, FRAP value of 66.80μmol Fe^{2+}/g and phenolic content was 10.86mg GAE/g.

Nyctanthes arbor-tristis L. Family: Oleaceae

Nagavani *et al.* (2010) evaluated the free radical scavenging activity of flowers of *Nycatanthes arbor-tristis*. Methanol extracts of dry flowers exhibit high phenolic content and high antioxidant activities, interestingly aqueous extracts showed high enzymatic antioxidants.

Nymphaea caerulea Sav. Family: Nymphaeaceae

Agnihotri *et al.* (2008) isolated 20 constituents from *Nymphaea caerulea* flowers including two 2S,3S,4S-trihydroxypentanoic acid (1), and myricetin 3-O-(3''-O-acetyl)-α-L-rhamnoside (2), along with the known myricetin 3-O-α-L-rhamnoside (3), myricetin 3-O-β-D-glucoside (4), quercetin 3-O-(3''-O-acetyl)-α-L-rhamnoside (5), quercetin 3-O-α-L-rhamnoside (6), quercetin 3-O-β-D-glucoside (7), kaempferol 3-O-(3''-O-acetyl)-α-L-rhamnoside (8), kaempferol 3-O-β-D-glucoside (9), naringenin (10),

(S)-naringenin 5-O-β-D-glucoside (11), isosalipurposide (12), β-sitosterol (13), β-sitosterol palmitate (14), 24-methylenecholesterol palmitate (15), 4α-methyl-5α-ergosta-7,24(28)-diene-3β,4β-diol (16), ethyl gallate (17), gallic acid (18), *p*-coumaric acid (19), and 4-methoxybenzoic acid (20). Compounds were tested for antioxidant activity and nine compounds 2-7, 11, 12 and 18 were considered active with IC(50) of 1.16, 4.1, 0.75, 1.7, 1.0, 0.34, 11.0, 1.7 and 0.95 microg/ml, respectively, while 1 was marginally active (IC(50)>31.25 µg/ml). The most promising activity was found in the EtOAc fraction (IC(50) 0.2 µg/ml). This can be attributed to the synergistic effect of the compounds present in it.

Nymphaea candida C. Presl. Family: Nymphaeaceae

Zhao *et al.* (2011) evaluated the antioxidant effects of flower of *Nymphaea candida* using DPPH radical scavenging assay and reducing power determination. The active extract was further purified to afford four fractions of which the ethyl acetate soluble fraction (NCE) exhibited the strongest antioxidant capacity with IC_{50} value of 12.6µg/ml for DPPH. Thirteen phenolic compounds were isolated from this fraction, and they all showed significan antioxidant activities in DPPH model system. Furthermore NCE showed potent antioxidant capacity with IC_{50} value of 59.32 µg/ml, 24.4µg/ml and 86.85µg/ml for O^{2-}, OH and H_2O_2 radicals, respectively.

Nymphaea hybrid Family: Nymphaeaceae

An all-round evaluation for the antioxidant activity of the ethanol extract of the flowers of *Nymphaea hybrid* was carried out by Ren *et al.* (2010). The evaluation indices included the clean-up rate towards superoxidized anion radicals, hydroxyl radicals, and the antioxidant activity towars the peroxidize of lipids, the reducing ability, total antioxidant capacity. Results indicate that the ethanol extract from the flower of *Nymphaea hybrid* possesses rather high and multiple antioxidant activity.

Nymphaea lotus L. Family: Nymphaeaceae

English Name: Lotus

Saleem *et al.* (2001) extracted flowers of *Nymphaea lotus* with 70 per cent acetone and analyzed for their total phenolic concentration and antioxidant potential. The IC_{50} results indicate that the extract was stronger antioxidant than α-tocopherol. Total phenolics concentration, expressed as gallic acid equivalents, showed close correlation with the antioxidant activity. HPLC analysis with diode array detection at 280 nm of the extract indicated the presence of hydroxybenzoic acid derivatives, hydroxycinnamic acid derivatives, flavonol aglycones and their glycosides as main phenolics compounds.

Nymphaea nouchali Burm. Family: Nymphaeaceae

Nagavani and Raghava Rao (2010) evaluated the enzymatic and nonenzymatic and antioxidant potential in ethanol, methanol and aqueous extract of *Nymphaea nouchali* dry and fresh flowers. Obtained results showed that the high levels of non-enzymatic antioxidants like phenols, flavonoids, tannins etc., as well as antioxidant potential was found to be more in methanol extracts of *Nymphaea nouchali* dry flowers.

It is very interesting that the levels of enzymatic antioxidants were found to be high in fresh flower aqueous extracts.

Nymphaea stellata Willd. Family: Nymphaeaceae

The free radical scavenging potential of the dried flower of plant *Nymphaea stellata* was studied by using different *in vitro* antioxidant models of screening. The hydroalcoholic extract at 500 2g/ml showed maximum scavenging of the *in vitro* riboflavin NBT assay (26.22 per cent), hydroxyl radical scavenging activity (26.06 per cent), DPPH (25.70 per cent) and nitric oxide radicals (25.11 per cent). However, the extract showed only moderate scavenging activity of total antioxidant capacity (15.25 per cent). DPPH- Radical scavenging activity, scavenging of superoxide radical, scavenging of nitric oxide radical, iron chelating activity, hydroxyl radical scavenging activity, rapid screening for antioxidant activity, total antioxidant capacity and Riboflavin NBT method were used as *in-vitro* antioxidant models for screening of antioxidant potential. The extract showed significant scavenging and IC_{50} value as compared to standard Ascorbic acid.

Ocimum americanum L. Family: Lamiaceae (Labiatae)

Chanwitheesuk *et al.* (2005) investigated the antioxidant activity of *Ocimum americanum* using a β-carotene bleaching method. The contents of plant chemicals such as vitamin C, Vitamin E, carotenoids, tannin and total phenolics, were also determined. Methanolic extract showed antioxidant activity. *Ocimum americanum* has antioxidant index of 3.67 Vitamin C 5.48 mg per cent, Vitamin E 0.0040mg per cent, Total carotenes 5.128 mg per cent, Total xanthophylls 9.52mg per cent, Tannins 11.1mg per cent and Total phenolics of 43.6 mg per cent.

Ocimum basilicum L. Family: Lamiaceae

English Name: Basil.

Ethanolic extracts of varieties of *Ocimum basilicum* were tested by Juliani and Simon (2002) for *in vitro* antioxidant activity using ABTS, TEAC, FRAP and AEAC. There was a strong relationship between the total phenolic content and the antioxidant capacity. In all the basils the essential oil contribution to the total antioxidant activity was low. These plants constitute a new source of antioxidant phenolics in the diet, providing 125mg of gallic acid equivalents, 85-125mg of Trolox or 106-140 mg of ascorbic acid equivalents per gram of dry weight.

Antioxidant activity and total phenolic content of 23 Iranian *Ocimum basilicum* accessions was investigated by Javanmardi *et al.* (2003). Total antioxidant capacity by TEAC assay varied from 10.8 ro 35.7μM Trolox, and total phenolic content ranged from 22.9 to 65.5 mg gallic acid/g dw in "Dezful I" and "Babol" accessions, respectively. A linear positive relationship existed between the antioxidant activity and total phenolic acids content of the tested basil accessions.

Souri *et al.* (2004) investigated the antioxidant activity of methanol extracts of basil, used as vegetable in Iranian diet against linoleic acid peroxidation. The extracts showed good antioxidant activity.

Hydrodistillated extracts from basil were assessed for their total phenol content, and antioxidant (iron(III) reduction, inhibition of linoleic acid peroxidation, iron (II) chelation, DPPH radical-scavenging and inhibition of hydroxyl radical-mediated 2-deoxy-D-ribose degradation, site and nonsite-specific activities by Hinneburg *et al.* (2005). The extracts from basil possessed the highest antioxidant activities except iron chelation. In linoleic acid peroxidation assay, 1 g of basil extract was effective as 177 of trolox.

Chanwitheesuk *et al.* (2005) investigated the antioxidant activity of *Ocimum basilicum* using a β-carotene bleaching method. The contents of plant chemicals such as vitamin C, Vitamin E, carotenoids, tannin and total phenolics, were also determined. Methanolic extract showed antioxidant activity. *Ocimum basilicum* has antioxidant index of 5.12, Vitamin C 6.79 mg per cent, Vitamin E 0.0101mg per cent, Total carotenes 1048 mg per cent, Total xanthophylls 13.3mg per cent, Tannins 3090mg per cent and Total phenolics of 83.3 mg per cent.

Wong *et al.* (2006) investigated the antioxidant properties of 25 edible tropical plants including *Ocimum basilicum* using DPPH and FRAP assays. Total phenolic content was also estimated. *Ocimum basilicum* showed good antioxidant activity. Abbas *et al.* (2006) evaluated the antioxidant activity of *Ocimum basilicum* using DPPH assay. Methanolic extract showed DPPH radical scavenging activity.

Maeda *et al.* (2006) examined the phenolic contents of *Ocimum basilicum* and their radical scavenging activities by using DPPH assay. High contents of phenolic compounds and strong DPPH scavenging activities were observed in extracts from this species. They also showed radical scavenging activities for the superoxide radical and tert-butylhydroperoxyl radical (t-BuOO·).

Wojcikowski *et al.* (2007) evaluated the antioxidant activity of aerial parts of *Ocimum basilicum*. Oxygen radical absorbance capacity of the ethyl extract, methanol extract and aqueous extracts were 38.44, 171.82 and 314.50μmol Trolox equivalent per g of dried starting material, respectively.

Gülçin *et al.* (2007) investigated the radical scavenging and antioxidant activity of water (WEB) and ethanol extracts (EEB) of basil using different antioxidant methodologies: DPPH free radical scavenging, scavenging of superoxide anion radical-generated non-enzymatic system, ferric thiocyanate method, reducing power, hydrogen peroxide scavenging and metal chelating activities. Experiments revealed that WEB and EEB have antioxidant effects which were concentration-dependent. The total antioxidant activity was performed according to the ferric thiocyanate method. At the 50 μg/ml concentration, the inhibition effects of WEB and EEB on peroxidation of linoleic acid emulsion were found to be 94.8 per cent and 97.5 per cent, respectively. On the other hand, the percentage inhibition of a 50 μg/ml concentration of BHA, BHT and α-tocopherol was found to be 97.1 per cent, 98.5 per cent and 70.4 per cent inhibition of peroxidation of linoleic acid emulsion, respectively. In addition, WEB and EEB had effective DPPH radical scavenging, superoxide anion radical scavenging, hydrogen peroxide scavenging, reducing power and metal chelating activities. Additionally, these various antioxidant activities were compared with BHA, BHT and α-tocopherol as reference antioxidants. The additional total phenolic content of these basil extracts were determined as the gallic acid equivalent and were found to be equivalent (Gülçin *et al.*, 2007).

Zakaria *et al.* (2008) studied the antioxidant properties of the methanol extract of leaves and stems of *Ocimum basilicum* by DPPH radical scavenging activity. The extracts showed antioxidant activity comparable to butylated hydroxytoulene (BHT).

Souri *et al.* (2008) evaluated the antioxidant activity of seed against linoleic acid peroxidation using 1,3-diethyl-2-thiobarbituric acid as reagent. Antioxidant activity (IC_{50}) against peroxidation of linoleic acid (2mg/ml) was 4.78 and phenolic content was 106.52 mg/100g dry weight. The results of this study showed that there is no significant correlation between antioxidant activity and phenolic content of the studied plant materials and phenolic content could not be a good indicator of antioxidant capacity.

Mariutti *et al.* (2008) evaluated the antioxidant activity of basil employing ABTS and DPPH radical assay. DPPH EC_{50} value was 43g/kg while DPPH-TEAC value was 6.5mM/g. ABTS-TEAC value was 8.6mM/g.

The significant hepatoprotective effects were obtained by ethanolic extract of leaves of *Ocimum basilicum* against liver damage induced by H_2O_2 and CCl_4 as evidenced by decreased levels of antioxidant enzymes (Meera *et al.*, 2009). The extract also showed significant anti lipid peroxidation effects *in vitro*, besides exhibiting significant activity in superoxide radical and nitric oxide radical scavenging, indicating higher potent antioxidant effects.

Cioroi and Dumitriu (2009) investigated the total polyphenols content and antioxidant activity of aqueous extract of *Ocimum basilicum*. The total phenol content was 516.352 mg/100g dried species. Antioxidant activity was directly correlated with the total amount of polyphenols in the extract.

Twenty-five essential oils, including Basil oil, were tested by Wei and Shibamoto (2010) for antioxidant activities using a conjugated diene assay, the aldehyde/carboxylic acid assay, the DPPH free radical scavenging assay, and the malonaldehyde/gas chromatography (MA/GC) assay. They were also tested for lipoxygenase inhibitory activities using the lipoxygenase inhibitor-screening assay. Basil oil exhibited a strong effect in the DPPH assay (86 per cent) and moderate activities in the MA/GC assay (35 per cent).

Wei *et al.* (2010) reported that essential oil of *Ocimum basilicum* had appreciable antioxidant activities comparable to that of α-tocopherol, the reference chosen by the authors. In the basil oil linalool, isoanethole and eugenol predominated.

Narayanaswamy and Balakrishnan (2011) evaluated the antioxidant activity of 13 medicinal plants including leaf of *Ocimum basilicum*. The aqueous extract showed highest 95 per cent inhibition of DPPH radical while ethanolic extract showed 95 per cent inhibition of DPPH radical. The total phenolic contents of herbs were also determined by Folin-Ciocalteu method. *Ocimum basilicum* contained maximum amount of phenols (16.8 mg catechol/g of plant tissue)

Ocimum x citriodorum Family: Lamiaceae (Labiatae)

Ethanolic extracts of varieties of *Ocimum citriodorum* were tested by Juliani and Simon (2002) for *in vitro* antioxidant activity using ABTS, TEAC, FRAP and AEAC. There was a strong relationship between the total phenolic content and the antioxidant capacity. In all the basils the essential oil contribution to the total antioxidant activity was low. These plants constitute a new source of antioxidant phenolics in the diet, providing 125mg of gallic acid equivalents, 85-125mg of Trolox or 106-140 mg of ascorbic acid equivalents per gram of dry weight.

Ocimum gratissimum L. Family: Lamiaceae (Labiatae)

Oboh and Akindahunsi (2004) investigated the effect of preservation of green leafy vegetables of Nigeria on the antioxidant phytoconstituents (Vitamin C and Total phenol) and activity (reducing property and free radical scavenging ability) including *Ocimum gratissimum*. The result of the study revealed that sun-drying of green leafy vegetables cause a significant decrease in the vitamin C content (16.67-64.68 per cent loss). Conversely it leads to a significant increase in the total phenol content (6.45 -223.08 per cent grain) reducing property (16.00-362.50 per cent grain) and free radical scavenging ability (126.00-5757.00 per cent gain) of green leafy vegetables.

Chanwitheesuk *et al.* (2005) investigated the antioxidant activity of *Ocimum gratissimum* using a β-carotene bleaching method. The contents of plant chemicals such as vitamin C, Vitamin E, carotenoids, tannin and total phenolics, were also determined. Methanolic extract showed antioxidant activity. *Ocimum gratissimum* has antioxidant index of 10.8, Vitamin C 5.09 mg per cent, Vitamin E 0.0206mg per cent, Total carotenes 2.56 mg per cent, Total xanthophylls 7.94mg per cent, Tannins 24.3mg per cent and Total phenolics of 125 mg per cent.

Odukoya *et al.* (2005) evaluated the antioxidant activity of Nigerian dietary spices including *Ocimum gratissimum*. The antioxidant activity (expressed as per cent inhibition of oxidation) was 56.41 per cent, reducing power was 9.12 per cent and Total phenolic content was 138.21 mg/100g Tannic acid Equivalent. Antioxidant activity correlated significantly and positively with total phenolics while there was no linear correlation between total antioxidant activity and reducing power neither between reducing power and total phenolic content.

Akinmoladun *et al.* (2010) assessed the phytochemical constituents and antioxidant and free radical scavenging activities of *Ocimum gratissimum* using seven different antioxidant assay methods. The extract had low nitric oxide radical scavenging activity and high lipid peroxidation inhibitory activity. The results suggest that the methanolic extracts possess significant antioxidant and radical scavenging activity that may be due to the phytochemical content of the plant.

Adefegha and Oboh (2011) reported that cooking decreases the vitamin C contents in the leafy vegetable *Ocimum gratissimum*, while it increased the phenolic content and antioxidant activities.

Salawu *et al.* (2011) evaluated the cellular antioxidant activities of green leafy vegetables including *Ocimum gratissimum* against peroxyl radical-induced oxidation in HepG2 cells. HPLC/DAD/MS investigation of the ethanolic extract of the vegetal matter revealed the presence of 6 phenolic compounds. At a concentration of 1 mg/ml the quenching of peroxyl radical in HepG2 cells revealed that *Ocimum gratissimum* has the best inhibition capacity.

Ocimum micranthum Willd. Family: Lamiaceae (Labiatae)

English Name: Amazonian basil.

Sacchetti *et al.* (2004) analyzed the composition and functional properties of the essential oil of Amazonian basil, *Ocimum micranthum* in comparison with commercial essential oils. Radical scavenging activity was evaluated employing DPPH assay.

Ocimum minimum Lour. Family: Lamiaceae (Labiatae)

The antioxidant activity of the essential oils, BHA, BHT, and α-tocopherol was evaluated in a series of *in vitro* test: DPPH radical scavenging, ferric thiocyanate method, reducing power, and scavenging of superoxide anion radical generatd non-enzymatic systems (Telci *et al.*, 2009). On the basis of results obtained the reducing power of essential oil and standard compound was in order of BHT> α-tocopherol>essential oil.

Ocimum sanctum L. Family: Lamiaceae (Labiatae)

English Name: Holy Basil.

Devi and Ganasoundari (1999) investigated the effect of aqueous extract of *Ocimum sanctum* to protect against radiation induced lipid peroxidation in liver and to determine the role, if any, of the inherent antioxidant system in radiation protection by *Ocimum sanctum.* The results indicate that *Ocimum* extract protects against radiation induced lipid peroxidation and that GSH and the antioxidant enzymes appear to have an important role in the protection.

Ethanolic extracts of varieties of *Ocimum sanctum* were tested by Juliani and Simon (2002) for *in vitro* antioxidant activity using ABTS, TEAC, FRAP and AEAC. There was a strong relationship between the total phenolic content and the antioxidant capacity. In all the basils the essential oil contribution to the total antioxidant activity was low. These plants constitute a new source of antioxidant phenolics in the diet, providing 125mg of gallic acid equivalents, 85-125mg of Trolox or 106-140 mg of ascorbic acid equivalents per gram of dry weight.

The leaves of Holy Basil, *Ocimum sanctum*, commonly known as *Tulsi* is traditionally used for many health problems. Current research works have reported strong protective effect of aqueous extract of leaves against radiation injury. Flavonoids isolated from aqueous extract namely orientin and vicenin have been found to be very effective in protecting against radiation induced lipid peroxidation in mouse liver. Flavonoids also scavenged free radicals *in vitro* and showed antilipoperoxidant activity *in vivo* at a very low concentration. In view of these reports Geetha *et al.* (2003) investigated superoxide and hydrogen peroxide scavenging action of leaf extracts and their fractions and compared with known antioxidant ascorbic acid. The results revealed that both extracts and their fractions are good scavengers of superoxide and hydrogen peroxide comparable to ascorbic acid. Thus tulsi leaves can be exploited for its impressive free radical scavenging activities (Geetha *et al.*, 2003).

A significant decrease in malondialdehyde (MDA) levels and a significant increase in antioxidant activity were observed by Chandra *et al.* (2003) when *Ocimum sanctum* was administered to diabetic rats. Sethi *et al.* (2004) reported that the leaves possess both superoxide and hydroxyl free radical scavenging action.

Chanwitheesuk *et al.* (2005) investigated the antioxidant activity of *Ocimum sanctum* using a β-carotene bleaching method. The contents of plant chemicals such as vitamin C, Vitamin E, carotenoids, tannin and total phenolics, were also determined. Methanolic extract showed antioxidant activity. *Ocimum sanctum* has antioxidant index of 4.23, Vitamin C 9.23 mg per cent, Vitamin E 0.0202mg per cent, Total carotenes 5.13 mg per cent, Total xanthophylls 3.18mg per cent, Tannins 40.8mg per cent and Total phenolics of 91.8 mg per cent.

The cardioprotective efficacy of *Ocimum sanctum* was examined by Arya *et al.* (2006) in the experimental model of isoproterenol-induced myocardial necrosis. Tulsi pretreatment augmented the basal endogenous antioxdants and restored the antioxidant status of the heart.

The leaves are used to prevent free radical formation and have been found useful in treating arthritis, muscular pains and rheumatism. The main constituents responsible for antioxidative property of basil are Ascorbic acid, β-carotene, β-sitosterol, eugenol, palmitic acid and tannin (Mishra *et al.*, 2007). Leelapornpisid *et al.* (2007) reported antioxidant activity of the volatile oil of *Ocimum sanctum* (IC_{50} = 0.6294mg/ml).

Lukkmanul Hakkim *et al.* (2007) investigated the chemical composition and antioxidant property of holy basil leaves, stems, and inflorescence and their *in vitro* callus cultures. They reported that the callus extracts exhibited higher antioxidant activity than the extract obtained from field-grown plant parts.

The effect of an aqueous leaf extract of *Ocimum sanctum* was examined by Halder *et al.* (2009) against H_2O_2 induced cytotoxic changes in human lens epithelial cells (HLEC). The results indicated that extracts have protective role against H_2O_2 injury in HLEC by maintaining the normal cellular architecture. The protection could be due to its ability to reduce H_2O_2 through its antioxidant property and thus reinforcing the concept that the extracts can penetrate the HLEC membrane.

The methanolic crude extracts of *Ocimum sanctum* were screened by Veeru *et al.* (2009) for their free radical scavenging properties using ascorbic acid as standard antioxidant. Free radical scavenging activity was evaluated using DPPH free radical. The overall antioxidant activity of *Ocimum sanctum* was found to be strongeer with IC_{50} value of 0.05 mg/l.The ascorbic acid level was 8.6 mg/100g and the carotenoids content were 9.53 mg/100g in plant extracts. The total phenol content was 48.93 mg/g.

Suanarunsawat *et al.* (2010) have investigated the anti-hyperlipidemic and antioxidant activities of essential oils extracted from *Ocimum sanctum* leaves in rats fed with high cholesterol diet. The oils were able to decrease the high levels of TBARS either in cardiac or liver tissues, protecting them against stress induced oxidation.

Oenathe javanica (Bl.) DC. Family: Apiaceae (Umbelliferae)

Huda-Faujan *et al.* (2007, 2009) reported antioxidative activities of aqueous and methanolic extracts of *Oenanthe javanica* and four other plants. The analysis carried out was total phenolic content, ferric reducing power, ferric thiocyanate (FTC) and TBA tests. The phenolic content was 19.96 mg (aqueous extract) 7.41 mg (methanolic extract) TAE/100 g fresh weight and antioxidant activity of 52.1 per cent (aqueous extract) and 65.41 per cent (methanolic extract) using FTC method. Increasing the concentration of the extracts resulted in increased ferric reducing antioxidant power. Total phenolic content had positive correlation with antioxidant capacity.

Oenanthe stolonifera Wall. ex DC. Family: Apiaceae (Umbelliferae)

English Name: Chinese celery.

Chanwitheesuk *et al.* (2005) investigated the antioxidant activity of leaves of *Oenanthe stolonifera* using a β-carotene bleaching method. The contents of plant chemicals such as vitamin C, Vitamin E, carotenoids, tannin and total phenolics, were also determined. Methanolic extract showed antioxidant activity. *Oenanthe stolonifera* has antioxidant index of 7.92, Vitamin C 9.11 mg per cent, Vitamin E 0.0051mg per cent, Total carotenes 3.83 mg per cent, Total xanthophylls 14.8mg per cent, Tannins 34.2mg per cent and Total phenolics of 329mg per cent.

Oenothera biennis L. Family: Onagraceae

English Name: Evening Primrose.

A study was carried out by Budinevi *et al.* (1995) on the antioxidative activity of *Oenothera biennis*. Using the ethanolic extracts of plant material the best results were obtained. Antioxidative activity of powdered plant was studied on the Rancimat apparatus at 100°C, and increased induction periods were observed. The extract of *Oenothera biennis* was also in the emulsion of linoleic acid and the course of oxidation was followed by measuring the decoloration rate of β-carotene emulsion. In addition extract was studied at 60°C(determination of peroxide number) and 100°C (Rancimat) using lard as substrate.

Knorr *et al.* (2004) reported antioxidant activity of triterpenoid esters in *Oenothera biennis*.

Borchardt *et al.* (2008) evaluated the antioxidant activity of methanol extracts of seeds of *Oenothera biennis* using DPPH antioxidant assay.The antioxidant value reported in µM Trolox/100g (TE) from DPPH radical scavenging activity of crude seeds extract was 98,563.

Oenothera depressa Greene Family: Onagraceae

A study was carried out by Budinevi *et al.* (1995) on the antioxidative activity of *Oenothera depressa*. Using the ethanolic extracts of plant material good results were obtained.

Oenothera lamarckiana L. Family: Onagraceae

Knorr *et al.* (2004) reported antioxidant activity of triterpenoid esters from *Oenothera lamarckiana*.

Oenothera multicaulis Family: Onagraceae

Lock *et al.* (2005) evaluated the *in vitro* antioxidant activity of 40 Peruvian plants using DPPH assay. *Oenothera multicaulis* showed highest antioxidant activity with EC_{50} value of 16.89µg/ml.

Oldenlandia diffusa (Willd.) Roxb. Family: Rubiaceae

Li *et al.* (2008) evaluated the antioxidant properties of 45 medicinal plants including *Oldenlandia diffusa* using FRAP and TEAC assays and the total phenolic contents of these plants were measured. *Oldenlandia diffusa* showed 26.90 µmol Fe(II)/g FRAP value, 20.66µmol Trolox/g of TEAC value and 9.04 mg GAE/g of Phenolic content.

Olea europaea L. Family: Oleaceae

The antioxidative activity of different butanol extract fractions of olive cake was investigated by Amro *et al.* (2003). Four fractions showed marked antioxidative activity in comparison with BHT. Fractions tested also showed good hydrogen donating abilities, indicating that they had effective activities as radical scavengers. Coumaric, ferulic, cinnamic acids and oleuropein were isolated (Amro *et al.*, 2003).

The dichloromethane and ethanol extracts of *Olea europaea* wood (picual olive cultivar) were screened by Altarejos *et al.* (2005) for antioxidant activity, determined by the DPPH free radical scavenging assay. The ethanol extract displayed potent antioxidant activity.

A method has been proposed by Efmorfopoilou and Rodis (2005) for the extraction of the phenolic compounds from olive oil obtained from *Olea europaea* with as mild conditions as possible. It is based on the ability of cyclodextrins to form stable inclusion complexes with phenolic compounds.

Aldini *et al.* (2006) studied the antioxidant activity of polyphenols from solid olive residues. The results indicate a greater antioxidant activity for the purified extract, due to a cooperative antioxidant interaction among its polyphenol constituents.

Wojcikowski *et al.* (2007) evaluated the antioxidant activity of aerial parts of *Olea europaea*. Oxygen radical absorbance capacity of the ethyl extract, methanol extract and aqueous extracts were 158.20, 860.85 and 261.53µmol Trolox equivalent per g of dried starting material, respectively.

The leaf extract of olive contains α-tocopherol, apigenin, β-carotene, γ-tocopherol, kaempferol and luteolin which is immune modulatory, maintains sugar and cholesterol level in the body (Mishra *et al.*, 2007).

Marwah *et al.* (2007) investigated the antioxidant activity of wound-healing herbs using *in vitro* DPPH and phosphomolybdenum assay methods. Out of the 19 plants screened, the aqueous ethanol extract of *Olea europaea* showed the best DPPH scavenging activities with per cent IP value 89.8 per cent, after 15 min of incubation at an effective test concentration of 50µg/ml. The Total Antioxidant Capacity as gallic acid equivalent of 913mg/g of ethanol extract was obtained in the phosphomolybdenum assay. Total phenolic content was 144 mg GAE/g of ethanol extract.

Bouziz *et al.* (2010) investigated the effect of maturation process of the olive fruit on oil content, phenolic profile and antioxidant activity of four Tunisian cultivars. The total phenolic content varies from 3.46 to 4.30 g/kg at the first stage of maturation and from 8.71 to 11.52 g/kg of fruit fresh weight at the last maturation phase. Total flavonoid content reached 432.80 mg/kg. The antioxidant activity of the extract was evaluated by DPPH and ABTS assays. The IC_{50} values of the olive extracts ranged from 2.69 to 10.96µg/l and from 2.15 to 3.03 mmol/l trolox equivalent at the last stage of maturation.

Haloui *et al.* (2010) reported antioxidant activity of the essential oil of *Olea europaea*.

Olea europaea L. subsp. *africana* Family: Oleaceae

Somova *et al.* (2003) have demonstrated the antioxidant potential of triterpenoids isolated from the leaves of *Olea europea* subsp. *africana* and wild African olive leaves.

Ononis spinosa L. Family: Fabaceae

Souri *et al.* (2004) evaluated the antioxidative activity of *Ononis spinosa* by linoleic acid peroxidation test using 1,3-diethyl-2-thiobarbituric acid as the reagent. Methanolic extract showed 91.06 per cent inhibition at 40µg concentration with IC_{50} value of 2.37µg.

Onosma argenteum Hub.-Mor. Family: Boraginaceae

The n-Hexane-dichloromethane (1:1) extract of the roots was tested *in vitro* for its antioxidant activity by Ozgen *et al.* (2004). The highest antioxidant activity (98 per cent) was observed at 0.1 per cent concentration for the roots.

Ophiorrhiza discolor R.Br. ex G.Don Family: Rubiaceae

Ahmad *et al.* (2010) evaluated the antioxidant potential of 22 species of medicinal plants from Malaysian Rubiaceae including *Ophiorrhiza discolor*. FTC, TBA, TPC and the DPPH assays were employed. The tested extracts showed strong antioxidant potential with percent inhibition of 95.4 per cent in the FTC, and 85.4 per cent in TBA assays. The TPC of the extract was 13.11mg GAE/g PE. A good correlation was observed between total phenolic content and radical-scavenging activities.

Ophiorrhiza major Ridl. Family: Rubiaceae

Ahmad *et al.* (2010) evaluated the antioxidant potential of 22 species of medicinal plants from Malaysian Rubiaceae including *Ophiorrhiza major*. FTC, TBA, TPC and the DPPH assays were employed. The tested extracts showed strong antioxidant potential with percent inhibition of 97.8 per cent in the FTC, and 95.3 per cent in TBA assays. The TPC of the extract was 20.41mg GAE/g PE. A good correlation was observed between total phenolic content and radical-scavenging activities.

Ophryosporus heptanthus (Schultz.-Bip. ex Wedd.) R. King and H.Robins. Family: Asteraceae (Compositae)

The aerial parts of 17 Bolivian plants including *Ophryosporus heptanthus* were screened by Rosas-Romero and Saavedra (2005) to determine antioxidant activity. A methanol extract of each plant was prepared and portioned sequentially with hexane, chloroform, and ethyl acetate, leaving an aqueous solution. All extracts and their 5 fractions for a total of 102 samples, were evaluated using two techniques: an adaptation of the β-carotene bleaching technique using an emulsion of linoleic acid in water as the oxidizable substrate, and the DPPH free radical trapping technique. The results of the β-carotene bleaching technique were more discriminating and better related to the rancidity process under normal conditions; with this assay, 11 species provided at least one fraction with highly promising antioxidant activity. All species gave good results under the DPPH technique, and in most cases they performed better than BHA, which was used as a reference antioxidant (Rosas-Romero and Saavedra, 2005).

The antioxidant activity of 14 compounds isolated from the ether and butanolic extracts of the aerial parts of *Ophryosporus heptanthus* has been assayed using a β-carotene bleaching method and the DPPH technique (Barrero *et al.*, 2006). Compounds 1 to 13 showed the most potent antioxidant activity.

Opuntia dillenii Haw. Family: Cactaceae

The antioxidant activity and inhibitory effect of extracts from *Opuntia dilleni* fruit and its active compounds on low-density lipoprotein (LDL) peroxidation were investigated by Chang su-Feng *et al.*, 2008). The results indicated that the antioxidant activity of methanolic extracts in Trolox equivalent antioxidant capacity and oxygen-radical absorbance capacity assays were in the order of seed>peel>pulp. The results also demonstrated that seeds contained the highest amounts of polyphenols and flavonoids (212.8 and 144.1mg/100g fresh seed, respectively), such as gallic acid, catechin, sinapic acid, epicatechin, *p*-coumaric acid, quercetin and ferulic acid, but no betanin, isobetanin and ascorbic acid as determined by HPLC. However, the peel and pulp contained high amounts of betanin, isobetanin and ascorbic acid, but with lower contents of phenolics and flavonoids as compared to the seed. These findings suggest that phenolics and flavonoids may directly contribute to the antioxidant activity of the seeds.

Opuntia ficus-indica Mill. Family: Cactaceae

English Name: Prickly pear, Cacti-Nea.

Dehydrated extract of the prickly pear fruit *Opuntia ficus-indica*, Cacti-Nea, was evaluated by Bisson *et al.* (2009) for its chronic diuretic and antioxidant effect in wistar rats. Chronic administration of Cacti-Nea significantly increased blood globular levels of glutathione peroxidase in comparison with that of hydrochlorothiazide treated rats The prickly pear fruit extract Cacti-Nea demonstrated chronic diuretic and antioxidant effects in rats with respect to the excretion of the metabolites.

Opuntia vulgaris Mill. Family: Cactaceae

The *in vitro* antioxidant activity of stems of *Opuntia vulgaris* had been investigated by Pal and Mitra (2010) by estimating degree of non-enzymatic haemoglobin glycosylation measured colorimetrically at 520 mm. It was found that benzene and chloroform extract of *O. vulgaris* had better antioxidant activity than petroleum ether, ethyl acetate, ethanol and aqueous extract. The antioxidant activity of the extracts was concentration dependent and comparable to that of D-α- tocopherol (vitamin E) and ascorbic acid (vitamin C), standard antioxidant compounds used.

Orbignya speciosa (Mart. ex Spreng.) Barb. Rodr. Family: Arecaceae (Palmae)

Silva *et al.* (2005) evaluated the antioxidant activity of ethanol and butanol extracts from endocarp/mesocarp of fruit of *Orbignya speciosa*.They evaluated the capacity of the extracts to inhibit the reduction of free radical DPPH and to protect *Saccharomyces cerevisae* cells against the lethal oxidative stress caused by tert-butylhydroperoxide (TBH). Ethanol and butanol extract of leaves and mesocarp of fruits of *Orbignya speciosa* were able to increase the tolerance of *Saccharomyces cerevisae* to TBH and showed to be active as DPPH radical scavengers, thus indicating that this plant extracts could be considered as potential sources of antioxidants. The extracts exhibited a DPPH radical scavenging activity higher than that obtained from *Ginkgo biloba*, a reference plant with well documented antioxidant activity.

Oriciopsis glaberrima Engl. Family: Rutaceae

The acridones C, F and 1,3,5-trihydroxy-4-prenylacridone isolated from stem bark of *Oriciopsis glaberrima* showed potent activity against α-glucosidase, while the C-F acridones showed moderate free radical scavenging activity against DPPH (Wansi *et al.*, 2006).

Origanum acutidens (Hand.-Mazz.) Ietswaart. Family: Lamiaceae (Labiatae)

The essential oil from *Origanum acutidens* exhibited moderate antioxidant capacity in DPPH and β-carotene/linoleic assays. MeOH extracts obtained from herbal parts showed better antioxidative effect than that of BHT, where as callus cultures also exhibited interesting antioxidant patterns (Sokmen *et al.*, 2004).

Alviano and Alviano (2009) evaluated free radical scavenging activity of *Origanum acutidens* essential oil and this was obviously related to the high content of carvacrol (72 per cent, v/v), main oxidant constituents of the oils isolated from several *Origanum* species.

The radical scavenging activities and the composition of *Origanum acutidens* essential oil were evaluated *in vitro* by Goze *et al.* (2010). The sample was subjected to a screening for antioxidant activity by using the DPPH and β-carotene/linoleic acid assays. The essential oil exhibited strong antioxidant activity.

Origanum dictamnus L. Family: Lamiaceae (Labiatae)

Couladis *et al.* (2003) screened Greek aromatic plants from the Lamiaceae family for the antioxidant activity. Of the 21 plants they tested they found ethanol extracts prepared from *Origanum dictamnus*, as well as 9 other plants exhibited the same antioxidant activity as α-tocopherol in their ability to inhibit bleomycin-Fe(II) complex-induced arichidonic acid superoxidation to MDA.

Origanum ehrenbergii Boiss. Family: Lamiaceae (Labiatae)

O. ehrenbergii demonstrated interesting scavenging effects of DPPH with IC_{50} value of 0.99 µg/ml. Essential oil inhibited oxidation of linoleic acid after 30 min of incubation, as well as after 60 min of incubation with IC_{50} values of 42.1 and 33.6µg/ml (Loizzo *et al.*, 2009).

Origanum floribundum Munby Family: Lamiaceae (Labiatae)

Hazzit *et al.* (2006) reported that essential oil of *Origanum floribundum* possessed antioxidant activity.

Origanum glandulosum Desf. Family: Lamiaceae (Labiatae)

Hazzit *et al.* (2006) reported that essential oil of *Origanum glandulosum* possessed antioxidant activity.

Origanum heracleoticum L. Family: Lamiaceae (Umbelliferae)

A total of 27 extracts from non-cultivated and weedy vegetables traditionally consumed by ethnic Albanians in southern Italy were tested for their free radical scavenging activity (FRSA) in the DPPH screening assay, for their *in vitro* non-enzymatic inhibition of bovine brain lipid peroxidation and for their inhibition of xanthine oxidase. In the lipid peroxidation assay, extracts from leaves of *Origanum heracleoticum* showed a remarkable inhibitory activity (>50 per cent). This activity was comparable to quercetin (at a concentration of 50µM) and *Rhodiola rosea* extract (Pieroni *et al.*, 2002).

Origanum majorana L. Family: Lamiaceae (Labiatae)

A purified compound with antioxidant properties (28 mg) T3b, was isolated from a methanol extract (10g) of *Origanum majoranum* (Jun *et al.*, 2001). The *in vitro* scavenging activity of T3b on superoxide anion radical was investigated and compared to those of seven commercially available synthetic or naural antioxidants. Of these, the strongest scavenging action was observed in T3b with and IC_{50} of 1.44µg/ml. The T3b also exhibited significant inhibitory effects on TPA-induceed generation of hydrogen peroxide formation in differentiated HL-60 cells.

Vagi *et al.* (2005) studied the effect of phenols and triterpenoid antioxidant activities from *Origanum majorana* herb and extracts obtained using different solvents and antioxidants were quantified with high performance liquid chromatography.

The antioxidant activity of *Origanum majorana* oil fromTunisia was tested against two refined vegetable oils (soybean and colza) and an animal far lard. The antioxidant effect was confirmed by measuring peroxide values and compared to α-tocopherol. *Origanum majorana* oil showed a high degree of activity at 1000 ppm (Nejla and Moncef, 2006).

Origanum majorana essential oil exhibited concentration-dependent inhibitory effects on DPPH, hydroxyl radical, hydrogen peroxide, reducing power and lipid peroxidation with IC_{50} values of 58.67, 67.11, 91.25, 78.67 and 68.75µg/ml, respectively, while the IC_{50} values for the standard trolox were noted to be 23.95, 44.97, 51.30, 42.22 and 52.72µg/ml, respectively.

Origanum minutiflorum O.Schwarz et P.H.Davis Family: Lamiaceae (Labiatae)

English Name: Toka oregano.

Kosar *et al.* (2003) prepared water extracts from steam distilled essential oil-extracted from *Origanum minutiflorum*. The HPLC-DPPH on-line method was applied to the qualitative and quantitative analysis of this plant extract. There was a strong correlation between the scavenging (negative) peak area and the concentration of the radical scavenging reference substances used. The radical scavenging compounds within the extracts were determined as benzoic acid and hydroxycinnamic acid derivatives, flavonoids and diterpenoids according to their retention time and UV spectral data. Rosmarinic acid and carnosic acid were identified as the dominant radical scavengers in these extracts by this method.

Water-soluble extract from *Origanum minutiflorum* was screened by Dorman *et al.* (2004) for antioxidant properties in a battery of six *in vitro* assays. The extract demonstrated varying degrees of efficacy in each screen. The extract contained Folin-Ciocalteu reagent-reactive substances, which was confirmed by the presence of polar phenolic analytes (*i.e.*, hydroxybenzoates, hydroxycinnamates, and flavonoids).

Origanum onites L. Family: Lamiaceae (Labiatae)

English Name: Oregano, Turkish oregano.

Kosar *et al.* (2003) prepared water extracts from steam distilled essential oil-extracted from *Origanum onites*. The HPLC-DPPH on-line method was applied to the qualitative and quantitative analysis of this plant extract. There was a strong correlation between the scavenging (negative) peak area and the concentration of the radical scavenging reference substances used. The radical scavenging compounds within the extracts were determined as benzoic acid and hydroxycinnamic acid derivatives, flavonoids and diterpenoids according to their retention time and UV spectral data. Rosmarinic acid and carnosic acid were identified as the dominant radical scavengers in these extracts by this method.

Water-soluble extract from *Origanum onites* was screened by Dorman *et al.* (2004) for antioxidant properties in a battery of six *in vitro* assays. The extract demonstrated varying degrees of efficacy in each screen.Turkish oregano was the most effective inhibitors of nonsite-specific hydroxyl radical-mediated 2-deoxy-d-ribose degradation.The extract contained Folin-Ciocalteu reagent-reactive substances, which was confirmed by the presence of polar phenolic analytes (*i.e.*, hydroxybenzoates, hydroxycinnamates, and flavonoids).

The antioxidant activity of oregano essential oil, ethanol extract, hexane extract and deodorized ethanol extract, deodorized hexane extract from oregano leaves were evaluated by Kunduraci *et al.* (2010) in refined sunflower oil using Schaal oven and Rancimat method. The main component in oregano essential oil, carvacrol was also evaluated. Among the extracts, hexane extract at 0.1 per cent concentration exhibited stronger antioxidant activity than deodorized ethanol extract (0.1 per cent), deodorized hexane extract (0.1 per cent) and ethanol extract (0.1 per cent) showed in Schaal oven test at 60°C. The effect of the extracts was less significant in Rancimat test. Only hexane and ethanol extracts showed activity, however the other extracts didn't show any activity in Rancimat test.

Ozkan *et al.* (2010) investigated the influence of harvest time on essential oil composition, phenolic constituents and antioxidant properties of Turkish oregano. Oil distilled from early-season (June) harvested leaves had the highest antioxidant ability expressed as low concentration providing 50 per cent inhibition of free radical scavenging activity and high levels of reducing/antioxidant capacity. Total phenolic contents, free radical scavenging activities and reducing/antioxidant capacities were found to be highest in the July harvest.

Origanum syriacum L. Family: Lamiaceae (Labiatae)

English Name: Syrian Oregano.

Chemical composition and *in vitro* antioxidant and antimicrobial activities of the essential oils from *Origanum syriacum* growing in Turkey were evaluated by Alma *et al.* (2003). The reducing power, DPPH radical scavenging activities and antioxidant activities were investigated. The results showed that the activities were concentration dependent. The antioxidant activities of the oil were slightly lower than those of ascorbic acid or BHT.

Kosar *et al.* (2003) prepared water extracts from steam distilled essential oil-extracted *Origanum syriacum*. The HPLC-DPPH on-line method was applied to the qualitative and quantitative analysis of this plant extract. There was a strong correlation between the scavenging (negative) peak area and the concentration of the radical scavenging reference substances used. The radical scavenging compounds within the extracts were determined as benzoic acid and hydroxycinnamic acid derivatives, flavonoids and diterpenoids according to their retention time and UV spectral data. Rosmarinic acid and carnosic acid were identified as the dominant radical scavengers in these extracts by this method.

Water-soluble extract from *Origanum syriacum* was screened by Dorman *et al.* (2004) for antioxidant properties in a battery of six *in vitro* assays. The extract demonstrated varying degrees of efficacy in each screen. The Syrian Oregano extract was the most effective chelator of iron(II). The extract contained Folin-Ciocalteu reagent-reactive substances, which was confirmed by the presence of polar phenolic analytes (*i.e.*, hydroxybenzoates, hydroxycinnamates, and flavonoids).

Tepe *et al.* (2004) examined the *in vitro* antioxidant properties of the essential oil and various extracts from the herbal parts of *Origanum syriacum* var. *bevani*. Polar subfractions of methanol extract from both deodorized and non-deodorised materials showed the highest DPPH radical-scavenging activity, with IC_{50} values of 21.40 and 26.98µg/ml respectively, where as the IC_{50} of the essential oil was 134.00µg/ml. The antioxidant potential of the extracts appeared to be closely related to the presence of polar phenolics. However, the inhibitive effects on linoleic acid oxidation might be promoted by the presence of non-polar phenolics, as both hexane and dichloromethane extracts showed high antioxidant activities.

Essential oil inhibited oxidation of linoleic acid after 30 min of incubation as well as after 60 min of incubation with IC_{50} values of 46.9 and 58.9 µg/ml (Loizzo *et al.*, 2009).

Origanum vulgare L. Family: Lamiaceae (Labiatae)

English Name: Oregano.

Kosar *et al.* (2003) prepared water extracts from steam distilled essential oil-extracted *Origanum vulgare*. The HPLC-DPPH on-line method was applied to the qualitative and quantitative analysis of this plant extract. There was a strong correlation between the scavenging (negative) peak area and the concentration of the radical scavenging reference substances used. The radical scavenging compounds within the extracts were determined as benzoic acid and hydroxycinnamic acid derivatives, flavonoids and diterpenoids according to their retention time and UV spectral data. Rosmarinic acid and carnosic acid were identified as the dominant radical scavengers in these extracts by this method.

Sahin *et al.* (2004) evaluated the antioxidant activities of the essential oil and methanol extract of *Origanum vulgare* ssp. *vulgare*. Antioxidant studies suggested that methanol extract behaved as a strong free radical scavenger providing IC_{50} at only 9.9 ug/ml, where as the oil showed weaker activity with IC_{50} at 8.9 mg/ml. Total phenolic constituents based on gallic acid equivalents revealed the presence of total soluble phenolics in the extract as 220 ug/mg dry extract (22 per cent, w/w). Ivanova *et al.* (2005) also reported high phenolics content and antioxidant properties in *Origanum vulgare*.

Thymol and rosmarinic acid in oregano prevents oxygen based damage to the cells in the body. It is found to be 42 times more antioxidant activity than apples, 30 times more antioxidant activity than potatoes, 4 times more antioxidant activity than blueberries, 12 times more antioxidant property than oranges and acts as immune booster. Thymol and rosmarinic acid are the antioxidative constituents in oregano (Mishra *et al.*, 2007).

Mariutti *et al.* (2008) evaluated the antioxidant activity of oregano employing ABTS and DPPH radical assay. DPPH EC_{50} value was 8.4g/kg while DPPH-TEAC value was 33.5mM/g. ABTS-TEAC value was 73.3mM/g.

Cioroi and Dumitriu (2009) investigated the total pophenols content and antioxidant activity of aqueous extract of *Origanum vulgare*. The total phenol content was 859.617 mg/100g dried species. Antioxidant activity was directly correlated with the total amount of polyphenols in the extract.

Using a multiple-method approach, the antioxidant activity of the essential oils from *Origanum vulgare* was tested by Viuda-Martos *et al.* (2010). The oregano essential oil had the highest antioxidant activity index in the Rancimat test. Antioxidant activity of essential oils of five spice plants, including *Origanum vulgare* used in the Mediterranean diet was evaluated by Viuda-Martos *et al.* (2010). All the essential oils tested had antioxidant activity.

Diverse oils obtained from *Origanum vulgare* subsp. *glandulosum* collected at different locations of Tunisia showed different but always good abilities for scavenging

DPPH radicals. Such results depended on the percentage of phenolic compounds (thymol) present in the oils (Mechergui *et al.*, 2010).

Motamed and Naghibi (2010) evaluated the antioxidant properties *Origanum vulgare* by four different methods: free radical scavenging using DPPH, evaluation of xanthine-oxidase activity, inhibition of lipid peroxidation by the ferric thiocyanate method, and the deoxyribose degradation assay. The extracts showed antioxidant activity.

Spiridon *et al.* (2011) evaluated the total phenolic content and antioxidant activity of plants used in Romanian herbal medicine. Total phenolic concentration was determined using Folin-Ciocalteu phenol reagent method, while total flavonoids were measured using the aluminium chloride colorimetric method. *Origanum vulgare* extract showed the highest antioxidant activity and total phenol content compared to the other plant extracts. A positive correlation was observed between total antioxidant activity and total phenolic content of the analyzed extracts.

Origanum vulgare L. subsp. *hirtum* Family: Lamiaceae (Labiatae)

Damien *et al.* (1995) reported the evaluation of the essential oils of *in vitro* plants like *Origanum vulgare* subsp. *hirtum* for their natural antioxidant properties.

Ornithogalum alpigenum Stapf. Family: Liliaceae

Makasci *et al.* (2010) studied that the total antioxidant and free radical scavenging activities of the extractions obtained from *Ornothogalum alpigenum* using β-carotenelinoleic acid model system and was found the highest antioxidant activity of 88.12 ± 0.9 per cent in methanol bulb extracts. Free radical scavenging activity of the extracts using the free radical DPPH was also determined. The overall results showed that some solutions have values (90.9 per cent > 90.5 per cent > 90.4 per cent > 88.4 per cent) very close to those of Butylated Hydroxytoluene (BHT) (90.0 per cent). The leaf extracts were more potent with respect to free radical scavenging activity.

Ornithogalum sintenisii L. Family: Liliaceae

The antioxidant potency of the freeze-dried methanol extract of *Ornithogalum sintenisii* bulbs and aerial parts were investigated by Ebrahimzadeh *et al.* (2010) by evaluating the following parameters: linoleic acid peroxidation, DPPH, scavenging of nitric oxide and hydrogen peroxide as well as reducing power and Fe^{2+} chelating activity, using standard methods. Phenol and flavonoid contents were determined as gallic acid and quercetin equivalents, respectively. The aerial parts contained higher phenol and flavonoid contents than the bulbs. IC_{50} for DPPH radical-scavenging activity was 368 and 669 µg/ml for aerial parts and bulbs, respectively. The reducing power of the extracts was less than that of vitamin C with the aerial parts showing stronger activity than the bulbs. The extracts did not show any activity in the peroxidation test but displayed good H_2O_2 radical scavenging activity compared with quercetin (IC_{50}=52.0 µg/ml) which was used as positive control. The bulb and aerial parts of *O. sintenisii* aerial parts (at flowering stage) exhibited good but varying levels of antioxidant activities in nearly all the models studied.

Orobanche caerulescens Steph. Family: Orobanchaceae

Two new phenylpropanoid glycosides, caerulescenoside and 3'-methyl crenatoside as well as five known phenylpropanoid glycosides were isolated by Lin *et al.* (2004) from the whole plant. All these compounds showed antioxidative activity.

Orostachys japonicus A.Berger Family: Crassulaceae

Antioxidant activities of ethanol and water extracts from *Orostachys japonicus* leaf, stem, and root were determined by Lee *et al.* (2009) by rancimat method, DPPH radical scavenging effect, chelating effect, and reducing power analysis. The highest total phenolic compound (TPC) as 14.6 mg/g of dry sample and the strongest antioxidant activity in rancimat method (value of AI 1.98), DPPH radical scavenging effect (96 per cent in 4 mg/ml), and reducing power (1.50 in 4 mg/ml) were observed in ethanol extracts from *Orostachys japonicus* leaf. Heat and pH stabilities on antioxidant activity of *Orostachys japonicus* leaf extract were studied through TPC and DPPH radical scavenging effect. As a result, the extracts from *Orostachys japonicus* leaf showed high stability.

Oroxylum indicum (L.) Vent. Family: Bignoniceae

Antioxidant capacities of 56 selected Chinese medicinal plants, including *Oroxylum indicum* were evaluated by Song *et al.* (2010) using the TEAC and FRAP assays, and their total phenolic content was measured by the Folin-Ciocalteu method. *Oroxylum indicum* showed moderate antioxidant activity with TEAC value of 85.20μmol Trolox/g, FRAP value of 45.64μmol Fe^{2+}/g and phenolic content was 8.15mg GAE/g.

Orthosiphon aristatus (Blume) Miq. Family: Lamiaceae (Labiatae)

Antioxidant and anti-inflammatory effects of methanol, ethanol and water extracts from *Orthosiphon aristatus* (abbreviated as MEOA, EEOA and WEOA, respectively) was investigated by Hsu *et al.* (2010). The evaluation of antioxidant activity was determined by total phenolics, Trolox equivalent antioxidant capacity (TEAC), oxygen-radical absorbance capacity (ORAC) and cellular antioxidant activity (CAA) assays. These assays demonstrated a relatively high antioxidant activity for MEOA and EEOA. These results revealed that EEOA had the most prominent inhibitory effect on lipopolysaccharide (LPS)-stimulated nitric oxide (NO), prostaglandin E_2 (PGE_2) and intracellular reactive oxygen species (ROS) production in RAW 264.7 cells. A high performance liquid chromatography profile indicated that MEOA and EEOA contained both ursolic acid and oleanolic acid. Moreover, ursolic acid significantly reduced NO production in LPS-stimulated RAW 264.7 cells. Both EEOA and ursolic acid inhibited LPS-stimulated protein and mRNA expression of both inducible nitric oxide synthase (iNOS) and cyclooxygenase-2 (COX-2) in these cells.

Orthosiphon grandiflorus Bondingh Family: Lamiaceae (Labiatae)

Chanwitheesuk *et al.* (2005) investigated the antioxidant activity of *Orthosiphon grandiflorus* using a β-carotene bleaching method. The contents of plant chemicals such as vitamin C, Vitamin E, carotenoids, tannin and total phenolics, were also determined. Methanolic extract showed antioxidant activity. *Orthosiphon grandiflorus* has antioxidant index of 1.99, Vitamin C 7.69 mg per cent, Vitamin E 0.0011mg per cent, Total carotenes 3.20 mg per cent, Total xanthophylls 25.4mg per cent, Tannins 30.9mg per cent and Total phenolics of 145 mg per cent.

Orthosiphon stamineus Benth. Family: Lamiaceae (Labiatae)

Akowuah *et al.* (2003) evaluated the *in vitro* antioxidant activity of water extracts of *Orthosiphon stamineus* leaves, taken from plants cultivated in different places of Malaysia. The antioxidant activity was assessed by using β-carotene bleaching method. All the water extracts of the leaf samples from various places showed antioxidant activity comparable to that of pure quercetin and butylated hydroxylanisole (BHA) thus presenting an alternative source for natural additives.

Zakaria *et al.* (2008) studied the antioxidant properties of the methanol extract of leaves and stems of *Orthosiphon stamineus* by DPPH radical scavenging activity. The extracts showed antioxidant activity comparable to butylated hydroxytoulene (BHT). Akowuah *et al.* (2003) also reported antioxidant activity.

Ho *et al.* (2010) evaluated *in vitro* antibacterial and antioxidant activities of extracts of *Orthosiphon stamineus*. This study showed that the highest concentration of rosmarinic acid had the best antibacterial and free radical scavenging activities. This suggest that rosmarinic acid content is closely associated with antibacterial and free radical scavenging activities of *O. stamineus* extracts.

Oryza sativa L. Family: Poaceae (Graminae)

English Name: Rice.

From the black colored rice bran of *Oryza sativa* cv. Heugjinjubyeo, a new 2-arylbenzofuran, 2-(3,4-dihydroxyphenyl)-4,6-dihydroxybenzofuran-3-carboxylic acid methyl ester, oryzafuran was isolated by Han *et al.* (2004). Its structure was elucidated on the basis of spectral data. This compound showed strong antioxidant activity in a DPPH free radical scavenging assay (Han *et al.*, 2004).

Antioxidative and compositional effects of methanolic crude long grain rice bran extracts (0.025, 0.05 and 0.1 per cent w/w) on oxidative stability of tuna oil were determined by Chatchawan *et al.* (2008). Oxidative stability of tuna oil was increased with increasing concentration of crude long grain rice bran extracts during 30 days of storage.

Sreeramulu *et al.* (2009) evaluated the antioxidant activity of milled and parboiled rice by DPPH, scavenging assay, FRAP assay and reducing power. Phenolic content was 47.64 mg/100g for rice and 50.87 mg/100g for parboiled rice. DPPH radical scavenging activity expressed as trolox equivalent, FRAP (μmol/g) and reducing power (mg/g) were 1.23, 61 and 0.79, respectively in milled rice and 0.75, 67.5 and 0.84 respectively in parboiled rice.

Qiu *et al.* (2010) investigated the antioxidant properties of commercial wild rice. The antioxidant activity of wild rice methanol extract was found to be up to 10 times greater than that of white rice (control sample) according to their DPPH radical scavenging activity and ORAC. Ferulic acid was found as the most abundant phenolic acid (up to 355mg/kg) followed by sinapic acid in wild rice. They both occurred in the insoluble form. Other monomeric phenolic acids present in wild rice consisted of p-coumaric, vanillic, syringic, and p-hydroxybenzoic acids, along with two phenolic acid aldehydes. They were present in both soluble and insoluble forms. DPPH free radical scavenging activities of soluble and insoluble fractions suggest that the antioxidant activity of wild rice is partially attributed to its phenolic acid profile.

Osbeckia aspera Bl. Family: Melastomataceae

Osbeckia aspera plant extract was shown to inhibit significantly in a dose-dependent manner the activities of DPPH free radical (EC_{50} of 27.5µg/ml), xanthine oxidase (EC_{50} of 1.16µg/ml) and demonstrated a scavenging effect on hydroxyl radical mediated damage to deoxyribose (Thabrew *et al.*, 1998). The plant extract possessed some prooxidant activity from the effect on bleomycin-induced DNA damage, but this was less than that shown by comparable concentrations of (+)-catechin or silymarin.

Osmanthus fragrans Lour. Family: Oleaceae

Wu *et al.* (2009) evaluated the antioxidant activity of the acetonic extract of *Osmanthus fragrans*. The antioxidation activities, measured in terms of EC_{50} values using DPPH and ABTS assays, were 304.9 mg ascorbic acid equivalent/g extract and 516.3 mg Trolox equivalent/g of extract, respectively.

In the study by Wang *et al.* (2010) a significant inhibitory effect (EC_{50} = 50.29µg/ml) on the ABTS free radicals was detected in the crude ethanol extracts (CEE) of *Osmanthus fragrans* pulp. Then CEE was separted into methanol soluble fraction (MSF) and a methanol insoluble fraction (MISF), and their scavenging activities on the ABTS free radicals ascended in the order MSF>CEE>MISF, respectively. Accordingly, the total phenolic (TP) contents of MSF, CEE and MISF were 40.53, 35.37 and 29.62 mg/g (dry weight) respectively. Furthermore, salidroside (tyrosol 8-O-β-glucopyranoside), which showed powerful antioxidant activities in all test systems was isolated from MSF. The antioxidant activity of salidroside (81.54µg/ml) on DPPH free radicals was found to be nearly twice that of MSF (144.78µg/ml), and its total antioxidant activity was superior to that of the synthetic antioxidant BHT.

Osyris lanceolata Hochst. et Steud. ex A.DC. Family: Santalaceaee

The root bark of *Osyris lanceolata* was screened by Yeboah and Majinda (2009) for its antioxidant potential using three variations of the DPPH radical scavenging method *viz.* a TLC-autographic assay, a semi-quantitative TLC assay, and a spectrometric assay. The total phenolic content was evaluated as GAE using the Folin-Ciocalteu method. The radical scavenging ability, as measured by IC_{50} values, was found to be 48.4 and 49.5µg/ml for the 90 per cent methanol/water and methanol extracts, respectively, while those of chloroform, n-Hexane and Supercritical fluid extract (SFE) were found to be >250µg/ml. Under the same conditions, the values for the standards, ascorbic acid and gallic acid, were found to be 38.70 and 2.86µg/ml, respectively. The radical scavenging power of the five extracts correlated positively with the total phenol content and was ranked in the following decreasing order. 90 per cent methanol/water > methanol > chloroform > SFE > n-Hexane with total phenolic content of 290.2, 271.0, 74.4, 62.5 and 49.5 mg GAE/g of dry extract, respectively. The 90 per cent methanol/water extract and methanol extracts showed several components with high antioxidant activity displaying best kinetics in both TLC assays, while chloroform, SFE, and n-Hexane extracts exhibited slow kinetics antioxidant activity.

Otostegia limbata Benth. Family: Lamiaceae (Labiatae)

Ahmed *et al.* (2010) evaluated the antioxidant activity of water, n-hexane, chloroform, ethyl acetate, n-butanol and methanol extracts of leaves of *Otostegia limbata*. Some extracts showed strong antioxidant activity. DPPH free radical scavenging activity was determined and EC_{50} (microg) values were found to be in between 61 (ethyl acetate fraction) and 96 (n-butanol fraction). The total phenolic content varied from 489-1273 mg GAE/100 g. Total flavonoid content was 198-3018 mg QE/100 g, TEAC values ranged from 30-139 (micromole/g) using ABTS assay, while FRAP values ranged from 5-41 (mmol/g).

Otostegia michauxii Briq. Family: Lamiaceae (Labiatae)

Firuzi *et al.* (2010) evaluated the antioxidant activity and total phenolic content of 24 Lamiaceae species growing in Iran. *Otostegia michauxii* showed high antioxidant activity with 26.7µM/g DW FRAP value and IC_{50} = 465.4µg/ml of DPPH radical scavenging activity. It also showed high phenolic content of 10.7 mg CE/g. FRAP and DPPH assay results showed good correlations with the total phenolic contents of the plant measured by the Folin-Ciocalteu assay.

Otostegia persica Boiss. Family: Lamiaceae (Labiatae)

The antioxidant activity of different extracts of aerial parts of *Otostegia persica*, were evaluated by Yassa *et al.* (2005) using β-carotene bleaching and lipid peroxidation models. The inhibitory activity of these extracts on the peroxidation of linoleic acid was measured by ammonium thiocyanate in comparison to green tea, *Camellia sinensis*, *Ginkgo biloba* and BHA. A methanol extract of the plant exhibited the highest antioxidant activity. Five compounds were separated and purified from the methanol extract by column and paper chromatography, respectively. Three isolated flavonols showed significant antioxidant activity comparable to BHA and vitamin E in both methods. These active compounds were identified as morin, kaempferol and quercetin. The only identified C-glucoflavone (isovitexin) also exhibited activity, but weaker than the flavonols. Trans-cinnamic acid showed no activity in these methods (Yassa *et al.*, 2005).

Oxalis corniculata L. Family: Oxalidaceae

Kathiroya *et al.* (2010) evaluated the antitumor and antioxidant activity of *Oxalis corniculata*. This study revealed that the ethanol extract of *Oxalis corniculata* (EEOC) showed significant antitumor and antioxidant activities in Ehrlich Ascites carcinoma bearing mice. A significant decrease in liver MDA levels and increase in catalase and reduced glutathione levels were observed in EEOC treated animals.

Senthilkumar and Rajkapoor (2010) investigated the effect of methanolic extract of *Oxalis corniculata* (MEOC) on antioxidant enzymes in rat brain after induction of seizures by MES and PTZ. SOD, glutathione peroxidase, glutathione reductase and Catalase were decreased in rat brain due to seizure and it was retored significantly by administration of ethanol extract of *Oxalis corniculata* treated rats. Similar dose dependent results were obtained in PTZ model also. Whereas MEOC significantly decreased lipid peroxidation in both models.

Alam *et al*. (2011) evaluated the antioxidant activities, both *in vitro* and *in vivo*, of the crude methanolic extracts of the whole plant of *Oxalis corniculata* along with its various orgnic fractions. The different assay methods including total antioxidant activity, scavenging free radical, authentic peroxynitrite, nitric oxide and reducing power assessment were used to evaluate the antioxidant potential of the crude extract and its orgnic fractions. The EtOAC fraction, showed strong activity in all the model systems tested and in peroxynitrite model this fraction (IC_{50} value of 2.29µg/ml) exerted three-fold stronger activity than standard penicillamine (IC_{50} value of 6.20 µg/ml)). The reducing power of the extract was found to be concentration dependent. The administration of the extract/fractions at a dose of 250 and 500 mg/kg body weight to the male wistar rats increased the percentgee inhibition of reduced glutathione, superoxide dismutase and catalase significantly. Whereas, lipid peroxidation level in hepatotoxic rats markedly decreased at a dose of 500 mg/kg body weight after 7 days. The total phenol and flavonoid content were also measured in the crude methanolic extract along with its organic fractions

Oxalis tuberosa Molina Family: Oxalidaceae

Penarrieeta *et al*. (2005) measured the Total Antioxidant Capacity (TAC) in some Andean foods. Eight Andean foods, including *Oxalis tuberosa* were analyzed by two methods ABTS and FRAP to assess TAC. TAC value of *Oxalis tuberosa* was good in both water-soluble fraction and water-insoluble fraction.

Campos *et al*. (2006) and Salluca *et al*. (2008) evaluated the phenolic compounds and antioxidant capacity of tuber of *Oxalis tuberosa*. Total antioxidant capacity was measured using ABTS and FRAP methods. *Oxalis tuberosa* showed moderate antioxidant acitivity. Data from FRAP displayed a linear correlation with TPH data.

Oxystelma esculentum (L.f.) R.Br. ex Schult. Family: Asclepiadaceae

Ashokkumar *et al*. (2008, 2009) evaluated the antioxidant and free radical scavenging properties of *Oxystelma esculentum* (MEOE). The total antioxidant, reducing power, DPPH radical, superoxide anion radical, nitric oxide radical, hydroxyl radical, hydrogen peroxide scavenging and total phenol content in MEOE was carried out. The total antioxidant activity of MEOE increases with increasing concentration. The reducing capability and free radical scavenging activity in DPPH also increased in a dose dependent manner. The MEOE was found to scavenge the superoxide generated by PMS/NADH/NBT system. Moreover, MEOE was found to inhibit the nitric oxide radical generated from sodium nitropruside. MEOE was also found to inhibit the hydroxyl radical generated by Fe^{3+}/ascorbate/EDTA/H_2O_2 system. The MEOE scavenges the hydrogen peroxide in a dose dependent manner. The amount of total phenolic compounds was also determined.

Oxytropis falcata Bunge Family: Fabaceae (Leguminosae)

The antioxidant properties of the various extracts and flavonoids prepared from *Oxytropis falcata* were investigated by Jiang *et al.* (2008) by DPPH radical-scavenging assay. In the chloroform, ethyl acetate and n-butanol extracts, the ethyl acetate extract exhibited the highest antioxidant activity (IC_{50} = 2.05 mg/ml). Kaempferol (IC_{50} = 0.11 mg/ml), rhamnetin (IC_{50} = 0.14 mg/ml) and rhamnocitrin (IC_{50} = 0.15 mg/ml), isolated from the extract exhibited considerable antioxidant activities.

Paederia foetida Family: Rubiaceae

The antioxidant activity of fresh and dried plant extracts of *Paederia foetida* were studied by Osman *et al.* (2009) using β-carotene bleaching and the ABTS radical cation assay. The percentage of antioxidant activity for all extract samples using both assays was between 58 and 80 per cent. The fresh samples had higher antioxidant activity than the dried samples. The results of β-carotene bleaching assay were correlated with those of the ABTS assay.

Paeonia emodi Wall. ex Hook.f. Family: Paeoniaceae

Paeoniol and paeonin C, oligostilbene and monoterpene galactoside, have been isolated from the methanolic extract of the fruits of *Paeonia emodi*. In addition, 4-hydroxybenzoic acid, gallic acid and methylgallate have also been reported. Paeoninol and paeonin have displayed potent inhibitory potential against enzyme lipoxygenase in a concentration-dependent fashion with the IC_{50} values 0.77 and 99.5muM, along with ABTS+ radical quenching activity with IC_{50} values of 147.5 and 498.2muM, respectively (Riaz *et al.*, 2004).

Paeonia lactiflora Pall. Family: Paeoniaceae

Cai *et al.* (2004) evaluated the antioxidant activity and phenolic compounds in traditional Chinese medicinal plants associated with anticancer. The improved ABTS.+ method was used to systematically assess the total antioxidant capacity (Trolox equivalent antioxidant capacity, TEAC) of the medicinal extracts. The TEAC values and total phenolic content for methanolic extract of root and seed of *Paeonia lactiflora* were 407.3 μmol Trolox equivalent/100g dry weight (DW), and 1.14 g of gallic acid equivalent/100g DW respectively. A positive significant linear relationship between antioxidant activity and total phenolic content showed that phenolic compounds were the dominant antioxidant components in the tested herbs. Major types of phenolic compounds in *Paeonia lactiflora* include flavonols (astragalin), tannins (gallotannin), stilbenes (resveratrol).

Li *et al.* (2008) evaluated the antioxidant capacities of 45 selected medicinal plants using ferric reducing antioxidant power (FRAP) and Trolox equivalent antioxidant capacity (TEAC) assays respectively, and the total phenolic contents of these plants were measured by the Folin-Ciocalteau method. It was found that *Paeonia lactiflora* possessed the highest antioxidant capacities with 345.02 μmol Fe(II)/g FRAP value, 219.42μmol Trolox/g of TEAC value and 26.75 mg GAE/g of Phenolic content. A strong correlation between TEAC values and those obtained from FRAP assay implied that antioxidants in these plants were capable of scavenging free radicals and reducing oxidants. A high correlation between antioxidant capacities and their total phenolic contents indicated that phenolic compounds were a major contributor of antioxidant activity of these plants.

Paeonia suffruticosa Andrl. Family: Paeoniaceae

Methanol aqueous extract of *Paeonia suffruticosa* was screened by Kim *et al.* (1997) for anti-oxidative activity using Fenton's reagent/ethyl lineolate system. The extract showed antioxidant activity.

Methanol extract of *Paeonia suffruticosa* showed relatively high levels of DPPH radical scavenging activity (IC_{50} <6.0 µg/ml) (Lee *et al.*, 2003). The activities of superoxide dismutase, catalase and glutathione peroxidase were dose-dependently enhanced in V79-4 cells treated with the plant extract.

Cai *et al.* (2004) evaluated the antioxidant activity and phenolic compounds in traditional Chinese medicinal plants associated with anticancer. The improved ABTS.+ method was used to systematically assess the total antioxidant capacity (Trolox equivalent antioxidant capacity, TEAC) of the medicinal extracts. The TEAC values and total phenolic content for methanolic extract of root-bark of *Paeonia suffruticosa* were 876.1 µmol Trolox equivalent/100g dry weight (DW), and 2.90 g of gallic acid equivalent/100g DW respectively. A positive significant linear relationship between antioxidant activity and total phenolic content showed that phenolic compounds were the dominant antioxidant components in the tested herbs. Major types of phenolic compounds in *Paeonia suffruticosa* include phenolic acid (gallic acid), hydrolysable tannins (gallotannin), other phenolics (paeonol).

Li *et al.* (2008) evaluated the antioxidant capacities of 45 selected medicinal plants using ferric reducing antioxidant power (FRAP) and Trolox equivalent antioxidant capacity (TEAC) assays respectively, and the total phenolic contents of these plants were measured by the Folin-Ciocalteau method. It was found that *Paeonia suffruticosa* possessed the highest antioxidant capacities with 328.44 µmol Fe(II)/g FRAP value, 221.10µmol Trolox/g of TEAC value and 24.51 mg GAE/g of Phenolic content. A high correlation between antioxidant capacities and their total phenolic contents indicated that phenolic compounds were a major contributor of antioxidant activity of these plants.

Paepalanthus bromelioides Silv. Family: Eriocaulaceae

Stasi *et al.* (2004) studied the intestinal anti-inflammatory activity of Paepalantine an isocoumarin isolated form the capitula of *Paepalanthus bromelioides*, in the tri nitrobenzenesulphonic acid model of rat colitis and this is probably related with its antioxidant properties.

Devienne *et al.* (2009) investigated the antioxidant activity of isocoumarins (1-50µM) paepalantine, 8,8'-paepalantine dimer, and vioxanthin isolated from *Paepalanthus bromelioides* on mitochondria. The paepalantine, 8,8'-paepalantine dimer, but not vioxanthin, were effective at scavenging both DPPH and superoxide radicals in non-mitochondrial systems and protected mitochondria from tert-butylhydroperoxide-induced H_2O_2 accumulation and Fe^{2+}citrate-mediated mitochondrial membrane lipid peroxidation, with almost the same potency as quercetin. These results point towards paepalantine, followed by paepalantine dimer, as being a powerful agent affording protection apparently via O^*_2-scavenging, from oxidative stress conditions imposed on mitochondria, the main intracellular source and target of those reactive oxygen species. This strong antioxidant action of paepalantine was reproduced in HepG2 cells exposed to oxidative stress conditions induced by H_2O_2.

Pajanelia longifolia (Willd.) K.Schum. Family: Bignoniaceae

Phytochemical characterization and bioactivity evaluation of the Indian medicinal plant *Pajanelia longifolia* was carried out by Choudhury (2010). Three new compounds were isolated and characterized from the bark of the plant. The different crude extracts were evaluated for hepatoprotective activity and as such various biochemical parameters were measured. The 70 per cent ethanolic extracts showed significant antioxidant activity as demonstrated by the DPPH and OH-radical scavenging activity as compared to other two extracts. The extracts also showed significant activity by reducing the level of lipid peroxidation with increase in SOD and CAT levels.

Palicourea acetosoides Wernham Family: Rubiaceae

Mosquera *et al.* (2007) studied the antioxidant activity of the extracts using DPPH free-radical scavenging method. The percentage of the antioxidant activity of the n-Hexane fraction was 19.995, Dichloromethane fraction was 40.92 and methanol fraction was 39.671.

Palicourea angustifolia Kunth. Family: Rubiaceae

Mosquera *et al.* (2009) evaluated the antioxidant activity of forty-six methanol plant extracts using DPPH radical scavenging assay. The plant extracts that showed greatest antioxidant activity were *Palicourea angustifolia* with 36.1 per cent of antioxidant activity.

Palisota hirsuta (Thumb.) K. Schum. Family: Commelinaceae

Mamyrbekova-Bekro *et al.* (2008) evaluated the phytocompounds of the extract and assessed their antioxidant potential of *Palisota hirsuta* leaves and root. The extracts showed the presence of polyphenols, flavonoids, tannins and coumarins. The extracts showed DPPH scavenging activity. This activity is probably due to the presence of the phenolic compounds, flavonoids and tannins.

Paliurus spina-christi Miller Family: Rhamnaceae

Souri *et al.* (2004) evaluated the antioxidative activity of *Paliurus spina-christi* by linoleic acid peroxidation test using 1,3-diethyl-2-thiobarbituric acid as the reagent. Methanolic extract showed 85.38 per cent inhibition at 40µg concentration with IC_{50} value of 1.83µg, which is approximately in the range of α-tocopherol.

Kirca and Arslan (2008) investigated the antioxidant capacity and total phenol contents of *Paliurus spina-christi*. The antioxidant capacity of methanolic extracts prepared from various parts of plants was evaluated by both trolox equivalent antioxidant capacity (TEAC) and DPPH assays, while the total phenolics were determined using the Folin-Ciocalteu method. *Paliurus spina-christi* showed higher antioxidant activities in both the TEAC and DPPH method.

Panax ginseng C.A.Meyer Family: Araliaceae

English Name: Ginseng.

A standardized aqueous extract of *Panax ginseng* root was tested for its antioxidant effect on primary astrocytes culture on an antioxidant stress model generated by H_2O_2 (Naval *et al.,* 2007). The results indicated that this extract had a significant effect on the reduction of astrocytic death induced by H_2O_2. Dose-response experiments revealed that this ginseng extract increased cell viability at a wide range of concentrations. Exposure of astrocytes to H_2O_2 decreased the activities of antioxidant enzymes, and increased ROS formation. Ginseng root extract reversed the effect of almost all of these parameters in H_2O_2-injured primary cultures of rat astrocytes.

Sun ginseng (SG) is heat-processed *Panax ginseng* steamed at 120°C, which has ginsenoside-Rg3, -RK1 and –Rg5 as its main ginsenoside components. The effect of SG on lipopolysaccharide (LPS)-induced liver injury in rats was investigated by Kang *et al.* (2007). Injection of LPS (i.v.) induced excessive nitric oxide (NO) generation in serum and increased the hepatic mitochondrial TBARS level. However the elevated TBARS level was significantly lowered by 15 consecutive days of SG administration. In addition, upregulated hepatic inducible nitric oxide synthase and heme oxygenase 1 levels in LPS-treated control rats were significantly lowered and increase, respectively, by 100mg/kg body weight/day of SG administration. The antioxidant effects were thought to be partially related to the deactivation of nuclear factor-kappaB by SG administration (Kang *et al.,* 2007).

Wojcikowski *et al.* (2007) evaluated the antioxidant activity of aerial parts of *Panax ginseng.* Oxygen radical absorbance capacity of the ethyl extract, methanol extract and aqueous extracts were 1.22, 20.79 and 16.17µmol Trolox equivalent per g of dried starting material, respectively.

Mahour and Saxena (2010) studied free radical scavenging activity of *Panax ginseng* under *in vitro* models *viz.,* DPPH, nitric oxide and Lipid peroxidation method. IC_{50} values in *Panax ginseng* were 199.22, 7.56 and 130.72 µg/ml with DPPH, nitric oxide and lipid peroxidation groups, respectively.

Panax notoginseng (Burk.) F.H.Chen Family: Araliaceae

Ng *et al.* (2004) reported antioxidant effects of aqueous and organic extracts of *Panax notoginseng.*

Gan *et al.* (2010) evaluated the antioxidant activity of 40 medicinal plants. *Panax notoginseng* showed very low antioxidant activity of 8.05 µmol Fe(II)/g in FRAP assay and 6.80µmol in TEAC assay. Total phenolic contet was also very low, 0.46 mg GAE/g.

Panax quinquefolius L. Family: Araliaceae

English Name: American ginseng.

Ng *et al.* (2004) reported antioxidant effects of aqueous and organic extracts of *Panax quinquefolius.*

A North American ginseng extract (NAGE) containing known principal ginsenosides for *Panax quinquefolius* was assayed by Kitts *et al.* (2010) for metal chelation, affinity to scavenge DPPH-stable free radical, and peroxyl (LOO*) and hydroxyl (OH) free radicals for the purpose of characterizing the mechanisms of antioxidant activity. These results indicate that NAGE exhibits effective antioxidant activity in both lipid and aqueous medium by both chelation of metal ions and scavenging of free radicals.

Pandanus amaryllifolius Roxb. Family: Pandanaceae

The potential uses of *Pandanus amaryllifolius* leaf extract as a natural antioxidant were evaluated in refined, bleached and deodorized palm olein. The extracts (optimum concentration 0.2 per cent) significantly retarded oil oxidation and deterioration, comparably to 0.02 per cent BHT in tests such as peroxide value, aniside value, iodine value, free fatty acid, oxidative stability index, polar and polymer compound contents. *Pandanus amaryllifolius* leaf extract, which had a polyphenol contet of 102 mg/g exhibited an excellent heat-stable antioxidant property (Nor *et al.*, 2008).

Pandanus odoratissimus L. Family: Pandanaceae

The methanolic and aqueous extracts of *Pandanus odoratissimus* L. root were examined by Sasikumar *et al.* (2009) for their contents of phenolics and flavonoids as well as for their *in vitro* antioxidant activities using 1, 1-diphenyl-2-picryl hydrazyl (DPPH) radical quenching assay and reducing power. The methanolic root extract yielded 56.24 mg/g GAE of phenolic content and 81.25 mg/g CE of flavonoid content. Higher antioxidant potential was observed in both DPPH scavenging assay (EC = 48.3 µg mLG) and reducing capacity 501(OD at 1000 µg mLG1 = 0.787) by the methanolic root extract than by the aqueous extract. A positive correlation was found between phenolics and flavonoid contents and antioxidant properties of the extracts.

Panigrahi *et al.* (2011) evaluated the antitumor effect and antioxidant role of *Pandanus odoritassimus* whole plant in animal model. The Acetone fraction of *Pandanus odoritassimus* (AFPO) was administered at 200 and 400 mg/kg b.w. once a day for 14 days, after 24 hours of tumor inoculation. The effect of AFPO on the growth of tumor, life span of EAC bearing mice, hematological profile, liver biochemical parameters (lipid peroxidation, antioxidant enzymes) were estimated. AFPO decrease, the tumor volume, viable cell count and increasing the life span of EAC bearing mice and brought back the hematological parameter more or less normal level. The effect of AFPO also decreased the levels of lipid peroxidation and increased the levels of glutathione (GSH), superoxide dismutase (SOD) and catalase (CAT).

Pandanus sanderi Sander ex M.T. Mast. Family: Pandanaceae

Rahman *et al.* (2010) investigated the phytochemical composition and antioxidant activity of *Pandanus sanderi* leaves extracts. In this study the yellow stiped leaves extract showed the most potent antioxidant activity. However, it was less effective in chelating ferrous ion, compared to EDTA.

Papaver bracteatum Lindl. Family: Papaveraceae

Souri *et al.* (2004) evaluated the antioxidative activity of *Papaver bracteatum* by linoleic acid peroxidation test using 1,3-diethyl-2-thiobarbituric acid as the reagent.

Methanolic extract showed 92.21 per cent inhibition at 40µg concentration with IC_{50} value of 3.51µg.

Papaver sominferum L. Family: Papaveraceae

English Name: Opium poppy.

Wojcikowski *et al.* (2007) evaluated the antioxidant activity of aerial parts of *Papaver somiferum.* Oxygen radical absorbance capacity of the ethyl extract, methanol extract and aqueous extracts were 3.35, 10.72 and 2.05µmol Trolox equivalent per g of dried starting material, respectively.

Souri *et al.* (2008) evaluated the antioxidant activity of seed against linoleic acid peroxidation using 1,3-diethyl-2-thiobarbituric acid as reagent. Antioxidant activity (IC_{50}) against peroxidation of linoleic acid (2mg/ml) was 49.75 and phenolic content was 44.42 mg/100g dry weight. The results of this study showed that there is no significant correlation between antioxidant activity and phenolic content of the studied plant materials and phenolic content could not be a good indicator of antioxidant capacity.

Parabarium huaitingii Chun et Tsiang Family: Apocynaceae

Three new phenylpropanoid-substituted epicatechin glycosides, namely parabosides A-C together with three known compounds, 5-caffeoylquinic acid, 5-caffeoylshikimic acid and 3,4-dicaffeoylquinic acid were obtained from the plant *Parabarium huaitingii* (Tang *et al.*, 2009). The oxygen radical absorbance capacity assay was applied to evaluate their antioxidative capacity *in vitro* which revealed that all the compounds possess strong antioxidant properties.

Parinari curatelifolia Planch. ex Benth. Family: Chrysobalanaceae

The phenolic content and antioxidant capacities of *Parinari curatelifolia, Strychnos spinosa* and *Adansonia digitata* were determined and compared to orange juice and baobab nectar, a commercial beverage (Nhukarume *et al.*, 2010). Methanolic extracts were investigated for their ability to scavenge free radicals by the 1, 1-diphenyl-2 picrylhydrazyl radical and superoxide radical scavenging assays whilst the β-Carotene Linoleic acid model system and inhibition of phosholipid peroxidation were used as model systems. The reducing power assay was used to determine the reducing potential of the extracts. Results showed that the beverages in this investigation were capable of acting as antioxidant sources as they displayed significant radical scavenging properties. The total phenolic, ascorbic acid, proantocyanidin and flavonoid contents ranged between 12 and 58 mg GAE/100 ml, 0.00 to 51.26 mg/100 ml, 0.35–1.071 per cent and 18.3–124 mg/100 ml, respectively. There was a positive correlation between antioxidant activities and phenolic compounds content but there was no clear relationship between proanthocyanidin content and antioxidant activity.

Parinari macrophylla Sabine Family: Chrysobalanaceae

Using Trolox assay, Cook *et al.* (1998) estimated the antioxidant activity of 17 wild edible plants of Niger Republic used for food and traditional medicine. They observed that *Parinari macrophylla* possessed lowest antioxidant activity.

Parkia biglobosa (Jacq.) R. Br. ex G. Don Family: Mimosaceae

Polyphenol content and antioxidant activity of fourteen wild edible fruits from Burkina Faso were investigated by Lamien-Meda *et al.* (2008). The data obtained show that the total phenolic and total flavonoid levels were significantly higher in the acetone than in the methanol extracts. *Parkia globosa* fruit had high flavonoid content and high antioxidant activity.

Parkia speciosa Hassk. Family: Mimosaceae

Wong *et al.* (2006) investigated the antioxidant properties of 25 edible tropical plants including *Parkia speciosa* using DPPH and FRAP assays. Total phenolic content was highest in *Parkia speciosa*. *Parkia speciosa* showed moderate antioxidant activity.

The phenolic compounds of five southern Thai indigenous vegetables including stink bean (*Parkia speciosa*) fruit were extracted using different solvents. The extracts were analyzed for total phenolic content using the Folin-Ciocalteu procedure, free radical scavenging capacity by using ABTS and DPPH methods, and reducing capacity by using FRAP assay. The acetone extract possessed the highest total phenolic content followed by methanolic and distilled water extracts, respectively. The acetone extract showed higher free radical scavenging capacity and reducing capacity than those of their methanolic and water extracts counterparts respectively (Panpipat *et al.*, 2010).

Parthenocissus tricuspidata (Sieb. et Zucc.) Planch. Family: Vitaceae

The ethyl acetate fraction of an aqueous alcoholic extract from the stem of *Parthenocissus tricuspidata* yielded 11 known compounds and two new stilbene dimers parthenostilbenins A and B (Kim *et al.*, 2005). Among them, pieatanol showed the strongest inhibitory activity in these assay systems. Two new compounds, parthenostilbenins A and B inhibited lipid peroxidation (IC_{50}=20.35 and 18.68µg/ml), respectively) in rat liver homogenate.

Passiflora alata Dryander Family: Passifloraceae

Passiflora alata extract was administered orally in rats at dose of 1000 mg/kg and it was observed an increase in high-density lipoprotein level (HDL-cholesterol). Two saponins and five C-glycosylflavones derived from apigenin, luteolin and chrysoeriol were identified (Doyama *et al.*, 2005).

Rudnicki *et al.* (2007a) verified the antioxidant activities of hydroalcoholic extracts of leaves of *Passiflora alata* and *Passiflora edulis*. *P. alata* showed a higher total reactive antioxidant potential than did *P. edulis*. The antioxidant activities of both extracts were significantly correlated with polyphenolic contents.

Rudnicki *et al.* (2007b) investigated the antioxidative activity of *Passiflora alata* leaf extract in carbon tetrachloride-treated rats. The pretreatment with *Passiflora alata* extract provided significant protection to liver, evidenced by lower degree of necrosis, decreased lipid peroxidation (TBARS) and higher catalase and superoxide dismutase activities. Additionally, pretreated rats with *P. alata* (5mg/kg) showed significantly decreased cardiac TBARS levels.

Passiflora edulis Sims Family: Passifloraceae

English Name: Passion fruit.

Chen *et al.* (2009) studied the antioxidant activity of leaves extract towards xanthine oxidase, lipoxygenase and tyrosinase activities. The inhibitory effect of acetone extract on xanthine oxidase activity was nil, tyrosine activity was more than 70 per cent and lipoxygenase activity was more than 40 per cent.

Rudnicki *et al.* (2007a) verified the antioxidant activities of hydroalcoholic extracts of leaves of *Passiflora alata* and *Passiflora edulis*. *P. alata* showed a higher total reactive antioxidant potential than did *P. edulis*. The antioxidant activities of both extracts were significantly correlated with polyphenolic contents.

Narayanaswamy and Balakrishnan (2011) evaluated the antioxidant activity of 13 medicinal plants including leaf of *Passiflora edulis*. The aqueous extract showed highest 95 per cent inhibition of DPPH radical while ethanolic extract showed 50 per cent inhibition of DPPH radical.

Passiflora liguralis Juss. Family: Passifloraceae

Considerable antioxidant activity was noted in the lipophilic extracts of *P. liguralis* fruits (Tommonaro *et al.*, 2007).

Pastinara sativa L. Family: Apiaceae

English Name: Parsnip.

A study was carried out by Budinevi *et al.* (1995) on the antioxidative activity of *Pastinara sativa*. Using the ethanolic extracts of plant material good results were obtained.

Patrinia scabiosaefolia Fisch. ex Link. Family: Valerianaceae

Cai *et al.* (2004) evaluated the antioxidant activity and phenolic compounds in traditional Chinese medicinal plants associated with anticancer. The improved ABTS.+ method was used to systematically assess the total antioxidant capacity (Trolox equivalent antioxidant capacity, TEAC) of the medicinal extracts. The TEAC values and total phenolic content for methanolic extract of aerial parts of *Patrinia scabiosaefolia* were 90.7 µmol Trolox equivalent/100g dry weight (DW), and 0.89 g of gallic acid equivalent/100g DW respectively. A positive significant linear relationship between antioxidant activity and total phenolic content showed that phenolic compounds were the dominant antioxidant components in the tested herbs. Major types of phenolic compounds in *Patrinia scabiosaefolia* include tannins.

Patrinia villosa Juss. Family: Valerianaceae

Li *et al.* (2008) evaluated the antioxidant properties of 45 medicinal plants including *Patrinia villosa* using FRAP and TEAC assays and the total phenolic contents of these plants were measured. *Patrinia villosa* showed 65.06 µmol Fe(II)/g FRAP value, 39.19µmol Trolox/g of TEAC value and 8.82 mg GAE/g of Phenolic content.

Paullinia cupana H.B.K. Family: Sapindaceae

English Name: Guarana.

Basile *et al.* (2005) evaluated the antibacterial and antioxidant activities of ethanol extract of seeds from *Paullinia cupana*. The antioxidant activity was determined by the malondialdehyde (MDA) test, measuring the MDA concentration in 3T3-L 1 cells after induced cellular damage using ferric ammonium citrate (FAC). The reduction of lipid peroxidation was 62.5 per cent using *Paullinia cupana* seeds extract with a concentration of 2 µg/ml. The effect was dose dependent. The concentration of catechol equivalent was 6.06 mg/g, while the total phenolic content was 8.43 mg/g. The correlation index between antioxidant activity and catechol content was 0.96.

The antioxidant activities of guarana seed extracts were determined by Majhenic *et al.* (2007). The seeds were extracted with water, methanol, 35 per cent acetone and 60 per cent ethanol at room and at boiling temperature of solvent. The guarana seed water extract obtained at room temperature contained higher amounts of caffeine and catechins than did alcoholic guarana seed extracts. The antioxidant and radical-scavenging activities of guarana seed extracts were evaluated using the β-carotene-linoleic acid emulsion system and the stable free radical DPPH. All the tested guarana seed extracts displayed strong antioxidant and radical-scavenging properties.

Paullinia pinnata L. Family: Sapindaceae

Jimoh *et al.* (2007) evaluated the antioxidant activities of the methanol extract of leaves of *Paullinia pinnata* using different testing systems, its scavenging activities on DPPH and ABTS radicals as well as its ferric ion reducing power. The results showed that *Paullinia pinnata* possessed strong scavenging activity and moderate reducing power. The total phenol, flavonoid and proanthocyanidin contents of the extracts were very close to those reported for most medicinal plants and showed good correlation with its antioxidant activities.

Paulownia tomentosa Steud. Family: Scrophulariaceae

Smejkal *et al.* (2007) evaluated the antioxidant activity of *Paulownia tomentosa*.

They identified and quantified acteoside (1) and isoacteoside (2) in the EtOAc and *n*-BuOH extracts; mimulone (3) and diplacone (4) in the MeOH extract. To determine the antiradical activity of extracts they used the anti DPPH and peroxynitrite assays. The activity was expressed as Trolox C equivalents, IC_{50} for DPPH scavenging and a time dependency course was established. The polyphenols content was determined; results were expressed as gallic acid equivalents. Using these methods they found the fractions of the *n*-BuOH, EtOAc and MeOH extracts that display antiradical activity.

Pedilanthus tithymaloides (L.) Poit. Family: Euphorbiaceae

A tincture from *Pedilanthus tithymaloides* in the scavenging assays showed to be effective against all the assayed ROS and RNS, specially for HO*, O_2*-, HOCl, ROO* and H_2O_2), reactive nitrogen species (RNS) (ONOO- and *NO) and DPPH* radical (Abreu *et al.*, 2006). The content of total phenolics was 17.4 mg GAE/g dry material.

A bioassay-guided methodology utilizing DPPH, the TEAC, and reducing power assays, as well as an assessment of scavenging properties against O(2)(.-), H_2O_2, HOCl, ROO., No and ONOO(-) were used to find the main antioxidant principles of *Pedilanthus tithymaloides* (Abreu *et al.*, 2008). The principles were identified as kaempferol 3-O-beta-D-glucopyranoside-6"-(3-hydroxy-3-methylglutarate), quercitrin, and scopoletin. The contents of total phenolics and flavonoids were found to be 76.0 of GAE/g extract and 9.8 mg of rutin equivalents/g extract, respectively.

Peganum harmala L. Family: Zygophyllaceae

Souri *et al.* (2004) evaluated the antioxidative activity of *Peganum harmala* by linoleic acid peroxidation test using 1,3-diethyl-2-thiobarbituric acid as the reagent. Methanolic extract showed 70.15 per cent inhibition at 40µg concentration with IC_{50} value of 1.98µg.

Pelargonium endlicherianum Willd. Family: Geraniaceae

Tepe *et al.* (2006) screened the antioxidative properties of the methanolic extracts of *Pelargonium endlicherianum* by two complementary test systems, namely DPPH free radical-scavenging and β-carotene/linoleic acid. The extract exerted greater antioxidant activity with IC_{50} value of 7.43 µg/ml. When compared to the synthetic antioxidant BHT (18.0µg/ml), the methanolic extract of *P. endlicherianum* exhibited more than two fold greater antioxidant activity. In the β-carotene/linoleic acid test system, *P. endlicherianum* showed 72.6 per cent inhibition rate. Antioxidant activities of curcumin and ascorbic acid were also determined as positive controls in parallel experiments.

Pelargonium odoratissimum (L.) L`Her ex. Ait. Family: Geraniaceae

The chloroform and methanolic extracts of 124 Egyptian plant species belonging to 56 families were investigated by Moussa *et al.* (2011) and compared their antioxidant activity by DPPH scavenging assay. Among the 124 plant species tested 18 exhibited extremely high antiradical activity (>80 per cent inhibition). The IC_{50} value of DPPH radical scavenging of *Pelargonium odoratissimum* was 26.0µg/ml, while total phenolic and flavonoid contents were 104.96 mg Tannic acid equivalent/g extract and 180.66 mg Rutin equivalent/g extract, respectively. Correlation coefficient between DPPH radical scavenging activity and the total phenolic and flavonoid contents suggest that phenolics and flavonoids in the extracts were partly responsible for the antiradical activities.

Pelargonium reniforme Spreng. Family: Geraniaceae

South African plant, *Pelargonium reniforme*, used locally for treatment of liver disorders, was assessed for antioxidant properties by Latte and Kolodziej (2004) using DPPH radical scavenging system and a luminal-dependent chemiluminiscence assay. Their work indicated that, in both assays, the polyphenols tested showed higher radical scavenging activities than the reference antioxidant ascorbic acid (IC_{50} 2.6-32.9µM vs 40.9µM in the DPPH test and 2-25 times stronger effects in the chemiluminiscence assay). Furthermore, they reported that the marked antioxidant potential than the flavonoids. They ascribed the marked antioxidant activities of the hydrolysable tannins to the presence of galloyl and hexahydroxydiphenoyl groups, and carbonyl (ester) functionalities in oxidatively modified hexahydroxydiphenoyl moieties and for the flavonoids, the catechol (3', 4' dihydroxy) element in the β ring were important determinants with the O-glucosides being more effective than the flavone-based C-glucosyl: their activities were potentiated by introduction of a galloyl group.

Pelargonium zonale L. Family: Geraniaceae

Demo *et al.* (1998) reported antioxidant α-tocopherol from the Hexane extract of *Pelargonium zonale*.

Peltophorum africanum Sond. Family: Mimosaceae

English Name: Weeping wattle.

Bizimenyera *et al.* (2006) evaluated antioxidant potential of *Peltophorum africanum* (weeping wattle). The dried leaves, bark and root of *P. africanum* were extracted with acetone. Thin layer chromatograms were sprayed with 0.2 per cent 2,2-diphenyl-1-picryl hydrazyl (DPPH) in methanol for screening for antioxidants. Quantification of antioxidant activity was assessed against 6-hydroxy-2, 5,7,8-tetramethylchromane-2-carboxylic acid (Trolox) and L-ascorbic acid (both standard antioxidants), using two free radicals, ABTS and DPPH, respectively. Results of their study showed that the bark and root extracts had higher antioxidant activity than L-ascorbic acid and Trolox, a synthetic vitamin-E analogue. The respective TEAC (Trolox Equivalent Antioxidant Capacity) values for the bark and root extracts, and Trolox were 1.08, 1.28 and 1.0. EC_{50} values for L-ascorbic acid (5.04 µg/ml) was more active than the leaf 6.54 (µg/ml), but much less active than the bark (4.37 µg/ml) and root (3.82 µg/ml) extracts.

Peltophorum ferrugineum Benth. Family: Caesalpiniaceae

Chanda *et al.* (2011) determined the antioxidant activity of twelve medicinal plants *Averrhoa carambola, Buchanania lanzan, Calophyllum inophyllum, Celastrus paniculatus, Clerodendron multiflorum, Luffa acutangula, Morinda citrifolia, Ocimum gratissimum, Peltophorum ferrugineum, Phyllanthus fraternus, Triumfetta rotundifolia, Ziziphus nummularia* belonging to different families. Antioxidant activity was determined by using different methods like DPPH free radical scavenging assay, hydroxyl radical scavenging assay, superoxide anion radical scavenging assay and reducing capacity assessment. The plants were extracted individually by cold percolation method using different organic solvents (petroleum ether, acetone and methanol) depending on their polarity. Ascorbic acid was used as standard to determine DPPH *free radical* scavenging activity and reducing capacity assessment. Gallic acid was used as standard to determine hydroxyl radical scavenging activity and superoxide anion radical scavenging activity. Amongst the twelve plants studied, acetone and methanolic extract of *Peltophorum ferrugineum* showed the best radical scavenging activity and reducing capacity assessment.

Pennisetum americanum (L.) Schum. (Syn: *Pennisetum typhoideum* (Burm.f.) Stapf et Hubb.; *Pennisetum glaucum* L.) Family: Poaceae (Graminae)

English Name: Pearl millet, Hindi: Bajra.

Sreeramulu *et al.* (2009) evaluated the antioxidant activity of pearl millet by DPPH, scavenging assay, FRAP assay and reducing power. Phenolic content was 133.63 mg/100g. DPPH radical scavenging activity expressed as trolox equivalent, FRAP (µmol/g) and reducing power (mg/g) were 1.26, 145 and 1.73, respectively in pearl millet.

Penstemon pallidus Small Family: Scrophulariaceae

Borchardt *et al.* (2008) evaluated the antioxidant activity of methanol extracts of seeds of *Penstemon pallidus* using DPPH antioxidant assay.The antioxidant value reported in µM Trolox/100g (TE) from DPPH radical scavenging activity of crude seeds extract was 47,320.

Pentacalia americana (L.f.) Cuatrec. Family: Asteraceae (Compositae)

Mosquera *et al.* (2007) studied the antioxidant activity of the extractsof *Pentacalia americana* using DPPH free-radical scavenging method. The percentage of the antioxidant activity of the n-Hexane fraction was 19.73, Dichloromethane fraction was 33.333 and methanol fraction was 49.531.

Pera glabrata (Schott.) Baill. Family: Euphorbiaceae

Santos *et al.* (2009) evaluated the antioxidant activity of extracts from nine plant species belonging to the Brazilian flora acting on the free radicals DPPH and TEMPOL. Results showed that the extracts of *Pera glabrata* display moderate antioxidant activities with DPPH scavenging IC_{50} value of 0.452 mg/ml while Hydroxyl scavenging IC_{50} value was 9.02 mg/ml.

Pereskia grandifolia Hort. ex Pfeiff. Family: Cactaceae

The antioxidant potential of *P. grandifolia* crude methanol and its fractionated extracts (hexane, ethyl acetate and water) have been investigated by Sim *et al.* (2010), employing three different established testing systems, such as scavenging activity on DPPH radicals, reducing power assay and β-carotene method. The total phenolic content of the *P. grandifolia* extracts was also assessed by the Folin-Ciocalteu's method. The ethyl acetate extract showed significantly the highest total phenolic content, DPPH scavenging ability and antioxidant activity in β-carotene bleaching assay while the hexane extract possessed significantly strongest reducing power.

Antioxidant activity of *Pereskia grandifolia* and four other species have been investigated by Adib *et al.* (2010) by DPPH assay. Methanolic extract of *P.grandifolia* was found to have the lowest antioxidant activity with an IC_{50} value of 450µg/ml.

Pergularia daemia Forsk. Family: Asclepiadaceae

Mirunalini and Karthiswaran (2010) evaluated the oxidant and antioxidant potential of *Pergularia daemia* methanolic extract on DMBA induced hamster buccal pouch carcinogenesis and in human KB cell line. Diminished lipid peroxidation in the HBP tumor was accompanied by decreased activities of superoxide dismutase and catalase with increase in reduced glutathione and glutathione peroxidase. However, in liver and erythrocytes of tumor bearing animals increased lipid peroxidation was associated with compromised antioxidant defenses. Oral administration of methanolic extracts of *P. daemia* aerial parts to DMBA painted animals restored the antioxidant status to near normal.

Perilla frutescens (L.) Britt. Family: Lamiaceae (Labiatae)

Two novel antioxidants (vinyl caffeate and trans-*p*-menth-8-en-7-yl caffeate) and seven known antioxidants (3,4-dihydroxybenzaldehyde, methyl 3,4-dihydroxy-benzoate, methyl caffeate, 3,4,5,7-tetrahydroxy-flavone, caffeic acid, 6,7-dihydroxycoumarin, and rosmarinic acid) were isolated by Masahiro *et al.* (1996) from *Perilla frutescens* var. *crispa.* The redox potentials of the novel isolated antioxidative compounds were comparable to those of known antioxidants. Rosmarinic acid has been isolated by Nakamura *et al.* (1998) in high yield (0.1 per cent from the leaves of *Perilla frutescens* var. *acuta* f. *viridis*, as a superoxide scavenger in a xanthine/xanthine oxidase system. The scavenging activity of Rosmarinic acid was significantly higher than that of ascorbic acid.

The antioxidant activity of roasted defatted *Perilla frutescens* seed was determined by Jung *et al.* (2001) by measuring its radical scavenging effect on DPPH radicals, inhibitory activity on total reactive oxygen species generation in kidney homogenates and scavenging effect on authentic peroxynitrites. The methanolic extract of roasted, defatted perilla seed showed strong scavenging activity in both DPPH and peroxynitrite radicals, and thus fractionated with several solvents. The antioxidant activity potential of the individual fraction was in the order of ethyl acetate>n-butanol>dichloromethane>water>n-hexane fraction. Luteolin was isolated as one of the active principles from the ethyl acetate fraction, together with the inactive chrysoeriol and apigenin.

Wojcikowski *et al.* (2007) evaluated the antioxidant activity of aerial parts of *Perilla frutescens.* Oxygen radical absorbance capacity of the ethyl extract, methanol extract and aqueous extracts were 1.130, 152.23 and 86.48μmol Trolox equivalent per g of dried starting material, respectively.

Zekonis *et al.* (2008) reported that the treatment of neutrophil leukocytes with the *Perilla* aqueous extract inhibited the release of reactive oxygen species, measured as luminal- and lucigenin-dependent chemiluminiscence, by about 30 per cent and more thant 90 per cent, respectively.

Meng *et al.* (2009) evaluated the antioxidant activities of polyphenols extracted from *Perilla frutescens* varieties based on inhibition of free radical DPPH. The DPPH radical scavenging activity was calculated as TEAC. The mean amount of total phenolics of the water extracts (4-29 μmol/100ml) and the TEAC value calculated (23-167 μmol TE/100ml) confirmed the high antioxidant activity of these leaf water extracts. These results were highly correlated within some O-dihydroxylated polyphenolic compounds and AA.

The antioxidant activities of *Perilla frutescens* were evaluated by Lin *et al.* (2010) by using DPPH radical scavenging methods. Reducing power, metal chelating ability, total phenolic content and total flavonoid content were also detected. In all the tested models, the methanolic extracts of stalk, leaf and seed showed their ability to scavenge the free radicals, reducing power and chelating capacity in a dose-dependent manner. The methanolic extract of stalk had higher antioxidant activity than of leaf and seed.

Antioxidant capacities of 56 selected Chinese medicinal plants, including *Perilla frutescens* were evaluated by Song *et al.* (2010) using the TEAC and FRAP assays, and their total phenolic content was measured by the Folin-Ciocalteu method. Leaf of *Perilla frutescens* showed moderate antioxidant activity with TEAC value of 36.47μmol Trolox/g, FRAP value of 46.8μmol Fe^{2+}/g and phenolic content was 7.17mg GAE/g while seed has TEAC value of 13.71μmol Trolox/g, FRAP value of 26.29μmol Fe^{2+}/g and phenolic content was 1.96mg GAE/g.

Perilla ocymoides L. Family: Lamiaceae (Labiatae)

Maeda *et al.* (2006) examined the phenolic contents of *Perilla ocymoides* and their radical scavenging activities by using DPPH assay. High contents of phenolic compounds and strong DPPH scavenging activities were observed in extracts from this species. They also showed radical scavenging activities for the superoxide radical and tert-butylhydroperoxyl radical (t-BuOO·).

Perilla pankinensis Decne Family: Lamiaceae (Labiatae)

At the concentrations of 30 and 45 μg/ml, *Perilla pankinensis* samples showed 91.9 and 96.4 per cent inhibition on lipid peroxidation of linoleic acid emulsion, respectively. In comparison, 45 μg/ml of standard antioxidant such as α-tocopherol and trolox exhibited 88.8 and 86.2 per cent inhibition on peroxidation of linoleic acid emulsion, respectively. The extract was found effective on DPPH scavenging, superoxide anion radical and hydrogen peroxide scavenging, total reducing power and metal chelating on ferrous ions activities (Gülçin *et al.*, 2005).

Periploca laevigata Ait. Family: Asclepiadaceae

Hajji *et al.* (2010) described the chemical composition and antimicrobial, antioxidant and angiotension 1-converting enzyme (ACE) inhibitory activities of essential oil from *Periploca laevigata* root barks, an aromatic plant widely distributed in Tunisia and used as a traditional medicinal plant. The root bark essential oil was found to possess antioxidant activities, as evaluated by the DPPH radical method, β-carotene bleaching and reducing power assays.

Persea americana Mill. Family: Lauraceae

English Name: Avacado

Chen *et al.* (2009) studied the antioxidant activity of fruit and leaves extract towards xanthine oxidase, lipoxygenase and tyrosinase activities. The inhibitory effect of acetone extract on xanthine oxidase activity was nil, tyrosine activity was less than 10 per cent and lipoxygenase activity of fruit was more than 20 per cent while that of leaf was more than 50 per cent.

Wang *et al.* (2010) determined the antioxidant capacities, total phenolic content and identified and quantified major antioxidant compounds in avacados of different strains and cultivars. For all varieties, seeds contained the highest antioxidant capacities, phenolic content and procyanidin, whereas the pulp had the lowest. Antioxidant capacities, phenolic contents and procyanidins in avacados were highly correlated, suggesting that procyanidins were the major phenolic compounds that contributed to antioxidant capacities.

Persea gamblei (King ex Hook.f.) Kosterm. Family: Lauraceae

Joshi *et al.* (2010) evaluated the antioxidant activity of seven Himalayan Lauraceae species. The essential oil of *Persea gamblei* was able to inhibit linoleic acid oxidation. Furanodienone and curzerenone constituted the major components of the oil.

Persicaria tenella (Blume) Hara Family: Polygonaceae

Abbas *et al.* (2006) evaluated the antioxidant activity of *Persicaria tenella* using DPPH assay and nitric oxide inhibition using Griess assay. Methanolic extract showed DPPH radical scavenging and nitric oxide inhibition in murine peritoneal macrophages.

Petasites formosanus Kitamura Family: Asteraceae (Compositae)

A novel phenylpropenoyl sulfonic acid, petasiformin-A was isolated from the leaves of *Petasites formosanus* which showed the significant antioxidant activity in DPPH radical scavenging assay (Lin *et al.*, 2004).

Petrorhagia velutina (Guss.) Ball. et Heyw. Family: Caryophyllaceae

Radical scavenging activity of the metabolites isolated from *Petrorhagia velutina* was assessed by D'Abrosca *et al.* (2010) measuring their ability to scavenge DPPH radical and ABTS radical cation. The dimeric metabolite and the monomeric trans isomers showed an effective reducing power on the oxidant probes.

Petroselenium crispum (Mill.) Nym. ex A.W. Hill. Family: Apiaceae (Umbelliferae)

English Name: Parsley; Japanese: Paseri.

Fejes *et al.* (2000) found considerable antioxidant activity of methanol extracts of parsley against lipid peroxidation.

Hydrodistillated extracts from parsley were assessed for their total phenol content, and antioxidant (iron(III) reduction, inhibition of linoleic acid peroxidation, iron (II) chelation, DPPH radical-scavenging and inhibition of hydroxyl radical-mediated 2-deoxy-D-ribose degradation, site and nonsite-specific) activities by Hinneburg *et al.* (2005). The extracts from parsley possessed the highest antioxidant activities in iron chelation assay. In linoleic acid peroxidation assay it was less effective.

Maeda *et al.* (2006) examined the phenolic contents of leaf of parsley and its radical scavenging activities for DPPH radical. The polyphenol content was 43.2 mg-gallic acid equivalent/100 g) and DPPH scavenging activity was 353. 9 μmol-Trolox eq./100 g).

Effects of parsley extract, versus glibornuride, on the liver of streptozotocin-induced diabetic rats were investigated by Ozsoy-Sacan *et al.* (2006). It was concluded that probably, due to its antioxidant property, parsley extract has a protective effect comparable to glibornuride against hepatoxicity caused by diabetes.

Wojcikowski *et al.* (2007) evaluated the antioxidant activity of aerial parts of *Petroselenium crispum*. Oxygen radical absorbance capacity of the ethyl extract, methanol extract and aqueous extracts were 8.83, 291.11 and 445.00µmol Trolox equivalent per g of dried starting material, respectively.

Mariutti *et al.* (2008) evaluated the antioxidant activity of parsley employing ABTS and DPPH radical assay. ABTS-TEAC value was 6.2mM/g.

Peucedanum japonicum Thunb. Family: Apiaceae (Umbelliferae)

Maeda *et al.* (2006) examined the phenolic contents of leaf of *Peucedanum japonicum* and their radical scavenging activities by using DPPH assay. High contents of phenolic compounds and strong DPPH scavenging activities were observed in extracts from this species. They also showed radical scavenging activities for the superoxide radical and tert-butylhydroperoxyl radical (t-BuOO·).

Study was conducted by Thuong *et al.* (2010) to evaluate the structure-activity relationships of the antioxidative of natural coumarins isolated from *Peucedanum japonicum*. The free radical scavenging and lipid peroxidation assays revealed that five phenolic coumarins, scopoletin (1), aesculetin (2), fraxetin (3), umbelliferone (18), and daphnetin (19), possessed considerable antioxidant activities. The coumarins having a catechol group 2, 3 and 19, showed significant free radical scavenging activity and inhibitory effects on lipid peroxidation indicating that the catechol group contributed to the antioxidant activities of coumarins. In contrast, the sugar moiety markedly reduced the activities of coumarin glycosides. The results also demonstrated that the α-pyrone ring of coumarins significantly enhanced the capacity of inhibiting oxidative reactions of coumarins (Thuong *et al.*, 2010).

Peucedanum longifolium Waldst. et Kit. Family: Apiaceae (Umbelliferae)

Tepe *et al.* (2011) evaluated the chemical composition and antioxidant activity of the essential oil of *Peucedanum longifolium* by four different test systems. The antioxidant activity of the essential oil was 70.73 per cent in β-carotene linoleic acid method, 41.87 per cent in DPPH method at 1mg/ml concentration, 24.12 per cent in chelating effect.

Peucedanum palimboides Boiss. Family: Apiaceae (Umbelliferae)

Tepe *et al.* (2011) evaluated the chemical composition and antioxidant activity of the essential oil of *Peucedanum palimboides* by four different test systems. The antioxidant activity of the essential oil was 70.73 per cent in β-carotene linoleic acid method, 41.87 per cent in DPPH method at 1mg/ml concentration, 24.12 per cent in chelating effect. The reducing power was 0.248 nm.

Peucedanum praeruptorum Dunn Family: Apiaceae (Umbelliferae)

Antioxidant capacities of 56 selected Chinese medicinal plants, including *Peucedanum praeruptorum* were evaluated by Song *et al.* (2010) using the TEAC and FRAP assays, and their total phenolic content was measured by the Folin-Ciocalteu method. *Peucedanum praeruptorum* showed very low antioxidant activity with TEAC value of 4.20µmol Trolox/g, FRAP value of 14.78µmol Fe^{2+}/g and phenolic content was 1.6mg GAE/g.

Peumus boldus Mol. Family: Monimiaceae

English Name: Boldo

Boldo leaves are rich in several aporphine-like alkaloids, of which boldine is the most abundant one. Research conducted during the early 1990s led to the discovery that boldine is one of the most potent natural antioxidants. Prompted by the latter, a large and increasing number of studies emerged, which have focused on characterizing some of the pharmacological properties that may arise from the free radical-scavenging properties of boldine. O'Brien *et al.* (2006) reviewed the pharmacological actions, which arise from its antioxidant properties (*e.g.*, cyto-protective, anti-tumour promoting, anti-inflammatory, anti-diabetic and anti-atherogenic actions), as well as those that do not seem to be associated with such activity (*e.g.*, vasorelaxing, anti-trypanocidal, immuno- and neuro-modulator, cholagogic and/or choleretic actions).

Boldine, an aporphine alkaloid extracted from the leaves and bark of boldo (*Peumus boldus*), has been shown to exhibit strong free-radical scavenger and antioxidant properties. Jimenez *et al.* (2000) reported the *in vitro* ability of boldine to protect intact red cells against the haemolytic damage induced by the free radical initiator 2, 2'-azobis-(2-amidinopropane) (AAPH). Boldine concentration-dependently prevented the AAPH-induced leakage of haemoglobin into the extracellular medium. Substantial and similar cyto-protective effects of boldine were observed whether the antioxidant was added 1 h prior to, or simultaneously with, the azo-compound. The delayed addition of boldine, by 1 h relative to AAPH, diminished but did not abolish its cytoprotective effect. However, negligible effects of boldine were observed after its addition to erythrocytes previously incubated with AAPH for 2 h. The data presented demonstrate that, in addition to its well-established antioxidant effects, boldine also displays time-dependently strong cytoprotective properties against chemically induced haemolytic damage.

Del Valle *et al.* (2004) reported that compounds with antioxidant activity of *Peumus boldus* varied depending on part plant, solvent and extraction conditions. Antioxidant activity (free radical DPPH method) was maximal in boldo leaf extracts with aqueous ethanol at 50°C (1648 TEAC/g dry leaf).

Preliminary assays showed free-radical scavenging activity in hot water extracts of boldo leaves (Krigstein and Cederbaum, 1995). Assay-guided isolation led to the active compounds. Catechin proved to be the main free-radical scavenger of the extracts. The IC_{50} for catechin and boldine in the lipid peroxidation test were 75.6 an 12.5 µg/ml, respectively. Lipid peroxidation in erythrocytes was inhibited by boldo extracts. The relative concentration of alkaloids and phenolics in boldo leaves and their activity suggest that the free-radical scavenging effect is mainly due to catechin and flavonoids and that the antioxidant effect is mainly related with the catechin content. The high catechin content of boldo leaves and its bioactivity suggest that quality control of boldo folium has to combine the analysis of catechin as well as their characteristic aporphine alkaloids (Schmeda-Hirschmann *et al.*, 2003).

Pfaffia glomerata (Spreng.) Pettersen Family: Amaranthaceae

English Name: Brazilian ginseng.

Daniel *et al.* (2005) evaluated the free radicals scavenging activity of methanolic (MeOH) extract, butanolic (BuOH) fraction and 20-hydroxyecdysone (EC) from the roots of *Pfaffia glomerata*. The free radical scavenging activity presented by EC was bigger than that one presented by the BuOH fraction. The MeOH extract showed a remarkable prooxidant activity. EC free radical reaction-inhibition was almost twice of that of the control α-tocopherol.

Phalaria macrocarpa (Scheff.) Boerl. Family: Thymeleaceae

Endrini (2010) evaluated the antioxidant activity of the methanol extract of *Phalaria macrocarpa*. The extract was found to have high antioxidant activity.

Phaseolus aureus Roxb. Family: Fabaceae (Papilionaceae)

English Name: Adzuki bean.

Lin *et al.* (2001) evaluated the anti-lipid peroxidation activity, free radical scavenger activity and anti-superoxide formation of hot water extracts of legumes incluing *Phaseolus arureus* to test the antioxidant activity. The results showed that the extract exhibited remarkable inhibition of $FeCl_2$-ascorbic acid-induced lipid peroxidation of mouse liver homogenate. Extract showed anti-lipid peroxidation activity and anti-superoxide formation. Adzuki bean extract demonstrated the strongest anti lipid peroxidation activity and highest superoxide anion scavenging activity.

Phaseolus calcaratus Roxb. Family: Fabaceae (Papilionaceae)

English Name: Rice bean.

Lin *et al.* (2001) evaluated the anti-lipid peroxidation activity, free radical scavenger activity and anti-superoxide formation of hot water extracts of legumes incluing *Phaseolus calcaratus* to test the antioxidant activity. The results showed that the extract exhibited remarkable inhibition of $FeCl_2$-ascorbic acid-induced lipid peroxidation of mouse liver homogenate. Extract showed anti-lipid peroxidation activity and anti-superoxide formation.

Phaseolus radiatus L. Family: Fabaceae (Papilionaceae)

English Name: Mung bean.

Lin *et al.* (2001) evaluated the anti-lipid peroxidation activity, free radical scavenger activity and anti-superoxide formation of hot water extracts of legumes incluing *Phaseolus radiatus* to test the antioxidant activity. The results showed that the extract exhibited remarkable inhibition of $FeCl_2$-ascorbic acid-induced lipid peroxidation of mouse liver homogenate. Extract showed anti-lipid peroxidation activity and anti-superoxide formation. Mung bean extract demonstrated the strongest anti lipid peroxidation activity and highest superoxide anion scavenging activity.

Phaseolus trilobus Ait. Family: Fabaceae (Papilionaceae)

Fursule and Patil (2010) evaluated the hepatoprotective and antioxidant activity of *Phaseolus trilobus* on bile duct ligation induced fibrosis in rats. Methanol and aqueous extracts reduced elevated level of ALT, AST, ALP and LDH. Both the extracts were found to reduce the elevated levels of serum TBARS and elevated superoxide radical scavenging activity providing antioxidant activity comparable to ascorbic acid. The reduced level of glutathione was found to be elevated in liver providing antioxidant activity comparable with silymarin.

Phaseolus vulgaris L. Family: Fabaceae (Papilionaceae)

English: Kidney bean, Green beans, common beans; Hindi: Rajmah.

Beans were pearled to evaluate the feasibility of increasing antioxidant activity and phenolic antioxidants. Phenolics were concentrated mostly in the hull fraction at about 56mg of catechin equivalents per gram of sample. The methanolic extracts of the pearled bean samples were screened for antioxidant potential using the β-carotene-lineolate and the DPPH *in vitro* model systems. The pearled material, also referred to as milled samples, exhibited antioxidant activity that correlated with phenolic content and inhibited DPPH significantly in a dose-dependent manner. Phenolics and antioxidant activities were also examined in chromatographic fractions of methanolic extracts of manually obtained hulls that represented a model used previously to ascertain antimutagenic activity. Fractions extracted with ethyl acetate/acetone and acetone displayed antioxidant activity (Cardador-Martinez *et al.*, 2002).

Jitaram and Liu (2004) reported antioxidant activity of green beans. The effects of soaking, boiling and steaming processes at atmospheric and high pressures on the phenolic components and antioxidant properties of pinto and black beans were investigsted by Xu and Chang (2008, 2009). In comparison to original raw beans all processing methods caused significant decreases in total phenolic content, total flavonoid content, condensed tannin content, amonomeric anthocyanin content, DPPH free-radical scavenging activity, FRAP, and ORAC values in both pinto and black beans. Steaming processing resulted in a greater retention of TPC, DPPH, FRAP and ORAC values than the boiling processes in both pinto and black beans.

Sreeramulu *et al.* (2009) evaluated the antioxidant activity of rajmah, by DPPH, scavenging assay, FRAP assay and reducing power. Phenolic content of the seed was 332.98 mg/100g. DPPH radical scavenging activity expressed as trolox equivalent, FRAP (μmol/g) and reducing power (mg/g) were 1.07, 373 and 4.89, respectively in rajmah.

The effect of fermentation on the polyphenol distribution and antioxidant activity of kidney bean was investigated by Oboh *et al.* (2009). The results of the study revealed that fermentation caused a significant decrease in the bound phenol content of the legume. Free soluble phenol from both the fermented and unfermented legumes had a significantly higher reducing power.

Phaseolus vulgaris L. var. *humulis* Alef. Family: Fabaceae (Papilionaceae)

English Name: Field snap bean; *Japanese Name*: Sayaingen.

Maeda *et al.* (2006) examined the phenolic contents of pod of field snap bean and its radical scavenging activities for DPPH radical. The polyphenol content was 42.9 mg-gallic acid equivalent/100 g) and DPPH scavenging activity was 205.9 μmol-Trolox eq./100 g).

Phaulopsis fascisepala C.B.Cl. Family: Acanthaceae

The antioxidant activities of crude extract of *Phaulopsis fascisepala* leaf were evaluated by Adesagun *et al.* (2009) and compared with α-tocopherol and BHT as synthetic antioxidants and ascorbic acid as natural-based antioxidant. *In vitro*, they studied its antioxidative activities, radical-scavenging effects, Fe^{2+}-chelating ability and reducing power. The total phenolic content was determined and expressed in gallic acid equivalent. The extract showed variable activities in all of these *in vitro* tests. The antioxidant effect of *P. fascisepala* was strongly dose dependent, increased with increasing leaf extract dose and then leveled off with further increase in extract dose. Compared to other antioxidants used in the study, α-Tocopherol, ascorbic acid and BHT, *P. fascisepala* leaf extract showed less scavenging effect on DPPH radical and less reducing power on Fe^{3+}/ferricyanide complex but better Fe^{2+}-chelating ability.

Phellodendron amurense Rupr. Family: Rubiaceae

Wu *et al.* (2003) isolated two new dihydroflavonols, three new coumarins, one new chlorophyll and one new phenyllactate in addition to 35 known compounds from the leaves of *Phellodendron amurense* var. *wilsonii*. The antioxidant and antityrosinase activities were also described.

Cai *et al.* (2004) evaluated the antioxidant activity and phenolic compounds in traditional Chinese medicinal plants associated with anticancer. The improved ABTS.+ method was used to systematically assess the total antioxidant capacity (Trolox equivalent antioxidant capacity, TEAC) of the medicinal extracts. The TEAC values and total phenolic content for methanolic extract of bark of *Phellodendron amurense* were 334.2 µmol Trolox equivalent/100g dry weight (DW), and 1.68 g of gallic acid equivalent/100g DW respectively. A positive significant linear relationship between antioxidant activity and total phenolic content showed that phenolic compounds were the dominant antioxidant components in the tested herbs. Major types of phenolic compounds in *Phellodendron amurense* include phenolic acids (phellodendrine, magnoflorine).

Li *et al.* (2008) evaluated the antioxidant properties of 45 medicinal plants including *Phellodendron amurense* using FRAP and TEAC assays and the total phenolic contents of these plants were measured. *Phellodendron amurense* showed 144.99 µmol Fe(II)/g FRAP value, 71.83µmol Trolox/g of TEAC value and 18.35 mg GAE/g of Phenolic content. A high correlation between antioxidant capacities and their phenolic contents indicated that phenolic compounds were a major contributor of antioxidant activity of these plants.

Phellodendron japonicum Maxim. Family: Rutaceae

Quercetin and Phellodenin isolated from the leaves of *Phellodendron japonicum* demonstrated significant DPPH radical scavenging activity (Chiu *et al.*, 2005).

Phillyria latifolia L. Family: Oleaceae

Demo *et al.* (1998) reported antioxidant α-tocopherol from the Hexane extract of *Phillyria latifolia*.

Philodendron bipinnatifidum Schott ex Endlicher Family: Araceae

Menezes *et al.* (2004) investigated the antioxidant activity of hydroalcoholic extracts of *Philodendron bipinnatifidum* by DPPH assay and phosphomolybdenum method. The relative activity in phosphomolybdenum complex was 0.37 while scavenging activity against DPPH in IC_{50} ($\mu g/ml$) was 39.91.

Philodendron solimoesense A.C. Smith Family: Araceae

Lock *et al.* (2005) evaluated the *in vitro* antioxidant activity of 40 Peruvian plants using DPPH assay. *Philodendron solimoesense* showed highest antioxidant activity with EC_{50} value of 8.8$\mu g/ml$.

Phlomis anisodonta Boiss. Family: Lamiaceae (Labiatae)

Souri *et al.* (2004) evaluated the antioxidative activity of *Phlomis anisodonta* by linoleic acid peroxidation test using 1,3-diethyl-2-thiobarbituric acid as the reagent. Methanolic extract showed 88.34 per cent inhibition at 40μg concentration with IC_{50} value of 3.14μg.

Phlomis bruguieri Desf. Family: Lamiaceae (Labiatae)

Firuzi *et al.* (2010) evaluated the antioxidant activity and total phenolic content of 24 Lamiaceae species growing in Iran. *Phlomis bruguieri* showed low antioxidant activity with 11.0$\mu M/g$ DW FRAP value and IC_{50} = 1029.7$\mu g/ml$ of DPPH radical scavenging activity. It also showed low phenolic content of 4.7 mg CE/g.

Phlomis elliptica Benth. Family: Lamiaceae (Labiatae)

Firuzi *et al.* (2010) evaluated the antioxidant activity and total phenolic content of 24 Lamiaceae species growing in Iran. *Phlomis elliptica* showed good antioxidant activity with 23.1$\mu M/g$ DW FRAP value and IC_{50} = 517.0$\mu g/ml$ of DPPH radical scavenging activity. It also showed high phenolic content of 7.9 mg CE/g. FRAP and DPPH assay results showed good correlations with the total phenolic contents of the plant measured by the Folin-Ciocalteu assay.

Phlomis lanata Willd. Family: Lamiaceae (Labiatae)

Couladis *et al.* (2003) screened Greek aromatic plants from the Lamiaceae family for the antioxidant activity. Of the 21 plants they tested they found ethanol extracts prepared from *Phlomis lanata*, as well as 9 other plants exhibited the same antioxidant activity as α-tocopherol in their ability to inhibit bleomycin-Fe(II) complex-induced arichidonic acid superoxidation to MDA.

Phlomis olivieri Benth. Family: Lamiaceae (Labiatae)

Firuzi *et al.* (2010) evaluated the antioxidant activity and total phenolic content of 24 Lamiaceae species growing in Iran. *Phlomis olivieri* showed good antioxidant activity with 22.3$\mu M/g$ DW FRAP value and IC_{50} = 417.1$\mu g/ml$ of DPPH radical scavenging activity. It showed phenolic content of 9.0 mg CE/g.

Phlomis persica Boiss. Family: Lamiaceae (Labiatae)

Firuzi *et al.* (2010) evaluated the antioxidant activity and total phenolic content of 24 Lamiaceae species growing in Iran. *Phlomis persica* showed low antioxidant activity with 11.6µM/g DW FRAP value and IC_{50} = 1199.2µg/ml of DPPH radical scavenging activity. It showed low phenolic content of 6.5 mg CE/g.

Phodococcum vitis-idaea (L.) Avror Family: Vacciniaceae

Myagmar and Aniya (2000) evaluated the free radical scavenging action of some medicinal herbs growing in Mongolia. The aqueous extract of nine herbs including *Phodococcum vitis-idaea* were used. The free radical scavenging action was determined *in vitro* and *ex vivo* by using ESR spectrometer and chemiluminiscence analyzer. The results showed that *Phodococcum vitis-idaea* extract possess strong scavenging action of DPPH, superoxide and hydroxyl radicals. *Phodococcum vitis-idaea* also depressed reactive oxygen production from polymorphonuclear leukocytes stimulated by phorbol-12-myristate *ex vivo*.

Phoenix dactylifera L. Family: Arecaceae (Palmae)

Tariq and Al-Yahya (1983) have reported an α-tocopherol rich fraction isolated from date palm seeds, which was found to have significant activity against endotoxin-induced myocardial damage.

Edible parts of date palm fruits from Iran were anyzed by Biglari *et al.* (2008) for their antioxidant activities using Trolox equivalent antioxidant capacity (TEAC) method, ABTS+ assays and ferric reducing antioxidant power method (FRAP assay). The study demonstrated the potential of Iranian dates and antioxidant functional food ingredients.

Phoradendron falcifrons Eichl. Family: Viscaceae

Menezes *et al.* (2004) investigated the antioxidant activity of hydroalcoholic extracts of leaves of *Phoradendron falcifrons* by DPPH assay and phosphomolybdenum method. The relative activity in phosphomolybdenum complex was 0.87 while scavenging activity against DPPH in IC_{50} (µg/ml) was 60.05.

Phoradendron piperoides (Kunth) Trel. Family: Viscaceae

Methanol extract, dichloromethane, ethyl acetate and methanol partitions of leaves of *Phoradendron piperoides* were evaluated for antioxidant and anti-inflammatory properties by Vasconellos *et al.* (2009). Methanol extract, dichloromethane, ethyl acetate and methanol partitions reacted with the DPPH radical and reduced the DPPH radical by 94.5, 37.2, 77.2 and 95.7 per cent, respectively. ME, ethyl acetate and methanol partitions exhibited low IC_{50} values.

Phragmites australis (Cav.) Trin. ex Steud. (Syn.: *Phragmites communis* (L.) Trin.) Family: Poaceae (Graminae)

English Name: Lake reed.

The antioxidative activity of phenolic extracts from leaf and stalk of lake reed was examined by autoxidation of methyl linoleate by Kahkonen *et al.* (1999). The phenolic content of the leaf was 5.7mg/g GAE while the antioxidant activity was 46 per cent at 500 ppm level. The stalk has lower phenolics and lower antioxidant activity.

Methanol aqueous extract of *Phragmites australis* (Syn: *P. communis*) was screened by Kim *et al.* (1997) for free radical scavenging activity using the DPPH free radical generating system. The extract showed free radical scavenging activity.

Li *et al.* (2008) evaluated the antioxidant properties of 45 medicinal plants including *Phragmites communis* using FRAP and TEAC assays and the total phenolic contents of these plants were measured. *Phragmites communis* showed 27.15 μmol Fe(II)/g FRAP value, 20.74μmol Trolox/g of TEAC value and 5.78 mg GAE/g of Phenolic content.

Phyllanthus acidus L. Family: Euphorbiaceae

Habib *et al.* (2011) evaluated the antioxidant activity of *Phyllanthus acidus*. The phenolics content was 159.601mg g GAE and the amount of flavonoid was 24.183 mg/g of quercetin equivalent. The DPPH radical scavenging activity was found to slight increase with increasing concentration of the extract and the IC_{50} value showed 1192.263μg/ml for plant extract compared to 13.37μg/ml which was the IC_{50} value for the reference ascorbic acid.

Usha Nagalakshmi (2011) reported antioxidant activity of *Phyllanthus acidus* fruits.

Phyllanthus amarus Schum. et Thonn. Family: Euphorbiaceae

P. amarus contains several active ingredients like tannins, phyllanthin, hypophyllanthin, polyphenols, quercetin, astragalin and some ellagitannins like catechin and epigallocatechin (Foo, 1995). Kumaran and Karunakaran (2007) have reported antioxidant activity in *Phyllanthus* species.

Kumar and Kuttan (2005) have shown that *Phyllathus amarus* possesses powerful scavenger properties of oxygen radicals. The compounds responsible for the antioxidant activity isolated from this species include a variety of tannins, several lignans and flavonoids such as quercetin and catechin (Kumar and Kuttan, 2005).

Phyllanthus amarus extract was found to show hepatoprotective effect by lowering down the content of TBARS and enhancing the reduced glutathione level and the activities of antioxidant enzymes, glutathione peroxidase, glutathione S-transferase, superoxide dismutase and catalase. Histopathological analysis of liver samples also confirmed the hepatoprotective value and antioxidant activity of the ethanolic extract of the herb, which was comparable to the standard antioxidant ascorbic acid (Naaz *et al.*, 2007).

The antioxidant activity of the principal constituents of *Phyllanthus amarus* namely amariin, 1-galloyl-2,3-dihydroxy diphenyl (DHHDP)-glucose, repandusinic acid, geraniin, corilagin, phyllanthusiin D, rutin and quercetin 3-O-glucoside were examined for their ability to scavenge free radicals in a range of systems including DPPH, ABTS/ferrylmyoglobin, FRAP and pulse analysis (Londhe *et al.*, 2008). In addition their ability to protect rat liver mitochondria against oxidative damage was determined by measuring the ROO radical induced damage to proteins and lipids and OH radical induced damage to plasmid DNA. The compounds showed significant antioxidant activities with different efficacy depending on the assays employed. Amariin, repandusinic acid and phyllanthusin D showed higher antioxidant activity among the ellagitannins and were comparable to the flavonoids rutin and quercetin 3-O-glucoside.

Patnibul *et al.* (2008) screened 18 vegetables for antioxidant activity using silica gel Thin Layer chromatography followed by spraying with DPPH. *Phyllanthus amarus* exhibited strong antioxidant activity.

Phyllanthus debilis Klein ex Willd. Family: Euphorbiaceae

P. debilis shows highest antioxidant property as compared to all the other plants belonging to the genus *Phyllanthus* as proved experimentally by Kumaran and Karunakaran (2007) using different antioxidant assays such as total antioxidant activity, free radical scavenging, superoxide anion radical scavenging, hypdrogen peroxide scavenging, nitric oxide scavenging, reducing power and metal ion chelating activities. The antioxidant property of *P. debilis* can be attributed to the presence of phenolic compounds, flavonoids and flavonols (Kumaran and Karunakaran, 2007).

Phyllanthus emblica L. (Syn.: *Emblica officinalis* Gaertn.) Family: Euphorbiaceae

English Name: Indian Gooseberry; Hindi: Amlaki.

Fruits of *Phyllanthus emblica* have been used in Ayurveda as a potent Rasayana and also for treatment of diseases of diverse etiology. It is an acknowledged rich source of vitamin C. The role of vitamin C as a potent antioxidant has been well documented. Unlike other citrus fruits, Amlaki fruits have ascorbic acid conjugated to gallic acid and reducing sugars forming a tannoid complex. Such complex of vitamin C is known to be more stable. It has been suggested that, due to an internal mechanism, tannoid complex liberates the nascent ascorbic acid into the body. The antioxidant activity of tannoid active principles of *Phyllanthus emblica* consisting of emblicanin A (37 per cent), emblicanin B (33 per cent), punigluconin (12 per cent) and pedunculagin (14 per cent), was investigated on the basis of their effects on rat brain frontal cortical and striatal concentrations of the oxidative free radical scavenging enzymes, superoxidase dismutase (SOD), catalase (CAT) and glutathione peroxidase (GPX), and lipid peroxidation, in terms of thiobarbituric acid-reactive products (Bhattacharya *et al.,* 1999, 2000). The active tannoids of *Phyllanthus emblica,* induced an increase in both frontal cortical and striatal SOD, CAT and GPX activity, with concomitant decrease in lipid peroxidation in these brain areas when administered daily for 7 days. Acute single administration of EOT had insignificant effects. The results indicate that the antioxidant activity of *Phyllanthus emblica* may reside in the tannoids which have vitamin-C like properties, rather than vitamin C itself.

The extract obtained from *Phyllanthus emblica* fruits inhibited lipid peroxidation. Scartezzini and Speroni (2000) are of the opinion that this plant contains antioxidant principles that can explain and justify their use in traditional medicine in the past as well as the present.

To assess the antioxidant activity of Amla, Khopde *et al.* (2001) examined the aqueous amla extract for its ability to inhibit γ-radiation-induced lipid peroxidation (LPO) in rat liver microsomes and SOD damage in rat liver mitochondria. It was observed that the amla extract acts as a very good antioxidant against γ-radiation-induced LPO. Similarly it was found to inhibit the damage to antioxidant enzyme SOD. The antioxidant activity of the amla extract was found to be both dose- and concentration-dependent. The amount of ascorbic acid in amla was found to be 3.25 to 4.5 per cent w/w. However, in microsomes containing this composition of pure ascorbic acid alone, no inhibition in LPO was observed.

Phyllanthus emblica is one of the three constituents of the famous Indian preparation Triphala, the other two being *Terminalia belerica* and *Terminalia chebula*. The methanolic extract of 5 commercial Triphala was evaluated by Jayajothi *et al.* (2004) for antioxidant activity by DPPH free radical scavenging method, total phenolic content by Folin-Ciocalteu method and gallic acid equivalents (GAE) by HPTLC method. All extracts exhibited antioxidant activity significantly. A clear correlation between IC_{50} and contents of GAE nor the total phenolic content could be observed. Consumption of Triphala has been suggested to exert several beneficial effects by virtue of its antioxidant activity.

The aqueous extract of the fruits of *Phyllanthus emblica* and their equiproportional mixture triphala were evaluated by Naik *et al.* (2005) for their *in vitro* antioxidant activity. The extracts were found to possess ability to scavenge free radicals such as DPPH and superoxide. The mixture, triphala, is expected to be more efficient due to the combined activity of the individual compounds.

Bajpai *et al.* (2005) investigated the phenolic contents and antioxidant activity of some food and medicinal plants including *Phyllanthus emblica*. The leaves and fruits were found to have high phenolic contents and antioxidant activity. The TPC of leaves and fruits of *Phyllanthus emblica* were 75.7 and 83.5 mg/g GAE, respectively. The antioxidant activity of *Phyllanthus emblica*, as assayed by auto-oxidation of β-carotene and linoleic acid and expressed as the per cent of inhibition relative to the control were, 72.4 per cent and 81.3 per cent for heart wood and leaves, respectively.

Scartezzini *et al.* (2006) investigated vitamin C content and antioxidant activity of the fruit and of the Ayurvedic preparation of *Phyllanthus emblica* (=*Emblica officinalis*). The data obtained show that the fruit contains ascorbic acid (0.40 per cent w/w), and that the Ayurvedic method of processing increases the healthy characteristics of the fruits, thanks to the higher antioxidant activity and higher content of ascorbic acid (1.28 per cent w/w). It has also been found that Vitamin C accounts for approximately 45-70 per cent of the antioxidant activity.

Om Prakash (2007) has shown that 95 per cent methanol extract of *Phyllanthus emblica* fruit has more antioxidant activity than other extact.

Shukla *et al.* (2009) prepared the aqueous and alcoholic extracts of fruits (amalaki) and analyzed for antioxidant vitamin content (vitamin C and E), total phenolic compounds. Antioxidant status, reducing power and effect of glutathione S-transferase (GST) activity were evaluated *in vitro*. Vitamin C content of crude amalaki powder was found to be 5.38mg/g. Amalaki was rich in vitamin E like activity, total phenolic content, reducing power and antioxidant activity. Total antioxidant activity of aqueous extract of amlaki was 7.78 mmol/l. At similar concentration the total antioxidant activity of alcoholic extract was 6.67 mmol/l. Amalaki was found to be rich source of phenolic compounds (241mg/g gallic acid equivalent). Both aqueous and alcoholic extracts of amalaki inhibited activity of rat liver glutathione S-transferase (GST) *in vitro* in dose dependent manner. Since GST acts as powerful drug metabolizing enzyme its inhibition by amalaki offers possibility of its use for lowering therapeutic dose of herbal preparations. The aqueous extract also showed protection against t-BOOH induced cytotoxicity and production of ROS in cultured C_6 glial cells (Shukla *et al.*, 2009).

Poltanov *et al.* (2009) carried out the chemical and antioxidant evaluation of Indian gooseberry supplements. The study investigated the chemistry and antioxidant properties of four extracts of fruit in order to determine if there are any qualitative-quantitative differences. The extracts demonstrated varying degrees of antioxidant efficacy.

Povichit *et al.* (2010) investigated phenolic content and free radical scavenging effect of *Phyllanthus emblica* by DPPH and ABTS assays, anti-lipid peroxidation activity by TBARS and for antiglycation activity. The results revealed that the total phenolic content showed good correlation with free radical scavenging by ABTS and anti-lipid peroxidation by TBARS, but showed no correlation with antiglycation. The extract of *Phyllanthus emblica* demonstrated a moderate antioxidant effect and also showed a moderate antiglycation effect. The IC_{50} were 0.06mg/ml for the DPPH method, TEAC value of 2.94 mg/Trolox/mg sample for the ABTS method, IC_{50} of 1.37 mg/ml for the TBARS method and IC_{50} of 0.02 for the antiglycation method.

Hazra *et al.* (2010) evaluated the *in vitro* antioxidant and reactive oxygen species scavenging activities of methanol extracts of *Terminalia chebula, Terminalia belerica* and *Phyllanthus emblica* (=*Emblica officinalis*). The ability of the extracts of the fruits in exhibiting their antioxidative properties follow the order *T. chebula>P.emblica>T.belerica*. The same order is followed for the flavonoid content, whereas in the case of phenolic content it becomes *P. emblica>T.belerica> T. chebula*.

Niwano *et al.* (2011) summarized their research for herbal extracts with potent antioxidant activity obtained from a large scale screening based on superoxide radical. Experiments with the Fenton reaction and photolysis of H_2O_2 induced by UV irradiation demonstrated that extract of *Phyllanthus emblica* (fruit) have potent ability to directly scavenge OH radicals. Furthermore, the scavenging activities aagainst O_2 and OH of extracts of *Phyllanthus emblica* (fruit) proved to be heat resistent.

Phyllanthus maderaspatensis L. Family: Euphorbiaceae

The hepatoprotective activities of different extracts (n-hexane, alcohol and water) of *P. maderaspatensis* (whole plant) and choleretic activity of the most active extract have been studied along with the hydroxyl radical scavenging and antilipid peroxidation activities of these extracts. *P. maderaspatensis* has been evaluated for its antihepatotoxic and choleretic activities. The plant extracts (n-hexane, ethyl alcohol or water) has been shown to possess a remarkable hepatoprotective activity against acetaminophen-induced hepatotoxicity as judged from the serum marker enzymes (Asha *et al.*, 2004). The water and ethyl alcohol extract, showed activity at low doses as well. The n-hexane extract is known to exhibit choloretic activity, *in vitro* hydroxyl radical scavenging activity and inhibition of lipid peroxidation (Asha *et al.*, 2004).

Phyllanthus niruri L. Family: Euphorbiaceae

Shamasundar *et al.* (1985) have shown that phyllanthin and hypophyllanthin protect cells against carbon tetrachloride cytotoxicity whereas triacontanal protects against galactosamine toxicity. *P. niruri* is used as one of the components of a multiherbal preparation for treating liver ailments (Kapur *et al.*, 1994). Methanolic and aqueous extract of leaves and fruits of *P. niruri* showed inhibition of membrane lipid peroxidation (LPO), scavenging of 1,1-diphenyl-2picrylhydrazyl (DPPH) radical and inhibition of reactive oxygen species (ROS) *in vitro* (Chatterjee *et al.*, 2006).

Antioxidant activity and hepatoprotective potential of *Phyllanthus niruri* were investigated by Harish and Shivanandappa (2006). Methanolic and aqueous extracts of leaves and fruits of *P. niruri* showed inhibition of membrane lipid peroxidation, scavenging of DPPH radical and inhibition of ROS *in vitro*. Antioxidant activities of the extracts were also demonstrable *in vivo* by the inhibition of the carbon tetrachloride induced formation of lipid peroxides in the liver of rats by pretreatment with the extracts (Harish and Shivanandappa, 2006). Apart from this protein fraction isolated from *P. niruri* partially prevented nimesulide induced hepatic disorder and possesses both preventive and curative activities against chemically induced oxidative stress in mice (Chatterjee *et al.*, 2006).

Mosquera *et al.* (2009) evaluated the antioxidant activity of forty-six methanol plant extracts using DPPH radical scavenging assay. *Phyllanthus niruri* showed 31.8 per cent antioxidant activity.

Phyllanthus polyphyllus Willd. Family: Euphorbiaceae

Methanolic extract of *Phyllanthus polyphyllus* was evaluated by BR *et al.* (2008) for hepatoprotective and antioxidant activities in rats. The plant extract (200 and 300 mg/kg, p.o.) showed a remarkable hepatoprotective and antioxidant activity against acetaminophen induced hepatotoxicity as judged from the serum marker enzymes and antioxidant levels in liver tissues. Treatment of rats with different doses of plant extract significantly altered serum marker ezymes and antioxidant levels to near normal against actaminopen treated rats.

Phyllanthus reticulatus Poir. Family: Euphorbiaceae

Antioxidant activity of the methanol crude extract of entire plant of *Phyllanthus reticulatus* was assessed by Maruthappana and Shreeb (2010) using DPPH, superoxide anion and metal chelating assays at different concentrations. The methanol extract exhibited potent antioxidant activity in hepatic malondialdehyde (330.70, 279.40 and 383.79µM/mg protein) with simultaneous improvement in hepatic glutathione (7.03, 18.16 and 6.88µg/mg protein) and catalase levels (678.10, 787.00 and 522.00µg/mg protein) respectively for 50, 100 mg.kg doses and control) compared to control group.

Phyllanthus sellowianus Muell.-Arg. Family: Euphorbiaceae

Aqueous, ethanolic and dichloromethane extracts of *Phyllanthus sellowianus* showed antioxidant activity (Hnatgyszyn *et al.*, 1996).

Phyllanthus urinaria L. Family: Euphorbiaceae

The antioxidant activity of *P. urinaria* in comparison with other plants of the genus *Phyllanthus* has been examined using different antioxidant assays such as total antioxidant activity, free radical scavenging (DPPH assay), superoxide anion radical scavenging, hydrogen peroxide scavenging, nitric oxide scavenging, reducing power and metal ion chelating activities in the study by Kumaran and Karunakaran (2007).

Phyllanthin (1), phyltetralin (2), trimethyl-3,4-dehydrochebulate (3), methylgallate (4) and rhamnocetrin (5), methylbrevifolincarboxylate(6), β-sitosterol-3-O-β-D-glucopyranoside (7), quercitrin (8) and rutin (9) were isolated from the ethanolic extract of *Phyllanthus urinaria* (Fang *et al.*, 2007). In the antioxidant assay, the isolates 3,4 and 6 exhibited significant DPPH radical scavenging activity with an IC_{50} value of 9.4, 9.8 and 8.9µM, respectively. All the isolates, except 7, significantly and dose-dependently inhibited the enhanced production of NO radicals.

Chen *et al.* (2009) studied the antioxidant activity of aerial parts extract towards xanthine oxidase, lipoxygenase and tyrosinase activities. The inhibitory effect of acetone extract on xanthine oxidase activity was less than 5 per cent, tyrosine activity was less than 40 per cent and lipoxygenase activity was about 50 per cent.

Phyllanthus virgatus G.Forst. Family: Euphorbiaceae

The antioxidant properties of *P. virgatus* were intermediate to those of the other species of the genus *Phyllanthus* as reported by Kumaran and Karunakaran (2007).

Phyllostachys bambusoides Sieb. et Zucc. Family: Poaceae (Graminae)

English Name: Moso bamboo.

Jun *et al.* (2004) evaluated the antioxidant activities of methanol extracts of culms of *Phyllostachys bambusoides* by DPPH assay, the inhibition activity for peroxidation of linoleic acid, and the reduction power. The methanol extracts of Moso bamboo culms presented stronger antioxidant activity compared with DPPH scavenging activity. Methanol-extract of moso bamboo culms was further fractionated by different solvents and n-butanol soluble fraction exhibited the most significant activity in the DPPH scavenging assay.

Phyllostachys edulis A. et C. Riviere Family: Poaceae (Graminae)

The butanol-soluble extract of the bamboo leaves was found to have a significant antioxidant activity, as measured by scavenging the stable DPPH free radical and the superoxide anion radical in the xanthine/xanthine oxidase assay system. Antioxidant activity-directed fractionation of the extract led to the isolation and characterization of three structural isomeric chlorogenic acid derivatives 3-O-(3'-methylcaffeoyl)quinic acid(1), 5-O-caffeoyl-4-methylquinic acid (2) and 3-O-caffeoyl-1-methylquinic acid (3). In the DPPH scavenging assay as well as in the iron-induced rat microsomal lipid peroxidation system, compounds 2 (IC_{50} = 8.8 and 19.2 µM) and 3 (IC_{50} = 6.9 and 14.6 µM) showed approximately 2-4 times higher antioxidant activity than did chlorogenic acid (IC_{50} = 12.3 and 28.3 µM) and other related hydroxycinnamates such as caffeic acid (IC_{50} = 13.7 and 25.5 µM) and ferulic acid (IC_{50} = 36.5 and 56.9µM).

Among the three compounds, compound 1 yielded the weakest antioxidant activity, and the DPPH scavenging and lipid peroxidation inhibitory activity (IC_{50} = 16.0 and 29.8 µM) was lower than those of chlorognic and caffeic acids. All three compounds exhibited both superoxide scavenging activities and inhibitory effects on Xanthine oxidase (Kweon *et al.*, 2001).

Phyllostachys nigra (Lodd. ex Lindl.) Munro Family: Poaceae (Graminae)

Luteolin 6-C-(6″-O trans-caffeoylglucoside), a flavone isolated from the leaves of *Phyllostachys nigra*, have antioxidant as well as aldose reductase and advanced glycation end products inhibitory effects (Jung *et al.*, 2007).

Phyllostachys pubescens Mazel. ex J. Houz. Family: Poaceae (Graminae)

English Name: Madake bamboo.

Jun *et al.* (2004) evaluated the antioxidant activities of methanol extracts of culms of *Phyllostachys pubescens* by DPPH assay, the inhibition activity for peroxidation of linoleic acid, and the reduction power. The methanol extracts of Madake bamboo leaves presented stronger antioxidant activity compared with DPPH scavenging activity.

Physalis alkekengi L. Family: Solanaceae

Antioxidant capacities of 56 selected Chinese medicinal plants, including *Physalis alkekengi* were evaluated by Song *et al.* (2010) using the TEAC and FRAP assays, and their total phenolic content was measured by the Folin-Ciocalteu method. *Physalis alkekengi* showed moderate antioxidant activity with TEAC value of 64.29µmol Trolox/g, FRAP value of 60.42µmol Fe^{2+}/g and phenolic content was 9.12mg GAE/g.

Physalis angulata L. Family: Solanaceae

Choi and Hwang (2005) studied the effect of flower extract of *Physalis angulata* on plasma oxidant system and lipid levels in rats. The study showed that the intake of the extract by rats results in an increase in antioxidant enzyme activity and HDL-cholesterol, and a decrease in malondialdehyde, which minimize the risk of inflammatory and heart disease.

Physalis peruviana L. Family: Solanaceae

Wu *et al.* (2005) evaluated the antioxidant activities of *Physalis peruviana*. In this study, the hot water extract (HWEPP) and extracts prepared from different concentrations of ethanol (20, 40, 60, 80 and 95 per cent EtOH) from the whole plant were evaluated for antioxidant activities. Results displayed that at 100 mug/ml, the extract prepared from 95 per cent EtOH exhibited the most potent inhibition rate (82.3 per cent) on $FeCl^2$-ascorbic acid induced lipid peroxidation in rat liver homogenate. At concentrations 10-100 µg/ml, this extract also demonstrated the strongest superoxide anion scavenging and inhibitory effect on xanthine oxidase activities. In general, the ethanol extracts revealed a stronger antioxidant activity than α-tocopherol and HWEPP. Compared to α-tocopherol, the IC_{50} value of 95 per cent EtOH PP extract was lower in thiobarbituric acid test (IC_{50}=23.74 µg/ml vs. 26.71 µg/ml), in cytochrome c test (IC$_{50}$=10.40 µg/ml vs. 13.39 µg/ml) and in xanthine oxidase inhibition test (IC$_{50}$=8.97 µg/ml vs. 20.68 µg/ml).

Physospermum verticillatum Waldst. et Kit. Family: Apiaceae (Umbelliferae)

Three triterpene saponins isolated from the roots of *Physospermum verticillatum* and identified as saikosaponin a (1), buddlejasaponin IV (2) and sangarosaponin D (3) were investigated inhibitory effect on nitric oxide production. Buddlejasaponin IV (2) and songarosaponin D (3) exerted significant inhibition of NO production in LPS induced RAW 264.7 macrophages with IC(50) of 4.2 nd 10.4 microM, respectively (Tundis *et al.*, 2009).

Phytolacca americana L. Family: Phytolaccaceae

Souri *et al.* (2004) evaluated the antioxidative activity of *Phytolacca americana* by linoleic acid peroxidation test using 1,3-diethyl-2-thiobarbituric acid as the reagent. Methanolic extract showed 68.77 per cent inhibition at 40µg concentration with IC$_{50}$ value of 2.08µg.

Extracts of *Phytolacca americana* were investigated by Hosseinimehr *et al.* (2007) for their total flavonoids, phenol contents and their radical scavenging activity using DPPH assays. Quercetin and butylated hydroxytoluene were used as standard reference with well documented antioxidant activity. Methanolic extract sample showed free radical scavenging activity. A correlation between radical scavenging capacity of the extract with total phenolic compounds content was observed.

Phytolacca rivinoides Kunth et Bouché Family: Phytolaccaceae

Ethanolic extracts from 15 plant species used in traditional medicine in Ecuador were evaluated by De las Heras *et al.* (1998) for anti-inflammatory and antioxidant activities. *Phytolacca rivinoides* extract was active as antioxidant.

Picrasma crenata (Vell.) Engl. Family: Simaroubaceae

Menezes *et al.* (2004) investigated the antioxidant activity of hydroalcoholic extracts of roots of *Picrasma crenata* by DPPH assay and phosphomolybdenum method. The relative activity in phosphomolybdenum complex was 0.67 while scavenging activity against DPPH in IC$_{50}$ (µg/ml) was 136.70.

Picrorrhiza kurroa Royle ex Benth. Family: Scrophulariaceae

Ethanol extract of rhizomes and roots showed antioxidant effect (Anandan and Devaki,1999). In D-galactosamine-induced hepatitis in rats, a significant increase of lipid peroxidation and a decrease in liver antioxidant enzymes levels were observed. Pretreatment with the ethanol extract of *Picrorrhiza kurroa* prevented these alterations (Anandan and Devaki, 1999).

The antioxidant activity of *Picrorhiza kurroa* extract was studied by lipid peroxidation assay using rat liver homogenate (Govindarajan *et al.*, 2003). The extract (1 mg/ml) showed marked protection (up to 66.68 per cent) against peroxidation of liver phospholipids. Besides, reduced glutathione showed very encouraging activity. The extract also exhibited significant scavenging activity.

Rajkumar *et al.* (2011) investigated the antioxidant and neoplastic activities of methanolic and aqueous extracts of *Picrorrhiza kurroa* rhizome. The antioxidant efficacies of the extracts were studied employing radical scavenging assays (DPPH and OH), FRAP and TBA assay for testing inhibition of lipid peroxidation. Both extracts exhibited promising antioxidant potentials.

Picrorhiza scrophulariflora Pennell. Family: Scrophulariaceae

Cai *et al.* (2004, 2006) evaluated the antioxidant activity and phenolic compounds in traditional Chinese medicinal plants associated with anticancer. The improved ABTS.+ method was used to systematically assess the total antioxidant capacity (Trolox equivalent antioxidant capacity, TEAC) of the medicinal extracts. The TEAC values and total phenolic content for methanolic extract of rhizome of *Picrorhiza scrophulariflora* were 518 μmol Trolox equivalent/100g dry weight (DW), and 3.92 g of gallic acid equivalent/100g DW respectively. A positive significant linear relationship between antioxidant activity and total phenolic content showed that phenolic compounds were the dominant antioxidant components in the tested herbs. Major types of phenolic compounds in *Picrorhiza scrophulariflora* include phenolic acids (ferulic acid, vanillic acid, coumaric acid).

Scrosides, isolated from the roots of *Picrorhiza scrophulariiflora* were evaluated for their scavenging effects on hydroxyl radicals and superoxide anion radicals. Scrosides showed potent antioxidative effects as those of ascorbic acid (Wang *et al.*, 2004).

Li *et al.* (2008) evaluated the antioxidant properties of 45 medicinal plants including *Picrorhiza scrophulariflora* using FRAP and TEAC assays and the total phenolic contents of these plants were measured. *Picrorhiza scrophulariflora* showed 168.04 μmol Fe(II)/g FRAP value, 96.57μmol Trolox/g of TEAC value and 31.24 mg GAE/g of Phenolic content. A high correlation between antioxidant capacities and their phenolic contents indicated that phenolic compounds were a major contributor of antioxidant activity of these plants.

Pilea melastemoides (Poir.) Bl. Family: Urticaceae

Extracts from 11 vegetables of Indonesian origin were screened for flavonoid content, total phenolics, and antioxidant activity by Andarwulan *et al.* (2010). Flavonoid content in *Pilea melastemoides* was 2.27mg/100g fresh weight.

Pilea microphylla (L.) Liebm. Family: Urticaceae

Prabhakar *et al.* (2007) evaluated antioxidant and radioprotective effect of the active fraction of *Pilea microphylla* ethanolic extract. The most active fraction reacts with free radicals, such as DPPH (50μM), ABTS (100μM) and OH (generated by Fenton reaction) with IC_{50} value of 23.15μg/ml, 3.0μg/ml and 310μg/ml, respectively. The most active fraction inhibited iron-induced lipid peroxidation in phosphatidyl choline liposomes with an IC_{50} of 13.74μg/ml. The kinetics of scavenging DPPH and ABTS radicals were followed at different concentrations of the fraction by employing stopped-flow studies. The observed first order decay rate constants at 200μg/ml and 50μg/ml of fraction with DPPH (50μM and ABTS(50μM) were found to be $0.4s^{-1}$ and 2.1^{-1}, respectively.

Piliostigma reticulatum (DC.) Hochst. Family: Fabaceae (Leguminosae)

Aderogba *et al.* (2004) evaluated the antioxidant activity of leaf extracts of *Piliostigma reticulatum*. The ethyl acetate and butanol fractions exhibited substantial inhibition of DPPH activity with EC_{50} values 19.70µg/ml and 10.27µg/ml, respectively.

The methanol and aqueous extracts of *Piliostigma reticulatum* barks were investigated by Zerbo *et al.* (2010) for its antioxidant and antibacterial activities. The antioxidant test using DPPH method demonstrated important radical scavenging activity for the methanol extract with $IC_{50} = 0.37$ µg/ml.

Piliostigma thonningii (Schum.) Milne-Redhead Family: Fabaceae (Leguminosae)

Aderogba *et al.* (2004) evaluated the antioxidant activity of leaf extracts of *Piliostigma thonningii*. The ethyl acetate and butanol fractions exhibited substantial inhibition of DPPH activity with EC_{50} values 18.51µg/ml and 14.70µg/ml, respectively.

Ajiboye (2011) investigated the *in vivo* antioxidant potentials of *Piliostigma thonnigii* in CCl_4-induced hepatic oxidative damage in rats. They concluded that *Piliostigma thonningii* leaves protected liver against hepatic and oxidative damage by CCl_4 possibly acting as an *in vivo* free radical scavenger, induction of antioxidant enzymes and lipid peroxidation.

Pimenta dioica (L.) Merr. Family: Myrtaceae

English Name: Allspice plant, Pimento.

A phenylpropanoid, threo-3-chloro-1-(4-hydroxy-3-methoxyphenyl) propane-1,2-diol together with five known compounds, eugenol, 3,4-dimethoxy-cinnamaldehyde, vanillin and 3-(-4-hydroxy-3-methoxyphenyl) propane 1,2-diol were isolated by Kikuzaki *et al.* (1999) from the berries of *Pimenta dioica*. The phenylpropanoids inhibited auoxidation of linoleic acid in a water-alcohol system.

Ramos *et al.* (2003) screened the reducing activity of *Pimenta dioica* on DPPH radical, OH radical scavenging potential, *in vitro* inhibition of lipid peroxidation and modulation of mutagenicity induced by ter-butyl hydroperoxide (TBH) in *Escherichia coli*. *Pimenta dioica* displayed IC_{50} less than 30µg/ml in DPPH radical reduction assay and IC_{50} less than 32µg/ml in lipid peroxidation inhibition testing. It also showed a 20 per cent inhibition of the *in vitro* induced OH attack of deoxyglucose. Eugenol, the main constituent of the essential oil of *Pimenta dioica*, also inhibited oxidative mutagenesis by TBH in *Escherichia coli*, at concentrations ranging from 150 to 400µg/plate.

The ethyl acetate-soluble part of allspice, berries of *Pimenta dioica*, showed strong antioxidant activity and radical-scavenging activity against DPPH radical. From the ethyl acetate-soluble part, two new compounds, 5-galloyloxy-3-4-dihydroxypentanoic acid and 5-(5-carboxmethyl-2-oxocyclopenty)3Z-pentany 6-O-galloy-beta-D-glucoside were isolated together with 11 known polyphenols by repeated column chromatography. All isolated compounds were evaluated for antioxidative effects on oxidation of methyl linoleate under aeration and heating and on peroxidation of liposome induced by AAPH as water-soluble initiator along with their radical-scavenging activity against DPPH. Quercetin and its glycoside showed remarkable activity for scavenging DPPH radical and inhibiting peroxidation of liposome. Two new compounds also exhibited strong DPPH radical-scavenging activity and inhibitory effect on the peroxidation of liposome as myricetin (Miyajima *et al.*, 2008).

Mariutti *et al.* (2008) evaluated the antioxidant activity of allspice employing ABTS and DPPH radical assay. DPPH EC_{50} value was 8.3g/kg while DPPH-TEAC value was 30.4mM/g. ABTS-TEAC value was 57.0mM/g.

Composition and antioxidant activity of essential oil of *Pimenta dioica* were investigated by Padmakumari *et al.* (2011). The antioxidant activities of the oils were evaluated in terms of their free radical-scavenging activity against DPPH, ABTS radical cation and superoxide anion. The antioxidant assays showed that the oil possess very high radical scavenging activities (DPPH IC_{50} = 4.82 µg/ml, ABTS IC_{50} = 2.27µg/ml, superoxide IC_{50} = 17.78µg/ml. The metal chelating capacities IC_{50} =83.62µg/ml and reducing power were also very high.

Pimpinella anisum L. Family: Apiaceae (Umbelliferae)

English Name: Aniseed.

Hydrodistillated extracts from aniseed were assessed for their total phenol content, and antioxidant (iron(III) reduction, inhibition of linoleic acid peroxidation, iron (II) chelation, DPPH radical-scavenging and inhibition of hydroxyl radical-mediated 2-deoxy-D-ribose degradation, site and nonsite-specific) activities by Hinneburg *et al.* (2005). The extracts performed significantly better and the total phenols were higher. In aniseed quercetin, luteolin and apigenin glycosides were found (Kunzemann and Herrmann, 1977).

Gülçin *et al.* (2003) reported that water and ethanol extracts (20µg/ml) of *Pimpinella anisum* seeds exhibited 99.1 and 77.5 per cent inhibition of peroxidation in linoleic acid system.

Souri *et al.* (2008) evaluated the antioxidant activity of fruit against linoleic acid peroxidation using 1,3-diethyl-2-thiobarbituric acid as reagent. Antioxidant activity (IC_{50}) against peroxidation of linoleic acid (2mg/ml) was 101.26 and phenolic content was 353.92 mg/100g dry weight. The results of this study showed that there was no significant correlation between antioxidant activity and phenolic content of the studied plant materials and phenolic content could not be a good indicator of antioxidant capacity.

Antioxidant activities of ethanol extracts from seven Apiaceae fruits have been studied by Nickavar and Abolhasani (2009) by the DPPH radical scavenging test. All the studied extracts, including *Pimpinella anisum,* showed antioxidant capability. For *Pimpinella anisum* IC_{50} value of the DPPH scavenging activity was 109.80µg/ml and total flavonoid content was 26.89µg/mg. A positive correlation was found between the antioxidant potency and flavonoid content of the fractions.

Pimpinella aurea DC. Family: Apiaceae (Umbelliferae)

Two phenylpropanoids were isolated from the aerial parts of *Pimpinella aurea* and their antioxidant properties were assessed by Delazar *et al.* (2006) by the DPPH assay.

Pimpinella saxifraga L. Family: Apiaceae (Umbelliferae)

A study was carried out by Budinevi *et al.* (1995) on the antioxidative activity of *Pimpinella saxifraga*. Using the ethanolic extracts of plant material good results were obtained.

Pinellia ternatea (Thunb.) Breitenb. Family: Araceae

Antioxidant capacities of 56 selected Chinese medicinal plants, including *Pinellia ternatea* were evaluated by Song *et al.* (2010) using the TEAC and FRAP assays, and their total phenolic content was measured by the Folin-Ciocalteu method. *Pinellia ternatea* showed low antioxidant activity with TEAC value of 0.61µmol Trolox/g, FRAP value of 0.46µmol Fe^{2+}/g and phenolic content was 0.12mg GAE/g.

Piper betel L. Family: Piperaceae

Dasgupta and De (2004) evaluated the antioxidant activities of aqueous extracts of three local varieties of *Piper betle* leaves by several *in vitro* systems – DPPH radical scavenging activity, superoxide radical scavenging activity in riboflavin/light/NBT system, hydroxyl radical scavenging activity was measured by the reduction of Mo(VI) to Mo(V), by the extract and subsequent formation of a green phosphate/Mo(V) complex at acid pH. The extracts were found to have different levels of antioxidant activity in the systems tested. The data indicate that the antioxidant activities differed in varieties.

Wong *et al.* (2006) investigated the antioxidant properties of 25 edible tropical plants including *Piper betel* using DPPH and FRAP assays. Total phenolic content was also estimated. *Piper betel* showed very high antioxidant activity.

The DPPH assay of the ethanol extract of three varieties (Bangla, Sweet and Mysore) of *Piper betel* revealed the Bangla variety to possess the best antioxidant activity that can be correlated with the total phenolic content and reducing powers of the respective extracts. Chevibetol, allylpyrocatechol and their respective glucosides were isolated from the leaves. Among the isolated compounds allylpyrocatechol showed the best results in all the *in vitro* experiments. It could prevent Fe(II)-induced LPO of liposomes and rat brain homogenates as well as gamma-ray-induced damage of pBR 322 plasmid DNA more efficiently than chevibetol (Rathee *et al.*, 2006).

Manigauha *et al.* (2009) evaluated the antioxidant activity of *Piper betel* leaves. Ethanolic extract of *Piper betel* leaf showed strong antioxidant activities like reducing power, DPPH radical, superoxide anion scavenging and deoxyribose degradation activities when compared with different standards such as ascorbic acid, DMSO and BHT.

Free radical scavenging activity of chloroform and ethyl acetate extracts of leaves of *Piper betle* were evaluated *in vitro* with spectrophotometric method based on the reduction of the stable DPPH free radical. The reducing power ability of these extracts was also determined. Both extracts of leaves and plant has shown significant antioxidant activity in all assay techniques. In DPPH method IC_{50} values of these chloroform and ethyl acetate extracts were found to be 9.187 µg/ml and 4.56 µg/ml (Islam *et al.*, 2010).

Chakraborty and Shah (2011) evaluated the antioxidative activity of betel leaf extracts by TBARS and DPPH method. In DPPH photometric assay DPPH was reduced in presence of extracts in concentration dependent manner. Petroleum ether extract showed highest amount of antioxidative activity followed by ethyl acetate extract, methanol extract and aqueous extratcts respectively. Reductive ability was maximum in case of water extract at 25µg/ml.

Piper cubeba L. Family: Piperaceae

English Name: Sweet Black pepper.

Choi and Hwang (2005) studied the effect of fruit extract on plasma oxidant system and lipid levels in rats. The study showed that the intake of the extract by rats resulted in an increase in antioxidant enzyme activity and HDL-cholesterol, and a decrease in malondialdehyde, which minimize the risk of inflammatory and heart disease.

Khalaf *et al.* (2008) reported high antioxidant activity in the methanolic crude extracts using DPPH assay.

Piper fulvescens DC. Family: Piperaceae

The antioxidant property of *Piper fulvescens* was studied using free radical-generating systems (Velazquez *et al.*, 2003). The methanol extract protected against enzymatic and non-enzymatic lipid peroxidation in microsomal membranes of rat.

Piper guineense Schumm. et Thonn Family: Piperaceae

Odukoya *et al.* (2005) evaluated the antioxidant activity of Nigerian dietary spices including *Piper guineense*. The antioxidant activity (expressed as per cent inhibition of oxidation) was 76.94 per cent, reducing power was 15.34 per cent and Total phenolic content was 169.52 mg/100g Tannic acid Equivalent. Antioxidant activity correlated significantly and positively with total phenolics while there was no linear correlation between total antioxidant activity and reducing power neither between reducing power and total phenolic content.

Piper kadsura (Choisy) Ohwi Family: Piperaceae

Gan *et al.* (2010) evaluated the antioxidant activity of *Piper kadsura* using the FRAP and TEAC assays, and its total phenolic content was measured by the Folin-Ciocalteu method. For FRAP assay the value was 147.41 µmol Fe(II)/g dry weight. For TEAC assay, the value was 103.41µmol Trolox/g dry weight. Total phenolic content was 8.94 mg gallic acid equivalent/g dry weight of plant material.

Piper leucophyllum C.DC. Family: Piperaceae

Ruiz-Terán *et al.* (2008) evaluated the antioxidant activity of aerial parts of *Piper leucophyllum* using DPPH and β-carotene-linoleic acid bleaching assay. The extract displayed antioxidant activity comparable to the commercial antioxidant BHA. For the methanolic extract analysed, a clear relation between the total phenolic content of the extracts and their antioxidant activity was found.

Piper longum L. Family: Piperaceae

Acetone and petroleum ether extract of *Piper longum* were evaluated for their antioxidant enzyme activity in testis of male albino rats by Krishnamoorthy *et al.* (2002). The antioxidant enzymes activity were increased and lipid peroxidation were decreased considerably in the *Piper longum* extract treated animal's testis.

Bajpai *et al.* (2005) investigated the phenolic contents and antioxidant activity of some food and medicinal plants including *Piper longum*. The fruits and leaves were found to have moderate phenolic contents and antioxidant activity. The TPC of fruits and leaves of *Piper longum* were 17.3 and 18.1 mg/g GAE, respectively. The antioxidant activity of *Piper longum*, as assayed by auto-oxidation of β-carotene and linoleic acid and expressed as the per cent of inhibition relative to the control were, 50.1 per cent and 51.2 per cent for fruits and leaves, respectively.

The methanolic crude extracts of *Piper longum* were screened by Veeru *et al.* (2009) for their free radical scavenging properties using ascorbic acid as standard antioxidant. Free radical scavenging activity was evaluated using DPPH free radical. The overall antioxidant activity of *Piper longum* was found to be strongeer with IC_{50} value of 0.1 mg/l.The ascorbic acid level was 7.36 mg/100g and the carotenoids content were 9.5 mg/100g in plant extracts. The total phenol content was 28.16 mg/g.

Piper nigrum L. Family: Piperaceae

English Name: Black pepper, Pepper.

Vijaykumar *et al.* (2004) investigated the antioxidant efficacy of black pepper and piperine in rats with high fat diet induced oxidative stress. Significantly elevated levels of TBARS, Conjugated dienes and significantly lowered activities of SOD, CAT, GPx, GST and GSH in the liver, heart, kidney, intestine and aorta were observed in rats fed with the high fat diet as compared to the control rats. Simultaneous supplementation with black pepper or the active principle of black pepper, piperine, can reduce high-fat diet induced oxidative stress to the cells.

Selvendian *et al.* (2004) investigated the impact of piperine, a principal ingredient of pepper (*Piper nigrum*) on alterations of mitochondrial antioxidant system and lipid peroxidation in Benzo(a)pyrene [B(a)p] induced experimental lung carcinogenesis. Oral supplementation piperine (50mg/kg body weight) effectively suppressed lung carcinogenesis in B(a)p induced mice as revealed by the decrease in the extent of mitochondrial lipid peroxidation and concomitant increase in the activities of enzymatic antioxidants (superoxide dismutase, catalase and glutathione peroxidase) and non-enzymatic antioxidant (reduced glutathione, vitamin E and vitamin C) levels when compared to lung carcinogenesis bearing antimals. Data suggests that piperine may extend its chemopreventive effect by modulating lipid peroxidation and augmenting antioxidant defense system.

The 75 µg/ml concentration of water and ethanolic extracts of *Piper nigrum* showed 95.5 and 93.3 per cent inhibition on peroxidation of linoleic acid emulsion, respectively (Gülçin, 2005).

Aqueous extract of *Piper nigrum* seeds were administered orally to alloxan induced diabetic rats once a day for 4 weeks (Kaleem *et al.*, 2005). These treatments led to significant lowering of blood sugar level and reduction in serum lipids. The levels of antioxidant enzymes, catalase and glutathione peroxidase decreased in alloxan induced diabetic rats, however these levels returned to normal in insulin, *P. nigrum* treated rats. There was no significant difference in superoxide dismutase activity compared to controls. Lipid peroxidation level was significantly higher in diabetic rats and it was slightly increased in insulin and *P. nigrum* treated rats as compared to control rat. The results of Kaleem *et al.* (2005) suggest that the treatment with *Piper nigrum* is useful in controlling not only the glucose and lipid levels but also in strengthening the antioxidant potential.

Khalaf *et al.* (2008) reported high antioxidant activity in the methanolic crude extracts of pepper using DPPH assay.

Singh *et al.* (2008) investigated the *in vitro* antioxidant activity of different fractions (R1, R2 and R5) obtained from pet ether extract of black pepper fruits. The fractions R1, R2 and R3 were eluted from pet ether and ethyl acetate in the ratio of 6:4, 5:5 and 4:6, respectively. DPPH radical, superoxide anion radical, nitric oxide radical and hydroxyl radical scavenging assays were carried out to evaluate the antioxidant potential of the extract. The free radical scavenging activity of the different fractions of pet ether extract of *P. nigrum* (PEPN) increased in concentration dependent manner. The R3 and R2 fraction of PEPN in 500µg/ml inhibited the peroxidation of a linoleic acid emulsion by 60.48 per cent, respectively. In DPPH free radical scavenging assay the activity of R3 and R2 were found to be almost similar. The R3 (100µg/ml) fraction of PEPN inhibited 55.68 per cent nitric oxide radicals generated from sodium nitroprusside, whereas curcumin in the same concentration inhibited 84.27 per cent. Moreover, PEPN scavenged the superoxide radical generated by Xanthine/Xanthine oxidase system. The fraction R2 and R3 in the doses of 1000µg/ml inhibited 61.04 per cent and 63.56 per cent respectively. The hydroxyl radical was generated by Fenton's reaction. The amounts of total phenolic compounds were determined and 5.98µg pyrocatechol phenol equivalents were detected in one mg of R3. Thus *Piper nigrum* could be considered as a potential source of natural antioxidant.

Mariutti *et al.* (2008) evaluated the antioxidant activity of black pepper employing ABTS and DPPH radical assay. DPPH EC_{50} value was 110g/kg while DPPH-TEAC value was 2.5mM/g. ABTS-TEAC value was 5.7mM/g.

The fruit of the plant is used in the treatment of arthritis, and effective against neuralgia, poor circulation, poor muscle tone, sprains and stiffness. Ascorbic acid, β-carotene, Lauric acid, Myristic acid, Palmitic acid and Piperine are the main constituents responsible for its antioxidative behaviour.

Piper retrofractum Vahl Family: Piperaceae

English Name: Long pepper. Japanese: Hihatsumodoki.

Chanwitheesuk *et al.* (2005) investigated the antioxidant activity of flowers of *Piper retrofractum* using a β-carotene bleaching method. The contents of plant chemicals such as vitamin C, Vitamin E, carotenoids, tannin and total phenolics, were also determined. Methanolic extract showed antioxidant activity. *Piper retrofractum* has antioxidant index of 3.36, Vitamin C 9.37 mg per cent, Vitamin E 0.0053mg per cent, Total carotenes 1.28 mg per cent, Total xanthophylls 5.31mg per cent, Tannins 7.78mg per cent and Total phenolics of 57.5mg per cent.

Maeda *et al.* (2006) examined the phenolic contents of Long Pepper and its radical scavenging activities for DPPH radical. The polyphenol content was 132.1 mg-gallic acid equivalent/100 g) and DPPH scavenging activity was 736.1 µmol-Trolox eq./ 100 g).

Piper sarmentosum Roxb. Family: Piperaceae

Chanwitheesuk *et al.* (2005) investigated the antioxidant activity of leaves of *Piper sarmentosum* using a β-carotene bleaching method. The contents of plant chemicals such as vitamin C, Vitamin E, carotenoids, tannin and total phenolics, were also determined. Methanolic extract showed antioxidant activity. *Piper sarmentosum* has antioxidant index of 13., Vitamin C16.6 mg per cent, Vitamin E 0.0100mg per cent, Total carotenes 3.82 mg per cent, Total xanthophylls 5.81mg per cent, Tannins 17.7mg per cent and Total phenolics of 123mg per cent.

Ethanol and aqueous extracts of different parts of *Piper sarmentosum* were investigated for antioxidant activity (β-carotene linoleate model and DPPH model) and total phenolic and amide contents. The extracts of the different parts exhibited different antioxidant activity, phenolic and amide contents. The ethanol extracts exhibited better antioxidant activity compared to the aqueous extracts. A positive correlation was found between antioxidant activity and total polyphenols, flavonoids and amides, in the β-carotene linoleate model and DPPH model (Hussain *et al.*, 2009, 2010).

The phenolic compounds of five southern Thai indigenous vegetables including Cha-plu (*Piper sarmentosum*) leaf were extracted using different solvents. The extracts were analyzed for total phenolic content using the Folin-Ciocalteu procedure, free radical scavenging capacity by using ABTS and DPPH methods, and reducing capacity by using FRAP assay. The acetone extract possessed the highest total phenolic content followed by methanolic and distilled water extracts, respectively. The acetone extract showed higher free radical scavenging capacity and reducing capacity than those of their methanolic and water extracts counterparts respectively (Panpipat *et al.*, 2010).

Piper umbellatum L. Family: Piperaceae

Agbor *et al.* (2005) evaluated the antioxidant capacity of 14 herbs from Cameroon. Methanol and HCl in methanol extracts were analyzed using two different antioxidant assay methods (Folin-Ciocalteu method and FRAP method). The 1.2M HCl in methanol extract of *Piper umbellatum* had significantly higher antioxidant capacity than the methanolic extract. The FRAP antioxidant values were significantly higher than the Folin antioxidant values. *Piper umbellatum* tops the folin free antioxidant list. Irrespective of the assay methods used, the samples were rich in antioxidants.

Pisonia alba Span. Family: Nyctaginaceae

Subhasree *et al.* (2009) carried out the antioxidant activity of leafy vegetable of *Pisonia alba* by measuring the ability of methanol extracts of this plant to scavenge radicals generated by *in vitro* systems and by their ability to inhibit lipid peroxidation. The levels of non-enzymatic antioxidants were also determined by standard spectrophotometric methods. Correlation and regression analysis established a positive correlation between some of these antioxidants and the *in vitro* free radical-scavenging activity of the plant extracts. The conclusions drawn from this study indicate that *in vivo* studies, isolation and analysis of individual bioactive compounds will reveal the crucial role that this plant may play in several therapeutic formulations.

Pistacia atlantica Desf. Family: Anacardiaceae

Souri *et al.* (2004) evaluated the antioxidative activity of *Pistacia atlantica* subsp. *mutica* by linoleic acid peroxidation test using 1,3-diethyl-2-thiobarbituric acid as the reagent. Methanolic extract showed 56.46 per cent inhibition at 40µg concentration with IC_{50} value of 7.51µg.

The quantification of the total phenolic compounds of *Pistacia atlantica* showed that the different parts of the tree are rich in natural phenolic compounds (Yousfi *et al.*, 2009). The antioxidant tests proved that the phenolic extracts have a strong antioxidant activity. The positive correlation between the TEAC and the amount of phenolic compounds confirmed their contribution to the antioxidant activity. Yousfi *et al.* (2009) isolated a new hispalone from antioxidant extracts of *Pistacia atlantica*.

The essential oils obtained from the leaves of *Pistacia atlantica* were screened by Gourine *et al.* (2010a,b) for their antioxidant activities *in vitro* using two different and complementary assays: the DPPH free radical scavenging and the FRAP. The result of the DPPH assay gave and IC_{50} range value of 8.8-27.48 mg/ml for all the samples studied. The antioxidant expressed as Ascorbic acid Equivalent Antioxidant Capacity has a range value of 2.8 – 11.1 mg/ml, *i.e.*, that the sample oils were almost 3-11 times much more active than ascorbic acid. The seasonal variation showed that most of the main components of the oils reached their highest values in September. The highest antioxidant capacity to scavenge free DPPH radicals was reached in the month of June for male oils and during the months of September – October for the female oils. The female oil was more active than the male oil. The antioxidant capacity of the female oil was almost ten times higher than ascorbic acid in the FRAP assay (Gourine *et al.*, 2010a).

Pistacia lentiscus L. Family: Anacardiaceae

Essential oil from aerial parts of *Pistacia lentiscus* exhibited good antioxidant activity (Barra *et al.*, 2007).

Pistacia terebinthus L. subsp. *palaestina* (Boiss.) Engler Family: Anacardiaceae

Demo *et al.* (1998) reported antioxidant α-tocopherol from the Hexane extract of *Pistacia terebinthus*. Serteser *et al.* (2009) studied the antioxidant activity of 50 per cent aqueous methanol extracts by various antioxidant assays, including free radical scavenging, hydrogen peroxide scavenging and metal (Fe^{2+}) chelating activities. The methanolic extract examined in the assay showed the strongest activities. The DPPH radical scavenging effects of leaves extract was 0.311 while those of fruit extracts showed 0.533. Fe^{2+} chelating activity (per cent) of leaves extracts is 19.26 while fruit extracts showed 30.65. Hydrogen peroxide inhibition activity of leaves extracts was 16.78 per cent while fruit extracts showed 24.65 per cent inhibition.

Pistacia terebinthus L. subsp. *terebinthus* Family: Anacardiaceae

Serteser *et al.* (2009) studied the antioxidant activity of 50 per cent aqueous methanol extracts by various antioxidant assays, including free radical scavenging, hydrogen peroxide scavenging and metal (Fe^{2+}) chelating activities. The methanolic extract examined in the assay showed the strongest activities. The DPPH radical scavenging effects of leaves extract was 0.336 while fruit extracts showed 0.523 antiradical activity. Fe^{2+} chelating activity (per cent) of leaves extracts was 24.65 while fruit extracts showed 29.42 per cent of inhibition. Hydrogen peroxide inhibition activity of leaves extracts was 21.67 per cent while fruit extracts showed 23.18 per cent inhibition.

Pistacia weinmannifolia J. Poisson ex Franch Family: Anacardiaceae

Zhao *et al.* (2005) evaluated the antioxidant properties of two gallotannins (Pistafolin A and Pistafolin B) isolated from the leaves of *Pistacia weinmannifolia*. Both Pistafolin A and Pistafolin B inhibited the peroxyl-radical induced lipid peroxidation of L-α-phosphatidylcholine liposomes dose-dependently and prevented the bovine serum albumin from peroxyl-induced oxidative damage. Both Pistafolin A and Pistafolin B scavenged the hydrophilic 2,2'-azinobis(3-ethylbenzothiozoline-6-sulphonic acid) diammonium salt-free radicals and the hydrophobic DPPH radicals effectively, suggesting they may act as hydrogen donating antioxidants. The protective effects of the two gallotannins against oxidative damage of biomacromolecules were due to their strong free radical scavenging ability. Pistafolin A with three galloyl moieties showed stronger antioxidant ability than Pistafolin B with two galloyl moieties.

Pisum sativum L. Family: Fabaceae (Papilionaceae)

English Name: Pea.

The antioxidative activity of phenolic extracts from seeds and fruits was examined by autoxidation of methyl linoleate by Kahkonen *et al.* (1999). Both phenolics and antioxidant activity of the fruits and seeds were low. The phenolics in the seeds were 0.4 mg/g GAE while antioxidant activity was 8 per cent at 500 ppm level. In the fruits the phenolics were 1.6 mg/g GAE and antioxidant activity 24 per cent at 500 ppm level.

The contents of phenolic compounds in seed coat of pea and their antioxidative properties were examined by Troszynska *et al.* (2002). The pea seed coat was extracted with acetone-water (7:3 v/v) mixture and the extract was separated into five (iV) fractions using a Sephadex LH-20 column chromatography. Antioxidative activity of extract and fractions was measured by the oxidation of phosphatidylcholine to hydroxyperoxidephosphatidylcholine in liposome model and by scavenging effects of superoxide radical anion in xanthine-xanthine oxidase system. Phenolic compounds of extract and fractions were determined by spectrophotometric methods and characterized by HPLC analysis. Strong antioxidative properties were noted for extract and its five fractions measured by liposome method. The extract and fractions I, IV and V also showed scavenging effects of superoxide radical anion. A statistically significant correlation was found between the inhibition of PC oxidation in the system tested and contents of either total phenols or tannins. However no statistically significant correlation was found between $O^{\bullet-}_2$ scavenging effect and contents of either total phenols or tannins. The HPLC analysis of phenolic compounds of extract and active fractions showed the presence of some phenolic acids (benzoic and cinnamic acids, and cinnamic acid derivatives), flavone and flavonol glycoside.

Troszynska and Ciska (2002) compared the composition and contents of phenolic acids and condensed tannins in the seed coats of white and coloured varieties of pea and to examine the antioxidant properties of methanol and acetone extracts containing these phenolic compounds. The sum of free phenolic acids was higher for coloured seed coat (78.53 g per g dry dry matter). Protocatechuic acid, genistic and vanillic acids were found dominant in the coloured seed coat, while ferulic and coumaric acids in the white seed coat. The content of condensed tannins was 1560 mg of catechin equivalent/100g of coloured seed coat. No condensed tannins were detected in the white seed coat. Strong antioxidant properties were observed in a crude tannin extract from the coloured seed coat.

Thirty-five varieties of the green pea (*Pisum sativum* L.) were analysed by Nilsson *et al.* (2004) for their total antioxidant capacity (TAC). After blanching and freezing, the water-soluble fraction had, on average, a TAC of 0.61(0.22) µmol/g (mean(SD)) and the water-insoluble fraction a value of 0.23(0.08) µmol/g. There was a significant correlation between the TAC in the water-soluble and water-insoluble fractions. Regarding the antioxidant capacity in both the water-soluble and the water-insoluble extract, there was a significant difference between the varieties but not between the harvest periods. A significant correlation between TAC in the water-soluble fraction and ascorbic acid content was found. Ascorbic acid accounted for a large part of the water-soluble antioxidant capacity.

Sreeramulu *et al.* (2009) evaluated the antioxidant activity of peas (dry) by DPPH, scavenging assay, FRAP assay and reducing power. Phenolic content of the whole seed was 126.63 mg/100g. DPPH radical scavenging activity expressed as trolox equivalent, FRAP (µmol/g) and reducing power (mg/g) were 0.88, 97 and 1.54, respectively in peas.

The amino acid composition and antioxidant activities of peptide fractions obtained from HPLC separation of a pea protein hydrolysate (PPH) were studied by Pownall *et al.* (2010). Thermolysin hydrolysis of pea protein isolate and ultrafiltration (3 kDa molecular weight cutoff membrane) yielded a PPH that was separated into five fractions (F1"F5) on a C_{18} reverse phase HPLC column. The fractions that eluted later from the column (F3"F5) contained higher contents hydrophobic and aromatic amino acids when compared to fractions that eluted early or the original PPH.

Fractions F3"F5 also exhibited the strongest radical scavenging and metal chelating activities; however, hydrophobic character did not seem to contribute to reducing power of the peptides. In comparison to glutathione, the peptide fractions had significantly higher ability to inhibit linoleic acid oxidation and chelate metals. In contrast, glutathione had significantly higher free radical scavenging properties than the peptide fractions.

Pithecellobium dulce (Roxb.)Benth. Family: Mimosaceae

The free radical-scavenging property and the inhibitory action on H+, K+-ATPase activity of aqueous and hydroalcoholic fruit extracts of *Pithecellobium dulce* were screened by Megala and Geetha (2010) in various *in vitro* models. Hydroalcoholic extract showed the greatest free radical-scavenging activity in all the experimental models. Hydroalcoholic extract was found to possess a good antioxidant capacity when compared to aqueous extract.

Pittosporum mannii Hook.f. Family: Pittosporaceae

Antioxidant activity test using two different methods namely DPPH and 2,2'-azinobis(3-ethylbenzothialozinesulfonate) diammonium salt free radical scavenging test has been carried out on three Cameroonian plants. Results of this study in the 2,2-diphenyl-1-picrylhydrazyl scavenging test show that the ethyl acetate extract of *P. mannii* exhibit high free radical scavenging activities with IC_{50} values of 177.74 µg/ml as compared to Trolox (939.19 µg/ml) used as standard. In the same manner, 2,2'-azinobis(3-ethylbenzothialozinesulfonate) diammonium salt radical scavenging test of these extracts was in accordance of the result of 2,2-diphenyl-1-picrylhydrazyl test.

Pittosporum moluccanum Miq. Family: Pittosporaceae

Lee *et al.* (2006) investigated the antioxidant activity of ethanol extract of leaves using DPPH, hydroxyl and superoxide radical scavenging and reducing power activities as well as the induction of heme oxygenase-1 (HO-1). The IC_{50} values against DPPH radical were 14.9µg/ml, OH radical 1.24µg, superoxide radical 223.3µg/ml and total phenolics was 19.7 mg of GAE/g.

Plantago afra L. Family: Plantaginaceae

English Name: Plantain.

Antioxidant activity of methanol extract of *P. afra* was characterized by the DPPH scavenging test and the inhibition of Fe^{2+}/ascorbate-induced lipid peroxidation on bovine brain liposomes. Extract showed antioxidant activity in both methods. Total phenolic compound were estimated by Folin-ciocalteu assay, flavonoids by $AlCl_3$ reagent, phenypropanoid glycosides (PPGs) by Arnow reagent, and iridioids by Trim-Hill assay. A high correlation was found between the scavenging potency and the total phenolic and phenylpropanoid content of the extracts but not betweem the lipid peroxidation potency and the extract composition (Galvez *et al.*, 2005a).

Khalil *et al.* (2007) studied the growth, phenolic compounds and antioxidant activity of *Plantago afra* grown under organic farming condition. Antioxidant activity of ethanol extract (inhibition per cent) was 87.61 at 200µl extract.

Plantago argentea Chaix. Family: Plantaginaceae

Beara *et al.* (2009) evaluated the free radical scavenging ability of *Plantago argentea* and four other species of *Plantago* by various assays: DPPH, hydroxyl radical, superoxide anion, and nitric oxide scavenger capacity tests, reucing power (FRAP) assay, and Fe^{2+}/ascorbate induced lipid peroxidation. In all of the tests extracts showed a potent antioxidant effect compared with BHT. Besides, in examined extracts the total phenolic amount (ranging from 38.43 to 70.97 mg of GAE/g of dw) and the total flavonoid content (5.31-13.10 mg of QE/g of dw) were determined. Furthermore, the presence and content of selected flavonoids (luteolin-7-O-glucoside, apigenin-7-O-glucoside, luteolin, apigenin, rutin, and quercetin) were studied using LC-MS/MS technique.

Plantago asiatica L. Family: Plantaginaceae

A potent antioxidative compound was isolated by Toda *et al.* (1985a) from *Plantago asiatica*. Cai *et al.* (2004) evaluated the antioxidant activity and phenolic compounds in traditional Chinese medicinal plants associated with anticancer. The improved ABTS.+ method was used to systematically assess the total antioxidant capacity (Trolox equivalent antioxidant capacity, TEAC) of the medicinal extracts. The TEAC values and total phenolic content for methanolic extract of seed of *Plantago asiatica* were 47.7 µmol Trolox equivalent/100g dry weight (DW), and 0.30 g of gallic acid equivalent/100g DW respectively. A positive significant linear relationship between antioxidant activity and total phenolic content showed that phenolic compounds were the dominant antioxidant components in the tested herbs. Major types of phenolic compounds in *Plantago asiatica* include flavones (plantaginin).

Gan *et al.* (2010) evaluated the antioxidant activity of *Plantago asiatica* using the FRAP and TEAC assays, and its total phenolic content was measured by the Folin-Ciocalteu method. For FRAP assay the value was 88.06 µmol Fe(II)/g dry weight. For TEAC assay, the value was 39.94µmol Trolox/g dry weight. Total phenolic content was 3.34 mg gallic acid equivalent/g dry weight of plant material.

Plantago bellardii All. Family: Plantaginaceae

Galvez *et al.* (2005b) evaluated the *in vitro* antioxidant activity of the methanol extract of *Plantago bellardii* aerial parts. This was assessed by two different tests, scavenging of DPPH radical, and inhibition of lipid peroxidation on liposomes prepared from bovine brain extract. In both tests the extract showed a potent antioxidant effect. The characterization of the major compounds in the extract as rutin, geniposide and verbascoside. The compounds that contribute most to the antioxidant activity were shown to be verbascoside and rutin.

Plantago coronopus L. Family: Plantaginaceae

Antioxidant activity of methanol extract of *P. coronopus* was characterized by the DPPH scavenging test and the inhibition of Fe^{2+}/ascorbate-induced lipid peroxidation on bovine brain liposomes. Extract showed antioxidant activity in both methods. Total phenolic compounds were estimated by Folin-ciocalteu assay, flavonoids by $AlCl_3$ reagent, phenypropanoid glycosides (PPGs) by Arnow reagent, and iridioids by Trim-Hill assay. A high correlation was found between the scavenging potency and the total phenolic and phenylpropanoid content of the extracts but not betweem the lipid peroxidation potency and the extract composition(Galvez *et al.*, 2005a).

Plantago holosteum Scop. Family: Plantaginaceae

Beara *et al.* (2009) evaluated the free radical scavenging ability of *Plantago holosteum* and four other species of *Plantago* by various assays: DPPH, hydroxyl radical, superoxide anion, and nitric oxide scavenger capacity tests, reducing power (FRAP) assay, and Fe^{2+}/ascorbate induced lipid peroxidation. In all of the tests extracts showed a potent antioxidant effect compared with BHT. Besides, in examined extracts the total phenolic amount (ranging from 38.43 to 70.97 mg of GAE/g of dw) and the total flavonoid content (5.31-13.10 mg of QE/g of dw) were determined. Furthermore, the presence and content of selected flavonoids (luteolin-7-O-glucoside, apigenin-7-O-glucoside, luteolin, apigenin, rutin, and quercetin) were studied using LC-MS/MS technique.

Plantago lagopus L. Family: Plantaginaceae

Antioxidant activity of methanol extract of *P. lagopus* was characterized by the DPPH scavenging test and the inhibition of Fe^{2+}/ascorbate-induced lipid peroxidation on bovine brain liposomes. Extract showed antioxidant activity in both methods. Total phenolic compound were estimated by Folin-ciocalteu assay, flavonoids by $AlCl_3$ reagent, phenypropanoid glycosides (PPGs) by Arnow reagent, and iridioids by Trim-Hill assay. A high correlation was found between the scavenging potency and the total phenolic and phenylpropanoid content of the extracts but not betweem the lipid peroxidation potency and the extract composition (Galvez *et al.*, 2005a).

Plantago lanceolata L. Family: Plantaginaceae

Antioxidant activity of methanol extract of *P. lanceolata* was characterized by the DPPH scavenging test and the inhibition of Fe^{2+}/ascorbate-induced lipid peroxidation on bovine brain liposomes. Extract showed antioxidant activity in both methods. Total phenolic compound were estimated by Folin-ciocalteu assay, flavonoids by $AlCl_3$ reagent, phenypropanoid glycosides (PPGs) by Arnow reagent, and iridioids by Trim-Hill assay. A high correlation was found between the scavenging potency and the total phenolic and phenylpropanoid content of the extracts but not betweem the lipid peroxidation potency and the extract composition (Galvez *et al.*, 2005a).

Plantago lanceolata L. var. *libor* Family: Plantaginaceae

Kardosova and Machova (2006) investigated the ability of polysaccharides, isolated from leaves, to inhibit peroxidation of soyabean lecithin liposomes by OH radicals. All polysaccharides exhibited antioxidant activity. The highest inhibition was found with glucuronoxylans of *P.lanceolata* var. *libor*.

Plantago major L. Family: Plantaginaceae

English Name: Great plantain

Pourmorad *et al.* (2006) carried out a systematic record of the relative antioxidant activity in selected Iranian medicinal plant species. The total phenol content of *Plantago major*, measured by Folin Ciocalteu method was 31 mg/g GAE. Flavonoid contents in terms of quercetin equivalent was 25.15 mg/g. DPPH radical scavenging effect of the extracts was detemined spectrophotometrically. The radical scavenging activity of *Plantago major* at 0.8mg/ml concentration was 89.3 per cent.

Souri *et al*. (2008) evaluated the antioxidant activity of seed against linoleic acid peroxidation using 1,3-diethyl-2-thiobarbituric acid as reagent. Antioxidant activity (IC_{50}) against peroxidation of linoleic acid (2mg/ml) was 16.77 and phenolic content was 672.79 mg/100g dry weight. The results of this study showed that there was no significant correlation between antioxidant activity and phenolic content of the studied plant materials and phenolic content could not be a good indicator of antioxidant capacity.

Beara *et al*. (2009) evaluated the free radical scavenging ability of *Plantago major* and four other species of *Plantago* by various assays: DPPH, hydroxyl radical, superoxide anion, and nitric oxide scavenger capacity tests, reducing power (FRAP) assay, and Fe^{2+}/ascorbate induced lipid peroxidation. In all of the tests extracts showed a potent antioxidant effect compared with BHT. Besides, in examined extracts the total phenolic amount (ranging from 38.43 to 70.97 mg of GAE/g of dw) and the total flavonoid content (5.31-13.10 mg of QE/g of dw) were determined. Furthermore, the presence and content of selected flavonoids (luteolin-7-O-glucoside, apigenin-7-O-glucoside, luteolin, apigenin, rutin, and quercetin) were studied using LC-MS/MS technique.

Gan *et al*. (2010) evaluated the antioxidant activity of *Plantago major* using the FRAP and TEAC assays, and its total phenolic content was measured by the Folin-Ciocalteu method. For FRAP assay the value was 137.23 µmol Fe(II)/g dry weight. For TEAC assay, the value was 37.77µmol Trolox/g dry weight. Total phenolic content was 6.62 mg gallic acid equivalent/g dry weight of plant material.

Plantago maritima L. Family: Plantaginaceae

Beara *et al*. (2009) evaluated the free radical scavenging ability of *Plantago maritima* and four other species of *Plantago* by various assays: DPPH, hydroxyl radical, superoxide anion, and nitric oxide scavenger capacity tests, reducing power (FRAP) assay, and Fe^{2+}/ascorbate induced lipid peroxidation. In all of the tests extracts showed a potent antioxidant effect compared with BHT. Besides, in examined extracts the total phenolic amount (ranging from 38.43 to 70.97 mg of GAE/g of dw) and the total flavonoid content (5.31-13.10 mg of QE/g of dw) were determined. Furthermore, the presence and content of selected flavonoids (luteolin-7-O-glucoside, apigenin-7-O-glucoside, luteolin, apigenin, rutin, and quercetin) were studied using LC-MS/MS technique.

Plantago media L. Family: Plantaginaceae

Beara *et al*. (2009) evaluated the free radical scavenging ability of *Plantago media* and four other species of *Plantago* by various assays: DPPH, hydroxyl radical, superoxide anion, and nitric oxide scavenger capacity tests, reducing power (FRAP) assay, and Fe^{2+}/ascorbate induced lipid peroxidation. In all of the tests extracts showed a potent antioxidant effect compared with BHT. Besides, in examined extracts the total phenolic amount (ranging from 38.43 to 70.97 mg of GAE/g of dw) and the total flavonoid content (5.31-13.10 mg of QE/g of dw) were determined. Furthermore, the presence and content of selected flavonoids (luteolin-7-O-glucoside, apigenin-7-O-glucoside, luteolin, apigenin, rutin, and quercetin) were studied using LC-MS/MS technique.

Plantago ovata L. Family: Plantaginaceae

English Name: Blond plantain

Souri *et al.* (2008) evaluated the antioxidant activity of seed against linoleic acid peroxidation using 1,3-diethyl-2-thiobarbituric acid as reagent. Antioxidant activity (IC_{50}) against peroxidation of linoleic acid (2mg/ml) was 126.56 and phenolic content was 249.4 mg/100g dry weight. The results of this study showed that there is no significant correlation between antioxidant activity and phenolic content of the studied plant materials and phenolic content could not be a good indicator of antioxidant capacity.

Plantago serraria L. Family: Plantaginaceae

Antioxidant activity of methanol extract of *P. serraria* was characterized by the DPPH scavenging test and the inhibition of Fe^{2+}/ascorbate-induced lipid peroxidation on bovine brain liposomes. Extract showed antioxidant activity in both methods. Total phenolic compound were estimated by Folin-ciocalteu assay, flavonoids by $AlCl_3$ reagent, phenypropanoid glycosides (PPGs) by Arnow reagent, and iridioids by Trim-Hill assay. A high correlation was found between the scavenging potency and the total phenolic and phenylpropanoid content of the extracts but not between the lipid peroxidation potency and the extract composition (Galvez *et al.*, 2005a).

Platostoma africanum P. Beauv. Family: Lamiaceae (Labiatae)

Hexane and dichloromethane extracts of *Platostoma africanum* were evaluated by Aladedunye *et al.* (2008) for anti-inflammatory and antioxidant activities. The hexane and dichloromethane extracts showed significant antioxidant activity with the dichloromethane extract exhibiting an IC_{50} comparable to that of BHT, a synthetic antioxidant. Phytochemical investigation of the extracts afforded eight acidic pentacyclic triterpenes, namely, ursolic acid, oleanolic acid, epimaslinic acid, maslinic acid, corsolic acid, heptadienic, euscaphic acid and tomentic acid, and a mixture of β-sitosterol and stigmasterol.

Platycodon grandiflorum A.DC. Family: Campanulaceae

Lee *et al.* (2004) reported antioxidant activity of phenylpropanoid esters isolated and identified from *Platycodon grandiflorum* whole root. The antioxidant activities of these two compounds, which were evaluated by DPPH, superoxide and nitric oxide radical scavenging capacity, were found to be as high as those of BHT or BHA.

Jeong *et al.* (2010) examined the antioxidant activities of solvent fractions from the ethanol extract of *Platycodon grandiflorum*. The butanol fraction showed the most potent antioxidant activities in each assay, showing 91.31 per cent in the DPPH radical scavenging method, 99.62 per cent in the ABTS radical scavenging method, 7.84 per cent in the reducing power method, and 1.29 per cent in the FRAP method at a concentration of 10mg/ml. The DPPH, ABTS, reducing power, and FRAP assay indicated that the butanol fraction of aerial parts of *P.grandiflorum* was the most potent scavengers and reducing agents compared to the other two extracts. Butanol fraction had strong antioxidant activities which were correlated with its high level of phenolics, particularly luteolin-7-O-glucoside and apigenin-7-O-glucoside.

Antioxidant capacities of 56 selected Chinese medicinal plants, including *Platycodon grandiflorus* were evaluated by Song *et al.* (2010) using the TEAC and FRAP assays, and their total phenolic content was measured by the Folin-Ciocalteu method. *Platycodon grandiflorus* showed low antioxidant activity with TEAC value of 6.42μmol Trolox/g, FRAP value of 5.26μmol Fe^{2+}/g and phenolic content was 1.15mg GAE/g.

Plazia daphnoides Wedd. Family: Asteraceae (Compositae)

The aerial parts of 17 Bolivian plants including *Plazia daphnoides* were screened by Rosas-Romero and Saavedra (2005) to determine antioxidant activity. A methanol extract of each plant was prepared and portioned sequentially with hexane, chloroform, and ethyl acetate, leaving an aqueous solution. All extracts and their 5 fractions for a total of 102 samples, were evaluated using two techniques: an adaptation of the β-carotene bleaching technique using an emulsion of linoleic acid in water as the oxidizable substrate, and the DPPH free radical trapping technique. The results of the β-carotene bleaching technique were more discriminating and better related to the rancidity process under normal conditions; with this assay, 11 species provided at least one fraction with highly promising antioxidant activity. All species gave good results under the DPPH technique, and in most cases they performed better than BHA, which was used as a reference antioxidant (Rosas-Romero and Saavedra, 2005)

Plectranthus amboinicus (Lour.) Spreng. (Syn.: *Coleus aromaticus* Benth.) Family: Lamiaceae (Labiatae)

English Name: Country borage.

Kumaran and Karunakaran (2005) reported that aqueous extract of *Coleus aromaticus* showed significant *in vitro* DPPH radical scavenging activity.

Palani *et al.* (2010) evaluated the nephroprotective, diuretic and antioxidant activities of *Plectranthus amboinicus* on acetaminophen-induced nephrotoxic rats. The data suggest that the ethanol extract possess nephroprotective antioxidant effects against acetaminophen-induced nephrotoxicity.

Aerial parts of *Plectranthus amboinicus* (Syn: *Coleus aromaticus*) extracted with ethanol and hot water, was analyzed by Anasuya and Gomathi (2010) for antioxidant activity.The ethanol extract was found to possess high antioxidant capacity as measured by reducing power ability and ability to scavenge DPPH stable free radicals, ABTS radical cations, hydroxyl, nitric and superoxide free radicals in the *in vitro* systems. However, the water extract chelated metal ions more effectively than the ethanol extract. Further, the extracts registered a moderate peroxidation inhibition in the β-carotene/linoleic acid emulsion system. Both the plant extracts contained considerable levels of non-enzymatic antioxidants such as vitamin C and E, total phenolics, tannins and flavonoids, estimated through standard spectrophotometric methods.

Plectranthus barbatus Andr. (Syn.: *Coleus forskohli* Briq.) Family: Lamiaceae (Labiatae)

Roots of the plant are used for fat lowering and prevent synthess of high cholesterol in body. Forskohlin is the main component in the tea that has antioxidative property.

Maioli *et al.* (2010) assessed the antioxidant activity of the aqueous extract of *Plectranthus barbatus* leaves on Fe^{2+}-citrate mediated membrane lipid peroxidation in isolated rat liver mitochondria as well as in non-mitochondrial systems: DPPH reduction, OH scavenging activity, and iron chelation by prevention of formation of Fe^{2+} bathophenanthroline disulphonic acid (BPS) complex. With all the tested concentrations (15-75µg/ml), *P. barbatus* extract presented significant free radical-scavenging activity (IC_{50} = 35.8µg/ml in the DPPH assay), and IC_{50} = 69.1µg/ml in the OH assay), and chelated iron (IC_{50} = 30.4µg/ml). Over the same concentration range, the plant extract protected mitochondria against Fe^{2+}/citrate-mediated swelling and malondialdehyde production, a property that persisted even after simulation of its passage through the digestive tract. These effects could be attributed to the phenoolic compounds, nepetoidin – caffeic acid esters, present in the extract.

Plectranthus hadiensis Forssk. Family: Lamiaceaae (Labiatae)

Methanolic extracts of *Plectranthus hadiensis* showed high free radical scavenging activity (Mothana *et al.*, 2010).

Pluchea arabica (Boiss.) Qaiser et Lack Family: Asteraceae (Compositae)

Marwah *et al.* (2007) investigated the antioxidant activity of wound-healing herbs using *in vitro* DPPH and phosphomolybdenum assay methods. Out of the 19 plants screened, the aqueous ethanol extract of *Pluchea arabica* showed good DPPH scavenging activities with per cent IP value 77 per cent, after 15 min of incubation at an effective test concentration of 50µg/ml. The Total Antioxidant Capacity as gallic acid equivalent of 343mg/g of ethanol extract was obtained in the phosphomolybdenum assay. Total phenolic content was 76.9 mg GAE/g of ethanol extract.

Pluchea carolinensis (Jacq.) G. Don Family: Asteraceae (Compositae)

Fernandez and Torres (2006) evaluated the antioxidant activity of extracts of *Pluchea carolinensis*. The crude phenol and 50 per cent ethanol extract obtained from leaves of *P. carolinensis* inhibited the L-epinephrine oxidation by hydroxyl radical generated in Fenton reaction in a dose-dependent manner.

Pluchea indica Less. Family: Asteraceae (Compositae)

Patnibul *et al.* (2008) screened 18 vegetables for antioxidant activity using silica gel Thin Layer chromatography followed by spraying with DPPH. *Pluchea indica* exhibited strong antioxidant activity.

Extracts from 11 vegetables of Indonesian origin were screened for flavonoid content, total phenolics, and antioxidant activity by Andarwulan *et al.* (2010). Flavonoid content in *Pluchea indica* was 6.39mg/100g fresh weight. The flavonoid content of the vegetables were mainly quercetin and kaempferol. *P. indica* had the highest antioxidant activity as measured by ferric cyanide reducing power, DPPH and ABTS scavenging and inhibition of linoleic acid oxidation.

Plumbago zeylanica L. Family: Plumbaginaceae

The herb is known for inhibitory effects against superoxide and nitric oxide production (Natarajan *et al.*, 2006; Seo *et al.*, 2001). Natarajan *et al.* (2006) found antioxidant property of the plant using ABTS assay. It is used as chemopreventive agent against intestinal neoplasia (Natarajan *et al.*, 2006; Sugie *et al.*, 1998). Rajarajan *et al.* (2010) also reported free radical scavenging activity of *Plumbago zeylanica* root extract on acetaminophen induced hepatic damage in rats.

Tilak *et al.* (2004) evaluated the antioxidant properties of *Plumbago zeylanica* and its active ingredient plumbagin. In FRAP/DPPH assays boiled ethanolic extracts were the most effective, while in the ABTS assay boiled aqueous extracts were the most efficient. These extracts also significantly inhibited lipid peroxidation induced by cumene hydroperoxide, ascorbatee-Fe(2+) and peroxynitrite and contained high amounts of polyphenols and flavonoids.

Methanolic extracts of *Plumbago zeylanica* root was evaluated for their antioxidant activity by ferric thiocyanate assay and campared with TBA method. The extract showed good antioxidant activity. 73.41 per cent decrease of DPPH standard solution was recorded for the extract at a concentration of 100µg/ml (Zahin *et al.*, 2009).

Plumeria alba L. Family: Apocynaceeae

Leelapornpisid *et al.* (2007) reported antioxidant activity of the absolutes (obtained by solvent extraction) of *Plumeria alba* (IC_{50} = 1.0766 mg/ml).

Plumeria rubra L. Family: Apocynaceae

Ruiz-Terán *et al.* (2008) evaluated the antioxidant activity of aerial parts of *Plumeria rubra* using DPPH and β-carotene-linoleic acid bleaching assay. The extract displayed antioxidant activity comparable to the commercial antioxidant BHA. For the methanolic extract analysed, a clear relation between the total phenolic content of the extracts and their antioxidant activity was found.

Podophyllum hexandrum Royle Family: Berberidaceae

Antioxidant activities of fractionated extract of *Podophyllum hexandrum* was studied by Arora *et al.* (2004, 2005). The fraction extract reduced both iron ascorbate and radiation induced lipid peroxidation and exhibited metal chelation activities, thereby rendering radiation protection.

Chawla *et al.* (2005) evaluated the antioxidant activity of fractionated extracts of rhizomes of high altitude *Podophyllum hexandrum* and discussed their role in radiation protection. Chloroform extract showed maximal metal chelation activity (41.59 per cent), evaluated using 2,2'-bipyridyl assay, followed by alcohol extract (31.25 per cent), which exhibited maximum antioxidant potentialin the reducing power assay. Alcohol extract also maximally inhibited iron/ascorbate-induced and radiation-induced LPO (99.76 and 92.249 per cent, respectively, at 2000µg/ml) in mouse liver homogenate. Under conditions of combined stress (radiation 250Gy + iron/ascorbate) at a concentration of 2000µg/ml, hydroalcohol extract exhibited higher percentage of inhibition (93.05 per cent) of LPO activity. Hydroalcoholic extract was found to be effective in significantly lowering LPO activity over a wide range of concentrations as compared to alcohol extract.

A semi-purified extract of low-altitude *Podophyllum hexandrum* containing a relatively low content of podophyllotoxin (3,25 per cent) exhibited potent antioxidant activity in lipid media (at 1000µg/ml against 0.25kGy) and significant hydroxyl ion scavenging potential (78.83 per cent at 500µg/ml) (Sagar *et al.*, 2006).

Poincianella pyramidalis (Tul.) L. Family: Fabaceae (Leguminosae)

De Melo *et al.* (2010) evaluated the antioxidant capacity and tannin content in *Poincianella pyramidalis*. IC_{50} value of DPPH radical scavenging was 42.95µg/ml while tannin content was 2.35 mg/g.

Polycarpaea corymbosa Lam. Family: Caryophyllaceae

Singh *et al.* (2009) evaluated the antioxidant activity of dichloromethane and methanolic extracts of 12 arid zone medicinal plants. *Polycarpaea corymbosa* showed appreciable antioxidant activity and per cent of inhibition of DPPH where RC_{50} (µg/ml) was 7.5 (Dichloromethane extract) and 6.5 (MeOH extract). per cent of inhibition of DPPH was 94.4 per cent in methanol extract of *Polycarpaea corymbosa*. Key phytochemicals include α–1-Barringenol, stigmasterol and camelligenin.

Polygala alpestris (L.) Reichenb. Family: Polygalaceae

Members of Polygalaceae are known to contain a variety of different polyphenolic compounds such as xanthones, flavonoids, and biphenyl derivatives. Cervellati *et al.* (2004) reported the isolation and structural characterization of two new phenol derivatives, named alpestrin (= 3,3',5'-trimethoxy[1,1'-biphenyl]-4-ol; 10) and alpestriose A (= 6-O-benzoyl-1-O-{6-O-acetyl-3-O-[(4-hydroxy-3,5-dimethoxy-phenyl)prop-2-enoyl]-beta-D-fructofuranosyl}-alpha-D-glucopyranoside; 11), and of four known compounds (12-15) from the MeOH extract of *Polygala alpestris*. The relative *in vitro* antioxidant activities of these compounds, in comparison with other phenolic substances from *Polygala vulgaris*, were evaluated by means of the Briggs-Rauscher (BR) oscillating reaction, a method based on the inhibitory effects of antioxidant free-radical scavengers. The experimental antioxidant-activity values (relative to resorcinol as a standard) were compared with those calculated on the basis of the bond-dissociation enthalpies. The structure/activity relationships for the compounds examined were also discussed.

Polygala chinensis L. Family: Polygalaceae

Gurav *et al.* (2000) assessed the free radical scavenging activity of alcoholic extract of *Polygala chinensis* by different analytical methods like DPPH free radical scavenging assay, nitric oxide scavenging assay and hydrogen peroxide scavenging assay. The alcoholic extract of *Polygala chinensis* showed concentration dependent free radical scavenging activity in all models. IC_{50} was found to be 56.09 µg/ml, 47.63 µg/ml and 41.003 µg/ml in DPPH assay, NO radical scavenging assay and H_2O_2 scavenging assay respectively. Total phenolic content and total flavonoids were evaluated according to Folin–Ciocalteu reagent and colorimetric method respectively which revealed a high amount of polyphenols (88.2±9.3 mg/100gm) as well as flavonoids 36.7±3.9 mg/100gm, suggesting a possible role of these compounds in the antioxidant properties.

Polygala hongkongensis Hemsl. Family: Polygalaceae

Two new xanthone *O*-glycosides, polyhongkongenosides A (*1*) and B (*2*), together with four known xanthone glycosides, were isolated by Wu *et al.* (2008) from the herbs of *Polygala hongkongensis*. The antioxidant *in vitro* activities of *1–6* were determined by the scavenging activities against DPPH radicals and hydroxyl radicals and their reductive activities to Fe^{3+}. Mangiferin, one of the four known xanthone glycosides, showed potential scavenging effect on DPPH and hydroxy radicals and reductive activity to Fe^{3+} with IC_{50} values of 4.7, 13.9, 23.7 µM, respectively.

Polygala senega L. Family: Polygalaceae

English Name: Senega.

Ethanolic extracts of roots of senega were prepared and evaluated by Amarowicz *et al.* (2004) for their free-radical scavenging capacity and their antioxidant activity, by a number of chemical assays. Assays employed included β-carotene linoleic acid (lineolate) model system, reducing power, scavenging effect on the DPPH free radical and capacity to scavenge hydroxyl free radical (HO), by use of electron paramagnetic resonance (EPR) spectroscopy.

Polygala tenuifolia Willd. Family: Polygalaceae

Cai *et al.* (2004) evaluated the antioxidant activity and phenolic compounds in traditional Chinese medicinal plants associated with anticancer. The improved ABTS.+ method was used to systematically assess the total antioxidant capacity (Trolox equivalent antioxidant capacity, TEAC) of the medicinal extracts. The TEAC values and total phenolic content for methanolic extract of root of *Polygala tenuifolia* were 133.8 µmol Trolox equivalent/100g dry weight (DW), and 0.82 g of gallic acid equivalent/100g DW respectively. A positive significant linear relationship between antioxidant activity and total phenolic content showed that phenolic compounds were the dominant antioxidant components in the tested herbs. Major types of phenolic compounds in *Polygala tenuifolia* include lignans (chinensin), phenolic acids (cinnamic acid).

The constituents of the ethanol extract from the root of *Polygala tenuifolia* were investigated by Liu *et al.* (2010) for antioxidant activity in senescence-accelerated mice. Consequently, two relevant samples were obtained, a fraction separated by macroporous resin (YZ-OE), and a major pure crystal of 3,6'-disinapoyl sucrose (DISS). Based on HPLC-ESI-MS analysis, the most constituents in the YZ-OE fraction from the extract of *P. tenuifolia* were oligosaccharide esters. The antioxidant activities of these two samples were evaluated using the accelerated senescence-prone, short-lived mice (SAMP) *in vivo*. The activities of superoxide dismutase (SOD) and glutathione peroxidase (GSH-PX) were increased significantly in SAMP mice fed oligosaccharide esters (YZ-OE 50 mg/kg) and its constituents (DISS 50 mg/kg). However, the content of malondialdehyde (MDA) was increased in the blood and liver of SAMP mice. But when given YZ-OE, it could be decreased, by 44.3 per cent and 47.5 per cent, respectively, compared with the SAMP model. Results from the analyses indicated that the oligosaccharide esters (YZ-OE) from roots of *P. tenuifolia* had a high *in vivo* antioxidant activity.

Polygonatum multiflorum All. Family: Polygonaceae

English Name: Solomorib seal, Lady's seals, Drop berry.

Chen *et al.* (1999) identified DPPH radical scavenging active compounds from *Polygonum multiflorum*. Dried root was extracted with 95 per cent ethanol and then separated into water, ethyl acetate and hexane fractions. Among these only the ethyl acetate phase showed strong antioxidant activity by the DPPH test when compared with water and hexane phases. Three compounds showing strong antioxidant activity were identified.

Crude extract from *Polygonatum multiflorum* was screened by Sengul *et al.* (2009) for its *in vitro* antioxidant and antimicrobial properties. Total phenolic content of extract of this plant was also determined. β-carotene bleaching assay and Folin-Ciocalteu reagent were used to determine total antioxidant activity and total phenols of plant extract. Antioxidant activity was 75.85 per cent and the total phenolic content was 4.07 mgGAE/g DW).

Polygonum amphibium L. Family: Polygonaceae

The antioxidative activity of the 80 per cent ethanol extract obtained from eleven commonly consumed wild edible plants was determined according to the phosphomolybdenum method, reducing power, metal chelating, superoxide anion and free radical scavenging activity (Özen, 2010). *Polygonum amphibium* had the highest antioxidant capacities. The results indicated that the antioxidant activity of vegetables seems to be due to the presence of polyphenols, flavonoid and anthocyanoside. Total phenolics in *Polygonum amphibium* were 57.7mg pyrocatechol equivalent/g dw, Flavonoids were 14.7 mg quercetin equivalent/g dw and anthocyanins 6.4 mg cyanidin glycoside equivalent/g dw.

Polygonum aviculare L. Family: Polygonaceae

Cai *et al.* (2004) evaluated the antioxidant activity and phenolic compounds in traditional Chinese medicinal plants associated with anticancer. The improved ABTS.+ method was used to systematically assess the total antioxidant capacity (Trolox equivalent antioxidant capacity, TEAC) of the medicinal extracts. The TEAC values and total phenolic content for methanolic extract of whole plant of *Polygonum aviculare* were 856.8 μmol Trolox equivalent/100g dry weight (DW), and 4.13 g of gallic acid equivalent/100g DW respectively. A positive significant linear relationship between antioxidant activity and total phenolic content showed that phenolic compounds were the dominant antioxidant components in the tested herbs. Major types of phenolic compounds in *Polygonum aviculare* include flavonols (avocilarin, quercetin), phenolic acids (gallic acid, caffeic acid), anthraquinones (emodin), tannins.

Ethanol extract of *Polygonum aviculare* exhibited antioxidant activity (Hsu, 2006). The antioxidant activities of extract powder were examined by free radical scavenging assays, superoxide radical scavenging assays, lipid peroxidation assays and hydroxyl radical-induced DNA strand scission assays. The results showed that IC_{50} value of the extract is 50µg/ml in free radical scavenging assays, 0.8µg/ml in superoxide radical scavenging assays, and 15µg/ml in lipid peroxidation assays, respectively. The total phenolics and flavonoid content of the extract was 677.4 µg/g.

Corush *et al.* (2007) reported antioxidant activity in *Polygonum aviculare*. There was a low correlation between total phenolic content and antioxidant activity in the plant samples.

Gan *et al.* (2010) evaluated the antioxidant activity of *Polygonum aviculare* using the FRAP and TEAC assays, and its total phenolic content was measured by the Folin-Ciocalteu method. For FRAP assay the value was 263.19 µmol Fe(II)/g dry weight. For TEAC assay, the value was 171.65µmol Trolox/g dry weight. Total phenolic content was 18.00 mg gallic acid equivalent/g dry weight of plant material.

Polygonum bistorta L. Family: Polygonaceae

English Name: Bistort, Snake root.

Demiray *et al.* (2009) evaluated the phenolic profiles and antioxidant activities of Turkish medicinal plant *Polygonum bistorta* roots. The amount of total phenolics in *P. bistorta* root ranged between 2336 to 3887 µg/g dw. Protocatechuic acid, ferulic, coumaric and gallic acids were major phenolic compounds. Catechin was not found in the extracts. 70 per cent acetone extract had the highest phenolic content (4.19ng GAE/g dw). The antioxidant activity was very low (Demiray *et al.*, 2009).

Polygonum cuspidatum Sieb. et Zucc. Family: Polygonaceae

English Name: Japanese knot weed, Mexican bamboo.

Cai *et al.* (2004, 2006) evaluated the antioxidant activity and phenolic compounds in traditional Chinese medicinal plants associated with anticancer. The improved ABTS.+ method was used to systematically assess the total antioxidant capacity (Trolox equivalent antioxidant capacity, TEAC) of the medicinal extracts. The TEAC values and total phenolic content for methanolic extract of rhizome of *Polygonum cuspidatum* were 1849 µmol Trolox equivalent/100g dry weight (DW), and 8.08 g of gallic acid equivalent/100g DW respectively. A positive significant linear relationship between antioxidant activity and total phenolic content showed that phenolic compounds were the dominant antioxidant components in the tested herbs. Major types of phenolic compounds in *Polygonum cuspidatum* include anthraquinones (emodin, rhein, physcion, anthraglycosides), stilbenes (resveratrol), phenolic acids (coumaric acid), tannins.

Polygonum cuspidatum was extracted with 95 per cent ethanol by Pan *et al.* (2007) to obtain a crude antioxidant extract. *P. cuspidatum* extract had a very high content of total phenol, which was 104.83 mg/g dry weight, expressed as pyrocatechol equivalent. The extract exhibited excellent antioxidant activity, as measured using DPPH and total antioxidant assays. It also showed a high lipid antioxidant activity and hydroxyl radicals scavenging activity. A positive correlation was found between the reducing power and the antioxidant activity of the extract.

Gan *et al.* (2010) evaluated the antioxidant activity of 40 medicinal plants associated with prevention and treatment of cardiovascular and cerebrovascular diseases using FRAP, TEAC assays. A high correlation between antoxidant capacity and total phenolic content indicated that phenolic compounds could be the main contributor of the antioxidant activity of these plants. *Polygonum cuspidatum* showed the highest antioxidant activity of 520.78 µmol Fe(II)/g in FRAP assay and 590.51µmol in TEAC assay. Total phenolic contet was also very high, 34.91 mg GAE/g.

Polygonum hydropiper L. Family: Polygonaceae

English Name: Pepper wort, Water pepper.

Wong *et al.* (2006) investigated the antioxidant properties of 25 edible tropical plants including *Polygonum hydropiper* using DPPH and FRAP assays. Total phenolic content was also estimated. *Polygonum hydropiper* showed very high antioxidant activity.

Polygonum hyrcanicum Rech. f. Family: Polygonaceae

Extracts of *Polygonum hyrcanicum* were investigated by Hosseinimehr *et al.* (2007) for their total flavonoids, phenol contents and their radical scavenging activity using DPPH assays. Quercetin and butylated hydroxytoluene were used as standard reference with well documented antioxidant activity. Methanolic extract sample showed free radical scavenging activity.The highest antioxidant activity was found in *P. hyrcanicum* with an IC_{50} equal to 0.036 mg/ml that is higher than BHT (IC_{50} = 0.054). A correlation between radical scavenging capacity of the extract with total phenolic compounds content was observed.

Polygonum minus Huds. Family: Polygonaceae

Huda-Faujan *et al.* (2007, 2009) reported antioxidative activities of aqueous and methanolic extracts of *Polygonum minus* and four other plants. The analysis carried out was total phenolic content, ferric reducing power, ferric thiocyanate (FTC) and TBA tests. The phenolic content was 44.35mg (aqueous extract) 16.73 mg (methanolic extract) TAE/100 g fresh weight and antioxidant activity of 63.2 (aqueous extract) 63.66 per cent (methanolic extract) using FTC method. Increasing the concentration of the extracts resulted in increased ferric reducing antioxidant power. Total phenolic contet had positive correlation with antioxidant capacity.

Polygonum multiflorum Thunb. Family: Polygonaceae

Ip *et al.* (1997) have demonstrated the presence of both *in vivo* and *in vitro* antioxidant activities in the crude water-extract of the root of *Polygonum multiflorum*, as indicated by its ability to protect against CCl_4-induced hepatotoxicity in rats and to reduce ferr-heme oxidants generated in an *in vitro* system. Activity-directed fractionation of the extract indicated that the antioxidant activities were contained in the ethylacetate fraction (Ip *et al.*, 1997).

An activity-directed fractionation and purification process was used by Chen *et al.* (1999) to identify the antioxidative components of *Polygonum multiflorum* (PM). Dried root of PM was extracted with 95 per cent ethanol and then separated into water, ethyl acetate and hexane fractions. Among these only the ethyl acetate phase showed strong antioxidant activity by the DPPH test when compared with water and hexane phases. Three compounds showing strong antioxidant activity were identified. These were gallic acid, catechin and 2,3,5,4'-tetrahydroxystilbene 2-O-beta-D-glucopyranoside.

The extract of the root of *Polygonum multiflorum* exhibited a significant antioxidant activity assessed by the DPPH radical scavenging activity *in vitro* (Ryu *et al.*, 2002). The bioassay-guided fractionation of the extract yielded a stilbene glucoside, (E)-2,3,5,4'-tetrahydroxystilbene-2-O-beta-D-glucopyranoside as an active constituent responsible for the antioxidant property. This compound demonstrated a moderate DPPH radical scavenging activity (IC_{50}, 40 microM), while the corresponding deglucosylated stilbene 2 exhibited a much higher activity (IC_{50}, 0.38 microM).

Cai *et al.* (2004, 2006) evaluated the antioxidant activity and phenolic compounds in traditional Chinese medicinal plants associated with anticancer. The improved ABTS.+ method was used to systematically assess the total antioxidant capacity (Trolox equivalent antioxidant capacity, TEAC) of the medicinal extracts. The TEAC values and total phenolic content for methanolic extract of root of *Polygonum multiflorum* were 1811.9 µmol Trolox equivalent/100g dry weight (DW), and 5.67 g of gallic acid equivalent/100g DW respectively. A positive significant linear relationship between antioxidant activity and total phenolic content showed that phenolic compounds were the dominant antioxidant components in the tested herbs. Major types of phenolic compounds in *Polygonum multiflorum* include anthraquinones (chrysophanol, emodin, rhein, physcion), tannins, stilbenes (resveratrol).

The antioxidant activity and total phenolic contents of stem and root of *Polygonum multiflorum* was evaluated by Wong *et al.* (2006) using the ferric reducing antioxidant power assay and the Folin-Ciocalteu method, respectively. The plant was extracted by the traditional method, boiling in water and also in 80 per cent methanol. A significant correlation coefficient between the antioxidant activity and the total phenolic content was found in both aqueous and methanol extracts. Phenolic compounds were thus a major contributor of antioxidant activity. Comparing the extraction efficiency of the two methods, the boiling water method extracted phenolic compounds more efficiently, and the antioxidant activity of the extracts was higher.

The crude polysaccharide (PMTP) was isolated from *Polygonum multiflorum* through hot water extraction followed by ethanol precipitation. Through antioxidant assay *in vitro*, the results exhibited PMTP has powerful scavenging abilities especially on ABTS, DPPH and Hydroxyl radicals (Luo *et al.*, 2011).

Polygonum orientale L. Family: Polygonaceae

Cai *et al.* (2004) evaluated the antioxidant activity and phenolic compounds in traditional Chinese medicinal plants associated with anticancer. The improved ABTS.+ method was used to systematically assess the total antioxidant capacity (Trolox equivalent antioxidant capacity, TEAC) of the medicinal extracts. The TEAC values and total phenolic content for methanolic extract of fruit of *Polygonum orientale* were 255.3 µmol Trolox equivalent/100g dry weight (DW), and 1.23 g of gallic acid equivalent/100g DW respectively. A positive significant linear relationship between antioxidant activity and total phenolic content showed that phenolic compounds were the dominant antioxidant components in the tested herbs. Major types of phenolic compounds in *Polygonum orientale* include flavonols (quercetin), flavonones (taxifolin).

Gan *et al.* (2010) evaluated the antioxidant activity of 40 medicinal plants. *Polygonum orientale* showed very low antioxidant activity of 2.34 µmol Fe(II)/g in FRAP assay and 9.43µmol in TEAC assay. Total phenolic content was also very low, 0.65 mg GAE/g.

Polygonum paleaceum H.Gross. Family: Polygonaceae

The crude aqueous acetone extract exhibited high antioxidant activity (IC_{50} 16.72 µg/ml) in DPPH radical scavenging assay. With bioactivity-guided fractionation, 14 antioxidant phenolic compounds were obtained from the air-dried rhizomes of this plant. Of them, determined to be 3,5-dihydroxyl hexanoic acid 1,5-lactone 3-(6'-O-galloyl)-O-β-D-glucopyranoside by detailed spectroscopic analysis (Wang *et al.*, 2005).

Polyscias fruticosa (L.) Harms. Family: Araliaceae

Chanwitheesuk *et al.* (2005) investigated the antioxidant activity of *Polyscias fruticosa* using a β-carotene bleaching method. The contents of plant chemicals such as vitamin C, Vitamin E, carotenoids, tannin and total phenolics, were also determined. Methanolic extract showed antioxidant activity. *Polyscias fruticosa* has antioxidant index of 2.92, Vitamin C 9.11mg per cent, Vitamin E 0.0201mg per cent, Total carotenes 2.52 mg per cent, Total xanthophylls 2.13mg per cent, Tannins 24.3mg per cent and Total phenolics of 46.3 mg per cent.

Polyscias pinnata J.R.Forster et G.Forster Family: Araliaceae

Extracts from 11 vegetables of Indonesian origin were screened for flavonoid content, total phenolics, and antioxidant activity by Andarwulan *et al.* (2010). Flavonoid content in *Polyscias pinnata* was 52.19mg/100g fresh weight.

Pongamia pinnata L. (Syn.: *P. glabra*) Family: Fabaceae (Papilionaceae)

Antioxidant activity of ethanol and water extract of *Pongamia pinnata* leaves was studied by Tenpe *et al.* (2008) in three *in vitro/ex vivo* models, *viz.*, radical scavenging activity by DPPH reduction, nitric oxide radical scavenging activity in sodium nitroprusside/Greiss reagent system and reducing power assay. Ethanolic extract have good antioxidant activity in all the three models. Scavenging DPPH radical extracts activity was IC_{50} = 22.72 µg/ml while in scavenging nitric oxide radical, the activity was IC_{50} = 13781 µg/ml. The reducing ability of the extract was found to increase with increasing concentration where as maximum absorbance of 0.613 was obtained at a concentration of 1000 µg/ml.

The root bark of *Pongamia pinnata* afforded a new chalcone (karanjapin) and two known flavonoids, a pyranoflavonoid (karanjachromene) and a furanoflavonoid (karanjin) The structure of karanjapin has been established from extensive 2D NMR spectral studies as β,2'-dihydroxy-a,4'-dimethoxy-3,4-methylenedioxychalcone. Karanjapin and karanjachromene were found to possess significant antioxidant activity (Ghosh *et al.*, 2009).

Porwal (2010) evaluated the antioxidant properties of *Pongamia pinnata*. The antioxidant activity of isolated compound and various extracts of *Pongamia pinnata* was evaluated *in vitro* by DPPH free radical scavenging activity. Their results showed that *Pongamia pinnata* displayed potent antioxidant properties.

Antioxidant activity of the methanol extract of *Pongamia pinnata* was determined by Gupta and Sharma (2011) by DPPH free radical, Nitric oxide scavenging assays, Superoxide ion scavenging assays, ABTS and iron chelating methods. Preliminary phytochemical screening revealed that the extract of *Pongamia pinnata* leaves possesses phenols, flavonoids, steroids, glycosides, saponins, tannins, and triterpenoids. The extract showed significant activities in all antioxidant assays compared to the standard antioxidant (ascorbic acid) in a dose dependent manner and remarkable activities to scavenge reactive oxygen species (ROS) may be attributed by the presence of the above active compounds in the leaves.

Populus davidiana Dode Family: Salicaceae

Phytochemical study by Zhang *et al.* (2006) of *Populus davidiana* resulted in the isolation of 10 phenolic glycosides. These compounds were tested for their radical scavenging activity against an azoradical ABTS+. Of these, populosides A-C (1-3), populoside (4), grandidenttatin (8), salireposide (9) and coumaryl-β-D-glucosde (10) exhibited antioxidant activity in their assay.

Populus nigra L. Family: Salicaceae

English anme: Poplar

Mainar *et al.* (2008) evaluated the antioxidant activity of supercritical extracts of *Populus nigra* buds by DPPH assay. It was concluded that buds of *Populus nigra* were a better source of natural antioxidants than those of MC.

Dudonne *et al.* (2011) evaluated the antioxidant activity of poplar buds (*Populus nigra*) extract. It presented a high total phenolic content, and moderate antioxidant properties as determined by ORAC assay. The main phenolic compounds identified were phenolic acids and flavonoids aglycones. Caffeic and *p*-coumaric acids were identified as the major antioxidant compounds. Representing only 3.5 per cent of its dry weight, these compounds represented together about 50 per cent of the total antioxidant activity of the extract.

Populus tremula L. Family: Salicaceae

English Name: Aspen.

The antioxidative activity of phenolic extracts from leaf and bark of aspen was examined by autoxidation of methyl linoleate by Kahkonen *et al.* (1999). Both phenolics and antioxidant activity of the flowers were low. Remarkable high antioxidant activity and high total phenolic content were found in the leaf and bark extracts of aspen.

Populus tremuloides Michx Family: Salicaceae

The antioxidant properties of the bark of *Populus tremuloides* hot water extracts and its fractions were determined by three *in vitro* experiments. DPPH assay, phosphomolybdenum assay and Canola oil thermoxidation assay by DSC analysis.

Most of the results of the reported tests showed that the crude hot water extract and its fractions exhibited a strong antioxidant activity, higher than the synthetic antioxidant BHT. Among the fractions from the aqueous extracts of bark, n-butanol, tert-butyl-methyl ether, and ethyl acetate soluble fractions exhibited the best antioxidant efficiency, in phosphomolybednum assay, DPPH assay and Canola oil thermoxidation, respectively. Total phenol, flavonoids and flavonol contents were also evaluated and the results confirmed that polyphenols contained in the hot water extract of this bark are mainly composed of non-flavonoid compounds (Diouf *et al.*, 2009).

Populus ussuriensis Kom. Family: Salicaceae

The methanol (MeOH) extract of *Populus ussuriensis* bark was analyzed by Si *et al.* (2011) for antioxidant assessing by DPPH radical scavenging potential. Among fractions using several solvents, the ethyl acetate (EtOAc) soluble fraction, which showed strong antioxidant activity (IC_{50} 2.02 µg/ml), was further purified by Thin layer chromatography (TLC) guided Sephadex LH-20 column chromatography. Three known phenolic glucosides, picein (*I*), salicortin (*II*), grandidentatin (*III*), and that of a new, 2-hydroxycyclohexyl-42 -*O*-*p*-coumaroyl-β-D-glucopyranoside (isograndidentatin A) (*IV*), were isolated and their structures were elucidated on the basis of physiochemical and spectroscopic methods. Phenolic glucosides *III* and *IV* exhibited strong antioxidant activities, with IC_{50} values 6.73 and 6.69 µM, respectively, comparable to the control (α-tocopherol, IC_{50} 6.80 µM).

Portulaca grandiflora Hook. Family: Portulacaceae

Singh *et al.* (2009) evaluated the antioxidant activity of dichloromethane and methanolic extracts of 12 arid zone medicinal plants. *Portulaca grandiflora* showed appreciable antioxidant activity and per cent of inhibition of DPPH where RC_{50} (µg/ml) was 10 (Dichloromethane extract) and 8 (MeOH extract). Per cent of inhibition of DPPH was 80.8 per cent in methanol extract of *Portulaca grandiflora*. Key phytochemicals include betalins, portulacaxanthins.

Portulaca oleracea L. Family: Portulacaceae

English Name: Purslane, Common purslane; Japanese: Suberihiyu.

Arruda *et al.* (2004) reported that ingestion of Purslane leaves have a protective effect against oxidative stress caused by vitamin A-deficiency. Maeda *et al.* (2006) examined the phenolic contents of leaf and stalk of *Portulaca oleracea* and its radical scavenging activities for DPPH radical. The polyphenol content was 47.5 mg-gallic acid equivalent/100g and DPPH scavenging activity was 784 µmol-Trolox eq./100 g.

Abbas *et al.* (2006) evaluated the antioxidant activity of *Portulaca oleracea* using DPPH assay and nitric oxide inhibition using Griess assay. Methanolic extract showed DPPH radical scavenging and nitric oxide inhibition in murine peritoneal macrophages.

Souri *et al.* (2008) evaluated the antioxidant activity of seed against linoleic acid peroxidation using 1,3-diethyl-2-thiobarbituric acid as reagent. Antioxidant activity

(IC$_{50}$) against peroxidation of linoleic acid (2mg/ml) was 11.74 and phenolic content was 33.66 mg/100g dry weight.

Li *et al.* (2008) evaluated the antioxidant properties of 45 medicinal plants using FRAP and TEAC assays and the total phenolic contents of these plants were measured. *Portulaca oleracea* showed 54.12 µmol Fe(II)/g FRAP value, 27.28µmol Trolox/g of TEAC value and 8.96 mg GAE/g of phenolic content.

Potentilla alba L. Family: Rosaceae

Oszmianski *et al.* (2007) investigated the antioxidant tannins from *Potentilla alba* roots. The predominant constitutive units of the procyanidins of roots were (-)epicatechin and the concentration was close to 80g/kg. The antioxidant activity, measured by the DPPH method was very high.

Using spectrophotometric methods, a H$_2$O-soluble *Potentilla alba* rhizome extract was evaluated phytochemically, i.e., the total phenol, flavonoid, flavonol, flavanone, and proanthocyanidin contents were determined, and its antioxidant and pro-oxidant properties, i.e., the Fe^{III} reductive and the Fe^{II} chelating properties, the 1,1-diphenyl-2-picrylhydrazyl radical (DPPH·), N,N-dimethyl-p-phenylenediamine (DMPD·+), and superoxide anion radical (O$_2^-$)-scavenging activities, the capacity to inhibit hydroxyl radical (HO·)-mediated deoxy-D-ribose and phospholipid degradation, and the interaction with the Cu-catalyzed HO·-mediated DNA degradation, were determined by Dorman *et al.* (2011). The extract was found to contain a range of phenolic compounds recognized to possess strong antioxidant-like properties. Moreover, the extract demonstrated dose-dependent activities in all the antioxidant assays with the exception of the DNA-degradation assay, where the components within the extract interfered with the assay components at concentrations ≥1.00mg/ml.

Potentilla anserina L. Family: Rosaceae

English Name: Silver weed.

The antioxidative activity of phenolic extracts from herb of silverweed was examined by autoxidation of methyl linoleate by Kahkonen *et al.* (1999). Phenolic content of the herb was low, however the antioxidant activity was high (58 per cent at 500 ppm level).

Potentilla argentea L. Family: Rosaceae

Tomczyk and Latte (2009) reported antioxidant activity of the herb of *Potentilla argentea*. Main antioxidant constituents were tannins.

Potentilla arguta Pursh. Family: Rosaceae

English Name: Prairie cinquefoil.

Borchardt *et al.* (2008) evaluated the antioxidant activity of methanol extracts of seeds of *Potentilla arguta* using DPPH antioxidant assay.The antioxidant value reported in µM Trolox/100g (TE) from DPPH radical scavenging activity of crude seeds extract was 59,635.

Potentilla atrosanguinea Lodd. Family: Rosaceae

The effects of different solvent systems (methanol, ethanol, acetone, and their 50 per cent aqueous concentrations) and extraction procedures (microwave, ultrasound, Soxhlet and maceration) on the antioxidant activity of aerial parts of *Potentilla atrosanguinea* were investigated by Kalia *et al.* (2008) by three different bioassays: ABTS, DPPH radical scavenging assays and FRAP. The 50 per cent aqueous ethanol extracts exhibited strong antioxidant activity measured in terms of Trolox equivalent antioxidant capacity (TEAC) [(54.34 to 122.96, 29.82 to 101.22 and 13.64 to 41.43) mg of Trolox/g] with ABTS (*+), DPPH (*) and FRAP assays, respectively. In general, TEAC of Soxhlet extracts was found to be 1.8 and 3 times higher than ultrasound and maceration but slightly (1.2 times) higher than microwave. A positive correlation was observed between total polyphenol (TPC) and total flavonoid (TFC) contents which ranged between 26.7 to 30.7 mg/g gallic acid equivalent and 16.8 to 20.8 mg/g quercetin equivalent respectively, with antioxidant activity. In addition, some of its bioactive phenolic constituents which contribute largely toward antioxidant potential such as chlorogenic acid, catechin, caffeic acid, *p*-coumaric acid and quercetin were also quantified in different extracts by RP-HPLC.

Potentilla chinensis Ser. Family: Rosaceae

Wei *et al.* (2010) isolated trans-tilioroside from *Potentilla chinensis* and the compound showed potent antioxidant activity.

Potentilla erecta (L.) Rausch. Family: Rosaceae

Tomczyk and Latte (2009) reported antioxidant activity of the herb of *Potentilla erecta*. Main antioxidant constituents were tannins.

Potentilla freyniana Bornm. Family: Rosaceae

The antioxidant activity of *Potentilla freyniana* has been investigated by Chen *et al.* (2005) using both enzymatic and non-enzymatic *in vitro* antioxidant assays. Extracts inhibited xanthine oxidase and lipoxygenase activities and were scavengers of the ABTS radical cation using the Trolox equivalent antioxidant capacity assay.

Potentilla fruticosa L. Family: Rosaceae

Miliauskas *et al.* (2004) investigated the radical scavenging activity of acetone, ethyl acetate and methanol extracts of flowers using DPPH and ABTS assays. They found that methanol extracts were most effective DPPH radical scavengers. Ethyl acetate and acetone extracts were considerably less effective radical scavengers. High content of phenolics in the extracts correlates with its antiradical activity, confirming that phenolic compounds are likely to contribute to the radical scavenging activity of this plant extracts.

Potentilla recta L. Family: Rosaceae

Tomczyk and Latte (2009) reported antioxidant activity of the herb of *Potentilla recta*. Main antioxidant constituents were tannins.

Potentilla reptans L. Family: Rosaceae

Avci *et al.* (2006) evaluated the hypercholesterolaemic and antioxidant activities of *Potentilla reptans*. The ethanol extract increased the serum HDL concentration.The ethanolic extract increased NOx.

Pouteria cambodiana (Pierre ex Dubard) Baehni Family: Sapotaceae

The methanolic stem bark extract from *Pouteria cambodiana* was evaluated for immunomodulating activity on BALB/c mice. The antioxidant effect was also assessed. The extract showed low free radical scavenging activity (Manosroi *et al.*, 2005).

Prangos ferulacea Lindl. Family: Apiaceae (Umbelliferae)

Antioxidant and radical scavenging activities, reducing powers and the amount of total phenolic compounds of aqueous and methanolic extracts of *Prangos ferulacea* were studied by Mavi *et al.* (2004). The extract showed moderate antioxidant activity.

Prangos uloptera DC. Family: Apiaceae (Umbelliferae)

A known furanocoumarin, 8-geranyloxy psoralen was isolated from n-hexane extract of *Prangos uloptera* roots by Razavi *et al.* (2009). The antioxidant activity of the compound was evaluated by DPPH assay. The compound exhibited weak antioxidant potential with an IC_{50} value of 0.262 mg/ml.

Primula elatior (L.) Hill. Family: Primulaceae

Katalinic *et al.* (2005) reported that *Primula elatior* possessed potent antioxidant activity. Main antioxidant constituents were saponins and tannins.

Primula veris L. Family: Primulaceae

Katalinic *et al.* (2005) reported that *Primula veris* possessed potent antioxidant activity. Main antioxidant constituents were saponins and tannins.

Prosopis juliflora (Sw.) DC. Family: Mimosaceae

A large amount of flavonoid has been extracted and isolated from the heartwood of *Prosopis julifora* by Sirmah *et al.* (2009). Structural and physicochemical elucidation clearly demonstrated the presence of (-)-mesquitol as the sole compound without any noticeable impurities. The product was able to slow down oxidation of methyl lineoleate induced by AIBN. The important amount and high purity of (-)-mesquitol present in the acetonic extract of *P. juliflora* could therefore be of valuable interest as a potential source of antioxidants from a renewable origin.

Sathiya and Muthuchelian (2010) evaluated the antioxidant and antitumor potential of the total phenolic extract of *Prosopis juliflora* leaves. The DPPH scavenging effect of the extract was increased with the increasing dosage and the IC_{50} estimations for the extract and the ascorbic acid standard were found to be 33.43µg/ml and 1.5µg/ml respectively.

Prosopis stephaniana (M.B.) Kunth. ex Spreng. Family: Mimosaceae

Souri *et al.* (2004) evaluated the antioxidative activity of *Prosopis stephaniana* by linoleic acid peroxidation test using 1,3-diethyl-2-thiobarbituric acid as the reagent.

Methanolic extract showed 57.37 per cent inhibition at 40µg concentration with IC_{50} value of 18.61µg.

Prunella vulgaris L. Family: Lamiaceae (Labiatae)

Cai *et al.* (2004) evaluated the antioxidant activity and phenolic compounds in traditional Chinese medicinal plants associated with anticancer. The improved ABTS.+ method was used to systematically assess the total antioxidant capacity (Trolox equivalent antioxidant capacity, TEAC) of the medicinal extracts. The TEAC values and total phenolic content for methanolic extract of inflorescence of *Prunella vulgaris* were 910.1 µmol Trolox equivalent/100g dry weight (DW), and 4.73 g of gallic acid equivalent/100g DW respectively. A positive significant linear relationship between antioxidant activity and total phenolic content showed that phenolic compounds were the dominant antioxidant components in the tested herbs. Major types of phenolic compounds in *Prunella vulgaris* include flavonols (rutin), anthocyanins (cyanidin, delphinidin), phenolic acids (caffeic acid), tannins.

Shyur *et al.* (2005) examined antioxidant activities of 26 medicinal herbal extracts that have been popularly used as folk medicines in Taiwan. The results of scavenging DPPH radical activity show that, among the 26 tested medicinal plants, *Prunella vulgaris* exhibited strong activities and its IC_{50} value for Superoxide anion scavenging activity was 113µg/ml.

Two natural antioxidant preparations, namely Rosmol (liquid) and Rosmol-P (powder) were produced by Lugasi *et al.* (2006) by extraction of a mixture of medicinal plants belonging to the Lamiaceae family, such as *Rosmarinus officinalis*, *Prunella vulgaris*, *Hyssopus officinalis* and *Melissa officinalis*. The main active compound of the extract was supposed to be a phenolic (caffeic) acid derivative. The total polyphenol content of the preparation is very high, 8.72g/l for Rosmol and 93.7 g/kg for Rosmol-P. The products acted as primary and secondary antioxidants chelatng transitional metal ions and inhibiting the autoxidation of linoleic acid. Rosmol and Romol-P scavenged free radicals formed during Fenton type reaction measured by chemiluminometry and also exhibited strong antioxidant property in Randox TAS measurement. The antioxidant activity of the products was unchanged after six months storage.

Li *et al.* (2008) evaluated the antioxidant properties of 45 medicinal plants using FRAP and TEAC assays and the total phenolic contents of these plants were measured. *Prunella vulgaris* showed 56.08 µmol Fe(II)/g FRAP value, 27.23µmol Trolox/g of TEAC value and 5.84 mg GAE/g of Phenolic content.

Mrudula *et al.* (2010) assessed the antioxidant activity of *Prunlla vulgaris* by using DPPH, nitric oxide scavenging activity and reducing power methods. Under *in vitro* DPPH free radical and nitric oxide free radicals were considerably inhibited in a dose-dependent manner and there is increasing in absorbing power, indicates increase in reducing power.

Prunus americana Marsh Family: Rosaceae

English Name: American plum.

The EtOAc extract obtained from *Prunus americana* was tested in the DPPH free radical assay. High antioxidant activity was obtained from the extract of fruit (Acuna *et al.*, 2002).

Prunus armeniaca L. Family: Rosaceae

English Name: Apricot.

Nectars of Apricot in Samsun markets in Turkey were analyzed by Tosun and Ustun (2003) for total antioxidant capacity, total phenolic content, ascorbic acid content and total carotenoids. Ferric reducing/antioxidant power (FRAP) assay was used to measure the total antioxidant power. Apricot nectar had moderate concentration of ascorbic acid (105.2 mg/kg). Total phenols were 457.51 mg/l and total carotenoids 3.32 mg/l. FRAP value of fruit nectar was 6µmol/ml.

The antioxidant properties of peeled, defatted, roasted apricot kernel flours were evaluated by Durmez and Alpasan (2005) by determining radical scavenging power (RSP), anti-lipid peroxidative activity (APLA), reducing power (RP), total phenolic content (TPC), assessed by DPPH test, β-carotene bleaching method, iron (III to II) reducing test and Folin method, respectively. Browning degree of the samples was also measured and found to increase almost linearly with the roasting time. Contrary to the browning degree, RSP, RP and TPC did not increase linearly but showed a maximum for 10 minutes of roasting. Roasting reduced ALPA value, thus unroasted sample showed the highest ALPA value, RSP, RP and TPC measurements of all samples, were in high correlation.

The total antioxidant status of several apricot cultivars differing in ripening season, pomological traits and geographical origin was determined by Leccese *et al.* (2007) by TEAC assay and total phenolic content by Folin-Ciocalteu (F-C) method. Among the cultivars analysed, the variability on the antioxidant capacity and total phenol content was consistent, showing an increasing amount of antioxidants in the late ripening genotypes. These genotypes exhibited the best combination of pomological and nutraceutical traits with an excellent fruit qualitative profile.

Madrau *et al.* (2009) studied the effect of drying temperature on polyphenolic content and antioxidant activity of apricot cultivars (Pelese and Cafona). Analysis of the data showed that the decrease in chlorogenic and neochlorogenic acid in Cafona cultivar was higher at the lower drying temperature. Catechin showed the same behaviour of hydroxycinnamic acids in both cultivars, while the decrease in the other compounds was significantly more marked in the sample dried at 75°C. The antioxidant activity increased significantly in Cafona fruits and this increase was confirmed by a diminution of the redox potential.

Yiit *et al.* (2009) evaluated the antioxidant activity of the methanol and water extracts of sweet and bitter apricot (*Prunus armeniaca*) kernels by determining radical scavenging power, lipid peroxidation inhibition activity and total phenol content measured with a DPPH test, the thiocyanate method and Folin method, respectively. In contrast to extracts of bitter kernels, both the water and methanol extracts of sweet kernels have antioxidant potential. The highest percent inhibition of lipid peroxidation (69 per cent) and total phenolic content (7.9µg/ml) were detected in the methanol extract of sweet kernels (Hasanbey) and in the water extract of the same cultivar, respectively.

Antioxidant capacities of 56 selected Chinese medicinal plants, including *Prunus armeniaca* were evaluated by Song *et al*. (2010) using the TEAC and FRAP assays, and their total phenolic content was measured by the Folin-Ciocalteu method. *Prunus armeniaca* showed low antioxidant activity with TEAC value of 4.18µmol Trolox/g, FRAP value of 0.41µmol Fe^{2+}/g and phenolic content was 0.58mg GAE/g.

Prunus avium L. Family: Rosaceae

English Name: Sweet cherry.

Goncalves *et al*. (2004) investigated the storage affects of the phenolic profiles and antioxidant activitites of cherry on human low-density lipoproteins. Murtaza *et al*. (2007) investigated the effect of storage on phenolic profiles and antioxidant activity of sweet cherry varities of Kashmir valley. Samples of five sweet cherry cultivars were harvested at maturity and stored at 4°C and 25°C for 20 and 7 days respectively. Total phenols in the extracts of the freshly harvested cherry samples varied from 180-350 mg/100g fresh weight and slightly increased during storage (180-388 mg/100g fresh weight) at room temperature (25°C) mostly attributable to slight increases in anthocyanins. Freshly harvested Siah Gole cherries exhibited the highest antioxidant activity (446µmole/g fr. wt) compared to the other varieties. Cool storage decreased phenol levels and antioxidant activities of these five varieties. A strong correlation was observed between total phenolic content and total antioxidant activity of extracts from fresh varieties.

Sweet cherry cultivars were evaluated by Usenik *et al*. (2008) for their phenolic content and antioxidant activity. The total phenolic content ranged from 44.3 to 87.9 mg gallic acid equivalent/100 g FW and antioxidant activity ranged from 8.0 to 17.2 mg ascorbic acid equivalent antioxidant capacity mg/100g FW. The correlation of antioxidant activity with total phenolics content and content of anthocyanins was cultivar dependent.

Six wild growing sweet cherry (*Prunus avium*) genotypes with different fruit skin colour were analyzed by Karlidag *et al*. (2009) for their antioxidant activity, ascorbic acid contents, total anthocyanins, total phenolic and total soluble solid contents. Antioxidant activity was relatively higher in blackish skin coloured fruits than in light skin coloured ones. Among the six genotypes, total antioxidant activity ranged from 51.13 per cent to 75.33 per cent while the total phenolic content was between 148 and 321 mg GAE/100g FW. The two parameters had a positive correlation. The vitamin was in general highest in blackish coloured fruits (21-27 mg/100ml).

The extract obtained with CO_2: EtOH (90:10, v/v) exhibited the highest antioxidant activity (181.4µmol TEAC/g). Polyphenols, in particular Sakuranetin and Sakurenin seemed to be the major contributors of the antioxidant capacity (Serra *et al*., 2010).

Prunus campanulata Maxim. Family: Rosaceae

Chen *et al*. (2009) studied the antioxidant activity of leaves extract towards xanthine oxidase, lipoxygenase and tyrosinase activities. The inhibitory effect of acetone extract on xanthine oxidase activity was nearly 80 per cent, tyrosine activity was less than 50 per cent and lipoxygenase activity was about 60 per cent.

Prunus cerasus L. Family: Rosaceae

English Name: Tart cherry, Sour cherry.

Nectars of sour cherry in Samsun markets in Turkey were analyzed by Tosun and Ustun (2003) for total antioxidant capacity, total phenolic content, ascorbic acid content and total carotenoids. Ferric reducing/antioxidant power (FRAP) assay was used to measure the total antioxidant power. Sour cherry nectar had moderate concentration of ascorbic acid (25.3 mg/kg) and highest total phenolic content of 1475.69mg/l. FRAP value of fruit nectar was 8µmol/ml.

Tart cherry produces various kinds of polyphenolics in its fruits. Kirakosyan *et al.* (2010) analysed the antioxidant capacities of these constituents using TEAC antioxidant assay. Kaempferol, quercetin, isorhamnetin 3-rutinoside, cyanidin 3-rutinoside, and melantonin showed significant antioxidant capacities; of these kaempferol proved to be the most active. The most important and new finding here was that pairs of compounds with the highest antioxidant capacity (*e.g.*, kaempferol and melatonin at a dose ratio of 2:1, respectively, and cyanidin 3-rutinoside and isorhamnetin 3-rutinoside at a dose ratio of 1:4, respectively) showed strong synergistic types of interactions.

Prunus davidiana (Carr.) Franch. Family: Rosaceae

The methanol extract of the stem of *Prunus davidiana* exerted inhibitory/ scavenging activities on DPPH radicals, total ROS and peroxynitrites (Jung *et al.,* 2003).

Prunus divaricata Ledeb. subsp. *divaricata* Family: Rosaceae

Serteser *et al.* (2009) studied the antioxidant activity of 50 per cent aqueous methanol extracts of fruits by various antioxidant assays, including free radical scavenging, hydrogen peroxide scavenging and metal (Fe^{2+}) chelating activities. The DPPH radical scavenging effect of fruit extract was 0.591. Fe^{2+} chelating activity (per cent) of fruit extract was 34.29 and Hydrogen peroxide inhibition activity of fruit extract was 32.65 per cent.

Prunus domestica L. Family: Rosaceae

English Name: Plum, Prune

The total antioxidant activity of plum was measured by Wang *et al.* (1996) using automated oxygen radical absorbance capacity (ORAC) assay. On the basis of the wet weight of the fruits (edible portion), plum had the ORAC activity (9.49µmoles of Trolox equivalents per gram). On the basis of dry weight of the fruits plum had the highest ORAC activity (79.1µmoles/gTrolox equivalent). The contribution of the fruit pulp fraction (extracted with acetone) to the total ORAC activity of a fruit was usually less than 10 per cent.

Donovan *et al.* (1998) investigated the phenolic compounds and antioxidant activity of prunes and prune juice. The mean concentrations of phenolics were 1840 mg/kg, 1397 mg/kg, and 441 mg/l in pitted prunes, extra large prunes with pits, and prune juice, respectively. Hydroxycinnamates, especially neochlorogenic acid, and chlorogenic acid predominated, and these compounds, as well as the prune and prune juice extracts, inhibited the oxidation of LDL. The pitted prune extract inhibited LDL oxidation by 24, 82, and 98 per cent at 5, 10, and 20 µM gallic acid equivalents (GAE). The prune juice extract inhibited LDL oxidation by 3, 62, and 97 per cent at 5, 10, and 20 µM GAE.

Kayano *et al.* (2002) reported that MeOH eluate of *Prunus domestica* exhibited the strongest antioxidant activity among the separated fractions evaluated by ORAC. The antioxidant activity of the MeOH eluate was highly dependent on the major prune components. Kayano *et al.* (2004) reported that prunes contain relatively high amount of 4-O-caffeoylquinic acid (CQA). The contribution of CQA isomers to the activity of prunes was revealed to be 28.4 per cent on the basis of ORAC. Total of 28 compounds were isolated, hydroxycinnamic acids, benzoic acids, coumarins, lignans and flavonoid showed high ORAC values.

Effect of polyphenolics on antioxidant capacities of plums, the amounts of total phenolics, total flavonoids and individual phenolic compounds, and vitamin C equivalent antioxidant capacity (VCEAC) of eleven plum cultivars was determined by Chun *et al.* (2003). There was a good linear relationship between the amount of total phenolics and total antioxidant capacity. The amount of total flavonoids and total antioxidant capacity also showed a good correlation. Chlorogenic acids and glycosides of cyanidin, peonidin, and quercetin were major phenolics among eleven plum cultivars. Chlorogenic acids were a major source of antioxidant activity in plums, and the consumption of one serving (100g) of plums can provide antioxidants equivalent to 144.4-889.6mg of vitamin C.

Four new abscisic acid related compounds (1-4), together with (+)abscisic acid (5), (+)-β-D-glucopyranosyl abscisate (6), (6S,9R)-roseoside (7), and two lignan glucosides (+)-pinoresinol mono- β-D-glucopyranoside (8) and 3-(- β-D-glucopyranosyloxymethyl)-2-(4-hydroxy-3-methoxyphenyl)-5-(3-hydroxypropyl)-7-methoxy(2R3S)-dihydrobenzofuran (9) were isolated from the antioxidative ethanol extract of prunes *Prunus domestica*. The antioxidant activities of these isolated compounds were evaluated on the basis of oxygen radical absorbance capacity (ORAC). The ORAC values of abscisic acid related compounds (1-7) were very low. Two lignans (8 and 9) were more effective antioxidants whose ORAC values were 1.09 and 2.33 micro mol of Trolox equiv./micromole, respectively (Kikuzaki *et al.*, 2004).

Rop *et al.* (2009) evaluated that the antioxidant activity of selected cultivars and subspecies of plum trees. Regional cultivars showed outstanding nutritional properties especially as far as the total content of phenolic substances concerned (3.48 – 4.95 mg GA/g FM), this parameter was highly correlated with the total antioxidant capacity of the fruit.

Prunus dulcis (Mill.) D.A.Webb. Family: Rosaceae

English Name: Almond.

The phenolic composition of almond skins was carried out by Monagas *et al.* (2007) in order to evaluate their potential applications as a potential food ingredient. Flavonols and flavonol glycosides were the most abundant phenolic compounds in almond skins, representing up to 38-57 per cent and 14-35 per cent of the total phenolics, respectively. Due to their antioxidant properties measured as ORAC at 0.398-0.500 mmol Trolox/g, almond skins can be considered as a value added byproduct for elaborating dietary antioxidant ingredients.

Prunus insititia L. Family: Rosaceae

English Name: Damson.

The antioxidative activity of phenolic extracts from herb of damson was examined by autoxidation of methyl linoleate by Kahkonen *et al.* (1999). The antioxidant activity and the phenolics were high. The phenolics were 23 mg/g GAE while the antioxidant activity was 48 per cent at 500 ppm level and 86 per cent at 5000 ppm level.

Prunus mahaleb L. Family: Rosaceae

English Name: White mahalab.

Mariod *et al.* (2010) investigated the antioxidant activities of phenolic rich fractions (PRFs) from crude methanolic extract and its fractions using ethyl acetate, hexane and water of white mahaleb seed cake. The total phenolic compounds were found to be higher in white mahaleb seed cakes. The seed cake exhibited the antioxidant activity determined by the DPPH method.

Prunus mume (Sieb.) Sieb. et Zucc. Family: Rosaceae

Methanol aqueous extract of *Prunus mume* was screened by Kim *et al.* (1997) for free radical scavenging activity using the DPPH free radical generating system. The extract showed free radical scavenging activity.

Shi *et al.* (2009) investigated the phenolic compounds and antioxidant activity of ethanolic extract from flowers of *Prunus mume* in China. The total phenol content was estimated as gallic acid equivalents by the Folin-Ciocalteu reagent method. The antioxidant activity was measured by DPPH, ABTS and OH free radicals scavenging and ferric-reducing antioxidant power (FRAP). Three chlorogenic acid isomers, namely, 3-O-caffeoylquinic, 4-O-caffeoylquinic and 5-O-caffeoylquinic acids, were isolated and purified. The ethanolic extract demonstrated activity to some degree in all the antioxidant assays. In all tested assays, all of the isolated chlorogenic acid isomers exhibited strong antioxidant activities, which were almost the same. The results showed that chlorogenic acid isomers were the key phenolic compounds which are responsible for antioxidant activity of the ethanolic extract from Chinese *Prunus mume* flowers.

Prunus persica Batsch. Family: Rosaceae

English Name: Peach.

Nectars of peach in Samsun markets in Turkey were analyzed by Tosun and Ustun (2003) for total antioxidant capacity, total phenolic content, ascorbic acid content and total carotenoids. Ferric reducing/antioxidant power (FRAP) assay was

Encyclopaedia of Herbal Antioxidants

used to measure the total antioxidant power. Peach nectar had moderate concentration of ascorbic acid (83.6 mg/kg). Total phenols were 413.72 mg/l and total carotenoids 1.90 mg/l.

Kardosova and Machova (2006) investigated the ability of polysaccharides, isolated from leaves, to inhibit peroxidation of soyabean lecithin liposomes by OH radicals. All polysaccharides exhibited antioxidant activity.

Gan *et al*. (2010) evaluated the antioxidant activity of 40 medicinal plants. *Prunus persica* showed very low antioxidant activity of 0.78 µmol Fe(II)/g in FRAP assay and 7.81µmol in TEAC assay. Total phenolic content was also very low, 0.46 mg GAE/g.

Prunus serrulata Lindl. var. *spontanea* Family: Rosaceae

Jung *et al*. (2002) evaluated the antioxidant activity of methanol extract and the organic solvent-soluble fractions, including the dichloromethane, ethyl acetate, n-butanol fractions and water layer of *Prunus serrulata* var. *spontanea* leaves. Jung *et al*. (2002) isolated eleven known flavonoids from EtOAc soluble fraction of MeOH extract exhibiting strong antioxidant activity and characterized as prunetin (1), genistein (2), quercetin (3), prunetin 4-O-β-glucopyranoside (4), kaempferol 3-O-α-arabinofuranoside (5), prunetin 5-O-β-glucopyranoside (6), kaempferol 3-O-β-xylopyranoside (7), genistin (8), kaempferol 3-O-β-glucoopyranoside (9), quercetin 3-O-β-glucopyranoside (10) and kaempferol 3-O-β-xyloopyranosyl-(1→2)-β-glucopyranoside (11). Compounds 3 and 10 showed good activities in DPPH and ONOO-tests, while compounds 5,7,9 and 11 were active in the ONOO- and ROS tests. On the other hand compounds 1, 4 and 6 did not show any activities in the tested model systems.

Prunus spinosa L. Family: Rosaceae

The antioxidant activity of fresh juice from *Prunus spinosa* L. fruit growing wild in Urbino (central Italy) was assessed by Fraternale *et al*. (2009) by using different cell-free *in vitro* analytical methods: 5-lipoxygenase test, 1,1-diphenyl-2-picrylhydrazyl (DPPH) free radical scavenging, and oxygen radical absorbance capacity (ORAC). Trolox was used as the reference antioxidant compound. In the 5-lipoxygenase and DPPH tests the fresh fruit juice of *P. spinosa* showed good antioxidant activity when compared with Trolox, while the ORAC value was 36.0 micromol eq. Trolox/g of fruit. These values are in accord with data reported in the literature for small fruits such as *Vaccinium*, *Rubus* and *Ribes*. The antioxidant capacity in cell-free systems of *P. spinosa* juice has been compared with its cytoprotective - bona fide antioxidant activity in cultured human promonocytes (U937 cells) exposed to hydrogen peroxide. The antioxidant activity of red berries has been correlated with their anthocyanin content. The results of this study indicate that the three most representative anthocyanins in *P. spinosa* fruit juice (cyanidin-3-rutinoside, peonidin-3-rutinoside and cyanidin-3-glucoside) are likely to play an important role in its antioxidant properties.

Psammogeton canescens (DC.) Vatke Family: Apiaceae (Umbelliferae)

Chemical composition and *in vitro* antioxidant activities of the essential oil and methanol extracts of *Psammogeton canescens* were investigated by Gholivand *et al*. (2010). Antioxidant activities of the samples were determined by three various systems, namely DPPH, β-carotene/linoleic acid and reducing power assay. In DPPH system, the highest radical-scavenging activity was seen by the polar subfraction of methanol extract (49.5µg/ml). Furthermore, in the second case the inhibition capacity(per cent) of the polar subfraction (92.40 per cent) found to be the stronger one. However, in the reducing power assay, a reverse activity pattern more than in the first two systems was observed, and the essential oil was stronger radical reducer than was the methanol extract in all the concentration tested.

Pseudarthria viscida (L) Wight et Arn. Family: Fabaceae (Papilionaceae)

Antioxidant activity of the crude methanol extract of *Pseudarthria viscida* stem and root was performed by DPPH radical quenching assay and reducing power test models. Both stem and root extracts exhibited potential antioxidant activity in both the assays (Mathew and Sasikumar, 2007).

Pseudobombax ellipticum H.B.K. Family: Bombacaceae

Ruiz-teran *et al.* (2008) evaluated the antioxidant activity of *Pseudobombax ellipticum*. Methanolic extract contained highest total phenolic contents. Hexane extract of *Pseudobombax ellipticum* had comparatively lower antioxidant activity while acetone extract had the highest antioxidant activity.

Pseudostellaria heterophylla (Miq.) Pax Family: Caryophyllaceae

Ng *et al.* (2004) evaluated the antioxidant activity of aqueous extract and organic fractions of *Pseudostellaria heterophylla*. The extract showed weak antioxidant activity.

Psidium guajava L. Family: Myrtaceae

English Name: Guava.

Huang *et al.* (2004a) evaluated the antioxidant activities of various fruits and vegetables produced in Taiwan. Guava seeds showed the highest antioxidant activity. At the level of 1 g fresh sample, low-density lipoprotein peroxidation was inhibited by at least 90 per cent by guava meat. The total phenolic content was significantly correlated with antioxidant activities measured by TBA and Iodometric assays.

Tachakittirungrod *et al.* (2007) isolated three compounds quercetin, quercetin-3-O-glucopyranoside and morin from the methanolic extract of *Psidium guajava* leaves which showed significant DPPH radical scavenging activity in terms of IC_{50} values 1.2, 3.58 and 5.41µg/ml, respectively.

Recent pharmacological studies have demonstrated the ability of this plant to exhibit antioxidant (the antioxidant capacity of the skin is 10 times higher than that of the pulp) activities. Alfredo (2007) also reported antioxidant activity of guava.

Chen *et al.* (2007) examined the antioxidant activities and free radical scavenging effects of extracts of aqueous leaves of *Psidium guajava* (PE), *Camellia sinensis* (GABA tea; CE), *Toona sinensis* (TE) and *Rosmarinus officinalis* (RE). Among the four extracts, PE exhibited the strongest efficiency and showed over 50 per cent scavenging effect on DPPH radicals at the concentration of 100 µg/ml. The reducing power of four nutraceutical herbs was in the order of PE > RE > CE > TE. The antioxidant activities of nutraceutical herbs were evaluated in a liposomes oxidation system promoted by

Fe^{3+}/ascorbic acid/H_2O_2. PE still showed the strongest antioxidant activity and exhibited over 95 per cent inhibition at concentration of 50 µg/ml. The antioxidant activity of TE was still lower than that of other herbal plants; however, it also displayed 89 per cent inhibition at concentration of 250 µg/ml. RE exhibited well inhibitory effects on the UVB-induced oxidation of erythrocyte ghosts at lower concentration (100 µg/mL). However, the protection of PE on the UVB-induced oxidation was significantly raised with increasing the concentrations and reached 95.4 per cent inhibitory effects at concentration of 500 µg/ml.

Ayoola *et al.* (2008) carried phytochemical screening and antioxidant activities of the ethanolic extract of leaves of *Psidium guajava*. The crude extract appeared to be as potent as Vitamin C with a maximum inhibition of 91 per cent at 0.5mg/ml which was comparable to 95 per cent for vitamin C at the same concentration. They are of the opinion that the antioxidant activity of this plant may contribute to their claimed antimalarial activity.

The antioxidant capacity and phenol content of Thai seedless guava (*Psidium guajava*) was studied by Alothman *et al.* (2009). Three solvent systems were used (methanol, ethanol and acetone) at three different concentrations (50, 70 and 90 per cent) and with 100 per cent distilled water. The antioxidant capacity of the fruit extracts was evaluated using a ferric reducing/antioxidant power assay and the free radical-scavenging capacity was evaluated using 2,2-diphenyl-1-picrylhydrazyl radical-scavenging assays. The polyphenol content was 123 to 191 Gallic Acid Equivalent (GAE)/100g. High phenol content was significantly correlated with high antioxidant capacity.

Akinmoladun *et al.* (2010) reported that the methanolic extract of *Psidium guajava* possess significant antioxidant and radical scavenging activities that may be due to the phytochemical content of the plant. *P. guajava* contained highest amount of total phenolics (380.08 mg/l gallic acid equivalents) and the flavonoids content was 269,72 quercetin equivalent. Percentage of radical scavenging activity was 82.79 and compared with values obtained for ascorbic acid and gallic acid. The reductive potential was 0.79.

Thirty-eight types of fruits commonly consumed in Singapore were systematically analyzed by Isabelle *et al.* (2010) for their hydrophilic oxygen radical absorbance capacity (H-ORAC), total phenolic content (TPC), ascorbic acid (AA) and various lipophilic antioxidants. Guava had the highest Ascorbic acid per gram fresh weight.

Psoralea corylifolia L. Family: Fabaceae (Papilionaceae)

Bakuchiol, a meroterpene isolated from *Psoralea corylifolia*, prevented mitochondrial lipid peroxidation. Inhibition of oxygen consumption originating in lipid peroxidation was time-dependent. Bakuchiol protected mitochondrial respiratory enzyme activities against both NADPH-dependent and dihydroxyfumarate-induced peroxidation injury. Bakuchiol was shown to be effective to protect mitochondrial functions against oxidative stress (Haraguchi *et al.*, 2000).

A meroterpene and four flavonoids were isolated by Haraguchi *et al.* (2002) from seeds of *Psoralea corylifolia* as antioxidative components. Their structures were elucidated by spectral data and identified as bakuchiol, bavachin, isobavachin and isobavachinin. In particular, meroterpene bakuchiol and flavonoids 4 and 5 showed broad antioxidative activities in rat liver microsomes and mitochondria. They inhibited NADPH-, ascorbate-, t-BUOOH- and CCl_4-induced lipid peroxidation in microsomes. They also prevented NADPH dependent and ascorbate-induced mitochondrial lipid peroxidation. Bakuchiol was the most potent antioxidant in microsomes, the inhibition of oxygen consumption induced by lipid peroxidation was time-dependent. Furthermore, bakuchiol protected red blood cells against oxidative haemolysis. These phenolic compounds in *P.corylifolia* were shown to be effective in protecting biological membranes against various oxidatives stresses.

Cai *et al.* (2004) evaluated the antioxidant activity and phenolic compounds in traditional Chinese medicinal plants associated with anticancer. The improved ABTS.+ method was used to systematically assess the total antioxidant capacity (Trolox equivalent antioxidant capacity, TEAC) of the medicinal extracts. The TEAC values and total phenolic content for methanolic extract of fruit of *Psoralea corylifolia* were 541.3 µmol Trolox equivalent/100g dry weight (DW), and 2.43 g of gallic acid equivalent/100g DW respectively. A positive significant linear relationship between antioxidant activity and total phenolic content showed that phenolic compounds were the dominant antioxidant components in the tested herbs. Major types of phenolic compounds in *Psoralea corylifolia* include coumarin (psorulen, psoralidin), flavonoids (bavachin, bavachinin, bavachalcone).

Tang *et al.* (2004) carried out characterization of antioxidant and antiglycation properties and isolation of active ingredients from traditional Chinese medicines. Extract of *Psoralea corylifolia* showed higher antioxidant activity.

Park *et al.* (2005) investigated the protective effect of (S)-bakuchiol isolated from the seed of *Psoralea corylifolia*, on liver injury. Treatment with bakuchiol significantly inhibited lipid peroxidation and intracellular glutathione depletion in hepatocytes induced by tBH, CCl_4 of D-galactosamine. They concluded that bakuchiol has a protective effect against tBH, CCl_4 or D-galactosamine-induced hepatotoxicity *in vitro* or *in vivo*.

Relation between isoflavones production and antioxidant activity in *Psoralea corylifolia* cell cultures was studied by Shinde *et al.* (2010). The antioxidant activity of extracts was determined using DPPH radical scavenging assay and phosphomolybdenum assay.

Psoralea esculenta Pursh. Family: Fabaceae (Papilionaceae)

Borchardt *et al.* (2008) evaluated the antioxidant activity of methanol extracts of seeds of *Psoralea esculenta* using DPPH antioxidant assay.The antioxidant value reported in µM Trolox/100g (TE) from DPPH radical scavenging activity of crude seeds extract was 43,182.

Psychotria brachyceras Muell.-Arg. Family: Rubiaceae

Nascimento *et al.* (2007) tested the antioxidant capacity of the crude foliar extract and the alkaloid brachycerine of *Psychotria brachyceras* using hypoxanthine/xanthine oxidase assay. The results showed that brachycerine and the crude foliar extract have antioxidant effects in scavenging OH radicals.

Psychotria griffithii Hook.f. Family: Rubiaceae

Ahmad *et al*. (2010) evaluated the antioxidant potential of 22 species of medicinal plants from Malaysian Rubiaceae including *Psychotria griffithii*. FTC, TBA, TPC and the DPPH assays were employed. The tested extracts showed strong antioxidant potential with percent inhibition of 98.2 per cent in the FTC, and 93.8 per cent in TBA assays. The TPC of the extract was 102mg GAE/g PE and strong DPPH radical-scavenging activity with IC_{50} values of 13.99µg/ml. A good correlation was observed between total phenolic content and radical-scavenging activities.

Psychotria lasiocarpa Family: Rubiaceae

Ahmad *et al*. (2010) evaluated the antioxidant potential of 22 species of medicinal plants from Malaysian Rubiaceae including *Psychotria lasiocarpa*. FTC, TBA, TPC and the DPPH assays were employed. The tested extracts showed strong antioxidant potential with percent inhibition of 96.9 per cent in the FTC, and 92.8 per cent in TBA assays. The TPC of the extract was 13.9mg GAE/g PE. A good correlation was observed between total phenolic content and radical-scavenging activities.

Psychotria ophirensis I.M.Turner Family: Rubiaceae

Ahmad *et al*. (2010) evaluated the antioxidant potential of 22 species of medicinal plants from Malaysian Rubiaceae including *Psychotria ophirensis*. FTC, TBA, TPC and the DPPH assays were employed. The tested extracts showed good antioxidant potential with percent inhibition of 97.6 per cent in the FTC, and 87.6 per cent in TBA assays. The TPC of the extract was 21.76mg GAE/g PE and strong DPPH radical-scavenging activity with IC_{50} value of 116.22µg/ml. A good correlation was observed between total phenolic content and radical-scavenging activities.

Psychotria rostrata Blume Family: Rubiaceae

Methanol exteracts of seven Malaysian medicinal plants, including *Psychotria rostrata* were screened by Saha *et al*. (2004) for antioxidant and nitric oxide inhibitory activities. Antioxidant activity was measured by using FTC, TBA and DPPH free radical scavenging methods and Greiss assay was used for the measurement of nitric oxide inhibition in LPS and interferon-γ-treated RAW264.7 cells. All the extracts showed strong antioxidant activity comparable to or higher than that of α-tocopherol, BHT and quercetin in FTC and TBA methods. Extracts from *Psychotria rostrata* also inhibited NO production but this was due to their cytotoxic effects upon cells during culture.

Pterocarpus erinaceus Poir. Family: Fabaceae (Papilionaceae)

Karou *et al*. (2005) evaluated the polyphenol content and antioxidant activity of *Pterocarpus erinaceus* by ABTS assay. Polyphenols in the lyophilized extract of leaves was 28.42 per cent, while in bark 40.80 per cent. The antioxidant activity of leaves was 1.88µmol Trolox/µg in the Phosphomolybdenum assay and 8.08 µmol Trolox/µg in ABTS assay, while bark showed 1.89µmol Trolox/µg in the

Phosphomolybdenum assay and 22.20 µmol Trolox/µg in ABTS assay. The total phenolic compounds were highly correlated with the antioxidant activities.

Lagnika *et al.* (2011) evaluated the antioxidant activity of *Pterocarpus erinaceus* using DPPH radical scavenging assay. The extract showed moderate radical scavenging activity with IC_{50} value of 3.37µg/ml while the standard L-Ascorbic acid had IC_{50} value of 1.1µg/ml.

Pterocarpus marsupium Roxb. Family: Fabaceae (Papilionaceae)

Bajpai *et al.* (2005) investigated the phenolic contents and antioxidant activity of some food and medicinal plants including *Pterocarpus marsupium*. The heartwood and leaves were found to have high phenolic contents and antioxidant activity. The TPC of heartwood and leaves of *Pterocarpus marsupium* were 62.6 and 20.6 mg/g GAE, respectively. The antioxidant activity of *Pterocarpus marsupium*, as assayed by auto-oxidation of β-carotene and linoleic acid and expressed as the per cent of inhibition relative to the control were, 78.9 per cent and 52.3 per cent for heart wood and leaves, respectively.

The heartwood of *Pterocarpus marsupium* used in Ayurvedic system of medicine as an antidiabetic was analysed for its phenol, flavonoid and flavonol contents and antioxidant activity. Methanol extract, ethyl acetate extract and isolated marsupin from heartwood of *Pterocarpus marsupium* showed significant antioxidant activity when evaluated by different methods such as DPPH free radical scavenging, reducing power and nitric oxide scavenging tests. IC_{50} values of isolated marsupin for DPPH and nitric acid scavenging activity were found to be 47.51 µg/ml and 22.42 µg/ml respectively. Therefore marsupin present in *Pterocarpus* wood can serve as a good source of antioxidant (Kasar Rahul *et al.*, 2006).

The antioxidant activities of water extract of heartwood of *Pterocarpus marsupium* were evaluated, by Yeole *et al.* (2008) in different systems, *viz.*, radical scavenging activity by DPPH reduction, nitric oxide radical scavenging in sodium nitroprusside/ Griess ragent system. The results indicate that the reducing power of a substance may be indicator of its potential antioxidant activity, but it is not necessarily a linear correlation between these two activities. The extract was found to have different level of antioxidant properties in the models tested. In scavenging DPPH and nitric oxide, its activity was intense (EC_{50}=182.74 and 66.028 µg/ml respectively). The free radical scavenging property may be one of the mechanism by which this drug is effective in several free radical mediated disease conditions.

The methanolic extract of the bark of *Pterocarpus marsupium* was screened by Radhika *et al.* (2010) for antioxidant activity *in vitro*. The *in vitro* antioxidant activity of the bark extract was evaluated using the DPPH assay, and the results were expressed as IC_{50}. Ascorbic acid, used as a standard, had an IC_{50} = 34.0µg/ml, whereas the bark extract of *P. marsupium* had an IC_{50} = 53.0µg/ml.

Pterocarpus mildbraedii Harms. Family: Fabaceae (Papilionaceae)

Odukoya *et al.* (2007) evaluated the antioxidant activity and total phenolic contents of Nigerian green leafy vegetables including *Pterocarpus mildbraedii*.

Pterocarpus santalinoides DC. Family: Fabaceae (Papilionaceae)

Odukoya *et al.* (2007) evaluated the antioxidant activity and total phenolic contents of Nigerian green leafy vegetables including *Pterocarpus santalinoides*.

Pterocarpus santalinus L. Family: Fabaceae (Papilionaceae)

English Name: Red Sanders, Red Sandalwood.

The ethanol extract of *Pterocarpus santalinus* (PS) was evaluated by Narayan *et al.* (2005) for gastroprotection in rats using ibuprofen as the induction model. The extract had the ability to increase the antioxidant enzymes SOD, CAT and GPx when compared with the untreated but induced rats.

Halim and Mishra (2011) evaluated the antioxidant, hypoglycemic, hypolipidemic and nephroprotective effects of the aqueous extract of *Pterocaarpus santalinus* alone and in combination with vitamin E supplementation in streptozotocin-induced diabetic rats. The antioxidant effect of the red sandal wood extract was also evident, as it caused a reduction in malondialdehyde (MDA) in the brain, liver and muscle tissues. The extract also caused a decrease in the formation of lipid peroxidase, estimated by Thiobarbituric Acid Reactive Substance (TBARS) and increased antioxidants, Superoxide Dismutase (SOD), Catalase (CAT), glutathione peroxidase and gluthathione transferase in erythrocytes.

Anti-inflammatory, analgesic and antioxidant activity of methanolic wood extract of *Pterocarpus santalinus* was investigated by Kumar (2011). The methanolic wood extract was found to contain glycosides, essential oils, flavonoids and polyphenolic compounds. The extract showed significant antioxidant activity and the reducing power was found to increase with increasing concentrations. The scavenging effect of the extract on the DPPH radical was 76.6 per cent at a concentration of 500µg/ml as compared to the scavenging effects of BHA at 500µg/ml of 88.4 per cent, respectively.

Pterocarya fraxinifolia (Poir.) Spach. Family: Juglandaceae

Souri *et al.* (2004) evaluated the antioxidative activity of *Pterocarya fraxinifolia* by linoleic acid peroxidation test using 1,3-diethyl-2-thiobarbituric acid as the reagent. Methanolic extract showed 93.99 per cent inhibition at 40µg concentration with IC_{50} value of 2.60µg.

The potential antioxidant activities of *Pterocarya fraxinifolia* bark and leaves were investigated employing six *in vitro* assay systems. IC_{50} for DPPH radical-scavenging activities were 3.89 for leaves and 41.57µg/ml for bark respectively. The leaf extract exhibited a good reducing power at 2.5 and 80µg/ml that comparable with vitamin C. The extracts also showed a weak nitric oxide-scaavenging activity and Fe^{2+} chelating ability. The peroxidation inhibition of extracts exhibited values from 92 ro 93 per cent at 72nd h, almost at the same pattern of vitamin C activity. Based on higher total phenol and flavonoid contents in leaves, higher antioxidant activities were observed in leaf extract (Ebrahimzadeh *et al.*, 2009).

Pterocaulon cordobense Kuntze Family: Asteraceae (Compositae)

Borneo *et al.* (2008) evaluated the antioxidant activity of ethanolic extract of *Pterocaulon cordobense.* The FRAP value was 536.5µmol of Fe(II)/g and DPPH radical scavenging activity was EC_{50} = 258.8.

Pterogyne nitens Family: Fabaceae (Leguminosae)

Santos *et al.* (2009) evaluated the antioxidant activity of extracts from nine plant species belonging to the Brazilian flora acting on the free radicals DPPH and TEMPOL. Results showed that the extracts of *Pterogyne nitens* display moderate antioxidant activities with DPPH scavenging IC_{50} value of 0.159 mg/ml while Hydroxyl scavenging IC_{50} value was 2.063 mg/ml.

Pteropyrum aucheri Jaub.& Spach. Family: Polygonaceae

Souri *et al.* (2004) evaluated the antioxidative activity of *Pteropyrum aucheri* by linoleic acid peroxidation test using 1,3-diethyl-2-thiobarbituric acid as the reagent. Methanolic extract showed 89.82 per cent inhibition at 40µg concentration with IC_{50} value of 0.94µg, which is approximately in the range of α-tocopherol (IC_{50}=0.60µg).

Pterospermum semisagittatum Buch.-Ham. ex Roxb. Family: Sterculiaceae

Kshirsagar and Upadhyay (2009) reported antioxidant activity of the methanol extract of stem. Its ability of scavenging radicals was measured by DPPH reduction spectrophotometric assay. The extract has the same activity profile as that of Curcumin.

Ptychopetalum olacoides Benth. Family: Olacaceae

Antioxidant effect of *Ptychopetalum olacoides* roots ethanol extract were examined *in vivo* by Siqueira *et al.* (2007). The study suggests that the extract contains compounds able to improve the cellular antioxidant network efficacy in the brain, ultimately reducing the brain damage caused by oxidative stress.

Pueraria lobata (Willd.) Ohwi Family: Fabaceae (Papilionaceae)

Cai *et al.* (2004, 2006) reported antioxidant activity in *Pueraria lobata*.

The roots of *Pueraria thomsonii* and *Pueraria lobata* are officially recorded in Chinese Pharmacopoeia under the name Radix Puerariae. However, the aqueous root extract of *Pueraria lobata* showed more potent antioxidant activity than that of *Pueraria thomsonii* (Jiang *et al.*, 2005).

The effect of the complex preparation (called Sheng), made of puerarin (isolated from *Pueraria lobata*) and Danshensu (isolated from the Chinese herb *Salvia miltiorrhiza*) were studied on the acute ischemic myocardinal injury in rats. Shenge exerted significant cardioprotective effects against acute ischemic myocardial injury in rats, likely through its antioxidant and antilipid peroxidation properties, and thus may be an effective medicine for both prophylaxis and treatment of ischemic heart disease (Wu *et al.*, 2007).

A water soluble glucan, PLB-2C, was isolated from the water extract of the root using anion-exchange and gel permeation chromatography by Cui Hengxian *et al.* (2008). *In viro* cell viability assay by MTT method, its sulfated derivative PLB-2CS which was substituted at 2-O,3-O, 4-O positions, at 0.1, 1 and 5 mg/ml, could attenuate PC12 cell damage significantly caused by hydrogen peroxide.

Antioxidant capacities of 56 selected Chinese medicinal plants, including *Pueraria lobata* were evaluated by Song *et al.* (2010) using the TEAC and FRAP assays, and their total phenolic content was measured by the Folin-Ciocalteu method. *Pueraria lobata* showed moderate antioxidant activity with TEAC value of 8.51µmol Trolox/g, FRAP value of 13.87µmol Fe^{2+}/g and phenolic content was 3.11mg GAE/g.

Pueraria mirifica Airy Shaw et Suvatab. Family: Fabaceae (Papilionaceae)

Chanwitheesuk *et al*. (2005) investigated the antioxidant activity of *Pueraria mirifica* using a β-carotene bleaching method. The contents of plant chemicals such as vitamin C, Vitamin E, carotenoids, tannin and total phenolics, were also determined. Methanolic extract showed antioxidant activity. *Pueraria mirifica* has antioxidant index of 6.54, Vitamin C 5.63 mg per cent, Vitamin E 0.0011mg per cent, Total carotenes 0.63 mg per cent, Total xanthophylls 0.52mg per cent, Tannins 1.18mg per cent and Total phenolics of 15.8mg per cent.

Phansawan and Poungbangpho (2007) reported antioxidant capacities of *Pueraria mirifica* by ABTS method. Ethanol was the best solvent for extraction to give the highest antioxidant capacity, followed by acetone, methanol, distilled water and acetic acid.

Pueraria thomsonii Benth. Family: Fabaceae (Papilionaceae)

The roots of *Pueraria thomsonii* and *Pueraria lobata* are officially recorded in Chinese Pharmacopoeia under the name Radix Puerariae. However, the aqueous root extract of *Pueraria lobata* showed more potent antioxidant activity than that of *Pueraria thomsonii* (Jiang *et al.*, 2005).

Pulicaria crispa (Forssk.) Benth. Family: Asteraceae (Compositae)

Marwah *et al*. (2007) investigated the antioxidant activity of wound-healing herbs using *in vitro* DPPH and phosphomolybdenum assay methods. Out of the 19 plants screened, the aqueous ethanol extract of *Cordia perrottettii* showed the best DPPH scavenging activities with per cent IP value 92.2 per cent, after 15 min of incubation at an effective test concentration of 50µg/ml. The highest Total Antioxidant Capacity as gallic acid equivalent of 1790g/g of ethanol extract was obtained in the phosphomolybdenum assay. Total phenolic content was 96.6 mg GAE/g of ethanol extract.

Pulmonaria officinalis L. Family: Boraginaceae

Ivanova *et al*. (2005) reported high phenolics content and antioxidant properties in *Pulmonaria oficinalis*.

Pulsatilla chinensis (Bge.) Regel. Family: Ranunculaceae

Cai *et al*. (2004) evaluated the antioxidant activity and phenolic compounds in traditional Chinese medicinal plants associated with anticancer. The improved ABTS.+ method was used to systematically assess the total antioxidant capacity (Trolox equivalent antioxidant capacity, TEAC) of the medicinal extracts. The TEAC values and total phenolic content for methanolic extract of root of *Pulsatilla chinensis* were 319.3 µmol Trolox equivalent/100g dry weight (DW), and 1.77 g of gallic acid equivalent/100g DW respectively. A positive significant linear relationship between antioxidant activity and total phenolic content showed that phenolic compounds were the dominant antioxidant components in the tested herbs. Major types of phenolic compounds in *Pulsatilla chinensis* include lignans (+)-pinoresinol).

Li *et al.* (2008) evaluated the antioxidant properties of 45 medicinal plants using FRAP and TEAC assays and the total phenolic contents of these plants were measured. *Pulsatilla chinensis* showed 67.81 μmol Fe(II)/g FRAP value, 33.34μmol Trolox/g of TEAC value and 9.72 mg GAE/g of Phenolic content.

Punica granatum L. Family: Punicaceae

English Name: Pomegranate.

Antioxidant-rich fractions were extracted from pomegranate peels and seeds using ethyl acetate, methanol and water by Singh *et al.* (2002). The extracts were screened for their potential as antioxidants using various *in vitro* models such as β-carotene-linoleate and DPPH model systems. The methanol extract of peels showed 83 and 81 per cent antioxidant activity at 50 ppm using the - β-carotene-linoleate and DPPH model systems. Similarly, the methanol extract of seeds showed 22.6 and 23.2 per cent antioxidant activity at 100ppm using the β-carotene-linoleate and DPPH model systems, respectively. As the methanol extract of pomegranate peel showed the highest antioxidant activity among all the extracts, it was selected for testing of its effect on lipid peroxidation, hydroxyl radical scavenging activity, and human low-density lipoprotein (LDL) oxidation. The methanol extract showed 56, 58 and 93.7 per cent inhibition using the thiobarbituric acid method, hydroxyl radical scavenging activity and LDL oxidation respectively, at 100 ppm.

Cai *et al.* (2004) evaluated the antioxidant activity and phenolic compounds in traditional Chinese medicinal plants associated with anticancer. The improved ABTS.+ method was used to systematically assess the total antioxidant capacity (Trolox equivalent antioxidant capacity, TEAC) of the medicinal extracts. The TEAC values and total phenolic content for methanolic extract of peel of *Punica granatum* were 6240.2 μmol Trolox equivalent/100g dry weight (DW), and 22.56 g of gallic acid equivalent/100g DW respectively. A positive significant linear relationship between antioxidant activity and total phenolic content showed that phenolic compounds were the dominant antioxidant components in the tested herbs. Major types of phenolic compounds in *Punica granatum* include tannins (ellagitannins), flavonols (quercetin), phenolic acids (gallic acid, chlorogenic acid).

Wang *et al.* (2004) isolated two new compounds, coniferyl 9-O-β-D-apiofuranosyl (1 to 6-O- β-D-glucopyranoside (1) and sinapyl 9-O- β-D-apiofuranosyl (1 to 6)-O- β-D-glucopyranoside (2) from the seeds of *Punica granatum*, together with five known compounds, 3,3-di-O-methylellagic acid (3), 3,3',4'-tri-O-methylellagic acid (4), phenethylrutinoside, icariside D1, and daucosterol. Compounds 1-4 exhibited antioxidant activity, which was evaluated by measurement of low-density lipoprotein (LDL) susceptibility to oxidation and by determination *in vitro* of malondialdehyde (MDA) levels in the rat brain (Wang *et al.*, 2004).

Ricci *et al.* (2006) investigated the antioxidant activity of *Punica granatum* fruits. Arils, juice and rinds of fruits and their aqueous and ethy acetate extracts displayed good antioxidant activity. Chemical and antioxidant properties of pomegranate cultivars grown in Mediterranean region of Turkey were investigated by Özgen *et al.* (2008).

Al-Mustafa and Al-Thunibat (2008) evaluated the antioxidant activity of *Punica granatum* fruit peel. The level of antioxidant activity was determined by DPPH and ABTS assay in relation to total phenolic contents of the medicinally used part. They concluded that *Punica granatum* fruit peel possessed high antioxidant capacity (DPPH-TEAC> or = 80mg/g).

Bagri *et al*. (2009) investigated the effects of pomegranate flowers aqueous extract (PgAq) on streptozotocin induced diabetic rats by measuring fasting blood glucose, lipid profiles (atherogenic index), lipid peroxidation (LPO) and activities of both non-enzymatic and enzymatic antioxidants. Diabetes was induced by single intraperitoneal injection of STZ (60mg/kg) to albino Wistar rats. The increase in blood glucose level, total cholesterol (TC), triglycerides (TG), low density lipoprotein cholesterol (LDL-C), very low density lipoprotein (VLDL), LPO level with decrease in high density lipoprotein cholesterol (HDL-C), reduced glutathione (GSH) content and antioxidant enzymes namely glutathione peroxidase, glutathione reductase, glutathione transferase, superoxide dismutase (SOD) and catalase were the salient features observed in diabetic rats. Oral administration of PgAq at doses of 250mg/kg and 500mg/kg for 21 days resulted in a significant reduction in fasting blood glucose, TC, TG, LDL-C, VLDL-C and tissue LPO levels coupled with elevation of HDL-C, GSH content and antioxidant enzymes in comparison with diabetic control group. The results suggest that pomegranate could be used, as a dietary supplement, in the treatment of chronic diseases characterized by atherogenous lipoprotein profile, aggravated antioxidant status and impaired glucose metabolism and also in their prevention.

Ismail *et al*. (2009) studied the hepatoprotective role and antioxidant capacity of pomegranate flowers. The study revealed that constituents present in the pomegranate flowers impart protection against carcinogenic chemical induced oxidative injury that may result in development of cancer during the period of a 52-day protective exposure.

Hydromethanol extracts of 15 Bangladeshi medicinal plants, including *Punica granatum*, were evaluated by Hasan *et al*. (2009) for antioxidant potential using DPPH radical scavenging assay. *Punica granatum* fruit peel exhibited radical scavenging activity with an IC$_{50}$ value of 10.82 µg/ml compared to the IC$_{50}$ value of 5.15µg/ml as shown by the reference antioxidant ascorbic acid, in a dose dependent fashion.

Sundararaju *et al*. (2010) screened the *Punica granatum* fruit rind extracts for antioxidant activity. All the extracts except hexane extract have shown more potent antioxidant activity compared to BHT, BHA, Vitamin C and E.

Antioxidant, metal chelating and ferric ion reducing and antiglycation property was checked by Kokila *et al*. (2010) for polysaccharide isolated from pomegranate fruit juice. At 20µg/ml sugar concentration, it gave 87 per cent and 85 per cent inhibition in hydroxyl radical scavenging assay and antiglycation assay respectively. It also had metal chelating and ferric ion reducing activity. Antioxidant-rich fractions were extracted from pomegranate peels and seeds using ethyl acetate, methanol and water by Singh *et al*. (2002). The extracts were screened for their potential as antioxidants using various *in vitro* models such as β-carotene-linoleate and DPPH model systems. The methanol extract of peels showed 83 and 81 per cent antioxidant activity at 50 ppm using the - β-carotene-linoleate and DPPH model systems. Similarly, the methanol extract of seeds showed 22.6 and 23.2 per cent antioxidant activity at 100ppm using the β-carotene-linoleate and DPPH model systems, respectively. As the methanol extract of pomegranate peel showed the highest antioxidant activity among all the extracts, it was selected for testing of its effect on lipid peroxidation, hydroxyl radical scavenging activity, and human low-density lipoprotein (LDL) oxidation. The methanol extract showed 56, 58 and 93.7 per cent inhibition using the thiobarbituric acid method, hydroxyl radical scavenging activity and LDL oxidation respectively, at 100 ppm.

Niwano *et al.* (2011) summarized their research for herbal extracts with potent antioxidant activity obtained from a large scale screening based on superoxide radical. Experiments with the Fenton reaction and photolysis of H_2O_2 induced by UV irradiation demonstrated that extract of *Punica granatum* (peel) have potent ability to directly scavenge OH radicals. Furthermore, the scavenging activities against O_2 and OH of extracts of *Punica granatum* (peel) proved to be heat resistent.

The chloroform and methanolic extracts of 124 Egyptian plant species belonging to 56 families were investigated by Moussa *et al.* (2011) and compared their antioxidant activity by DPPH scavenging assay. Among the 124 plant species tested 18 exhibited extremely high antiradical activity (>80 per cent inhibition). The IC_{50} value of DPPH radical scavenging of *Punica granatum* was 18.68µg/ml, while total phenolic and flavonoid contents were 54.36 mg Tannic acid equivalent/g extract and 242.26 mg Rutin equivalent/g extract, respectively. Correlation coefficient between DPPH radical scavenging activity and the total phenolic and flavonoid contents suggest that phenolics and flavonoids in the extracts were partly responsible for the antiradical activities.

Putoria calabrica Pers. Family: Rubiaceae

Tasdemir *et al.* (2004) reported antioxidant activity of the root extracts.

Pyracantha koidzumii (Hayata) Rehder Family: Fagaceae

Hou *et al.* (2003) evaluated the antioxidant activity of 70 per cent aqueous acetone extract of *Pyracantha koidzumii* by various assays including DPPH, hydroxyl radicals and reducing power assay. The extract exhibited stronger activity against DPPH radicals with IC_{50} value of 21.2 µg/ml and inhibited formation of OH generated in the Fenton reaction system with IC_{50} value of 0.5 µg/ml). The amount of phenolics was 2.6 mg of GAE/g.

Pyrola calliantha (Choisy) Ohwi Family: Pyrolaceae

Gan *et al.* (2010) evaluated the antioxidant activity of *Pyrola calliantha* using the FRAP and TEAC assays, and its total phenolic content was measured by the Folin-Ciocalteu method. For FRAP assay the value was 160.96 µmol Fe(II)/g dry weight. For TEAC assay, the value was 115.77µmol Trolox/g dry weight. Total phenolic content was 9.31 mg gallic acid equivalent/g dry weight of plant material.

Pyrus communis L. Family: Rosaceae

English Name: Pear.

The total antioxidant activity of pear was measured by Wang *et al.* (1996) using automated oxygen radical absorbance capacity (ORAC) assay. On the basis of the wet weight of the fruits (edible portion), pear had moderate ORAC activity (1.34 micromoles of Trolox equivalents per gram). On the basis of dry weight of the fruits pear had the highest ORAC activity (9.6μmoles/g Trolox equivalent). The contribution of the fruit pulp fraction (extracted with acetone) to the total ORAC activity of a fruit was usually less than 10 per cent.

Cai *et al.* (2004) evaluated the antioxidant activity and phenolic compounds in traditional Chinese medicinal plants and fruits associated with anticancer. The improved ABTS.+ method was used to systematically assess the total antioxidant capacity (TEAC) of the medicinal extracts. The TEAC values and total phenolic content for methanolic extract of fruit of *Pyrus communis* were 21.5 μmol Trolox equivalent/100g dry weight (DW), and 0.12 g of gallic acid equivalent/100g DW respectively.

Pyrus elaeagnifolia Pall. subsp. *elaeagnifolia* Family: Rosaceae

Serteser *et al.* (2009) studied the antioxidant activity of 50 per cent aqueous methanol extracts of fruits by various antioxidant assays, including free radical scavenging, hydrogen peroxide scavenging and metal (Fe^{2+}) chelating activities. The DPPH radical scavenging effect of fruit extract was 0.661. Fe^{2+} chelating activity (per cent) of fruit extract was 41.83 and Hydrogen peroxide inhibition activity of fruit extract was 54.23 per cent.

Pyrus elaeagnifolia Pall. subsp. *kotschyana* (Boiss.) Browicz Family: Rosaceae

Serteser *et al.* (2009) studied the antioxidant activity of 50 per cent aqueous methanol extracts of fruits by various antioxidant assays, including free radical scavenging, hydrogen peroxide scavenging and metal (Fe^{2+}) chelating activities. The DPPH radical scavenging effect of fruit extract was 0.682. Fe^{2+} chelating activity (per cent) of fruit extract was 43.74 and Hydrogen peroxide inhibition activity of fruit extract was 55.28 per cent.

Pyrus taiwanensis Iketani et Oashi Family: Rosaceae

Hou *et al.* (2003) evaluated the antioxidant activity of 70 per cent aqueous acetone extract of *Pyrus taiwanensis* by various assays including DPPH, hydroxyl radicals and reducing power assay. The extract exhibited moderate activity against DPPH radicals with IC_{50} value of 13.6 μg/ml and inhibited formation of OH generated in the Fenton reaction system with IC_{50} value of 0.6 μg/ml). The amount of phenolics was 2.6 mg of GAE/g.

Quercus alba L. Family: Fagaceae

English Name: White willow.

McCune and Johns (2002) evaluated antioxidant activity in medicinal plants associated with the symptoms of diabetes mellitus used by the indigenous people of the north American boreal forest. They evaluated antioxidant activities by three different assays (DPPH, NBT/xanthine oxidase and DCF/APPH). *Quercus alba* showed antioxidant activity similar to vitamin C, and green tea in all the assays.

Quercus aliena Blume Family: Fagaceae

Jin *et al.* (2005) investigated the protective effect of *Quercus aliena* acorn extracts against CCl_4-induced hepatotoxicity in rats. Aqueous extracts of *Quercus aliena* acorn had higher superoxide radical scavenging activity than other types of extracts. The *Quercus aliena* acorn extracts displayed dose-dependent superoxide radical scavenging activity (IC_{50} = 4.92 µg/ml), as assayed by the electron spin resonance (ESR) spin-trapping technique. Pretreatment with *Quercus aliena* acorn extracts reduced the increase in serum aspartate aminotransferase (AST) and serum alanine aminotransferase (ALT) levels. The aqueous extracts reversed CCl_4-induced liver injury and had an antioxidant action in assays of $FeCl^2$- ascorbic acid induced lipid peroxidation in rats. Expression of cytochrome P450 2E1 (CYP2E1) mRNA, as measured by RT-PCR, was significantly decreased in the livers of *Quercus aliena* acorn-pretreated rats compared with the livers of the control group.

Quercus calliprinos Webb. Family: Fagaceae

Al-Mustafa and Al-Thunibat (2008) evaluated the antioxidant activity of Jordanian medicinal plant *Quercus calliprinos*. The level of antioxidant activity was determined by DPPH and ABTS assay in relation to total phenolic contents of the medicinally used part. They concluded that *Quercus calliprinos* leaves and fruit have high antioxidant capacity (DPPH-TEAC> or = 80mg/g).

Quercus eduardii Trel Family: Fagaceae

English Name: Red Oak

Rivas-Arreola *et al.* (2010) evaluated the antioxidant capacity and cardioprotective potential of leaves infusions and partially purified fractions of *Quercus sideroxyla* and *Q. eduardii* (red oaks) and *Q. resinosa* (white oak). Infusions from Oak leaves were obtained and probed for total phenolics by Folin-Ciocalteu, DPPH and hydroxyl radicals scavenging by DPPH test and Deoxy-D-ribose method, the antioxidant capacity was evaluated by FRAP and ORAC tests, inhibitions of Low Density Lipoproteins (LDL) oxidation and Angiotensin Converting Enzyme (ACE) activity were measured. Bioactive polyphenols such as gallic and ellagic acids, catechin, quercetin and derivatives: naringenin and naringin were detected in *Quercus* infusions. A distinctive HPLC profile was observed among the red and white oak samples. *Q. resinosa* infusions have exhibited the highest antioxidant activity in comparison with the other species, although in the inhibition of LDL oxidation no differences were observed. In the inhibition of the ACE, *Q. resinosa* was more effective (IC_{50}, 18 ppm) than *Q. sideroxyla*, showing same effect as the control Captopril. From the results it is possible to postulate that not only chelating activity is important in these infusions, especially in *Q. resinosa*.

Quercus infectoria Oliver Family: Fagaceae

Kaur *et al.* (2008) evaluated the antioxidant activity of ethanolic extract of *Quercus infectoria* galls employing several established *in vitro* model systems. Their protective efficacy on oxidative modulation of murine macrophages was also explored. Gall extract was found to contain a large amount of polyphenols and possess a potent reducing power. HPTLC analysis of the extract suggested it to contain 19.925 per cent tannic acid (TA) and 8.75 per cent gallic acid (GA). The extract potently scavenged free radicals including DPPH (IC_{50}~0.5 µg/ml), ABTS (IC_{50}~1 µg/ml), hydrogen peroxide (H_2O_2) (IC_{50}~2.6 µg/ml) and hydroxyl (OH) radicals (IC_{50}~6 µg/ml). Gall extract also chelated metal ions and inhibited Fe^{3+}–ascorbate-induced oxidation of protein and peroxidation of lipids. Exposure of rat peritoneal macrophages to tertiary butyl hydroperoxide (*t*BOOH) induced oxidative stress in them and altered their phagocytic functions. These macrophages showed elevated secretion of lysosomal hydrolases, and attenuated phagocytosis and respiratory burst. Activity of macrophage mannose receptor (MR) also diminished following oxidant exposure. Pretreatment of macrophages with gall extract preserved antioxidant armory near to control values and significantly protected against all the investigated functional mutilations. MTT assay revealed gall extract to enhance percent survival of *t*BOOH exposed macrophages.

Quercus macrocarpa Michx Family: Fagaceae

He *et al.* (2011) reported the isolation of 12 known compounds along with their antioxidant activity from *Quercus macrocarpa*.

Quercus pubescens Willd. Family: Fagaceae

Demo *et al.* (1998) reported antioxidant α-tocopherol from the Hexane extract of *Quercus pubescens*.

Quercus resinosa Liebm. Family: Fagaceae

Rivas-Arreola *et al.* (2010) evaluated the antioxidant capacity and cardioprotective potential of leaves infusions and partially purified fractions of *Quercus sideroxyla* and *Q. eduardii* (red oaks) and *Q. resinosa* (white oak). For details refer *Quercus eduardii* mentioned above.

Quercus salicina Bl. Family: Fagaceae

Cytoprotective effect on gamma-ray radiation induced oxidative stress from forty one Korean plant extracts was screened by Jung *et al.* (2006). *Quercus salicina* was found to scavenge DPPH radical and intracellular reactive oxygen species. As a result, extract of this plant reduced cell death of Chinese hamster lung fibroblast cells induced by H_2O_2 treatment. In addition, these extracts scavenged ROS generated by radiation.

Five phenolic compounds – D-threo-guaiacylglycerol 8-O-beta-D-(6'-O-galloyl) glucopyranoside (1), 9-methoxy-D-threo-guaiacylglycerol 8-O-beta-D-(6'-O-galloyl) glucopyranoside (2), 6"-O-galloyl salidroside (3), methyl gallate (4), quercetin (5) were isolated from the MeOH extract from the stem of *Quercus salicina* (Kim *et al.*, 2008). They measured radical scavenging activity with the DPPH method and anti-lipid peroxidative efficacy on human LDL with TBARS assay, with the result that all these compounds exhibited the antioxidative activity.

Quercus sideroxyla Humb. et Bonpl. Family: Fagaceae

Rivas-Arreola *et al.* (2010) evaluated the antioxidant capacity and cardioprotective potential of leaves infusions and partially purified fractions of

Quercus sideroxyla and *Q. eduardii* (red oaks) and *Q. resinosa* (white oak). For details refer *Quercus eduardii* mentioned above.

Quisqualis indica L. Family: Combretaceae

Leelapornpisid *et al.* (2007) reported antioxidant activity of absolutes (Obtained by solvent extraction) of *Quisqualis indica*.

Rabdosia japonica Hara Family: Lamiaceae (Labiatae)

The antioxidant activities of phenolic compounds pedalitin, quercetin, rutin, isoquercetin and rosmarinic acid, isolated from the leaves of *Rabdosia japonica* were elucidated by Masuoka *et al.* (2006). All the phenolics tested exhibited superoxide scavenging activity and pedalin showed the most potent antioxidant activity. Pedaletin prevented generation of superoxide radicals in part by inhibiting xanthin oxidase competitively. Both pedaletin and quercetin inhibited the formation by xanthine oxidase and the inhibition kinetics analysed by Lineweaver-Burk plots found both flavonoids to be common inhibitors. On the other hand, isoquercetin, rutin and rosmarinic acid were effective in scavenging superoxide radicals generated in xanthine-xanthine oxidase system without inhibiting the enzyme.

Randia dumetorum Lamk. Family: Rubiaceae

Dharmishta *et al.* (2010) investigated the antioxidant potential of Aqueous extract of *Randia dumetorum* fruits using different models *viz.* DPPH radical scavenging, iron chelating activity and nitric oxide scavenging assay. The per cent scavenging activity at different concentrations was determined and the IC^{50} value of the extracts was compared with that of standard, ascorbic acid. Its antioxidant activity was estimated by IC^{50} value and the values are 45.02 µg/ml (reducing power assay), 66 µg/ml, (DPPH scavenging assay) and 79.09 µg/ml (nitric oxide scavenging assay). In all the testing, a significant correlation existed between concentrations of the extract and percentage inhibition of free radicals, metal chelation. According to these results, it may be hypothesized that antioxidant effect of fruits could be related to presence of polyphenolic compounds.

Randia echinocarpa Moc. et Sesse ex DC. Family: Rubiaceeae

Santos-cervantes *et al.* (2007) reported the antioxidant and antimutagenic activities of fractions from *Randia echinocarpa* fruit. The pulp of the fruit was sequentially extracted with solvents of different polarity (*i.e.* hexane, chloroform, methanol and water). A high extraction yield was obtained with methanol (72.17 per cent d.w.). The aqueous extract showed the highest content of phenolics (2.27 mg/g as ferulic acid equivalents) and the highest antioxidant activity based on the β-carotene bleaching method (486.15). The commercial antioxidant BHT was used as control (835.05).

Randia hebecarpa Benth. Family: Rubiaceae

The methanol extract of *Randia hebecarpa* leaves was evaluated by Nazari *et al.* (2006) for anti-inflammatory activity using carragenan- and dextran-induced rat paw edema models. Antioxidant activities of the methanol extract and of the fractions resulting from its partition were also measured using the 1,1-diphenyl-1-

picrylhydrazyl (DPPH) free radical scavenging assay and the linoleic acid peroxidation method. The methanol extract, ethyl acetate fraction, and hydromethanol fraction exhibited percent inhibition of lipid peroxidation comparable to that of commercial antioxidant butylated hydroxytoluene (BHT). Fractionation of the ethyl acetate and hydromethanol fractions through chromatographic methods yielded kaempferol-3, 7-*O*.-α.-L-dirhamnoside, kaempferol-3-*O*.-β.-D-galactoside, quercetin-3-*O*.-β.-D-galactoside, myricetin-3-*O*.-α.-L-rhamnoside, kaempferol-3-*O*.-α.-L-rhamnosyl-(1→6)-β.-D-galactosyl-7-*O*.-α.-L-rhamnoside, quinovic acid-3-*O*.-β.-D-quinovopyranosil-28-*O*.-β.-D-glucopyranoside, cincholic acid 3-*O*.-β.-D-quinovopyranosil-28-*O*.-β.-D-glucopyranoside, and the sugar D-mannitol.

Ranunculus marginatus d'Urv. var. *trachycarpus* (Fisch. and Mey) Azn. Family: Ranunculaceae

Hexane, ethyl acetate, methanol and aqueous extracts of *Ranunculus marginatus* var. *trachycarpus* were tested *in vitro* for their antioxidiant activities by DPPH radical scavenging and TEAC assays. Methanol extract showed the highest antioxidant activity in both assays. The results obtained in the antioxidant activity tests were in positive correlation with the total phenolic and flavonoid contents of the extracts.

Ranunculus repens L. Family: Ranunculaceae

English Name: Creeping buttercup.

The antioxidative activity of phenolic extracts from herb of creeping buttercup was examined by autoxidation of methyl linoleate by Kahkonen *et al.* (1999). The total phenolics were low (4.1 mg/g GAE) while the antioxidant activity was 46 per cent at 500 ppm level and 86 per cent at 5000 ppm level.

Ranunculus sprunerianus Boiss. Family: Ranunculaceae

Hexane, ethyl acetate, methanol and aqueous extracts of *Ranunculus sprunerianus* were tested *in vitro* for their antioxidiant activities by DPPH radical scavenging and TEAC assays. Methanol extract showed the highest antioxidant activity in both assays. The results obtained in the antioxidant activity tests were in positive correlation with the total phenolic and flavonoid contents of the extracts.

Raphanus sativus L. var. *niger* Family: Brassicaceae (Cruciferae)

English Name: Black radish.

Lugasi *et al.* (2005) in their *in vitro* studies reported that the squeezed juice from black radish root exhibited significant antioxidant properties.

Raphanus sativus L. var. *sativus* Family: Brassicaceae (Cruciferae)

English Name: Radish; Japanese: Hadaikon.

Methanol aqueous extract of *Raphanus sativus* was screened by Kim *et al.* (1997) for anti-oxidative activity using Fenton's reagent/ethyl lineolate system. The extract showed antioxidant activity.

Souri *et al.* (2004) investigated the antioxidant activity of methanol extracts of garlic, used as vegetable in Iranian diet against linoleic acid peroxidation. The extracts showed good antioxidant activity. Radish leaf had exceptionally high antioxidant activity even higher than that of quercetin.

Maeda *et al.* (2006) examined the phenolic contents of Radish and its radical scavenging activities for DPPH radical. The polyphenol content was 45.5 mg-gallic acid equivalent/100 g) and DPPH scavenging activity was 160.1 µmol-Trolox eq./ 100 g).

A study was carried out by Raghu *et al.* (2010) to determine the antioxidant activity of aqueous extracts of ten selected common vegetables. For *Raphanus sativus* the IC_{50} value for DPPH scavenging was 4.75mg/ml while standard was 2.45µg/ml. Reducing power of the extract was moderate.

Rauvolfia sellowii Muell.-Arg. Family: Apocynaceae

Menezes *et al.* (2004) investigated the antioxidant activity of hydroalcoholic extracts of *Rauvolfia sellowii* by DPPH assay and phosphomolybdenum method. The relative activity in phosphomolybdenum complex was 0.65 while scavenging activity against DPPH in IC_{50} (µg/ml) was 30.27.

Rauvolfia vomitoria Afzel. Family: Apocynaceae

Erasto *et al.* (2011) evaluated the antioxidant activity of roots of *Rauvolfia vomitoria*. The antioxidant activity of the alkaloids and ethanolic extracts was determined using DPPH and reducing capacity assays. The alkaloids and ethanolic extracts of *R. vomitoria* exhibited high antioxidant activity.

Rehmannia glutinosa Libosch. Family: Scrophulariaceae

Cai *et al.* (2004) evaluated the antioxidant activity and phenolic compounds in traditional Chinese medicinal plants associated with anticancer. The improved ABTS.+ method was used to systematically assess the total antioxidant capacity (Trolox equivalent antioxidant capacity, TEAC) of the medicinal extracts. The TEAC values and total phenolic content for methanolic extract of root of *Rehmannia glutinosa* were 74.5 µmol Trolox equivalent/100g dry weight (DW), and 0.53 g of gallic acid equivalent/100g DW respectively. A positive significant linear relationship between antioxidant activity and total phenolic content showed that phenolic compounds were the dominant antioxidant components in the tested herbs. Major types of phenolic compounds in *Rehmannia glutinosa* include phenolic acids (ferulic acid, caffeic acid).

Li *et al.* (2008) evaluated the antioxidant properties of 45 medicinal plants using FRAP and TEAC assays and the total phenolic contents of these plants were measured. *Rehmannia glutinosa* showed low antioxidant activity with 18.02 µmol Fe(II)/g FRAP value, 10.74µmol Trolox/g of TEAC value and 4.83 mg GAE/g of Phenolic content.

Remusatia vivipara Schott Family: Araceae

Marwah *et al.* (2007) investigated the antioxidant activity of wound-healing herbs using *in vitro* DPPH and phosphomolybdenum assay methods. Out of the 19 plants screened, the aqueous ethanol extract of *Remusatia vivipara* showed very low DPPH scavenging activities with per cent IP value 8.4 per cent, after 15 min of incubation at an effective test concentration of 50µg/ml. The Total Antioxidant

Capacity as gallic acid equivalent of 161mg/g of ethanol extract was obtained in the phosphomolybdenum assay. Total phenolic content was 14.9 mg GAE/g of ethanol extract.

Rennellia elliptica Korth. Family: Rubiaceae

Ahmad *et al.* (2010) evaluated the antioxidant potential of 22 species of medicinal plants from Malaysian Rubiaceae including *Rennellia elliptica*. FTC, TBA, TPC and the DPPH assays were employed. The tested extracts showed strong antioxidant potential with percent inhibition of 97.8 per cent in the FTC, and 96.8 per cent in TBA assays. The TPC of the extract was 19.48mg GAE/g PE and strong DPPH radical-scavenging activity with IC_{50} value of 227.42µg/ml. A good correlation was observed between total phenolic content and radical-scavenging activities.

Retama raetam Webb. et Berth. subsp. *gussonei* Family: Fabaceae (Leguminosae)

The antioxidant activity of methanol extract of *Retama raetam* subsp. *gussonei* was assessed by Conforti *et al.* (2004) by means of two different tests: (1) bleaching of the stable DPPH radical and (2) lipid peroxidation of liposomes which were prepared from bovine brain extract. In both tests used leaves extracts showed a significant antioxidant effect.

Flower oils of *Retama raetum* cultivated in Tunisia presented good antioxidant activity, as measured through the DPPH method where IC_{50} value was fortyfold superior to that of the reference BHT. The apparent good activity reported by the authors could be attributed to the relative high percentage of monoterpenes present in the essential oils (Edziri *et al.*, 2010).

Rhamnus cornifolia Boiss.& Hohen. subsp. *cornifolia* Family: Rhamnaceae

Souri *et al.* (2004) evaluated the antioxidative activity of *Rhamnus cornifolia* by linoleic acid peroxidation test using 1,3-diethyl-2-thiobarbituric acid as the reagent. Methanolic extract showed 88.40 per cent inhibition at 40µg concentration with IC_{50} value of 2.27µg.

Rhamnus lycioides L. Family: Rhamnaceae

Demo *et al.* (1998) reported antioxidant α-tocopherol from the Hexane extract of *Rhamnus lycioides*.

Rhaphidophora pertusa Schott. Family: Araceae

Sasikumar and Doss (2006) analyzed the methanol extract of stem for its antioxidant (DPPH, reducing power and Fe^{3+} metal chelation methods) activity. The extract was found effective against the three antioxidant test models.

Linnert *et al.* (2010) evaluated anti-inflammatory, analgesic and anti-lipid peroxidative effects of aerial parts of *Rhaphidophora pertusa*. The ethanol extract showed significant anti-lipid peroxidant effects *in vitro* in rat liver homogenate. There was a significant increase of malondialdehyde (MDA) in $FeCl_2$-AA treated rat liver homogenate, compared to normal control.

Rhaponticum carthamoides (Willd.) Iljin Family: Asteraceae

Miliauskas *et al.* (2004) investigated the radical scavenging activity of acetone, ethyl acetate and methanol extracts of leaves and stems at growing stage using DPPH and ABTS assays. They found that Methanol extracts were most effective DPPH radical scavengers. Ethyl acetate and acetone extracts were considerably less effective radical scavengers. High content of phenolics in the extracts correlates with its antiradical activity, confirming that phenolic compounds are likely to contribute to the radical scavenging activity of this plant extracts. Miliuskas *et al.* (2005) tested extracts of *Rhaponticum carthamoides* against DPPH radical and was found to be weaker than that of the reference antioxidants rosmarinic acid and Trolox.

Rhaponticum uniflorum (L.) DC. Family: Asclepiadaceae

Li *et al.* (2008) evaluated the antioxidant properties of 45 medicinal plants using FRAP and TEAC assays and the total phenolic contents of these plants were measured. *Rhaponticum uniflorum* showed 26.27 µmol Fe(II)/g FRAP value, 19.92µmol Trolox/g of TEAC value and 3.81 mg GAE/g of Phenolic content.

Rhazya stricta Decaisne Family: Apocynaceae

Iqbal *et al.* (2006) evaluated the antioxidant activity of water, 80 per cent methanol, 70 per cent ethanol, and diethyl ether extract of *Rhazya stricta*. The methanolic extract exhibited the highest total phenolic content among the extracts. The antioxidant potential of methanolic extracts of *Rhazya stricta* leaves was comparable with previously exploited potent antioxidants and is strongly concentration dependent.

Rheum emodi Wall. ex Meissn. Family: Polygonaceae

A sulfated emodin glucoside, emodin 8-O-β-D-glucopyranosyl-6-I-sulfate (1), was isolated from the roots of *Rheum emodi* in an investigation of the active constituents of this Nepalese medicinal plant. Additionally, two rare auronols, carpusin (2) and maesopsin (3), besides other anthraquinones and phenolics, were isolated and identified. Compounds 2 and 3 showed significant antioxidant activity in the DPPH assay, while chrysophanol physcion, and emodin and their 8-O-glucoside were found to be inactive (Kren *et al.*, 2003).

Rajkumar *et al.* (2010a) determined antioxidant and cytotoxic efficacies of methanolic and aqueous extracts of *Rheum emodi* rhizome. DPPH and hydroxyl radical scavenging activities, inhibitory effect on lipid peroxidation and Fe^{3+} reducing antioxidant property have been used to investigate antioxidant properties of the extracts. The aqueous extract, though inferior to methanolic extrct failed to protect the DNA. The methanolic extract contained higher polyphenolic contents than aqueous extract. Significant positive correlations were observed between results of phenolic content estimation and that of antioxidant assays.

Rheum officinale Baill. Family: Polygonaceae

Cai *et al.* (2004a,b, 2006) evaluated the antioxidant activity and phenolic compounds in traditional Chinese medicinal plants associated with anticancer. The improved ABTS.+ method was used to systematically assess the total antioxidant capacity (Trolox equivalent antioxidant capacity, TEAC) of the medicinal extracts.

The TEAC values and total phenolic content for methanolic extract of root of *Rheum officinale* were 2108.6 µmol Trolox equivalent/100g dry weight (DW), and 8.37 g of gallic acid equivalent/100g DW respectively. A positive significant linear relationship between antioxidant activity and total phenolic content showed that phenolic compounds were the dominant antioxidant components in the tested herbs. Major types of phenolic compounds in *Rheum officinale* include anthraquinones (chrysophanol, emodin, rhein, physcion and their glycosides), phenolic acids (gallic acid), hydrolysable tannins.

The aqueous extracts of 13 oriental medicinal plants were examined for their reducing power, scavenging ability towards superoxide and hydroxyl radicals and their inhibitory effect on lipid peroxidation. All extracts tested were found to be highly active on scavenging of superoxide radicals (Nam and Kang, 2004).

Rheum palmatum L. Family: Polygonaceae

Methanol aqueous extract of *Rheum palmatum* was screened by Kim *et al.* (1997) for anti-oxidative activity using Fenton's reagent/ethyl lineolate system. The extract showed antioxidant activity.

Wojcikowski *et al.* (2007) evaluated the antioxidant activity of aerial parts of *Rheum palmatum*. Oxygen radical absorbance capacity of the ethyl extract, methanol extract and aqueous extracts were 679.63, 336.98 and 178.68µmol Trolox equivalent per g of dried starting material, respectively.

Rheum ribes L. Family: Polygonaceae

English Name: Rhubarb.

The antioxidant activity of chloroform and methanol extract of roots and stems of *Rheum ribes* was studied by Öztürk *et al.* (2007). The antioxidant potential of both extracts of roots and stems were evaluated using different antioxidant tests, namely total antioxidant (lipid peroxidation inhibition activity), DPPH radical scavenging, superoxide anion radical scavenging, ferric reducing power, and cupric reducing power and metal chelating activities. Total antioxidant activity was also measured according to the β-carotene bleaching method, and all four extracts exhibited stronger activity than known standards, namely BHT and α-tocopherol. Particularly higher activity was exhibited by roots with 93.1 per cent and 84.1 per cent inhibitions of chloroform and methanol extracts, while 82.2 per cent and 82.0 per cent inhibitions by stem extracts, respectively. However, both methanol extracts exhibited higher DPPH radical scavenging activity than the corresponding chloroform extracts, moreover, methanol extract of the stems showed better activity than BHT. In addition, both root extracts showed more potent superoxide anion radical scavenging activity than BHT, and comparable with well known radical scavenger L-ascorbic acid. Except chloroform extract of the roots, the other three extracts exhibited better metal chelating activity than quercetin. Also total phenolic and flavonoid contents in both extracts of the roots and stems of *R. ribes* were determined as pyrocatechol and quercetin equivalents, respectively (Öztürk *et al.*, 2007).

Rhinacanthus nasutus L. Family: Acanthaceae

In vitro antioxidant activity of methanolic extract of *Rhinacanthus nasutus* leaves was determined by DPPH free radical scavenging assay and reducing power assays by Visweswara Rao and Dhananjaya Naidu (2010). Ascorbic acid was used as standard and positive control for both the analysis. The methanolic extract had shown very significant DPPH radical scavenging activity compared to standard antioxidant. In DPPH free radical scavenging assay IC_{50} value of leaves extract was found to be 34.4. µg/ml.

Rhizophora conjugata L. Family: Rhizophoraceae

Vadlapudi and Naidu (2009) evaluated the antioxidant potential of *Rhizophora conjugata*, a mangrove plant. The SOD was 1.31U/mg, Catalase was 0.95U/mg, Ascorbic acid was 70mg/100g, DPPH radical inhibition was 30 per cent and FRAP units were 730.

Rhizophora mangle L. Family: Rhizophoraceae

The antioxidant activity of *Rhizophora mangle* bark aqueous extract and its majority component and high molecular weight polyphenols' fraction were studied by Sanchez *et al.* (2006) using deoxyribose assay. The total extract and its fraction showed scavenging activity of hydroxyl radicals and ability to chelate iron ions.

Berenguer *et al.* (2006) reported protective and antioxidant effects of *Rhizophora mangle* against NSAID-induced gastric ulcers.

Rhizophora mucronata Poir. Family: Rhizophoraceae

Vadlapudi and Naidu (2009) evaluated the antioxidant potential of *Rhizophora conjugata*, a mangrove plant. The SOD was 0.25U/mg, Catalase was 0.46U/mg, Ascorbic acid was 270mg/100g, DPPH radical inhibition was 71.6 per cent and FRAP units were 576.

Suganthy *et al.* (2009) evaluated the radical scavenging activity of methanolic leaf extracts of *Rhizophora mucronata* using DPPH, nitric oxide, hydrogen peroxide, hydroxyl radical scavenging assay, reducing power, ferrous ion chelating and lipid peroxidation inhibition assay. *Rhizophora mucronata* extract (100microg/ml) showed significantly higher (p<0.05) activities for all antioxidant assays, and its IC_{50} values were 43.17, 116, 60.06 and 46.76 µg/ml for DPPH, hydrogen peroxide, nitric oxide and hydroxyl radical scavenging activity.

Rhodiola imbricata Edgew. Family: Crassulaceae

Cytoprotective and antioxidant activity of *Rhodiola imbricata* on U937 human macrophages was investigated by Hagat *et al.* (2004). Exposure of U 937 human cell to t-BHP caused cytotoxicity due to the generation of tert-butyoxyl radicals. But the cells treated with root extract showed protective effect of extract against t-BHP induced cytotoxicity and ROS production. As a result GSH levels decreased in cells leading to cell death and fall in Rh-123 uptake. Thus, *Rhodiola* inhibits t-BHP induced ROS production and prevent fall in GSH levels in pretreated cells and maintain cellular antioxidant level similar to that of normal cells.

Rhodiola rosea L. Family: Crassulaceae

Kim *et al.* (2008) reported that *Rhodiola rosea* extract significantly decreased on blood glucose activities of glutathione reductase, glutathione S-transferase, glutathione peroxidase, treatment also significantly decreased lipid peroxidation.

Chen *et al.* (2009) studied the antioxidant activity of leaves extract towards xanthine oxidase, lipoxygenase and tyrosinase activities. The inhibitory effect of acetone extract on xanthine oxidase activity was about 80 per cent, tyrosine activity was almost nil and lipoxygenase activity was more than 70 per cent.

Rhodiola sacra Fu Family: Crassulaceae

The antioxidant activity and total phenolic contents of *Rhodiola sacra* was evaluated by Wong *et al.* (2006) using the ferric reducing antioxidant power assay and the Folin-Ciocalteu method, respectively. The plant was extracted by the traditional method, boiling in water and also in 80 per cent methanol. A significant correlation coefficient between the antioxidant activity and the total phenolic content was found in both aqueous and methanol extracts. Phenolic compounds were thus a major contributor of antioxidant activity. Comparing the extraction efficiency of the two methods, the boiling water method extracted phenolic compounds more efficiently, and the antioxidant activity of the extracts was higher.

Rhodococcum vitis-idaea (L.) Avrorin Family: Ericaceae

Myagmar *et al.* (2004) evaluated antioxidant activity of medicinal herb *Rhodococcum vitis-idaea* on galactosamine-induced liver injury in rats. The results showed that the hepatotoxicity and oxidative stress induced by galactosamine (700mg/kg, s.c.) after 24 h evidenced by an increase in serum alanine aminotransferase and glutathione (GSH) S-transferase activities, and lipid peroxidation in liver homogenate were significantly inhibited, when 10 times diluted extract (5ml/kg, i.p.) was given to rats 12 and 1 h before galactosamine treatment demonstrating that the extract of *Rhodococcum vitis-idaea* is a potent antioxidant and protective against galactosamine-induced hepatotoxicity. The main antioxidant compound of the herb water extract used in the experiment was determined as arbutin, which possess 8 per cent of the dry weight of the herb. The electron spin resonance (ESR) spectrometer analysis revealed that the arbutin isolated from *Rhodococcum vitis-idaea* exhibited strong superoxide and hydroxyl radical scavenging ability.

Rhododendron arboreum Sm. Family: Ericaceae

Prakash *et al.* (2007d) evaluated the free radical scavenging activities of 21 species of *Rhododendron*. For *Rhododendron arboreum* Total phenolic content was 78.5 mg GAE/g, Flavonoids was 62.6mg quercetin equivalent/g and antioxidant activity was 81.6 per cent measured by auto-oxidation of β-carotene and linoleic acid-coupled reaction on dry weight basis. IC_{50} value of DPPH radical scavenging activity was 0.47 and ferric reducing antioxidant power expressed as ascorbic acid equivalent was 1.25 while quercetin standard was 0.20 and 0.5, respectively.

Rhododendron baileyi Balbour. Family: Ericaceae

Prakash *et al.* (2007d) evaluated the free radical scavenging activities of 21 species of *Rhododendron*. For *Rhododendron baileyi* Total phenolic content was 97.9 mg GAE/

g, Flavonoids was 32.0mg quercetin equivalent/g and antioxidant activity was 92.0 per cent measured by auto-oxidation of β-carotene and linoleic acid-coupled reaction on dry weight basis. IC_{50} value of DPPH radical scavenging activity was 0.14 and ferric reducing antioxidant power expressed as ascorbic acid equivalent was 1.09 while quercetin standard was 0.20 and 0.5, respectively.

Rhododendron barbatum Wall. ex G. Don Family: Ericaceae

Prakash *et al.* (2007d) evaluated the free radical scavenging activities of 21 species of *Rhododendron*. For *Rhododendron barbatum* Total phenolic content was 44.1 mg GAE/g, Flavonoids was 11.5mg quercetin equivalent/g and antioxidant activity was 66.5 per cent measured by auto-oxidation of β-carotene and linoleic acid-coupled reaction on dry weight basis. IC_{50} value of DPPH radical scavenging activity was 0.64 and ferric reducing antioxidant power expressed as ascorbic acid equivalent was 1.21 while quercetin standard was 0.20 and 0.5, respectively.

Rhododendron camellieflorum Glen doick Family: Ericaceae

Prakash *et al.* (2007d) evaluated the free radical scavenging activities of 21 species of *Rhododendron*. For *Rhododendron camellieflorum* Total phenolic content was 132.2 mg GAE/g, Flavonoids was 59.1mg quercetin equivalent/g and antioxidant activity was 93.6 per cent measured by auto-oxidation of β-carotene and linoleic acid-coupled reaction on dry weight basis. IC_{50} value of DPPH radical scavenging activity was 0.12 and ferric reducing antioxidant power expressed as ascorbic acid equivalent was 1.26 while quercetin standard was 0.20 and 0.5, respectively.

Rhododendron campanulatum D. Don Family: Ericaceae

Prakash *et al.* (2007d) evaluated the free radical scavenging activities of 21 species of *Rhododendron*. For *Rhododendron campanulatum* Total phenolic content was 123.9 mg GAE/g, Flavonoids was 24.5mg quercetin equivalent/g and antioxidant activity was 94.4 per cent measured by auto-oxidation of β-carotene and linoleic acid-coupled reaction on dry weight basis. IC_{50} value of DPPH radical scavenging activity was 0.13 and ferric reducing antioxidant power expressed as ascorbic acid equivalent was 0.71 while quercetin standard was 0.20 and 0.5, respectively.

Rhododendron ciliatum Hook. Family: Ericaceae

Prakash *et al.* (2007d) evaluated the free radical scavenging activities of 21 species of *Rhododendron*. For *Rhododendron ciliatum* Total phenolic content was 91.4 mg GAE/g, Flavonoids was 88.4mg quercetin equivalent/g and antioxidant activity was 71.5 per cent measured by auto-oxidation of β-carotene and linoleic acid-coupled reaction on dry weight basis. IC_{50} value of DPPH radical scavenging activity was 0.19 and ferric reducing antioxidant power expressed as ascorbic acid equivalent was 1.43 while quercetin standard was 0.20 and 0.5, respectively.

Rhododendron cinnabarinum Roylei Family: Ericaceae

Prakash *et al.* (2007d) evaluated the free radical scavenging activities of 21 species of *Rhododendron*. For *Rhododendron cinnabarinum* Total phenolic content was 93.9 mg GAE/g, Flavonoids was 26.3mg quercetin equivalent/g and antioxidant activity was 83.2 per cent measured by auto-oxidation of β-carotene and linoleic acid-coupled reaction on dry weight basis. IC_{50} value of DPPH radical scavenging activity was 0.16 and ferric reducing antioxidant power expressed as ascorbic acid equivalent was 1.57 while quercetin standard was 0.20 and 0.5, respectively.

Rhododendron dalhousiae Hook. f. Family: Ericaceae

Prakash *et al.* (2007d) evaluated the free radical scavenging activities of 21 species of *Rhododendron*. For *Rhododendron dalhousiae* Total phenolic content was 55.4 mg GAE/g, Flavonoids was 19.6mg quercetin equivalent/g and antioxidant activity was 57.2 per cent measured by auto-oxidation of β-carotene and linoleic acid-coupled reaction on dry weight basis. IC_{50} value of DPPH radical scavenging activity was 0.51 and ferric reducing antioxidant power expressed as ascorbic acid equivalent was 1.11 while quercetin standard was 0.20 and 0.5, respectively.

Rhododendron decipiens Lacaita Family: Ericaceae

Prakash *et al.* (2007d) evaluated the free radical scavenging activities of 21 species of *Rhododendron*. For *Rhododendron decipiens* Total phenolic content was 39.6 mg GAE/g, Flavonoids was 23.2mg quercetin equivalent/g and antioxidant activity was 55.0 per cent measured by auto-oxidation of β-carotene and linoleic acid-coupled reaction on dry weight basis. IC_{50} value of DPPH radical scavenging activity was 0.72 and ferric reducing antioxidant power expressed as ascorbic acid equivalent was 1.71 while quercetin standard was 0.20 and 0.5, respectively.

Rhododendron falconeri Hook. f. Family: Ericaceae

Prakash *et al.* (2007d) evaluated the free radical scavenging activities of 21 species of *Rhododendron*. For *Rhododendron falconeri* Total phenolic content was 39.2 mg GAE/g, Flavonoids was 21.1mg quercetin equivalent/g and antioxidant activity was 30.4 per cent measured by auto-oxidation of β-carotene and linoleic acid-coupled reaction on dry weight basis. IC_{50} value of DPPH radical scavenging activity was 0.82 and ferric reducing antioxidant power expressed as ascorbic acid equivalent was 1.63 while quercetin standard was 0.20 and 0.5, respectively.

Rhododendron grande Wight Family: Ericaceae

Prakash *et al.* (2007d) evaluated the free radical scavenging activities of 21 species of *Rhododendron*. For *Rhododendron grande* Total phenolic content was 77.3 mg GAE/g, Flavonoids was 41.4mg quercetin equivalent/g and antioxidant activity was 56.4 per cent measured by auto-oxidation of β-carotene and linoleic acid-coupled reaction on dry weight basis. IC_{50} value of DPPH radical scavenging activity was 0.86 and ferric reducing antioxidant power expressed as ascorbic acid equivalent was 1.61 while quercetin standard was 0.20 and 0.5, respectively.

Rhododendron griffithianum Wight Family: Ericaceae

Prakash *et al.* (2007d) evaluated the free radical scavenging activities of 21 species of *Rhododendron*. For *Rhododendron griffithianum* Total phenolic content was 165.4 mg GAE/g, Flavonoids was 77.3mg quercetin equivalent/g and antioxidant activity was 93.4 per cent measured by auto-oxidation of β-carotene and linoleic acid-coupled reaction on dry weight basis. IC_{50} value of DPPH radical scavenging activity was 0.10 and ferric reducing antioxidant power expressed as ascorbic acid equivalent was

0.64 while quercetin standard was 0.20 and 0.5, respectively. The IC_{50} values for inhibition of peroxidation measured by Ammonium thiocyanate assay was 0.84 mg/l and showed better inhibition of peroxide formation compared to reference standards, BHT (1.27mg./ml) and quercetin (1.85mg/ml).

Rhododendron lepidotum Wall. ex D. Don Family: Ericaceae

Prakash *et al*. (2007d) evaluated the free radical scavenging activities of 21 species of *Rhododendron*. For *Rhododendron lepidotum* Total phenolic content was 148.5 mg GAE/g, Flavonoids was 94.5mg quercetin equivalent/g and antioxidant activity was 87.2 per cent measured by auto-oxidation of β-carotene and linoleic acid-coupled reaction on dry weight basis. IC_{50} value of DPPH radical scavenging activity was 0.12 and ferric reducing antioxidant power expressed as ascorbic acid equivalent was 1.00 while quercetin standard was 0.20 and 0.5, respectively. The IC_{50} values for inhibition of peroxidation measured by Ammonium thiocyanate assay was 1.15 mg/l and showed better inhibition of peroxide formation compared to reference standards, BHT (1.27mg./ml) and quercetin (1.85mg/ml).

Rhododendron maddenii Hook. f. Family: Ericaceae

Prakash *et al*. (2007d) evaluated the free radical scavenging activities of 21 species of *Rhododendron*. For *Rhododendron baileyi* Total phenolic content was 87.3 mg GAE/g, Flavonoids was 54.4mg quercetin equivalent/g and antioxidant activity was 62.8 per cent measured by auto-oxidation of β-carotene and linoleic acid-coupled reaction on dry weight basis. IC_{50} value of DPPH radical scavenging activity was 0.30 and ferric reducing antioxidant power expressed as ascorbic acid equivalent was 1.31 while quercetin standard was 0.20 and 0.5, respectively.

Rhodendron mucronulatum Turcz. Family: Ericaceae

From the n-BuOH soluble fraction of the 70 per cent aqueous acetone extract of *Rhododendron mucronulatum* stem, twelve compounds were isolated by Lee *et al*. (2005). Two compounds (+)-taxifolin and quercetin showed the potent antioxidant activity.

Rhododendron niveum Hook. f. Family: Ericaceae

Prakash *et al*. (2007d) evaluated the free radical scavenging activities of 21 species of *Rhododendron*. For *Rhododendron niveum* Total phenolic was 106.6 mg GAE/g, Flavonoids was 57.1mg quercetin equivalent/g and antioxidant activity was 85.2 per cent measured by auto-oxidation of β-carotene and linoleic acid-coupled reaction on dry weight basis. IC_{50} value of DPPH radical scavenging activity was 0.13 and ferric reducing antioxidant power expressed as ascorbic acid equivalent was 0.72 while quercetin standard was 0.20 and 0.5 respectively.

Rhododendron pendulum Family: Ericaceae

Prakash *et al*. (2007d) evaluated the free radical scavenging activities of 21 species of *Rhododendron*. For *Rhododendron pendulum* Total phenolic content was 89.3 mg GAE/g, Flavonoids was 65.2mg quercetin equivalent/g and antioxidant activity was 76. 8 per cent measured by auto-oxidation of β-carotene and linoleic acid-coupled reaction on dry weight basis. IC_{50} value of DPPH radical scavenging activity was 0.22

and ferric reducing antioxidant power expressed as ascorbic acid equivalent was 1.72 while quercetin standard was 0.20 and 0.5 respectively.

Rhododendron ponticum L. Family: Ericaceae

Using TLC bioautography, a total of 58 extracts from various organs (aerial parts, leaves, flowers, fruits, roots) of 16 plants were tested by Tasdemir *et al.* (2004) using hexane, $CHCl_3$, CH_2Cl_2, Water and total MeOH extracts. β-carotene, β-carotene/ linoleic acid mixture, and DPPH solutions sprayed on to TLC plates were used for detecting antioxidant and radical scavenging properties of the crude extracts. These activities were found to be predominant in highly polar extracts. The water solubles of all *Rhododendron* species present the most significant activity.

Rhododendron sallignum Hook.f. Family: Ericaceae

Prakash *et al.* (2007d) evaluated the free radical scavenging activities of 21 species of *Rhododendron*. For *Rhododendron sallignum* Total phenolic content was 97.5 mg GAE/g, Flavonoids was 34.8mg quercetin equivalent/g and antioxidant activity was 88.4 per cent measured by auto-oxidation of β-carotene and linoleic acid-coupled reaction on dry weight basis. IC_{50} value of DPPH radical scavenging activity was 0.14 and ferric reducing antioxidant power expressed as ascorbic acid equivalent was 1.58 while quercetin standard was 0.20 and 0.5 respectively.

Rhododendron simsii Planch. Family: Ericaceae

Antioxidative substances were isolated from *Rhododendron simsii* by Takahashi *et al.* (2001) from the leaves of *Rhododendron simsii*. These were triterpene and flavone glycoside, together with known matteucinol and two known benzoic acid derivatives.

Rhododendron smirnovii Trautu Family: Ericaceae

Using TLC bioautography, a total of 58 extracts from various organs (aerial parts, leaves, flowers, fruits, roots) of 16 plants were tested by Tasdemir *et al.* (2004) using hexane, $CHCl_3$, CH_2Cl_2, Water and total MeOH extracts. β-carotene, β-carotene/ linoleic acid mixture, and DPPH solutions sprayed onto TLC plates were used for detecting antioxidant and radical scavenging properties of the crude extracts. These activities were found to be predominant in highly polar extracts. The water solubles of all *Rhododendron* species present the most significant activity.

Rhododendron thomsonii Hook. f. Family: Ericaceae

Prakash *et al.* (2007d) evaluated the free radical scavenging activities of 21 species of *Rhododendron*. For *Rhododendron thomsonii* Total phenolic content was 90.6 mg GAE/g, Flavonoids was 22.1mg quercetin equivalent/g and antioxidant activity was 90.0 per cent measured by auto-oxidation of β-carotene and linoleic acid-coupled reaction on dry weight basis. IC_{50} value of DPPH radical scavenging activity was 0.22 and ferric reducing antioxidant power expressed as ascorbic acid equivalent was 2.03 while quercetin standard was 0.20 and 0.5 respectively.

Rhododendron ungernii Trautv. Family: Ericaceae

Using TLC bioautography, a total of 58 extracts from various organs (aerial parts, leaves, flowers, fruits, roots) of 16 plants were tested by Tasdemir *et al.* (2004)

using hexane, $CHCl_3$, CH_2Cl_2, Water and total MeOH extracts. β-carotene, β-carotene/ linoleic acid mixture, and DPPH solutions sprayed onto TLC plates were used for detecting antioxidant and radical scavenging properties of the crude extracts. These activities were found to be predominant in highly polar extracts. The water solubles of all *Rhododendron* species present the most significant activity.

Rhododendron vaccinioides Hook. f. Family: Ericaceae

Prakash *et al.* (2007d) evaluated the free radical scavenging activities of 21 species of *Rhododendron*. For *Rhododendron vaccinioides* Total phenolic content was 67.8 mg GAE/g, Flavonoids was 14.7mg quercetin equivalent/g and antioxidant activity was 72.0 per cent measured by auto-oxidation of β-carotene and linoleic acid-coupled reaction on dry weight basis. IC_{50} value of DPPH radical scavenging activity was 0.45 and ferric reducing antioxidant power expressed as ascorbic acid equivalent was 1.16 while quercetin standard was 0.20 and 0.5 respectively.

Rhododendron virgatum Hook. f. Family: Ericaceae

Prakash *et al.* (2007d) evaluated the free radical scavenging activities of 21 species of *Rhododendron*. For *Rhododendron virgatum* Total phenolic content was 208.9 mg GAE/g, Flavonoids was 137.1mg quercetin equivalent/g and antioxidant activity was 97.4 per cent measured by auto-oxidation of β-carotene and linoleic acid-coupled reaction on dry weight basis. IC_{50} value of DPPH radical scavenging activity was 0.67 and ferric reducing antioxidant power expressed as ascorbic acid equivalent was 0.46 while quercetin standard was 0.20 and 0.5 respectively.

Rhododendron yedoense Rehder var. *poukhanense* Family: Ericaceae

One new flavonoid (myrcitrin-5-methyl ether (2), and three known flavonoids, quercetin-5-O-β-D-glucopyranoside (1), quercetin (3) and quercitrin (4) were isolated from the butanol and ethyl acetate extracts of the flower of *Rhododendron yedoense* var. *poukhanense* (Jung *et al.*, 2007). The flavonoids were evaluated for their antioxidant activities using DPPH, TBARS, and superoxide anion radical in the xanthine/ xanthine oxidase assay system. In the DPPH scavenging assay, the IC_{50} values were 4.5 µM for compound 2 and 9.7 µM for compound 3, which showed an antioxidant activity 1.5-2 times higher than the antioxidant activity of α-tocopherol (9.8 µM). Compound 2 showed high activity in both the inhibition of xanthine oxidase (1.1µM) and in the activation of superoxide scavenging (Jung *et al.*, 2007).

Rhoicissus digitata (L.f.) Gilg. et Brabndt. Family: Vitaceae

Analysis of methanol extracts of *Rhoicissus digitata*, a South African medicinal plant used by the Zulu traditional healers did not show inhibitory properties of the activities of NAPPH free radicals, xanthine oxidase except at very high concentrations (Fernandes *et al.*, 2004).

Rhoicissus rhomboidea (E.Mey ex Harv.) Planch. Family: Vitaceae

Analysis of methanol extracts of *Rhoicissus rhomboidea*, a South African medicinal plant used by the Zulu traditional healers inhibited the activities of NAPPH free radicals, xanthine oxidase and also prevented production of thiobarbituric acid reactive substances and also free-radical mediated sugar damage (Fernandes *et al.*, 2004).

Rhoicissus tomentosa (Lam.) Wild. et R.B.Drumm. Family: Vitaceae

Analysis of methanol extracts of *Rhoicissus tomentosa*, a South African medicinal plant used by the Zulu traditional healers did not show inhibitory properties of the activities of NAPPH free radicals, xanthine oxidase except at very high concentrations (Fernandes *et al.*, 2004).

Rhoicissus tridentata (L.f.) Wild. et R.B.Drumm. Family: Vitaceae

Analysis of methanol extracts of *Rhoicissus tridentata*, a South African medicinal plant used by the Zulu traditional healers inhibted the activities of NAPPH free radicals, xanthine oxidase and also prevented production of thiobarbituric acid reactive substances and also free-radical mediated sugar damage (Fernandes *et al.*, 2004).

Naidoo *et al.* (2006) quantified the antioxidant capacity of cultivated *Rhoicissus tridentata*. Acetone extracts of different plant parts had good antioxidant activity in DPPH and TEAC assay. HPLC analysis demonstrated that the activity was in part due to the presence of catechin, epicatechin, gallic acid and epigallo-catechin-gallate in the tuber acetone extracts.

Rhopalocnemis phalloides Jungh. Family: Balanophoraceae

She *et al.* (2010) reported that the 80 per cent aqueous acetone extract of the whole plant of *Rhopalocnemis phalloides* showed obvious radical scavenging activity (SC$_{50}$ = 32.1 mug/ml) on DPPH radical assay. Further chemical investigation of the extract led to the isolation of 12 phenolic compounds. The DPPH assay showed that flavan-3-ol dimers had remarkable free radical scavenging activity, while flavonoid aglycones presented stronger activities than those of their glycosides.

Rhus chinensis Mill. (=*R. potaninii* Maxim.) Family: Anacardiaceae

Cai *et al.* (2004, 2006) evaluated the antioxidant activity and phenolic compounds in traditional Chinese medicinal plants associated with anticancer. The improved ABTS.+method was used to systematically assess the total antioxidant capacity (Trolox equivalent antioxidant capacity, TEAC) of the medicinal extracts. The TEAC values and total phenolic content for methanolic extract of *Rhus chinensis* were 17323 µmol Trolox equivalent/100g dry weight (DW), and 50.3 g of gallic acid equivalent/100g DW respectively. A positive significant linear relationship between antioxidant activity and total phenolic content showed that phenolic compounds were the dominant antioxidant components in the tested herbs. Major types of phenolic compounds in *Rhus chinensis* include hydrolysable tannins (gallotannin) and phenolic acids (gallic acie).

Rhus coriaria L. Family: Anacardiaceae

Candan and Sikman (2004) evaluated the antioxidant potential of *Rhus coriaria*, a well-known spice, the methanolic pextract (soluble part) was prepared and investigated using free radical-generating systems *in vitro*. The data suggest that the methanolic extracts of *Rhus coriaria* fruits have considerable antioxidant activity against free radicals and lipid peroxidation *in vitro*.

Souri *et al.* (2004) evaluated the antioxidative activity of *Rhus coriaria* by linoleic acid peroxidation test using 1,3-diethyl-2-thiobarbituric acid as the reagent. Methanolic extract showed 93.81 per cent inhibition at 40µg concentration with IC_{50} value of 0.91µg, which is approximately in the range of α-tocopherol (IC_{50}=0.60µg).

Serteser *et al.* (2009) studied the antioxidant activity of 50 per cent aqueous methanol extracts of leaves and fruits by various antioxidant assays, including free radical scavenging, hydrogen peroxide scavenging and metal (Fe^{2+}) chelating activities. The DPPH radical scavenging effect of leaves extract was 0.424 and fruit extract was 0.591. Fe^{2+} chelating activity (per cent) of leaves extract was 26.34 while fruit extract was 34.29. Hydrogen peroxide inhibition activity of leaves extract was 24.41 per cent while fruit extract showed 32.65 per cent inhibition.

Rhus hirta (L.) Sudw. Family: Anacardiaceae

McCune (1999) evaluated antioxidant activities of indigenous medicinal plants of Canadian boreal forest by three different assays (DPPH, NBT/xanthine oxidase and DCF/APPH). *Rhus hirta* showed antioxidant activity similar to vitamin C, and green tea in all the assays.

Rhus verniciflua Stokes Family: Anacardiaceae

Methanol aqueous extract of *Rhus verniciflua* was screened by Kim *et al.* (1997) for antioxidative activity using Fenton's reagent/ethyl lineolate system. The extract showed antioxidant activity.

Treatment with the EtOAc fraction containing sulfuretin significantly decreased malondialdehyde formation and highly increased the activities of superoxide dismutase, catalase and glutathione peoxidase. Inhibition of xanthine oxidase and aldehyde oxidase in FCA-treated rats was also evident. Since treatment with sulfuretin and the EtOAc extract decreased the concentration of infiltrated mast cells in the rat knee exhibiting rheumatoid arthritis, it is suggested that the *Rhus verniciflua* extract, which contains sulfuretin as an active component, may prevent rheumatoid syndromes by inhibiting reactive oxygen species (Choi *et al.*, 2003).

Cai *et al.* (2006) reported structure-radical scavenging activity relationships of phenolic compounds from traditional Chinese medicinal plants including *Rhus verniciflua*.

The 80 per cent ethanol extract of *Rhus verniciflua* (RVS) was purified by ion exchange chromatography by Jung *et al.* (2006) Purified RVS extract contained a high amount of phenolics and flavonoids. Its antioxidant activities, such as DPPH, superoxde anion and hydroxyl radical scavenging activities were higher than the unpurified RVS extract. Purified RVS extract significantly reduced intracellular ROS formation caused by H_2O_2. Purified RVS extract also prevented the cell death of macrophage RAW 264.7 cells induced by H_2O_2.

Ribes grossularia L. Family: Grossulariaceae

English Name: Gooseberry.

The antioxidative activity of phenolic extracts from fruits of *Ribes nigrum* was examined by autoxidation of methyl linoleate by Kahkonen *et al.* (1999).The formation of MeLo-conjugated diene hydroperoxides was inhibited over 90 per cent by gooseberry extracts.

Ribes x nidigrolaria Bauer Family: Grossulariaceae

English Name: Jostaberry.

Moyer *et al.* (2002) analyzed the anthocyanins, phenolics and antioxidant capacity in diverse small fruits including *Ribes x nidigrolaria*. Total anthocynins was 74mg/100g, total phenolics was 309mg/100g, Oxygen radical absorbance capacity (ORAC) was 28.1µmol TE/g, FRAP was 39.0 µmol/g.

Ribes nigrum L. Family: Grosulariaceae

English Name: Black currant.

Remarkable high antioxidant activity and high total phenolic content were found in black currant. Extracts of black currants possessed a remarkably high scavenging activity toward chemically generated superoxide radicals (Constantino *et al.*, 1992). The antioxidative activity of phenolic extracts from fruits of *Ribes nigrum* was examined by autoxidation of methyl linoleate by Kahkonen *et al.* (1999).

Moyer *et al.* (2000) determined the range of total anthocyanin content, total phenolic content and antioxidant capacity in *Ribes nigrum*. Anthocyanin content was determined by the pH differential method. Total phenolics content was determined via the Folin-Cocalteu method. Antioxidant capacity was measured by oxygen radical absorbing capacity (ORAC). Total anthocyanin content of *Ribes nigrum* cultivars ranged from 128 to 411 mg/100 g fruit. Total anthocyanin content was highly correlated to TPH and ORAC.

Moyer *et al.* (2002) analyzed the anthocyanins, phenolics and antioxidant capacity in diverse small fruits including *Ribes nigrum* and hybrids. Total anthocynins was 229mg/100g, total phenolics was 799mg/100g, Oxygen radical absorbance capacity (ORAC) was 57.1µmol TE/g, FRAP was 92.0 µmol/g.

Inheritance of antioxidants in a New Zealand Black currant population was studied by Currie *et al.* (2006). The antioxidant capacity (AOC) of black currant was determined by Borges *et al.* (2010) using FRAP assay. In addition, the vitamin C content of the berries was determined and phenolic and polyphenolic compounds in the extracts were analyzed by reversed-Phase HPLC-PDA-MS and by reversed-phase HPLC-PDA with online antioxidant detection system. A complex spectrum of anthocyanins was the major contributor to the antioxidant capacity of black currants. Vitamin C was responsible for 18-23 per cent of HPLC-AOC of black currants.

Ribes rubrum L. Family: Grossulariaceae

English Name: Red currant.

The antioxidative activity of phenolic extracts from fruits of *Ribes rubrum* was examined by autoxidation of methyl linoleate by Kahkonen *et al.* (1999). Remarkable high antioxidant activity and high total phenolic content were found in red currant. Extracts of red currants possessed a remarkably high scavenging activity toward chemically generated superoxide radicals (Constantino *et al.*, 1992).

The antioxidant capacity (AOC) of red currant was determined by Borges *et al.* (2010) using FRAP assay. In addition, the vitamin C content of the berries was determined and phenolic and polyphenolic compounds in the extracts were analyzed by reversed-Phase HPLC-PDA-MS and by reversed-phase HPLC-PDA with online antioxidant detection system. AOC was less in Red currants was mainly due to a reduced anthocyanin content.

Ribes uva-crispa L. Family: Grossulariaceae

English Name: Gooseberry.

Moyer *et al.* (2002) analyzed the anthocyanins, phenolics and antioxidant capacity in diverse small fruits including *Ribes uva-crispa*. Total anthocynins was 14mg/100g, total phenolics was 191mg/100g, Oxygen radical absorbance capacity (ORAC) was 17.0µmol TE/g, FRAP was 25.2 µmol/g.

Ribes valdivianum Phil. Family: Grossulariaceae

Moyer *et al.* (2000) determined the range of total anthocyanin content, total phenolic content and antioxidant capacity in *Ribes valdivianum*. Anthocyanin content was determined by the pH differential method. Total phenolics content was determined via the Folin-Cocalteu method. Antioxidant capacity was measured by oxygen radical absorbing capacity (ORAC). *Ribes valdivianum* had the highest Total phenolic content (1790 mg gallic acid/100 g fruit) and ORAC 115 µm trolox equivalents/1g fruit. Total anthocyanin content was highly correlated to TPH and ORAC.

Moyer *et al.* (2002) analyzed the anthocyanins, phenolics and antioxidant capacity in diverse small fruits including *Ribes valdivianum*. Total anthocynins was 358mg/100g, total phenolics was 1790mg/100g, Oxygen radical absorbance capacity (ORAC) was 115.9µmol TE/g, FRAP was 219.3 µmol/g.

Ricinodendron heudelotii (Baill.) Pierre ex Pax Family: Euphorbiaceae

Antioxidant activity test using two different methods namely DPPH and 2,2'-azinobis(3-ethylbenzothialozinesulfonate) diammonium salt free radical scavenging test has been carried out by Momeni *et al.* (2010) on three Cameroonian plant extracts. Results of this study in the DPPH scavenging test show that the methanol/dichloromethane (1+1) extract of *R. heudelotii* showed weak free radical scavenging activities as compared to Trolox used as standard. In the same manner, 2,2'-azinobis(3-ethylbenzothialozinesulfonate) diammonium salt radical scavenging test of these extracts was in accordance of the result of 2,2-diphenyl-1-picrylhydrazyl test.

Ricinus commuis L. Family: Euphorbiaceae

English Name: Castor.

Antioxidant activity of different extracts of *Ricinus communis* were evaluated by Shahwar *et al.* (2010) through DPPH radical scavenging activity, phosphomolybdate

Encyclopaedia of Herbal Antioxidants

and ferric thiocyanate methods. The DPPH radical scavenging activity of ethyl acetate extract was 93.2 per cent, total antioxidant activity was 0.532 and inhibition of lipid peroxidation was 79.3 per cent while total phenols was 547.0 mg GAE/g of crude extract.

Ridolfia segetum (L.) Moris Family: Apiaceae (Umbelliferae)

Jabrane *et al.* (2010) reported good antioxidant activity for the root essential oil of *Ridolfia segetum* with IC_{50} values close to that of BHT.

Roemeria refracta DC. Family: Papaveraceae

Souri *et al.* (2004) evaluated the antioxidative activity of *Roemeria refracta* by linoleic acid peroxidation test using 1,3-diethyl-2-thiobarbituric acid as the reagent. Methanolic extract showed 93.96 per cent inhibition at 40μg concentration with IC_{50} value of 2.79μg.

Rorippa nasturtium-aquaticum (L.) Hayek. Family: Brassicaceae (Cruciferae)

English Name: Water cress, Indian Cress.

Lako *et al.* (2006) reported antioxidant activity in *Rorippa nasturtium-aquaticum*.

Rosa sp. Family: Rosaceae

Liu and Ng (2000) demonstrated for the first time the presence of both gallic acid derivative and polysaccharides as major antioxidant principles of the aqueous extract of rose flowers, reflected in the ability to inhibit lipid peroxidation in rat brain and kidney homogenates.

Rosa agrestis Savi Family: Rosaceae

Bitis *et al.* (2010) reported the isolation and characterization of seven flavonoids, the levels of phenolics, flavonoids and proanthocyanidins and the antioxidant activity of the leaf extract of *Rosa agrestis*. The results showed that the *R. agrestis* leaf extract exhibited significant antioxidant activity as measured by DPPH (EC_{50}+47.4 μg/ml), inhibited both β-carotene bleaching and deoxyribose degradation, quenched a chemically generated superoxide anion *in vitro* and showed high ferrous ion chelating activity. Reactivity towards ABTS radical cation and FRAP valued were equivalent to 2.30 mM/l Trolox, the water soluble α-tocopherol analogue and 1.91 mM/l Fe^{2+}, respectively. The high antioxidant activity of the extract appeared to be attributed to its high content of total phenolics, flavonoids and proanthocyanidins. The flavonoids isolated from *R. agrestis* leaves were diosmetin, kaempferol, quercetin, kaempferol 3-glucoside (astragalin), quercetin 3-rhamnoside (quercitrin), quercitrin 3-xyloside and quercetin 3-galactoside (hyperoside). Diosmetin (5,7,3'-trihydroxy-4'-methoxyflavone) was isolated for the first time from *Rosa* species (Bitis *et al.*, 2010).

Rosa canina L. Family: Rosaceae

Local name: Nourna Dog rose.

Bioactive flavonoid glycosides were isolated from the seeds of *Rosa canina* by Kumaraswamy *et al.* (2003). These compounds showed antioxidant activity.

Souri *et al.* (2004) evaluated the antioxidative activity of *Rosa canina* by linoleic acid peroxidation test using 1,3-diethyl-2-thiobarbituric acid as the reagent. Methanolic extract showed 91.79 per cent inhibition at 40µg concentration with IC_{50} value of 0.41µg, which is approximately in the range of α-tocopherol ($IC_{50} = 0.60µg$).

The antioxidant activity of leaf extracts of *Rosa canina* from diverse localities of Tunisia were evaluated by ABTS and DPPH methods, whereas those of essential oils and carotenoids extracts such activity was determined only by ABTS method (Ghazghazi *et al.*, 2010). Total phenols determined by the Folin method revealed that at Aindraham, samples showed a great variability of phenol content in contrast to those from Feija. All the samples possess antioxidant activity, nevertheless much more significant in phenol extracts in contrast to the carotenoid extracts, which possess the lowest activity.

Rosa chinensis Jacq. Family: Rosaceae

Cai *et al.* (2004) evaluated the antioxidant activity and phenolic compounds in traditional Chinese medicinal plants associated with anticancer. The improved ABTS.+ method was used to systematically assess the total antioxidant capacity (Trolox equivalent antioxidant capacity, TEAC) of the medicinal extracts. The TEAC values and total phenolic content for methanolic extract of flower of *Rosa chinensis* were 62206.9 µmol Trolox equivalent/100g dry weight (DW), and 18.75 g of gallic acid equivalent/100g DW respectively. A positive significant linear relationship between antioxidant activity and total phenolic content showed that phenolic compounds were the dominant antioxidant components in the tested herbs. Major types of phenolic compounds in *Rosa chinensis* include flavonoids (quercetin, catechin, anthocyanins), phenolic acids (gallic acid), tannins.

Gan *et al.* (2010) evaluated the antioxidant activity of 40 medicinal plants associated with prevention and treatment of cardiovascular and cerebrovascular diseases. A high correlation between antioxidant capacity and total phenolic content indicated that phenolic compounds could be the main contributor of the antioxidant activity of these plants. *Rosa chinensis* showed very high antioxidant activity of 660.34 µmol Fe(II)/g in FRAP assay and 758.65µmol in TEAC assay. Total phenolic content was also very high, 38.06 mg GAE/g.

Rosa damascena Mill. Family: Rosaceae

English Name: Damask rose.

A study conducted by Achuthan *et al.* (2003) on antioxidant and hepatoprotective effects of the fresh juice of Damask rose flower on rats resulted promising *in vitro* antioxidant potential. The partially purified acetone fraction from silica gel column chromatography was found to be the active fraction with antioxidant properties. The acetone fraction required for 50 per cent inhibition of superoxide radical production, hydroxyl radical generation and lipid peroxide formation were 13.75, 135 and 410µg/ ml, respectively. Oral administration of acetone fraction at 50mg/kg body weight significantly reduced the serum alkaline phosphatase, glutamine pyruvate transaminase and glutamine oxaloacetate transaminase activity and lipid peroxide levels in rats receiving an acute dose of CCl_4. This indicated that *R. damascena* could protect against CCl_4 induced hepatotoxicity, possibly by its free radical scavenging activity (Achuthan *et al.*, 2003).

Özkan *et al.* (2004) evaluated the antioxidant activity of *Rosa damascena* flower extracts. The total phenolic contents were 276.02 mg gallic acid equivalent (GAE)/g in fresh flower (FF) extract and 248.97 GAE/g in spent flower (SF) extract. FF and SF extracts showed 74.5 and 75.94 per cent antiradical activities at 100ppm. The antioxidant activity of FF extract (372.26mg/g) was higher than that of SF extract (351.36mg/g).

Kalim *et al.* (2010) evaluated the antioxidant activity of Methanol (50 per cent) extract of *Rosa damascena*. IC_{50} values for DPPH, ABTS, NO, OH, O_2 and ONOO were 10.36, 3.57, 23.01, 273.18, 42.10 and 637.57µg/ml. The extract contained significant amount of polyphenols.

Rosa dumalis Bechst. Family: Rosaceae

Gao *et al.* (2000) investigated the effects of temperature during drying and extraction on the total antioxidant activity (FRAP) in water and ethanol extracts of *Rosa dumalis* rose hips. Freeze-drying and drying at 50°C and 75°C appeared to preserve the antioxidant activity of rose hips at a high level. Degradation of antioxidant activity and ascorbic acid in water extracts was prevented by boiling for 2 and 10 minutes, and by microwave heating for 2 minutes. They also found that the decrease in total antioxidant activity occurring in extracts kept at ambient temperature was probably caused by enzymatic oxidation.

Rosa hybrida L. Family: Rosaceae

Choi and Hwang (2005) studied the effect of flower extract on plasma oxidant system and lipid levels in rats. The study showed that the intake of the extract by rats results in an increase in antioxidant enzyme activity and HDL-cholesterol, and a decrease in malondialdehyde, which minimize the risk of inflammatory and heart disease.

Rosa nutkana Presl. Family: Rosaceae

Yi *et al.* (2007) investigated antioxidant activity of rose hip extracts (*Rosa nutkana*) from wild British Columbia populations using liposome oxidation. The extracts exhibited strong antioxidant activity. *R. nutkana* pericarp extracts contained high phenolic concentrations and showed greater antioxidant activity than seed extracts.

Rosa pimpinellifolia L. Family: Rosaceae

Antioxidant and radical scavenging activities, reducing powers and the amount of total phenolic compounds of aqueous and methanolic extracts of *Rosa pimpinellifolia* were studied by Mavi *et al.* (2004). The aqueous extract showed highest peroxidation inhibition (IC_{50} : 20 mg/l), DPPH radical scavenging activity, reducing power and the amount of phenolic compounds.

Rosa pisocarpa A. Gray Family: Rosaceae

Yi *et al.* (2007) investigated antioxidant activity of rose hip extracts (*Rosa pisocarpa*) from wild British Columbia populations using liposome oxidation. The extracts exhibited strong antioxidant activity. *R. pisocarpa* pericarp extracts contained high phenolic concentrations and showed greater antioxidant activity than seed extracts.

Rosa rubiginosa L. Family: Rosaceae

English Name: Mosqueta rose.

Mosqueta rose (*Rosa rubiginosa* L.) meals were extracted with methanol, ethanol, acidified water, acetone, butanol, diethyl ether and ethyl acetate by Moure *et al.* (2001). The highest concentration of total polyphenols was found in the ethanolic and acetone extracts. The antioxidant activity of the extracts evaluated by the β-carotene assay and with hydrogen radical scavenging ability showed that the activity of the water extracts showed 80 per cent inhibition of BHT in respect to control achieved with the ethanol extracts.

Rosa rugosa Thunb. Family: Rosaceae

Recently, it was pointed that the petal of *Rosa rugosa* possess antioxidant activity through the condensed hydrolysable tannins, which may have strong antioxidant and anti-inflammatory properties (Cho *et al.*, 2003a,b; Ng *et al.*, 2005).

Park *et al.* (2009) investigated the antioxidative activities of white rose (*Rosa rugosa*) flower extract and pharmaceutical advantages of its hexane fraction via free radical scavenging effects. The hexane fraction contained the highest abundance of total phenols (330.3μg of gallic acid equivalent per gram of dry mass which is approximately 2-fold higher than that of the butanol fraction (177.3μg/g) and 2-6-fold higher than that of water fraction 123 μg/g. Hexane fraction was more active in DPPH and ABTS radical scavenging activity. The hexane fraction efficiently inhibited the degradation of BSA by free radicals in a dose-dependent manner with a maximum effect at a concentration of 10μg/ml. The hexane fraction drastically inhibited lipid peroxidation by almost as much as BHA, whereas the butanol fraction was not effective in inhibiting lipid peroxidation.

Rosa woodsii Lindl. Family: Rosaceae

Yi *et al.* (2007) investigated antioxidant activity of rose hip extracts (*Rosa woodsii*) from wild British Columbia populations using liposome oxidation. The extracts exhibited strong antioxidant activity. *R. woodsii* seed extracts had a higher phenolic concentration and greater antioxidant activity than pericarp extracts.

Rosmarinus officinalis L. Family: Lamiaceae (Labiatae)

English Name: Rosemary.

Four diterpenoids, carnosic acid (1), carnosol (2), rosmanol (3) and epirosmanol (4), were isolated as antioxidative agents from the leaves of *Rosmarinus officinalis* by bioassay-directed fractionation (Haraguchi *et al.*, 1995). These diterpenoids inhibited superoxide anion production in the xanthin/xanthin oxidase system. Mitochondrial and microsomal lipid peroxidation induced by NADH or NADPH oxidation were also completely inhibited by these diterpenes at the concentration of 3-30microM. Furthermore, carnosic acid protected red cells against oxidative hemolysis.

Sara Lopez *et al.* (1998) reported the treatment with super critical CO_2 with de odourizing antioxidant rosemary extracts obtained by steam distillation and soxhlet extraction. Evelyne *et al.* (2007) analyzed the essential oil of rosemary by means of HPLC and LC-MS in methanol extracts and identified and characterized two unknown compounds while screening for their antioxidant activity. Biljana *et al.* (2007) analyzed the essential oils of rosemary by means of gas chromatography-mass spectrometry and assayed for their antimicrobial and antioxidant activities.

Demo *et al.* (1999) reported antioxidant α-tocopherol from the Hexane extract of *Rosmarinus officinalis*. Ibanez *et al.* (1999) evaluated rosemary leaves from different sources in terms of antioxidant activity and essential oil yield and composition. Ibanez *et al.* (2003) studied the selectivity of sub critical water extraction at several temperatures to extract antioxidant compounds from rosemary leaves.

Patricia *et al.* (2003) reported the antioxidant, anticancer and antimicrobial activities of extracts from rosemary. The extracts were obtained using super critical CO_2 with and without ethanol or isopropyl alcohol as co-solvent.

Rosemary volatile oil exhibited antioxidant and radical scavenging activities on DPPH at various concentrations compared with the known synthetic antioxdants such as butylhydroxyanisol (BHA) (El-Ghorab, 2003).

The oleoresin extract of the plant scavenges superoxide radicals, inhibit lipid oxidation in food system. It contains oil soluble antioxidants. Carsonic acid, rosemarinic acid, β-sitosterol, caryophyllene oxide, eugenol and iso-eugenol in the plant have been found antioxidative in nature (Mishra *et al.*, 2007).

Two natural antioxidant preparations, namely Rosmol (liquid) and Rosmol-P (powder) were produced by Lugasi *et al.* (2006) by extraction of a mixture of medicinal plants belonging to the Lamiaceae family, such as *Rosmarinus officinalis, Prunella vulgaris, Hyssopus officinalis* and *Melissa officinalis*. The main active compound of the extract was supposed to be a phenolic (caffeic) acid derivative. The total polyphenol content of the preparation was very high, 8.72g/l for Rosmol and 93.7 g/kg for Rosmol-P. The products acted as primary and secondary antioxiants chelating transitional metal ions and inhibiting the autoxidation of linoleic acid. Rosmol and Romol-P scavenged free radicals formed during Fenton type reaction measured by chemiluminometry and also exhibited strong antioxidant property in Randox TAS measurement. The antioxidant activity of the products was unchanged after six months storage.

Mariutti *et al.* (2008) evaluated the antioxidant activity of garlic employing ABTS and DPPH radical assay. DPPH EC_{50} value was 3.86g/kg while DPPH-TEAC value was 83mM/g. ABTS-TEAC value was 89mM/g.

Rosmarinus officinalis possesses antioxidant activity and hepatoprotective effects, and so may provide a possible therapeutic alternative for chronic liver disease. The effect produced by a methanolic extract of *Rosmarinus officinalis* on CCl_4-induced liver cirrhosis in rats was investigated by Gutierrez *et al.* (2009) using both prevention and reversion models. On cirrhotic animals treated with CCl_4, histolocial studies showed a massive necrosis, periportal inflammation and fibrosis which were modified by *R. officinalis* (Guetierrez *et al.*, 2009).

The chemopreventive potential of rosemary on 7,12-dimethylbenz(a)anthracene (DMBA) initiated and promoted mouse skin tumorogensis was assessed. The modulatory effects of *R. officinalis* was monitored on the bases of the latency period, tumor incidence, tumor burden, tumor yield, tumor weight and diameter as wells as lipid peroxidation and glutathione level. The result indicated that *R. officinalis* leaves extract could prolong the latency period of tumor occurrence, decrease tumor incidence, tumor burden and tumor yield. The level of lipid peroxidation was significantly reduced in blood serum and liver. Further depleted levels of glutathione were restored in RE-administered animal groups. Thus, at a dose rate of 500mg/kg body weight oral administration of rosemary extract was found to be significantly protective against two-stage skin tumorogenesis.

Using a multiple-method approach, the antioxidant activity of the essential oils from rosemary was tested by Viuda-Martos *et al.* (2010). The essential oil was capable of chelating iron (76.06 per cent). Antioxidant activity of essential oils of five spice plants, including *Rosmarinus officinalis* used in the Mediterranean diet was evaluated by Viuda-Martos *et al.* (2010). All the essential oils tested had antioxidant activity. Yang *et al.* (2010) reported antioxidant activity of the essential oil from *Rosmarinus officinalis*. Spiridon *et al.* (2011) evaluated the total phenolic content and antioxidant activity of plants used in Romanian herbal medicine. Total phenolic concentration was determined using Folin-Ciocalteu phenol reagent method, while total flavonoids were measured using the aluminium chloride colorimetric method. *Rosmarinus officinalis* extracts showed good antioxidant activity.

Rotula aquatica Lour. Family: Boraginaceae

Patil *et al.* (2003) investigated the antioxidant potential of *Rotula aquatica* to understand their beneficial role in cancer therapy. The aqueous extract of *Rotula aquatica* showed a marked antioxidant activity when compared to ascorbic acid. The study revealed that polyphenols present in the polar extracts have a stronger antioxidant activity compared to the non-polar phytoconstituents.

Roylea cinerea (D.Don) Baill. Family: Lamiaceae (Labiatae)

Mathur *et al.* (2010) carried out phytochemical investigation and antioxidant activity of *Roylea cinerea*. Traditional solvent extraction (TSE) using four solvents – methanol, aqueous, hexane and petroleum ether were utilized to determine the content of antioxidants. DPPH assay and superoxide anion radical scavenging activity were used to determine the antioxidant capacity. The most antioxidant capacity was achieved using methanol as the solvent followed by aqueous extracts. The antioxidant capacity with in the specific extract of plant was found to be correlated with the total phenolic content. The radical scavenging activity by DPPH radical scavenging method (IC_{50}) was 31.07µg/ml (hexane extract), 37.10 µg/ml (petroleum ether extract), 22.14 (aqueous) and 25.00 (methanol extract).

Rubia cordifolia L. Family: Rubiaceae

English Name: Indian madder.

Antiperoxidative property of the solvent free alcoholic extract of *Rubia cordifolia* has been studied by Pandey *et al.* (1994) in rat liver homogenate. It prevents the cumene hydroperoxide induced malondialdehyde formation in the dose and time dependent manner. This effect was accompanied by the maintained reduced glutathione level even in the presence of above toxin.

Rubiadin, a dihydroxy anthraquinone, isolated from alcoholic extract of *Rubia cordifolia* possessed potent antioxidant property (Tripathi *et al.*, 1997, Tripathi and Sharma, 1998). It preveted lipid peroxidation induced by $FeSO_4$ and t-butylhydroperoxide in a dose dependent manner. The per cent of inhibition was more in the case of Fe^{2+} induced lipid peroxidation. The antioxidant property of the preparation has been found to be better than that of EDTA, Tris, mannitol, vitamin E and p-benzoquinone.

Joharapurkar *et al.* (2003) evaluated the *in vivo* antioxidant activity of alcoholic extract of the roots of *Rubia cordifolia*. Chronic administration of ethanol decreased the humoral and cell-mediated immune response, phagocytosis, phagocytosis index, TLC, GSH, CAT, SOD activities and increased the LPO. These influences of ethanol were prevented by concurrent daily administration of *Rubia cordifolia* and the effect was comparable with that of the combination of vitamin E and C.

R. cordifolia has been shown to exert cell/neuroprotective properties via preventing the depletion and increasing GSH (glutathione) levels by inducing GCLC (γ-glutamylcysteine ligase) expression, reducing oxidant levels by direct scavenging and decreasing iNOS expression. The protective ability may be attributed to the GSH and vitamin C content of the herb (Rawal *et al.*, 2004).

Cai *et al.* (2004a,b, 2006) evaluated the antioxidant activity and phenolic compounds in traditional Chinese medicinal plants associated with anticancer. The improved ABTS.+ method was used to systematically assess the total antioxidant capacity (Trolox equivalent antioxidant capacity, TEAC) of the medicinal extracts. The TEAC values and total phenolic content for methanolic extract of root of *Rubia cordifolia* were 623.2 µmol Trolox equivalent/100g dry weight (DW), and 2.11 g of gallic acid equivalent/100g DW respectively. A positive significant linear relationship between antioxidant activity and total phenolic content showed that phenolic compounds were the dominant antioxidant components in the tested herbs. Major types of phenolic compounds in *Rubia cordifolia* include anthraquinones (purpurin, alizarin, munjistin and their glycosides).

Extracts of the root bark of *Rubia cordifolia* were evaluated by Basu and Hazra (2005, 2006) for their effect on reactive nitric oxide (NO) levels. The NO released in presence of the extracts could be assayed by Griess reagent which interacts with the nitrite ions to form a purple azo dye estimated spectrophotometrically at 550 nm. The assessment of the same extract for their capacity to scavenge superoxide radicals was carried out. It was found that the NO-scavenging activity of the methanolic extract was significantly greater in comparison to that of the water extract, and was concentration dependent. This extract also exhibited the potency to quench reactive oxygen species at a much lower dose than that of the aqueous fraction.

Gan *et al.* (2010) evaluated the antioxidant activity of 40 medicinal plants. *Rubia cordifolia* showed moderate antioxidant activity of 116.67 µmol Fe(II)/g in FRAP assay and 89.49µmol in TEAC assay. Total phenolic contet was also very low, 5.55 mg GAE/g.

Rubia peregrina L. Family: Rubiaceae

Methanol extract of the underground parts (roots and rhizomes) of *Rubia peregrina* were tested *in vitro* for their antioxidant activity by Ozgen *et al.* (2004). The highest antioxidant activity (96 per cent) was observed at 0.25 per cent concentration on the ethyl acetate fraction.

Rubus caucasicus Focke Family: Rosaceae

Deiton *et al.* (2000) investigated the antioxidant capacity, total phenol, anthocyanin and ascorbic content of *Rubus caucasicus*. The antioxidant capacity of the fruit was 25.32µmol Trolox equivalent/g TEAC or 65669 µM/l FRAP. Ascorbic acid contributes only minimally to the antioxidant potential of *Rubus caucasicus* juice. There are apparent linear relationships between antioxidant capacity (assessed as both TEAC and FRAP) and total phenols (4.527 g/kg). Also, anthocyanin content has a minor influence on antioxidant capacity. *Rubus caucasicus* had the highest phenol content, but only a low percentage was represented by anthocyanins.

Rubus chamaemorus L. Family: Rosaceae

English Name: Cloudberry.

The antioxidative activity of phenolic extracts from fruits of *Rubus chamaemorus* was examined by autoxidation of methyl linoleate by Kahkonen *et al.* (1999). Remarkable high antioxidant activity and high total phenolic content were found in berry. In addition, berry wines made of berries showed stronger antioxidative activity in the oxidation of MeLo.

Rubus chingii Hu. Family: Rosaceae

Cai *et al.* (2004) evaluated the antioxidant activity and phenolic compounds in traditional Chinese medicinal plants associated with anticancer. The improved ABTS.+ method was used to systematically assess the total antioxidant capacity (Trolox equivalent antioxidant capacity, TEAC) of the medicinal extracts. The TEAC values and total phenolic content for methanolic extract of fruit of *Rubus chingii* were 946.1 µmol Trolox equivalent/100g dry weight (DW), and 4.54 g of gallic acid equivalent/100g DW respectively. A positive significant linear relationship between antioxidant activity and total phenolic content showed that phenolic compounds were the dominant antioxidant components in the tested herbs. Major types of phenolic compounds in *Rubus chingii* include phenolic acids (gallic acid), tannins (ellagic acid).

Rubus coreanus Miq. Family: Rosaceae

Deiton *et al.* (2000) investigated the antioxidant capacity, total phenol, anthocyanin and ascorbic content of *Rubus coreanus*. The antioxidant capacity of the fruit was from 0.03 to 7.98µmol Trolox equivalent/g TEAC) or 191 to 13078 µM/l FRAP. Ascorbic acid contributes only minimally to the antioxidant potential of *Rubus coreanus* juice. There are apparent linear relationships between antioxidant capacity (assessed as both TEAC and FRAP) and total phenols (0.267 to 1.259 g/kg). Also, anthocyanin content has a minor influence on antioxidant capacity.

The antioxidant properties of ten Korean medicinal plants were analyzed by Kim *et al.* (2010) using various antioxidant assays, including their ability to scavenge the DPPH free radical, ABTS radical, superoxide anion, and nitrite, and their reducing powers. The contents of total phenolic compounds and flavonoids were also determined. Of the ten plants, *Rubus coreanus* (RC) possessed strong antioxidant activity for all models tested. The half-maximal inhibitory concentration (IC_{50}) for DPPH free radicals of RC was 4.87 µg/ml. The IC_{50} for superoxide anion of RC 136.3 µg/ml. In addition, the ABTS radical scavenging activity of these materials was higher than that of Trolox. The total phenol content of RC was also significantly higher than in the other herbs. A correlation was observed between the antioxidant activity and total phenol content.

Rubus cyri Juz Family: Rosaceae

English Name: Blackberry.

Moyer *et al.* (2002) analyzed the anthocyanins, phenolics and antioxidant capacity in diverse small fruits including *Rubus cyri*. Total anthocynins was 143mg/100g, total phenolics was 545mg/100g, Oxygen radical absorbance capacity (ORAC) was 46.2µmol TE/g, FRAP was 71.4 µmol/g.

Seven wild and ten cultivated blackberries (Arapaho, Bartin, Black satin, Bursa 1, Bursa 2, Cherokee, Chester, Jumbo, Navaho and Ness) were analyzed for total anthocyanins, total phenolics, and antioxidant activity as ferric reducing antioxidant power (FRAP) in this study. The respective ranges of total anthocyanin and total phenolic contents of the tested samples were: black berries 0.95-1.97 and 1.73-3.79 mg/g. FRAP values varied from 35.05 to 70.41µmol/g for blackberries. Wild blackberries had the highest FRAP values. A linear relationship was observed between FRAP values and total phenolics for blueberries (Mishra *et al.*, 2007).

Rubus diversa Family: Rosaceae

Ivanova *et al.* (2005) reported high phenolics content and antioxidant properties in *Rubus diversa*. TEAC 4.23 mm/QE.

Rubus ellipticus Sm. Family: Rosaceae

Karuppusamy *et al.* (2011) evaluated the anthocyanins, ascorbic acid, total phenolics and flavonoids and antioxidant activity of methanol extracts of fruits of *Rubus ellipticus*. The anthocyanins were 1.71 CGE/100g, Ascorbic acid 44.0 AAE/100g, Total phenolics 72.0GAE/100g, Total flavonoids 86.4QE/100g and antioxidant activity (DPPH) 196.4µg/ml.

Rubus georgicus Focke Family: Rosaceae

English Name: Blackberry.

Moyer *et al.* (2002) analyzed the anthocyanins, phenolics and antioxidant capacity in diverse small fruits including *Rubus georgicus*. Total anthocynins was 89mg/100g, total phenolics was 561mg/100g, Oxygen radical absorbance capacity (ORAC) was 41.6µmol TE/g, FRAP was 96.3 µmol/g.

Rubus hunanensis Hand-Mazz. Family: Rosaceae

Deiton *et al.* (2000) investigated the antioxidant capacity, total phenol, anthocyanin and ascorbic content of *Rubus hunanensis*. The antioxidant capacity of the fruit was 10.06µmol Trolox equivalent/g TEAC or 20711 µM/l FRAP. Ascorbic acid contributes only minimally to the antioxidant potential of *Rubus hunanensis* juice. There are apparent linear relationships between antioxidant capacity (assessed as both TEAC and FRAP) and total phenols (1.702 g/kg). Also, anthocyanin content has a minor influence on antioxidant capacity.

Rubus idaeus L. Family: Rosaceae

English Name: Raspberry.

The antioxidative activity of phenolic extracts from fruits of *Rubus idaeus* was examined by autoxidation of methyl linoleate by Kahkonen *et al.* (1999). Raspberry showed a lower (23 per cent) antioxidant activity and high total phenolic content. Berry wines made of berries showed stronger antioxidative activity in the oxidation of MeLo. According to Hakkinen *et al.* (1998) redraspberry has an especially high content of ellagic acid. Extracts of red raspberries possessed also a remarkably high scavenging activity toward chemically generated superoxide radicals (Constantino *et al.*, 1992).

Deiton *et al.* (2000) investigated the antioxidant capacity, total phenol, anthocyanin and ascorbic content of *Rubus idaeus*. The antioxidant capacity of the fruit was 17.25µmol Trolox equivalent/g TEAC or 25569 µM/l FRAP. Ascorbic acid contributes only minimally to the antioxidant potential of *Rubus idaeus* juice. There are apparent linear relationships between antioxidant capacity (assessed as both TEAC and FRAP) and total phenols (2.112 g/kg). Also, anthocyanin content has a minor influence on antioxidant capacity.

Fruits and leaves from different cultivars of red raspberry plants were analyzed by Wang and Lin (2000) for total antioxidant capacity (ORAC) and total phenolic content. In addition, fruits were analyzed for total anthocyanin content. Red raspberries had the highest ORAC activity at the ripe stage. Total anthocyanin content increased with maturity of fruits. Compared with fruits, leaves were found to have higher ORAC values. Thes results showed a linear correlation between total phenolic content and ORAC activity of fruits and leaves. For ripe berries, a linear relationship existed between ORAC values and anthocyanin content.

High pigmented berry fruits like blackberry and raspberry have the highest antioxidant capacity (5100-5500 TE) (Türemis *et al.*, 2003). In a study carried out by Özgen and Scheerens (2006) antioxidant capacities of some black and red raspberries were compared. While no more difference was observed between types in terms of photochemical ingredients and antioxidant capacity, in respect to black ones, red raspberries. They found more than 80 per cent total phenolic matter and approximately 6 times more anthocyanin. Anti-oxidant capacities of black raspberries are two times more than red ones, especially, high anthocyanin (2176-2417 µg/g) and total phenolic (2853-3102µg/g).

Raspberry leaves, collected in different locations of Lithuania, were extracted with ethanol and the extracts were tested by Venskutonis *et al.* (2007) for their antioxidant activity by using ABTS+ decolourisation and DPPH scavenging methods. All extracts were active, with radical scavenging activity at the used concentrations from 20.5 to 82.5 per cent in DPPH reaction system and from 8.0 to 42.7 per cent in ABTS+ reaction. The total amount of phenolic compounds in the leaves varied from 4.8 to 12.0 mg of gallic acid equivalents in 1 g of plant extract. Quercetin glucuronide, quercetin-3-*O*-glucoside and rutin were identified in the extracts.

The antioxidant capacity (AOC) of raspberry was determined by Borges *et al.* (2010) using FRAP assay. In addition, the vitamin C content of the berries was determined and phenolic and polyphenolic compounds in the extracts were analyzed by reversed-Phase HPLC-PDA-MS and by reversed-phase HPLC-PDA with online antioxidant detection system. Raspberries had a lower anthocyanin content but there was only a slight decline in the AOC because of the presence of the ellagitannins sanguine H-6 and lambertianin C, which were responsible for 58 per cent of the HPLC-AOC of the berries. Vitamin C was responsible for 11 per cent.

Cekic and Ozgen (2010) investigated some of the chemical properties and antioxidant capacities of 14 wild red raspberry accessions selected from northern Turkey. Antioxidant capacity of fruits was determined by both FRAP and TEAC assays. The result of this study indicated that some of the wild accessions of red raspberries have higher antioxidant capacity and phytonutrient content than existing domesticated cultivars. The antioxidant capacity among samples averaged 14.6 and 14.1 µmol TE/gfw using FRAP and TEAC methods.

The effect of raspberry on early atherosclerosis in hamsters was investigated by Rouanet *et al.* (2010). They received an atherogenic diet at the same time a juice at a daily dose corresponding to the consumption of 275 ml by a 70kg human. After 12 weeks berry juices inhibited aortic lipid deposition by 79-96 per cent and triggered reduced activity of hepatic antioxidant enzymes, not accompanied by a lowered plasma cholesterol. These findings suggest that moderate consumption of berry juices and teas can help prevent the development of early atherosclerosis.

Zhang *et al.* (2010) evaluated the antioxidant properties and phenolic composition of seven primocane fall-bearing raspberry cultivars. The raspberry extracts possessed significant antioxidant activities with oxygen radical absorbance capacities (ORAC) ranging from 136.7 to 205.2µmol Trolox equivalents (TE)/g dry weight fruit and DPPH radical scavenging activities from 305 to 351 µmol TE/g. The total phenol content of raspberry cultivars varied significantly from 40.9 to 98.5 mg of gallic acid equivalent/g dry weight. Nine phenolic acids were quantified and their total amounts varied from 157.3 to 713.5µg/g. The enzyme inhibition and antioxidant properties of raspberry cultivars were not correlated with their total phenolic, anthocyanin and phenolic acid content.

Rubus innominatus S.Moore Family: Rosaceae

English Name: Raspberries.

Deiton *et al.* (2000) investigated the antioxidant capacity, total phenol, anthocyanin and ascorbic content of *Rubus innominatus*. The antioxidant capacity of the fruit ranged from 5.74 to 11.53µmol Trolox equivalent/g TEAC) or from 33436 to 44260 µM/l FRAP. Ascorbic acid contributes only minimally to the antioxidant potential of *Rubus innominatus* juice. There are apparent linear relationships between antioxidant capacity (assessed as both TEAC and FRAP) and total phenols (from 2.502 to 3.264 g/kg). Also, anthocyanin content has a minor influence on antioxidant capacity.

Moyer *et al.* (2002) analyzed the anthocyanins, phenolics and antioxidant capacity in diverse small fruits including *Rubus innominatus*. Total anthocynins was 52mg/100g, total phenolics was 126mg/100g, Oxygen radical absorbance capacity (ORAC) was 13.1µmol TE/g, FRAP was 19.9 µmol/g. The wild, red form of *R.innominatus* from China had the lowest ACY, TPH, ORAC and FRAP of the *Rubus* genotypes tested.

Rubus insularis F.Aresch Family: Rosaceae

English Name: Blackberry.

Moyer *et al.* (2002) analyzed the anthocyanins, phenolics and antioxidant capacity in diverse small fruits including *Rubus insularis*. Total anthocynins was 170mg/100g, total phenolics was 472mg/100g, Oxygen radical absorbance capacity (ORAC) was 51.4µmol TE/g, FRAP was 97.1 µmol/g.

Rubus lambertianus Focke Family: Rosaceae

Deiton *et al.* (2000) investigated the antioxidant capacity, total phenol, anthocyanin and ascorbic content of *Rubus lambertianus*. The antioxidant capacity of the fruit was 9.81µmol Trolox equivalent/g TEAC) or 51289 µM/l FRAP. Ascorbic acid contributes only minimally to the antioxidant potential of *Rubus lambertianus* juice. There are apparent linear relationships between antioxidant capacity (assessed as both TEAC and FRAP) and total phenols (3.173 g/kg). Also, anthocyanin content has a minor influence on antioxidant capacity.

Rubus leucodermis Douglas ex Torrey et A.Gray. Family: Rosaceae

Moyer *et al.* (2000) determined the range of total anthocyanin content, total phenolic content and antioxidant capacity in *Rubus leucodermis*. Anthocyanin content was determined by the pH differential method. Total phenolics content was determined via the Folin-Coicalteu method. Antioxidant capacity was measured by oxygen radical absorbing capacity (ORAC). Total anthocyanin content of *Rubus leucodermis* cultivars ranged from 230 to 627 mg/100 g fruit. Total anthocyanin content was highly correlated to TPH and ORAC.

Rubus niveus Thunb. Family: Rosaceae

English Name: Raspberries.

Moyer *et al.* (2000) determined the range of total anthocyanin content, total phenolic content and antioxidant capacity in *Rubus niveus*. Anthocyanin content was determined by the pH differential method. Total phenolics content was determined via the Folin-Ciocalteu method. Antioxidant capacity was measured by oxygen radical absorbing capacity (ORAC). Total anthocyanin content of *Rubus niveus* cultivars ranged from 230 to 627 mg/100 g fruit. Total anthocyanin content was highly correlated to TPH and ORAC.

Deiton *et al.* (2000) investigated the antioxidant capacity, total phenol, anthocyanin and ascorbic content of *Rubus niveus*. The antioxidant capacity of the fruit was 9.72μmol Trolox equivalent/g TEAC) or 40799 μM/l FRAP. Ascorbic acid contributes only minimally to the antioxidant potential of *Rubus niveus* juice. There are apparent linear relationships between antioxidant capacity (assessed as both TEAC and FRAP) and total phenols (2.991 g/kg). Also, anthocyanin content has a minor influence on antioxidant capacity.

Moyer *et al.* (2002) analyzed the anthocyanins, phenolics and antioxidant capacity in diverse small fruits including *Rubus niveus*. Total anthocynins was 230mg/100g, total phenolics was 402mg/100g, Oxygen radical absorbance capacity (ORAC) was 45.2μmol TE/g, FRAP was 69.4 μmol/g.

Rubus occidentalis L. Family: Rosaceae

English Name: Black raspberry.

Moyer *et al.* (2000) determined the range of total anthocyanin content, total phenolic content and antioxidant capacity in *Rubus occidentalis*. Anthocyanin content was determined by the pH differential method. Total phenolics content was determined via the Folin-Cocalteu method. Antioxidant capacity was measured by oxygen radical absorbing capacity (ORAC). Total anthocyanin content of *Rubus occidentalis* cultivars was 627 mg/100 g fruit. Total anthocyanin content was highly correlated to TPH and ORAC.

Fruits and leaves from different cultivars of black raspberry plants were analyzed by Wang and Lin (2000) for total antioxidant capacity (ORAC) and total phenolic content. In addition, fruits were analyzed for total anthocyanin content. Black raspberries had the highest ORAC activity during the green stages. Total anthocyanin content increased with maturity of fruits. Compared with fruits, leaves were found to have higher ORAC values. Thes results showed a linear correlation between total phenolic content and ORAC activity of fruits and leaves. For ripe berries, a linear relationship existed between ORAC values and anthocyanin content.

High pigmented berry fruits like blackberry and raspberry have the highest antioxidant capacity (5100-5500 TE) (Türemis *et al.*, 2003).

Phytonutrient accumulation and antioxidant capacity of eight developmental stages of black raspberry fruits were investigated by Özgen *et al.* (2008). Özgen *et al.* (2008) reported that antioxidant capacity and phenolic antioxidants of Midwestern black raspberries grown for direct markets are influenced by production site.

Rubus parvifolius L. Family: Rosaceae

Deiton *et al.* (2000) investigated the antioxidant capacity, total phenol, anthocyanin and ascorbic content of *Rubus parvifolius*. The antioxidant capacity of the fruit was 1.07μmol Trolox equivalent/g TEAC or 17436 μM/l FRAP. Ascorbic acid contributes only minimally to the antioxidant potential of *Rubus parvifolius* juice. There are apparent linear relationships between antioxidant capacity (assessed as both TEAC and FRAP) and total phenols (1.717 g/kg). Also, anthocyanin content has a minor influence on antioxidant capacity.

Shyur *et al.* (2005) examined antioxidant activities of 26 medicinal herbal extracts that have been popularly used as folk medicines in Taiwan. The results of scavenging DPPH radical activity showed that, among the 26 tested medicinal plants, *Rubus parvifolius* exhibited strong activities and its IC_{50} value for DPPH radical was 27µg/ml. Superoxide anion scavenging activity was IC_{50} = 151µg/ml.

The hepatoprotective and antioxidant activities of the n-butanol extract of *Rubus parvifolius* were evaluated by Gao *et al.* (2011). Treatment with the extract markedly attenuated the increase in serum alanine aminotransferase (ALT) and aspartase transferase (AST) levels caused by CCl_4 intoxication. It also significantly prevented the decrease in SOD and MDA content of liver tissue. Phytochemical analysis revealed the presence of various phenolic compounds, including caffeic acid conjugates, ellagic acid glucosides and flavonol glycosides which might be responsible for the hepatoprotective and antioxidant activities.

Rubus sumatranus Miq. Family: Rosaceae

Deiton *et al.* (2000) investigated the antioxidant capacity, total phenol, anthocyanin and ascorbic content of *Rubus sumatranus*. The antioxidant capacity of the fruit was 2.52µmol Trolox equivalent/g TEAC) or 13529 µM/l FRAP. Ascorbic acid contributes only minimally to the antioxidant potential of *Rubus sumatranus* juice. There are apparent linear relationships between antioxidant capacity (assessed as both TEAC and FRAP) and total phenols (2.287 g/kg). Also, anthocyanin content has a minor influence on antioxidant capacity.

Rubus tsangorum Hand-Mazz. Family: Rosaceae

Deiton *et al.* (2000) investigated the antioxidant capacity, total phenol, anthocyanin and ascorbic content of *Rubus tsangorum*. The antioxidant capacity of the fruit was 3.47µmol Trolox equivalent/g TEAC or 2681 µM/l FRAP. Ascorbic acid contributes only minimally to the antioxidant potential of *Rubus tsangorum* juice. There are apparent linear relationships between antioxidant capacity (assessed as both TEAC and FRAP) and total phenols (0.840 g/kg). Also, anthocyanin content has a minor influence on antioxidant capacity.

Rubus ulmifolius Schott. Family: Rosaceae

Deiton *et al.* (2000) investigated the antioxidant capacity, total phenol, anthocyanin and ascorbic content of *Rubus ulmifolius*. The antioxidant capacity of the fruit was 15.47µmol Trolox equivalent/g TEAC or 34137 µM/l FRAP. Ascorbic acid contributes only minimally to the antioxidant potential of *Rubus ulmifolius* juice. There are apparent linear relationships between antioxidant capacity (assessed as both TEAC and FRAP) and total phenols (2.362 g/kg). Also, anthocyanin content has a minor influence on antioxidant capacity.

Rubus ursinus Cham. et Schindl. Family: Rosaceae

English Name: Thornless blackberry.

Moyer *et al.* (2000) determined the range of total anthocyanin content, total phenolic content and antioxidant capacity in *Rubus ursinus*. Anthocyanin content was determined by the pH differential method. Total phenolics content was determined via the Folin-Cocalteu method. Antioxidant capacity was measured by oxygen radical absorbing capacity (ORAC). Total anthocyanin content of *Rubus ursinus* hybrid 'Marion' had the highest anthocyanin content of 230 mg/100 g fruit. Total anthocyanin content was highly correlated to TPH and ORAC.

Deiton *et al.* (2000) investigated the antioxidant capacity, total phenol, anthocyanin and ascorbic content of *Rubus ursinus*. The antioxidant capacity of the fruit ranged from 11.71 to 13.89µmol Trolox equivalent/g TEAC or from 27333 to 55529µM/l FRAP. Ascorbic acid contributes only minimally to the antioxidant potential of *Rubus ursinus* juice. There are apparent linear relationships between antioxidant capacity (assessed as both TEAC and FRAP) and total phenols (2.021 to 3.419 g/kg). Also, anthocyanin content has a minor influence on antioxidant capacity.

Fruits and leaves from different cultivars of thornless blackberry plants were analyzed by Wang and Lin (2000) for total antioxidant capacity (ORAC) and total phenolic content. In addition, fruits were analyzed for total anthocyanin content. Total anthocyanin content increased with maturity of fruits. Compared with fruits, leaves were found to have higher ORAC values. Thes results showed a linear correlation between total phenolic content and ORAC activity of fruits and leaves. For ripe berries, a linear relationship existed between ORAC values and anthocyanin content.

Moyer *et al.* (2002) analyzed the anthocyanins, phenolics and antioxidant capacity in diverse small fruits including *Rubus ursinus*. Total anthocynins was 211mg/100g, total phenolics was 629mg/100g, Oxygen radical absorbance capacity (ORAC) was 60.4µmol TE/g and FRAP was 79.8 µmol/g.

Rudbeckia fulgida (L.) Ait. var. *sullivantii* Family: Asteraceae (Compositae)

Kardosova and Machova (2006) investigated the ability of polysaccharides, isolated from leaves, to inhibit peroxidation of soyabean lecithin liposomes by OH radicals. All polysaccharides exhibited antioxidant activity. The highest inhibition of lipid peroxidation was observed with glucuronoxylans (29 per cent) from *R. fulgida*.

Ruellia tuberosa L. Family: Acanthaceae

Chen *et al.* (2006) investigated the antioxidant activity of *Ruellia tuberosa* using DPPH free radical-scavenging assay and the hydrogen peroxide-induced luminal-chemiluminiscence assay. The methanolic extract and its four fractions of water, ethyl acetate, chloroform and n-Hexane were prepared and subjected to antioxidant evaluation. The results of both methods revealed that *Ruellia tuberosa* possesses potent antioxidant activity.

Chen *et al.* (2009) studied the antioxidant activity of leaves extract towards xanthine oxidase, lipoxygenase and tyrosinase activities. The inhibitory effect of acetone extract on xanthine oxidase activity was nil, tyrosine activity was less than 20 per cent and lipoxygenase activity was about 50 per cent.

Rumex acetosa L. Family: Polygonaceae

English Name: Sorrel.

The antioxidative activity of phenolic extracts from herb of sorrel was examined by autoxidation of methyl linoleate by Kahkonen *et al.* (1999). The total phenolics and the antioxidant activity were low. The total phenolics were 3.5 mg/g GAE and antioxidant activity was 17 per cent at 500 ppm level.

Rumex acetosella L. Family: Polygonaceae

The antioxidative activity of the 80 per cent ethanol extract obtained from eleven commonly consumed wild edible plants was determined according to the phosphomolybdenum method, reducing power, metal chelating, superoxide anion and free radical scavenging activity (Özen, 2010). *Rumex acetosella* had the highest antioxidant capacities. The results indicated that the antioxidant activity of vegetables seems to be due to the presence of polyphenols, flavonoid and anthocyanoside. Total phenolics in *Rumex acetosella* were 76.6mg pyrocatechol equivalent/g dw, Flavonoids were 51.6 mg quercetin equivalent/g dw and anthocyanins 10.6 mg cyanidin glycoside equivalent/g dw.

Rumex bucephalophorus L. Family: Polygonaceae

The roots of *Rumex bucephalophorus* were analyzed for resveratrol and analogues. Two stilbene-O-methyl derivatives were identified, in addition to resveratrol by Kerem *et al.* (2003). The antioxidant capacities of all these stilbenes were determined.

Rumex crispus L. Family: Polygonaceae

English Name: Curly dock.

The antioxidant activities, reducing powers, DPPH scavenging activities and amount of total phenolic compounds of ether, ethanol and hot water extracts of leaves and seeds of *Rumex crispus* were studied by Yildirim *et al.* (2001). The antioxidant activities of extracts increase with increasing amount of extracts (50-150µg). However, the water extracts of both the leaves and seeds have shown the highest antioxidant activities. Thus, addition of 75µg of each of the above extracts of the linoleic acid emulsion caused the inhibition of peroxide formation by 96 and 94 per cent, respectively. Although the antioxidant activity of the ethanol extract of seeds was lower than the water extract, the difference between these was not statistically significant. Among all the extracts, the highest amount of total phenolic compound was found in the ethanol extract of seeds, whereas the lowest amount was found in the ether extract of seeds. Like phenolic compounds, the highest reducing power and the highest DPPH scavenging activity were found in the ethanol extract of seeds. However, the reducing activity of the ethanol extract of seeds was approximately 40 per cent that of ascorbic acid, whereas in the presence of 400µg of water and ethanol extracts of seeds scavenging activities were about 85 and 90 per cent, respectively. There were statistically significant correlations between amount of phenolic compounds and reducing power and between amount of phenolic compounds and per cent DPPH scavenging activities and also between reducing powers and per cent DPPH scavenging activities.

Borchardt *et al.* (2008) evaluated the antioxidant activity of methanol extracts of seeds of *Rumex crispus* using DPPH antioxidant assay.The antioxidant value reported in μM Trolox/100g (TE) from DPPH radical scavenging activity of crude seeds extract was 69,029.

Rumex hymenosepalus Torr. Family: Polygonaceae

Vander Jagt *et al.* (2002) evaluated the total antioxidant content of widely used medicinal plants of New Mexico. The plant *Rumex hymenosepalus* contained the higher amount of antioxidant.

Rumex induratus Boiss. et Reut. Family: Polygonaceae

Targeted metabolite analysis of aqueous extract of *Rumex induratus* leaves, in terms of phenolic compounds and organic acids, and the study of its antioxidant activity against the DPPH radical, a reactive oxygen species, hypochlorous acid, and a reactive nitrogen species, nitric oxide, were performed by Guerra *et al.* (2008). The aqueous extract exhibited a dose-related activity against all tested reactive species.

Rumex japonicus Houtt. Family: Polygonaceae

Elzaawely *et al.* (2005) evaluated total phenolic content, antioxidant activity, reducing power and antibacterial activity of ethanol, hexane, chloroform, ethyl acetate and aqueous extracts of aerial parts of *Rumex japonicus.* The ethyl acetate extract had the highest amount of phenolic compounds. It also exhibited highest reducing power and antioxidant activity when assayed by the DPPH, β-carotene bleaching and superoxide radical methods. HPLC analysis showed that pyrogallol was the predominant phenolic compound in this extract.

Rumex tuberosus L. Family: Polygonaceae

Motamed and Naghibi (2010) evaluated the antioxidant properties *Satureja mutica* and *Rumex tuberosus* by four different methods: free radical scavenging using DPPH, evaluation of xanthine-oxidase activity, inhibition of lipid peroxidation by the ferric thiocyanate method, and the deoxyribose degradation assay. The extracts showed antioxidant activity.

Rumex verticillatus L. Family: Lythraceae

Borchardt *et al.* (2008) evaluated the antioxidant activity of methanol extracts of seeds of *Rumex verticillatus* using DPPH antioxidant assay.The antioxidant value reported in μM Trolox/100g (TE) from DPPH radical scavenging activity of crude seeds extract was 123,423.

Rupechtia fusca Fernald. Family: Polygonaceae

Ruiz-Terán *et al.* (2008) evaluated the antioxidant activity of aerial parts of *Rupechtia fusca* using DPPH and β-carotene-linoleic acid bleaching assay. The extract displayed antioxidant activity comparable to the commercial antioxidant BHA. For the methanolic extract analysed, a clear relation between the total phenolic content of the extracts and their antioxidant activity was found.

Ruscus aculeatus L. Family: Ruscaceae

Wojcikowski *et al.* (2007) evaluated the antioxidant activity of aerial parts of *Ruscus aculeatus*. Oxygen radical absorbance capacity of the ethyl extract, methanol extract and aqueous extracts were 79.53, 183.65 and 138.53µmol Trolox equivalent per g of dried starting material, respectively.

Ruta chalepensis L. Family: Rutaceae

Ahmed *et al.* (2011) evaluated the antioxidant activities of 10 plants from Tunisia by the DPPH method. The essential oil of *Ruta chalepensis* presented a moderate DPPH scavenging effect. The essential oil was rich in 2-undecanone at 84.28 per cent.

Ruta graveolens L. Family: Rutaceae

Wong *et al.* (2006) investigated the antioxidant properties of 25 edible tropical plants including *Ruta graveolens* using DPPH and FRAP assays. Total phenolic content was also estimated. *Ruta graveolens* showed moderate antioxidant activity.

Pandey *et al.* (2011) evaluated the antioxidant and antidiabetic activity of ethanolic extract of *Ruta graveolens* leaves using various *in vitro* models. The alcoholic leaves extract of *Ruta graveolens* showed DPPH radical scavenging activity 08.48, 10.45, 11.15,13.01 and 19.37 per cent at respective concentrations of 10, 50, 100, 250 and 400 µg/ml and the results were compared with that of standard drug Butylated Hydroxyl Anisole (BHA). Also the extract showed relatively better hydroxyl radical scavenging activity of 42.10, 47.90, 58.71, 61.61and 72.67 per cent at respective concentrations. The total phenolic content of leaves extract was found to be 13µg/mg Catechol equivalent when assayed by Folin-Ciocalteau method. The extract significantly inhibited nitric oxide, superoxide anions and showed ferric reducing power.

Ruta montana L. Family: Rutaceae

The essential oil of aerial parts of *Ruta montana* growing in the Oran region in the west of Algeria was obtained by hydrodistillation with a 1.63 per cent yield on a dry weight basis. Gas chromatography (GC) and GC/mass spectrometry (MS) analyses were carried out to identify the chemical composition of *R. montana* essential oil. Moreover, spectrophotometric analyses were employed to highlight the scavenger capacity of this oil using the DPPH test. Twenty compounds were identified by GC and CG/MS analyses, and the bulk of the compounds of the oil were undecan-2-one (32.8 per cent), nonan-2-one (29.5 per cent), nonanol-2-acetate (18.2 per cent), and psoralen (3.5 per cent). The results obtained using the DPPH test show that *R. montana* essential oil possesses antiradical activity in a concentration-dependent manner. Thus, a linear correlation was found between the reduction of DPPH stable free radical and the concentration of *R. montana* essential oil (Kambouche *et al.*, 2008).

Saccharum officinarum L. Family: Poaceae (Graminae)

English Name: Sugarcane.

Sugarcane juice is widely consumed by people of the tropics and subtropics. It has been used to cure jaundice and liver-related disorders in Indian systems of medicine. It's possible mechanism of action was examined in terms of antioxidant availability by Kadam *et al.* (2008). The assays involved different levels of antioxidant action such as oxygen radical absorbance capacity (ORAC), radical scavenging abilities using DPPH, ABTS, FRAP and protection of membranes examined by inhibition of lipid peroxidation. In addition, the content of phenols and total flavonoids were measured. The extracts of three varieties showed good antioxidant properties. In conclusion the study revealed that the ability of sugarcane juice to scavenge free radicals reduce iron complex and inhibit lipid peroxidation, may explain possible mechanisms by which sugarcane juice exhibits its beneficial effects in relation to its reported health benefits.

From sugarcane juice, a flavone, identified as tricin-7-O-β-(6"-methoxycinnamic)-glucoside, was isolated in addition to orientin (Duarte-Almeida *et al.*, 2007). The tricin derivative was shown to have antioxidant activity higher than Trolox by means of the DPPH assay and lower by the β-carotene/linoleic acid system.

DPPH-scavenging antioxidant potential of *in vitro* regenerated tissues of *Saccharum officinarum* was carried out by Ahmad *et al.* (2011). Their study revealed that significantly higher activity was observed in callus of *Saccharum officinarum*.

Saccharum spontaneum L. Family: Poaceae (Graminae)

Ripa *et al.* (2009) investigated the *in vitro* antioxidant activity of chloroform extract of flower of *Saccharum spontaneum*. Antioxidant activity was determined by DPPH free radical scavenging assay. The crude chloroform extract and standard ascorbic acid showed antioxidant activity with the IC_{50} value of 51.04µg/ml and 43.04µg/ml respectively.

Sacoglottis gabonensis (Baill.) Urban. Family: Houmiriaceae

Maduka and Okoye (2002) investigated the antioxidant activity of *Sacoglottis gabonensis* by *in vivo* and they found the activity good.

Sageretia theezans (L.) Brongn. Family: Rhamnaceae

Myricetrin, europetin-3-O-α-L-rhamnoside and 7-O-methylquercetin 3-O- α-L-rhamnopyranoside isolated from the leaves of *Sageretia theezans* showed stronger antioxidant activities than ascorbic acid and α-tocopherol, 7-O-methylmenarnsitrin showed very weak antioxidant activities by ESR and LDL oxidation inhibition tests (Chung *et al.*, 2004).

Salacia oblonga Wall. ex Wight et Arn. Family: Celastraceae

Sabu and Kuttan (2003) investigated the antioxidant activity of Indian herbal drugs, including *Salacia oblonga*, in rats with alloxan-induced diabetes. The extract of *Salacia oblonga* showed antidiabetic and antioxidant activity.

Salacia pallescens Oliv. Family: Celastraceae

Sofidya *et al.* (2006) evaluated the DPPH radical scavenging, total antioxidant activities, reducing power and contents of phenolic compounds in methanolic leaf extracts of *Salacia pallescens*. The extract showed a dose-dependent effect both on the free radical scavenging DPPH and also on Fe^{3+} reducing power. *Hymenocardia acida* possess very high radical scavenging activity in both assays. Total antioxidant activities correlated with total phenols. A linear positive relationship existed between the reducing power and total phenolics.

Salicornea brachiata Roxb. Family: Chenopodiaceae

Vadlapudi and Naidu (2009) evaluated the antioxidant potential of *Salicornea brachiata*, a mangrove plant. The SOD was 0.921U/mg, Catalase was 0.28U/mg, Ascorbic acid was 180mg/100g, DPPH radical inhibition was 30.2 per cent and FRAP units were 180.

Salicornia herbacea (L.) L. Family: Chenopodiaceae

Ha *et al.* (2006) investigated the role of *Salicornia herbacea* in ovariectomy-induced oxidative stress. Antioxidative values in the ovariectomized group decreased, by those in the *Salicornia herbacea*-treated group increased due to free radical scavenging activity of *Salicornia herbacea*.

Jang *et al.* (2007) investigated the antioxidant activity of enzyme-treated *Salicornia herbacea* (SH) extracts by *in vitro* assays observing the inhibitory activity of rat liver microsomal lipid peroxidation and DPPH radical scavenging activity. The water and methanol extracts from enzyme-treated SH inhibited the lipid peroxidation in a dose-dependent manner over a concentration range of 0.1-10 mg/ml. The activity of enzyme-treated water and methanol extract was stronger than that of non-enzyme-treated water and methanol extracts. The inhitory activity of the water extract was higher at a concentration of 1.0mg/ml than that of methanol extract. The activity was the highest in the enzyme-treated water extract, and was approximately 1.08 times higher than α-tocopherol. The DPPH radical scavenging activities of SH extracts were similar to their lipid peroxidation inhibitory activity.

Salix aegyptiaca L. Family: Salicaceae

Sonboli *et al.* (2010) examined the *in vitro* antioxidant activities and total phenolic contents of the methanolic extracts from male inflorescence of *Salix aegyptiaca* grown in Iran. The methanolic extract (ME) and its three fractions including water (WF), butanol (BF) and chloroform (CF) were prepared and then their antioxidant activities, as well as total phenolic contents, were evaluated by DPPH free radical scavenging assay and the Folin–Ciocalteu method, respectively. Among the different fractions of methanol extract, BF indicated the most antioxidant activity with an IC_{50} value of 27.7 μg/ml and total phenols of 313.8 ppm, which was comparable with the synthetic antioxidant BHT (IC_{50} = 26.5μg/ml). The antioxidant activities of the other fractions decreased in the order of ME >WF > CF.

Salix alba L. Family: Salicaceae

English Name: Willow.

The antioxidative activity of phenolic extracts from bark and leaf was examined by autoxidation of methyl linoleate by Kahkonen *et al.* (1999). Remarkable high antioxidant activity and high total phenolic content were found in silver willow.

Salix caprea L. Family: Salicaceae

English Name: Willow.

The antioxidative activity of phenolic extracts from bark and leaf was examined by autoxidation of methyl linoleate by Kahkonen *et al.* (1999). Remarkable high antioxidant activity and high total phenolic content were found in willow. The antioxidative effect was high (96 per cent).

Alam *et al.* (2006) evaluated the antioxidant activity of an ethanol extract of *Salix caprea* flowers. They were found to possess a large amount of polyphenols and also exhibited a high reducing ability. The extract significantly and dose dependently scavenged DPPH, superoxide, hydrogen peroxide (H_2O_2) and nitric oxide (NO). Their study indicated that the flowers possess a significant antioxidant and hepatoprotective property, the former being implicated in the latter.

Salix warburgii O.Seem. Family: Salicaceae

Lee *et al.* (2006) investigated the antioxidant activity of ethanol extract of leaves using DPPH, hydroxyl and superoxide radical scavenging and reducing power activities as well as the induction of heme oxygenase-1 (HO-1). The IC_{50} values against DPPH radical were 12.7µg/ml, OH radical 0.67µg, superoxide radical 39.0µg/ml and total phenolics was 24.7 mg of GAE/g.

Salix sp. Family: Salicaceae

English Name: Willow.

Souri *et al.* (2008) investigated the antioxidant activity and radical scavenging activity of methanolic extract of bark of *Salix* sp. The antioxidant activity of this plant extract against linoleic acid peroxidation expressed as IC_{50} was 21.57ng/ml and radical scavenging activity against DPPH expressed as IC_{50} was 1.03 µg/ml.

Salvadora persica L. Family: Salvadoraceae

English Name: Meshwak

Souri *et al.* (2008) investigated the antioxidant activity and radical scavenging activity of methanolic extract of stem. The antioxidant activity of this plant extract against linoleic acid peroxidation expressed as IC_{50} was 52.35ng/ml and radical scavenging activity against DPPH expressed as IC_{50} was 37.19 µg/ml.

Vadlapudi and Naidu (2009) evaluated the antioxidant potential of *Salvadora persica*, a mangrove plant. The SOD was 0.26U/mg, Catalase was 0.63U/mg, Ascorbic acid was 220mg/100g, DPPH radical inhibition was 25.9 per cent and FRAP units were 840.

Salvia aethiopica L. Family: Lamiaceae (Labiatae)

Tepe *et al.* (2006) examined the *in vitro* antioxidant activities of the methanol extract of *Salvia aethiopica*. The extracts were screened for their possible antioxidant activities by two complementary test systems, namely DPPH free radical-scavenging and β-carotene/linoleic acid systems. Non-polar subfractions of the methanol extract did not show any antioxidant activity in both test systems. The polar subfractions showed antioxidant activity.

Salvia africana-caerulea L. Family: Lamiaceae (Labiatae)

The antioxidant and antiinflammatory activities of the methanol:chloroform (1:1) extracts of 16 *Salvia* species indigenous to South Africa were evaluated by Kamatou *et al.* (2010). Antioxidant activity was measured using ABTS and DPPH scavenging assays and compared to the control values obtained with Trolox. The solvent extract of *Salvia africana-caerulea* displayed antioxidant activity, with IC_{50} value of 33.4 µg/ml using DPPH whilst the IC_{50} value was 39.7µg/ml, when tested with ABTS. Total phenolic content was 115 mgGAE/g. Betulafolientriol oxide and rosmarinic acid were detected in all the species investigated and rosmarinic acid, carnosic acid, carnosol and oleanolic acid/ursolic acid were abundant in many species.

Salvia africana-lutea L. Family: Lamiaceae (Labiatae)

The antioxidant and antiinflammatory activities of the methanol:chloroform (1:1) extracts of 16 *Salvia* species indigenous to South Africa were evaluated by Kamatou *et al.* (2010). Antioxidant activity was measured using ABTS and DPPH scavenging assays and compared to the control values obtained with Trolox. The solvent extract of *Salvia africana-lutea* displayed antioxidant activity, with IC_{50} value of 47.6 µg/ml using DPPH whilst the IC_{50} value was 30.4µg/ml, when tested with ABTS. Total phenolic content was 67.8 mgGAE/g. Betulafolientriol oxide and rosmarinic acid were detected in all the species investigated and rosmarinic acid, carnosic acid, carnosol and oleanolic acid/ursolic acid were abundant in many species.

Salvia albicaulis Benth. Family: Lamiaceae (Labiatae)

The antioxidant and antioinflammatory activities of the methanol:chloroform (1:1) extracts of 16 *Salvia* species indigenous to South Africa were evaluated by Kamatou *et al.* (2010). Antioxidant activity was measured using ABTS and DPPH scavenging assays and compared to the control values obtained with Trolox. The solvent extract of *Salvia albicaulis* displayed antioxidant activity, with IC_{50} value of 19.9 µg/ml using DPPH whilst the IC_{50} value was 24.2µg/ml, when tested with ABTS. Total phenolic content was 100 mgGAE/g. Betulafolientriol oxide and rosmarinic acid were detected in all the species investigated and rosmarinic acid, carnosic acid, carnosol and oleanolic acid/ursolic acid were abundant in many species.

Salvia aramiensis Rech. f. Family: Lamiaceae (Labiatae)

Essential oils of *Salvia aramiensis* and two other species of *Salvia* were screened by Kelen and Tepe (2007) for their possible antioxidant properties as well as their chemical compositions. Antioxidant activity was employed by two complementary test systems namely DPPH free radical scavenging and β-carotene/linoleic acid systems. Antioxidant activity of *S. aramiensis* was found to be higher than those of the others for both the systems (12.26 and 92.46 per cent μg, respectively).

Salvia aucheri Boiss. *var. aucheri* Family: Lamiaceae (Labiatae)

Essential oils of *Salvia aucheri* var. *aucheri* and two other species of *Salvia* were screened by Kelen and Tepe (2007) for their possible antioxidant properties as well as their chemical compositions. Antioxidant activity was employed by two complementary test systems namely DPPH free radical scavenging and β-carotene/linoleic acid systems. Essential oil showed moderate antioxidant activity.

Salvia aurita L.f. Family: Lamiaceae (Labiatae)

The antioxidant and antiinflammatory activities of the methanol:chloroform (1:1) extracts of 16 *Salvia* species indigenous to South Africa were evaluated by Kamatou *et al.* (2010). Antioxidant activity was measured using ABTS and DPPH scavenging assays and compared to the control values obtained with Trolox. The solvent extract of *Salvia aurita* displayed antioxidant activity, with IC_{50} value of 16.6 μg/ml using DPPH whilst the IC_{50} value was 22.9μg/ml, when tested with ABTS. Total phenolic content was 119 mgGAE/g. Betulafolientriol oxide and rosmarinic acid were detected in all the species investigated and rosmarinic acid, carnosic acid, carnosol and oleanolic acid/ursolic acid were abundant in many species.

Salvia brachyantha (Bordz.) Probed. Family: Lamiaceae (Labiatae)

Esmaeili and Soboli (2010) assessed the antioxidant and free radical scavenging activities of *Salvia brachyantha in vitro* using DPPH radical, β-carotene linoleic acid, superoxide anion radical, hydroxyl radical, and reducing power assays. The plant extract exhibited antioxidant and radical scavenging activities at different magnitudes of potency. In addition, this study was undertaken to assess whether methanol extract could increase the endogenous antioxidant enzymes in cells, and where such increased cellular defence could provide protection against oxidative cell injury. Increased reactive oxygen species and cell apoptosis induced by xanthine/xanthane oxidase was dose-dependently prevented when cells were pretreated for 24 h with plant extract. These results indicated that *S. brachyantha* could protect against cell injury via induction of the antioxidant enzyme defences.

Salvia caespitosa Montbret et Aucher ex Benth. Family: Lamiaceae (Labiatae)

Tepe *et al.* (2006) examined the *in vitro* antioxidant activities of the methanol extract of *Salvia caespitosa*. The extracts were screened for their possible antioxidant activities by two complementary test systems, namely DPPH free radical-scavenging and β-carotene/linoleic acid systems. Non-polar subfractions of the methanol extract did not show any antioxidant activity in both test systems. The polar subfractions showed antioxidant activity.

Salvia candelabrum Boiss. Family: Lamiaceae (Labiatae)

Zupkó *et al.* (2001) evaluated the antioxidant activity of leaves of eleven *Salvia* species in enzyme-dependent and enzyme-independent systems of lipid peroxidation and their phenolic constituents. The 50 per cent aqueous methanolic extract of *Salvia candelabrum* displayed considerable concentration-dependent antioxidative effects that were comparable to those of *Salvia officinalis*

From the most active fractions of methanolic extract of the aerial parts of *Salvia candelabrum* seven abietane and seco-abietane diterpenes were isolated. All the identified compounds were evaluated for antioxidant activity in enzyme-dependent and enzyme-independent systems of lipid peroxidation. All compounds displayed marked concentration-dependent effects in both tests as compared with those of authentic ascorbic, rosmarinic and caffeic acids (Janicsak *et al.*, 2003).

Salvia chamelaeagnea Berg. Family: Lamiaceae (Labiatae)

The antioxidant and antiinflammatory activities of the methanol:chloroform (1:1) extracts of 16 *Salvia* species indigenous to South Africa were evaluated by Kamatou *et al.* (2010). Antioxidant activity was measured using ABTS and DPPH scavenging assays and compared to the control values obtained with Trolox. The solvent extract of *Salvia chamelaeagnea* displayed antioxidant activity, with IC_{50} value of 12.8 µg/ml using DPPH whilst the IC_{50} value was 14.6µg/ml, when tested with ABTS. Total phenolic content was 212 mgGAE/g. Betulafolientriol oxide and rosmarinic acid were detected in all the species investigated and rosmarinic acid, carnosic acid, carnosol and oleanolic acid/ursolic acid were abundant in many species.

Salvia cryptantha Montbret et Aucher ex Benth. Family: Lamiaceae (Labiatae)

The essential oils and methanolic extracts of *Salvia cryptantha* were examined by Tepe *et al.* (2004) for their potential antimicrobial and radical scavenging activities. Antioxidant activities of the polar subfraction and the essential oil were examined using DPPH, hydroxyl radical scavenging and lipid peroxidation assays. The essential oils, in particular, and the non-polar subfractions of methanol extracts, showed antioxidant activity.

Salvia disermas L. Family: Lamiaceae (Labiatae)

The antioxidant and antiinflammatory activities of the methanol:chloroform (1:1) extracts of 16 *Salvia* species indigenous to South Africa were evaluated by Kamatou *et al.* (2010). Antioxidant activity was measured using ABTS and DPPH scavenging assays and compared to the control values obtained with Trolox. The solvent extract of *Salvia disermas* displayed antioxidant activity, with IC_{50} value of 55.1 µg/ml using DPPH whilst the IC_{50} value was 33.1µg/ml, when tested with ABTS. Total phenolic content was 69 mgGAE/g. Betulafolientriol oxide and rosmarinic acid were detected in all the species investigated and rosmarinic acid, carnosic acid, carnosol and oleanolic acid/ursolic acid were abundant in many species.

Salvia dolomitica Codd. Family: Lamiaceae (Labiatae)

The antioxidant and antiinflammatory activities of the methanol:chloroform (1:1) extracts of 16 *Salvia* species indigenous to South Africa were evaluated by Kamatou *et al.* (2010). Antioxidant activity was measured using ABTS and DPPH scavenging assays and compared to the control values obtained with Trolox. The solvent extract of *Salvia dolomitica* displayed antioxidant activity, with IC_{50} value of >100 µg/ml using DPPH whilst the IC_{50} value was 49.9µg/ml, when tested with ABTS. Total phenolic content was 53 mgGAE/g. Betulafolientriol oxide and rosmarinic acid were detected in all the species investigated and rosmarinic acid, carnosic acid, carnosol and oleanolic acid/ursolic acid were abundant in many species.

Salvia eremophila Boiss. Family: Lamiaceae (Labiatae)

The essential oil composition and *in vitro* antioxidant and antimicrobial activity of the essential oil and methanol extract of *Salvia eremophila* were investigated by Ebrahimabadi *et al.* (2010). While the essential oil showed only weak antioxidant activities, its methanol extract was considerably active in both DPPH (IC_{50} = 35.19µg/ml) and β-carotene linoleic acid (inhibition percentage: 72.42 per cent) tests. Appreciable total phenolic content (101.25µg/g) was also detected for the plant methanol extract as gallic acid equivalent in the Folin-Ciocalteu test.

Firuzi *et al.* (2010) evaluated the antioxidant activity and total phenolic content of 24 Lamiaceae species growing in Iran. *Salvia eremophila* showed very high antioxidant activity with 45.3µM/g DW FRAP value and IC_{50} = 206.1µg/ml of DPPH radical scavenging activity. It also showed high phenolic content of 16.1 mg CE/g. FRAP and DPPH assay results showed good correlations with the total phenolic contents of the plant measured by the Folin-Ciocalteu assay.

Salvia euphratica Montbret and Aucher ex Benth. subsp. *euphratica* Family: Lamiaceae (Labiatae)

Tepe *et al.* (2006) examined the *in vitro* antioxidant activities of the methanol extract of *Salvia euphratica*. The extracts were screened for their possible antioxidant activities by two complementary test systems, namely DPPH free radical-scavenging and β-carotene/linoleic acid systems. Non-polar subfractions of the methanol extract did not show any antioxidant activity in both test systems. The polar subfractions showed antioxidant activity. Among the polar subfractions of the six species of *Salvia* studied, *Salvia euphratica* subsp. *euphratica* was most active with IC_{50} value of 20.7 µg/ml.

Salvia fruticosa Mill. Family: Lamiaceae (Labiatae)

Demo *et al.* (1998) reported antioxidant α-tocopherol from the Hexane extract of *Salvia fruticosa*. Al-Mustafa and Al-Thunibat (2008) evaluated the antioxidant activity of Jordanian medicinal plant *Salvia fruticosa* shoot. The level of antioxidant activity was determined by DPPH and ABTS assay in relation to total phenolic contents of the medicinally used part. They concluded that *Salvia fruticosa* shoot has moderate antioxidant capacity (DPPH-TEAC range 20-80mg/g).

Salvia garipensis L. Family: Lamiaceae (Labiatae)

The antioxidant and antiinflammatory activities of the methanol:chloroform (1:1) extracts of 16 *Salvia* species indigenous to South Africa were evaluated by Kamatou *et al.* (2010). Antioxidant activity was measured using ABTS and DPPH scavenging assays and compared to the control values obtained with Trolox. The solvent extract of *Salvia garipensis* displayed antioxidant activity, with IC_{50} value of 74.2 µg/ml using DPPH whilst the IC_{50} value was 41.7µg/ml, when tested with ABTS. Total phenolic content was 45.6 mgGAE/g. Betulafolientriol oxide and rosmarinic acid were detected in all the species investigated and rosmarinic acid, carnosic acid, carnosol and oleanolic acid/ursolic acid were abundant in many species.

Salvia glutinosa L. Family: Lamiaceae (Labiatae)

Zupkó *et al.* (2001) evaluated the antioxidant activity of leaves of eleven *Salvia* species in enzyme-dependent and enzyme-independent systems of lipid peroxidation and their phenolic constituents. The 50 per cent aqueous methanolic extract of *Salvia glutinosa* displayed considerable concentration-dependent antioxidative effects that were comparable to those of *Salvia officinalis*.

Miliauskas *et al.* (2009) investigated the radical scavenging activity of acetone, ethyl acetate and methanol extracts of leaves at full bloom stage using DPPH and ABTS assays. They found that methanol extracts were most effective DPPH radical scavengers. Ethyl acetate and acetone extracts were considerably less effective radical scavengers. High content of phenolics in the extracts correlated with its antiradical activity, confirming that phenolic compounds are likely to contribute to the radical scavenging activity of this plant extracts.

Salvia halophila Hedge Family: Lamiaceae (Labiatae)

Bozan *et al.* (2002) stated that methanol extracts of *S. halophila* and other *Salvia* species exhibited antioxidant and free radical scavenging activity.

The antioxidant activity of the ethanol extract of *Salvia halophila* was evaluated by Albayrak *et al.* (2008) by 2 antioxidant assays, including phosphomolybdenum reduction and free radical scavenging activity. The ethanol extract showed free radical scavenging activity with IC_{50} of 67.73µg/ml in DPPH assay. The antioxidant activity of the extract was found to be 84.87 mg ascorbic acid equivalent (AAE)/g extract in phosphomolybdenum assay. In addition to the antioxidant activity of the extract, the total phenolic content was measured. Total phenolic content of the ethanol extract was found to be 8.21 gallic acid equivalent (GAE)/g extract.

Salvia hydrangea DC. ex Benth. Family: Lamiaceae (Labiatae)

English Name: Mountain sage

Souri *et al.* (2008) investigated the antioxidant activity and radical scavenging activity of methanolic extract of aerial part. The antioxidant activity of this plant extract against linoleic acid peroxidation expressed as IC_{50} was 2.50ng/ml and radical scavenging activity against DPPH expressed as IC_{50} was 1.55 µg/ml.

Firuzi *et al.* (2010) evaluated the antioxidant activity and total phenolic content of 24 Lamiaceae species growing in Iran. *Salvia hydrangea* showed high antioxidant activity with 30.8µM/g DW FRAP value and IC_{50} = 301.9µg/ml of DPPH radical scavenging activity. It also showed high phenolic content of 10.9 mg CE/g. FRAP and DPPH assay results showed good correlations with the total phenolic contents of the plant measured by the Folin-Ciocalteu assay.

Salvia hypargeia Fischer et May Family: Lamiaceae (Labiatae)

Tepe *et al.* (2006) examined the *in vitro* antioxidant activities of the methanol extract of *Salvia hypargeia*. The extracts were screened for their possible antioxidant activities by two complementary test systems, namely DPPH free radical-scavenging and β-carotene/linoleic acid systems. Non-polar subfractions of the methanol extract did not show any antioxidant activity in both test systems. The polar subfractions showed antioxidant activity. In the β-carotene/linoleic acid systems, polar extract showed 69.2 per cent inhibition.

Salvia hypoleuca Benth. Family: Lamiaceae (Labiatae)

Souri *et al.* (2004) evaluated the antioxidative activity of *Salvia hypoleuca* by linoleic acid peroxidation test using 1,3-diethyl-2-thiobarbituric acid as the reagent. Methanolic extract showed 91.90 per cent inhibition at 40µg concentration with IC_{50} value of 5.27µg.

The radical scavenging activity of ethanolic extract of *Salvia hypoleuca* was evaluated *in vitro* by Nickavar *et al.* (2007) with the spectrophotometric method based on the reduction of the stable DPPH free radical. The extract exhibited concentration dependent radical scavenging activity and IC_{50} value was 36.81µg/ml. Total flavonoid content in rutin equivalent was 53.16µg/mg. However, a favourable correlation was not found between the radical scavenging potency and the total flavonoid content.

Salvia lanceolata L. Family: Lamiaceae (Labiatae)

The antioxidant and antiinflammatory activities of the methanol:chloroform (1:1) extracts of 16 *Salvia* species indigenous to South Africa were evaluated by Kamatou *et al.* (2010). Antioxidant activity was measured using ABTS and DPPH scavenging assays and compared to the control values obtained with Trolox. The solvent extract of *Salvia lanceolata* displayed antioxidant activity, with IC_{50} value of 68.1 µg/ml using DPPH whilst the IC_{50} value was 25.6µg/ml, when tested with ABTS. Total phenolic content was 54.2 mgGAE/g. Betulafolientriol oxide and rosmarinic acid were detected in all the species investigated and rosmarinic acid, carnosic acid, carnosol and oleanolic acid/ursolic acid were abundant in many species.

Salvia leriifolia Benth. Family: Lamiaceae (Labiatae)

Essential oil from the aerial parts of *Salvia leriifolia* exhibited a promising antioxidant activity by DPPH assay with an IC_{50} 2.26 mul/ml (Loizzo *et al.*, 2009). Loizzo *et al.* (2010) investigated the *in vitro* antioxidant properties of *Salvia leriifolia* extracts and fractions. Antioxidant activity was tested using DPPH, β-carotene bleaching test, and thiobarbituric acid assay. The ethyl acetate extract showed the highest activity, with IC_{50} values of 2 and 33µg/ml on β-carotene bleaching test, and thiobarbituric acid test, respectively.

Salvia macrosiphon Boiss. Family: Lamiaceae (Labiatae)

Souri *et al.* (2004) evaluated the antioxidative activity of *Salvia macrosiphon* by linoleic acid peroxidation test using 1,3-diethyl-2-thiobarbituric acid as the reagent. Methanolic extract showed 91.52 per cent inhibition at 40µg concentration with IC_{50} value of 2.96µg.

Salvia miltiorrhiza Bunge. Family: Lamiaceae (Labiatae)

English Name: Tanshen.

Gordon and Weng (1992) reported that powdered tanshen has strong antioxidant properties in lard. Powdered tanshen was less effective than rosemary but more effective than other plant materials with antioxidant properties including licorice, tea and ginseng. A hexane extract from tanshen was a very effective antioxidant both at 100°C and at 180°C. The tanshen extract acts as a primary antioxidant showing a synergistic effect with citric acid and also stabilizing lard containing tocopherols. Weng and Gordon (1992) isolated six compounds from *Salvia miltiorrhiza*. They proposed that the quinines act as primary antioxidants by addition of the lipid radical to the quinine to form a stabilized radical.

Cai *et al.* (2004) evaluated the antioxidant activity and phenolic compounds in traditional Chinese medicinal plants associated with anticancer. The improved ABTS.+ method was used to systematically assess the total antioxidant capacity (Trolox equivalent antioxidant capacity, TEAC) of the medicinal extracts. The TEAC values and total phenolic content for methanolic extract of root of *Salvia miltiorrhiza* were 761.5 μmol Trolox equivalent/100g dry weight (DW), and 4.26 g of gallic acid equivalent/100g DW respectively. A positive significant linear relationship between antioxidant activity and total phenolic content showed that phenolic compounds were the dominant antioxidant components in the tested herbs. Major types of phenolic compounds in *Salvia miltiorrhiza* include phenanthraquinones (tanshinone I, tanshinone II_A and II_B), flavonoids (baicalin), phenolic acids (protocatechuic acid).

Zhao *et al.* (2006) determined that *S. miltiorhiza* extract possessed high scavenging activities against three radicals including superoxide anion, hydroxyl and DPPH radicals.

Wojcikowski *et al.* (2007) evaluated the antioxidant activity of aerial parts of *Salvia miltiorrhiza*. Oxygen radical absorbance capacity of the ethyl extract, methanol extract and aqueous extracts were 6.98, 13.89 and 118.35μmol Trolox equivalent per g of dried starting material, respectively.

The effect of the complex preparation (called Sheng), made of puerarin (isolated from *Pueraria lobata*) and Danshensu (isolated from the Chinese herb *Salvia miltiorrhiza*) were studied on the acute ischemic myocardinal injury in rats. Shenge exerted significant cardioprotective effects against acute ischemic myocardial injury in rats, likely through its antioxidant and antilipid peroxidation properties, and thus may be an effective medicine for both prophylaxis and treatment of ischemic heart disease (Wu *et al.*, 2007). Matkowski *et al.* (2008) reported that the antioxidant activity correlates to the total polyphenol and, depending on the assay, to the hydroxycinnamic acids content.

The acetone and methanol extracts of leaves (AL and ML respectively) of *Salvia miltiorrhiza* were evaluated by Zhang *et al.* (2010) by various *in vitro* antioxidant assays. The total phenolic contents of AL and ML were 39.0 and 54.3 mg gallic acid equivalents/g extract tested, respectively. EC_{50} of ML was 7.0μg/ml in DPPH radical scavenging assay and 246.5 μg/ml in superoxide radical quenching assay. It was also found that ML has prominent effects on the inhibition of linoleic acid oxidation (93.2 per cent) which was equivalent to the positive control, BHT and was significantly higher than α-tocopherol. The reducing power of leaf extracts was as strong as roots. HPLC and correlation analysis show that salvionolic acid B and rosmarinic acid constitute the most abundant phenolic compounds. They are the major contributors to antioxidant activities.

Gan *et al.* (2010) evaluated the antioxidant activity of 40 medicinal plants associated with prevention and treatment of cardiovascular and cerebrovascular diseases. A high correlation between antioxidant capacity and total phenolic content indicated that phenolic compounds could be the main contributor of the antioxidant activity of these plants. *Salvia miltiorhiza* showed high antioxidant activity of 788.78 μmol Fe(II)/g in FRAP assay and 185.67μmol in TEAC assay. Total phenolic content was also very high, 29.60 mg GAE/g.

Salvia mirzayani Rech. Family: Lamiaceae (Labiatae)

Moein *et al.* (2007) investigated the free radical scavenging capacity, reducing power, and inhibition of β-carotene peroxidation of ethanol extract of *Salvia mirzayani*. The extract showed free radical scavenging capacity and reducing power and prevented β-carotene peroxidation. These results were confirmed by a protective effect on H_2O_2-induced cytotoxicity in human nonimmortalized fibroblasts.

Salvia muirii L. Family: Lamiaceae (Labiatae)

The antioxidant and antiinflammatory activities of the methanol:chloroform (1:1) extracts of 16 *Salvia* species indigenous to South Africa were evaluated by Kamatou *et al.* (2010). Antioxidant activity was measured using ABTS and DPPH scavenging assays and compared to the control values obtained with Trolox. The solvent extract of *Salvia muirii* displayed antioxidant activity, with IC_{50} value of 11.1 μg/ml using DPPH whilst the IC_{50} value was 11.9μg/ml, when tested with ABTS. Total phenolic content was 186 mgGAE/g. Betulafolientriol oxide and rosmarinic acid were detected in all the species investigated and rosmarinic acid, carnosic acid, carnosol and oleanolic acid/ursolic acid were abundant in many species.

Salvia multicaulis Vahl Family: Lamiaceae (Labiatae)

The essential oils and methanolic extracts of *Salvia multicaulis* were examined by Tepe *et al.* (2004) for their potential antimicrobial and radical scavenging activities. Antioxidant activities of the polar subfraction and the essential oil were examined using DPPH, hydroxyl radical scavenging and lipid peroxidation assays. The essential oils, in particular, and the non-polar subfractions of methanol extracts, showed antioxidant activity.

Erdemoglu *et al.* (2006) evaluated the antioxidant activity of the extract of *Salvia multicaulis* by using DPPH as well as by flow-injection analysis-luminol chemiluminiscence (FIA-CL). Extracts were shown to possess a significant scavenger activity against DPPH free radical and an inhibitory effect on H_2O_2- or HOCl-luminol chemiluminiscence. The results concluded that the extracts have a potential source of antioxidants of natural origin.

Firuzi *et al.* (2010) evaluated the antioxidant activity and total phenolic content of 24 Lamiaceae species growing in Iran. *Salvia multiacaulis* showed high antioxidant activity with 40.8µM/g DW FRAP value and IC_{50} = 246.6µg/ml of DPPH radical scavenging activity. It also showed high phenolic content of 13.0 mg CE/g. FRAP and DPPH assay results showed good correlations with the total phenolic contents of the plant measured by the Folin-Ciocalteu assay.

Salvia namaensis (L.) Schinz. Family: Lamiaceae (Labiatae)

The antioxidant and antiinflammatory activities of the methanol:chloroform (1:1) extracts of 16 *Salvia* species indigenous to South Africa were evaluated by Kamatou *et al.* (2010). The solvent extract of *Salvia namaensis* displayed antioxidant activity, with IC_{50} value of 10.6 µg/ml using DPPH whilst the IC_{50} value was 16.6µg/ml, when tested with ABTS. Total phenolic content was 191 mgGAE/g. Betulafolientriol oxide and rosmarinic acid were detected in all the species investigated and rosmarinic acid, carnosic acid, carnosol and oleanolic acid/ursolic acid were abundant in many species.

Salvia nemerosa L. Family: Lamiaceae (Labiatae)

Zupkó *et al.* (2001) evaluated the antioxidant activity of leaves of eleven *Salvia* species in enzyme-dependent and enzyme-independent systems of lipid peroxidation and their phenolic constituents. The 50 per cent aqueous methanolic extract of *Salvia nemerosa* displayed considerable concentration-dependent antioxidative effects that were comparable to those of *Salvia officinalis*.

Salvia officinalis L. Family: Lamiaceae

English Name: Sage.

Ten phenolic compounds were isolated by Wang *et al.* (1998) from a butanol fraction of sage extracts. Two test systems, DPPH free radical scavenging activity and radical cation ABTS scavenging activity, were used to evaluate their antioxidant activity. The most active compounds were found to be rosmarinic acid and luteolin-7-O-β-glucopyranoside.

The antioxidant effects of aqueous methanolic extracts from *Salvia officinalis* were investigated in enzyme-dependent and enzyme-independent lipid peroxidation systems. The extract caused a considerable concentration-dependent inhibition of lipid peroxidation. Phenolic components present in the plant extract were evaluated for antioxidant activity and were found effective in both tests (Hohmann *et al.*, 1999).

A new abietene diterpenoid, 12-O-methyl carnosol (2), was isolated from the leaves of sage together with 11 abietene diterpenoids, 3 apianane terpenoids, 1 anthraquinone and 8 flavonoids (Miura *et al.*, 2002). Antioxidant activity of these compounds was evaluated by the oil stability index method using a model substrate oil including methyl lineolate in silicone oil at 90°C. Carnosol, rosamarol, epirosmarol, isorosmarol, galdosol and carnosic acid exhibited remarkably strong activity which was comparable to that of α-tocopherol. The activity of miltirone, atuntzenin A, luteolin, 7-O-methyl luteolin, and eupafolin was comparable to that of BHT. The activity of these compounds was mainly due to the presence of ortho-dihydroxy groups. The DPPH radical scavenging activity of these compounds showed the similar result.

Miliauskas *et al.* (2004) investigated the radical scavenging activity of acetone, ethyl acetate and methanol extracts of leaves and stems at initial blossom stage using DPPH and ABTS assays. They found that Methanol extracts were most effective DPPH radical scavengers. Ethyl acetate and acetone extracts were considerably less effective radical scavengers. High content of phenolics in the extracts correlated with its antiradical activity, confirming that phenolic compounds are likely to contribute to the radical scavenging activity of this plant extracts.

The antioxidant activity of sage tea drinking was studied *in vitro* in a model using rat hepatocytes in primary culture. The replacement of drinking water with sage tea in the rats used as hepatocyte donors resulted in an improvement of the antioxidant status of rat hepatocytes in primary culture, namely a significant increase in GSH content and GST activity after 4th of culture (Lima *et al.*, 2005).

Biljana *et al.* (2007) analyzed the essential oils of sage by means of gas chromatography-mass spectrometry and assayed for their antimicrobial and antioxidant activities.

The antioxidant activities of sage polyphenols consisting of flavone glycosides and a range of rosmarinic acid derivatives, were evaluated for their capacity to scavenge DPPH and superoxide antion radicals and also to reduce Mo (VI) to Mo (V). The rosmarinic acid derivatives all showed potent antioxidant activity in three test systems and their capacity to reduce Mo (VI) to Mo (V) and their superoxide radical scavenging activities, with values ranging from 220 to 300 SID units/mg, in particular, were 4-6 and 15-20 times greater than trolox respectively. The high SOD activity of rosmarinic acid could be attributed to the radical scavenging catechols and the xanthine-oxidase inhibiting caffeic acid moieties contained in them. The antioxidant activity of flavonoids was variable and those with catechol B-ring (luteolin glycosides) were more active than those without (apigenin glycosides).

The radical scavenging activity of ethanolic extract of *Salvia officinalis* was evaluated *in vitro* by Nickavar *et al.* (2007) with the spectrophotometric method based on the reduction of the stable DPPH free radical. The extract exhibited concentration dependent radical scavenging activity and IC_{50} value was 30.67µg/ml. Total flavonoid content in rutin equivalent was 17.24µg/mg. However, a favourable correlation was not found between the radical scavenging potency and the total flavonoid content.

Mariutti *et al.* (2008) evaluated the antioxidant activity of garlic employing ABTS and DPPH radical assay. DPPH EC_{50} value was 1.65g/kg while DPPH-TEAC value was 192mM/g. ABTS-TEAC value was 202mM/g.

Cioroi and Dumitriu (2009) investigated the total pophenols content and antioxidant activity of aqueous extract of *Salvia officinalis*. The total phenol content was 577.46 mg/100g dried material. Antioxidant activity was directly correlated with the total amount of polyphenols in the extract.

Using a multiple-method approach, the antioxidant activity of the essential oil from *Salvia officinalis* was tested by Viuda-Martos *et al.* (2010). Their phenolic content was also determined. All the essential oils tested had antioxidant activity,

Spiridon *et al.* (2011) evaluated the total phenolic content and antioxidant activity of plants used in Romanian herbal medicine. *Salvia officinalis* extracts showed good antioxidant activity.

Salvia palestina Benth. Family: Lamiaceae (Labiatae)

The antioxidant activities of the methanolic extracts of 9 *Salvia* species and 15 other Lamiaceae plants growing in Iran were evaluated by Firuzi *et al.* (2010) using FRAP and DPPH free radical scavenging assays. *Salvia palestina* showed very high antioxidant activity with FRAP value 41.3µM quercetin equivalents/g dry weight, and IC_{50} values in the DPPH assay 207.9µg dry weight/ml. Total phenolic content was also higher with14.6 mg CE/gDW. FRAP and DPPH assays results showed good correlations with the total phenolic contents of the plants.

Salvia pilifera Montbr. et Auch. ex Benth. Family: Lamiaceae (Labiatae)

Essential oils of *Salvia pilifera* and two other species of *Salvia* were screened for their possible antioxidant properties as well as their chemical compositions. Antioxidant activity was employed by two complementary test systems namely DPPH free radical scavenging and β-carotene/linoleic acid systems. Essential oil of *S. pilifera* showed good antioxidant activity (Kelen and Tepe, 2007).

Salvia plebeia R. Br. Family: Lamiaceae (Labiatae)

Six compounds, hispidulin-glucuronide (1), hispidulin-7-O-D-glucoside (2), 6-methoxy-luteolin-7-glucoside (3), β-sitosterol (4), 2-hydroxy-5'-methoxybiochanin (5) and coniferyl aldehyde (6), were isolated from *Salvia plebeia* by Weng and Wang (2000). Their antioxidant activities were investigated individually and compared with BHT and α-tocopherol. Compounds 3,4 and 5 had strong antioxidant activities, but compounds 1,2 and 6 had low antioxidant activity at 0.02 and 0.04 per cent levels.

Three antioxidant components – royleanonic acid, hispidulin and supatorin- were isolated from *Salvia plebeia* (Gu and Weng, 2001). Though royleanonic acid and hispidulin prolonged the induction period significantly their antioxidant activities were much weaker than the crude extracts, implying that synergistic effects might be responsible for the high activity of the crude extracts.

β-Sitosterol, 2'-hydroxyl-5'-methoxybiochanin A and 6-methoxyluteolin-7-glycoside, isolated from *Salvia plebeia* exhibited strong antioxidant activities (Ai-Li and Wang, 2005).

Salvia pomifera L. Family: Lamiaceae (Labiatae)

Demo *et al.* (1998) reported antioxidant α-tocopherol in from the Hexane extract of *Salvia pomifera*. Couladis *et al.* (2003) screened Greek aromatic plants from the Lamiaceae family for the antioxidant activity. Of the 21 plants they tested they found ethanol extracts prepared from *Salvia pomifera*, as well as 9 other plants exhibited the same antioxidant activity as α-tocopherol in their ability to inhibit bleomycin-Fe(II) complex-induced arichidonic acid superoxidation to MDA.

Salvia pratensis L. Family: Lamiaceae (Labiatae)

English Name: Meadow sage.

Miliauskas *et al.* (2004) investigated the radical scavenging activity of acetone, ethyl acetate and methanol extracts of leaves and stems at full bloom stage using DPPH and ABTS assays. They found that methanol extracts were most effective DPPH radical scavengers. Ethyl acetate and acetone extracts were considerably less effective radical scavengers. High content of phenolics in the extracts correlates with its antiradical activity, confirming that phenolic compounds are likely to contribute to the radical scavenging activity of this plant extracts.

Salvia przewalskii Maxim. Family: Lamiaceae (Labiatae)

Matkowski *et al.* (2008) reported antioxidant activity in methanol extracts from roots and leaves of *Salvia przewalskii.* The antioxidant activity correlates to the total polyphenol and, depending on the assay, to the hydroxycinnamic acids content.

Salvia psidica Boiss. et Heldr. ex Benth. Family: Lamiaceae (Labiatae)

Total content of phenolic, flavanol and flavonol, antioxidant activities and antimicrobial activities of *Salvia psidica* extract and essential oil were assessed *in vitro*. Total phenolic, flavanol and flavonol contents in the extract were 54.57 mg GAE/g, 16.70 mg catechin equivalent/g and 18.19 mg rutin equivalent/g, respectively. Antioxidant activities (IC^{50} value) of the extract and essential oil were determined as 4.88 and 6.41 mg/ml, by DPPH assay, respectively (Ozkan *et al.*, 2010).

Salvia radula Benth. Family: Lamiaceae (Labiatae)

The antioxidant and antioinflammatory activities of the methanol:chloroform (1:1) extracts of 16 *Salvia* species indigenous to South Africa were evaluated by Kamatou *et al.* (2010). Antioxidant activity was measured using ABTS and DPPH scavenging assays and compared to the control values obtained with Trolox. The solvent extract of *Salvia radula* displayed antioxidant activity, with IC_{50} value of >100 µg/ml using DPPH whilst the IC_{50} value was 69.3µg/ml, when tested with ABTS. Total phenolic content was 55.7 mgGAE/g. Betulafolientriol oxide and rosmarinic acid were detected in all the species investigated and rosmarinic acid, carnosic acid, carnosol and oleanolic acid/ursolic acid were abundant in many species.

Salvia reflexa Hornem. Family: Lamiaceae (Labiatae)

The antioxidant properties of the wild growing sage species, *Salvia reflexa*, were investigated by Malencic *et al.* (2000). The presence of superoxide, and hydroxyl radicals, malonyldialdehyde (MDA) reduced glutathione and total flavonoids were observed in the above-ground parts of plant, as well as activities of the antioxidant enzymes SOD and peroxidase. The potential antioxidant activity of the methanol : water extract has been assessed based on scavenging activity of stable DPPH free radicals. Significant quantities of superoxide, hydroxyl radicals and MDA were observed. Thus this species exhibited high SOD and peroxidase activities as well as a content of total flavonoids. The dominant naturally occurring compound was rosmarinic acid (Malencic *et al.*, 2000).

Salvia repens Burch. ex Benth. Family: Lamiaceae (Labiatae)

The antioxidant and antiinflammatory activities of the methanol:chloroform (1:1) extracts of 16 *Salvia* species indigenous to South Africa were evaluated by Kamatou *et al.* (2010). The solvent extract of *Salvia repens* displayed antioxidant activity, with IC_{50} value of 15.5 µg/ml using DPPH whilst the IC_{50} value was 18.2µg/ml, when tested with ABTS. Total phenolic content was 178 mgGAE/g. Betulafolientriol oxide and rosmarinic acid were detected in all the species investigated and rosmarinic acid, carnosic acid, carnosol and oleanolic acid/ursolic acid were abundant in many species.

Salvia reuterana Boiss. Family: Lamiaceae (Labiatae)

The radical scavenging activity of ethanolic extract of *Salvia hypoleuca* was evaluated *in vitro* by Nickavar *et al.* (2007) with the spectrophotometric method based on the reduction of the stable DPPH free radical. The extract exhibited concentration dependent radical scavenging activity and IC_{50} value was 125.1µg/ml. Total flavonoid content in rutin equivalent was 46.97µg/mg. However, a favourable correlation was not found between the radical scavenging potency and the total flavonoid content.

Firuzi *et al.* (2010) evaluated the antioxidant activity and total phenolic content of 24 Lamiaceae species growing in Iran. *Salvia reuterana* showed high antioxidant activity with 24.4µM/g DW FRAP value and IC_{50} = 459.1µg/ml of DPPH radical scavenging activity. It also showed high phenolic content of 10.1 mg CE/g. FRAP and DPPH assay results showed good correlations with the total phenolic contents of the plant measured by the Folin-Ciocalteu assay.

Salvia ringens Sibth. et Sm. Family: Lamiaceae (Labiatae)

Zupkó *et al.* (2001) evaluated the antioxidant activity of leaves of eleven *Salvia* species in enzyme-dependent and enzyme-independent systems of lipid peroxidation and their phenolic constituents. The 50 per cent aqueous methanolic extract of *Salvia ringens* displayed considerable concentration-dependent antioxidative effects that were comparable to those of *Salvia officinalis*.

Couladis *et al.* (2003) screened Greek aromatic plants from the Lamiaceae family for the antioxidant activity. Of the 21 plants they tested they found ethanol extracts prepared from *Salvia ringens*, as well as 9 other plants exhibited the same antioxidant activity as α-tocopherol in their ability to inhibit bleomycin-Fe(II) complex-induced arichidonic acid superoxidation to MDA.

Salvia runcinata L.f. Family: Lamiaceae (Labiatae)

The antioxidant and antiinflammatory activities of the methanol:chloroform (1:1) extracts of 16 *Salvia* species indigenous to South Africa were evaluated by Kamatou *et al.* (2010). Antioxidant activity was measured using ABTS and DPPH scavenging assays and compared to the control values obtained with Trolox. The solvent extract of *Salvia runcinata* displayed antioxidant activity, with IC_{50} value of 19.3 µg/ml using DPPH whilst the IC_{50} value was 19.4µg/ml, when tested with ABTS. Total phenolic content was 149 mgGAE/g. Betulafolientriol oxide and rosmarinic acid were detected in all the species investigated and rosmarinic acid, carnosic acid, carnosol and oleanolic acid/ursolic acid were abundant in many species.

Salvia sahendica Boiss. et Buhse Family: Lamiaceae (Labiatae)

The antioxidant activity of *Salvia sahendica* extracts and its essential oils was evaluated in a series of *in vitro* tests. The free radical scavenging activity of methanol extract was found to be superior to all other extracts. Polar extracts exhibited stronger activity than nonpolar ones (Salehi *et al.*, 2007).

Salvia santolinifolia Boiss. Family: Lamiaceae (Labiatae)

The antioxidant activities of the methanolic extracts of 9 *Salvia* species and 15 other Lamiaceae plants growing in Iran were evaluated by Firuzi *et al.* (2010) using FRAP and DPPH free radical scavenging assays. *Salvia santolinifolia* showed highest antioxidant activity with 79.0μM/g DW FRAP value and IC_{50} = 115.7μg/ml of DPPH radical scavenging activity. It also showed high phenolic content of 28.6 mg CE/g. FRAP and DPPH assay results showed good correlations with the total phenolic contents of the plant measured by the Folin-Ciocalteu assay.

Salvia schlechteri Briq. Family: Lamiaceae (Labiatae)

The antioxidant and antiinflammatory activities of the methanol : chloroform (1:1) extracts of 16 *Salvia* species indigenous to South Africa were evaluated by Kamatou *et al.* (2010). The solvent extract displayed antioxidant activity, with IC_{50} value of 1.61μg/ml using DPPH whilst the IC_{50} values ranged from 17.5μg/ml, when tested with ABTS. The extract of *Salvia schlechteri*, with an IC_{50} value of 1.61μg/ml, was three times more active than the reference compound, Trolox (IC_{50} value: 2.51μg/ml). The total phenolic content based on gallic acid equivalents (GAE) confirmed the presence of total soluble phenolics in the extract at 209 mg of GAE per g dry sample and showed strong association with antioxidant activity. Betulafolientriol oxide and rosmarinic acid were detected in all the species investigated and rosmarinic acid, carnosic acid, carnosol and oleanolic acid/ursolic acid were abundant in many species.

Salvia sclarea L. Family: Lamiaceae

English Name: Clary sage.

It is commercially cultivated for its aromatic and widely used essential oil.

Miliauskas *et al.* (2004) investigated the radical scavenging activity of acetone, ethyl acetate and methanol extracts of leaves and stems at full bloom stage using DPPH and ABTS assays. They found that Methanol extracts were most effective DPPH radical scavengers. Ethyl acetate and acetone extracts were considerably less effective radical scavengers. High content of phenolics in the extracts correlates with its antiradical activity, confirming that phenolic compounds are likely to contribute to the radical scavenging activity of this plant extracts.

The plant prevents free radical formation and is effective in treatment against arthritis and rheumatism. γ-terpinene, linalyl acetate, myrcene, palmitic acid, androsemarinic acid are the main antioxidative components in the plant.

Tepe *et al.* (2006) examined the *in vitro* antioxidant activities of the methanol extract of *Salvia sclarea*. The extracts were screened for their possible antioxidant activities by two complementary test systems, namely DPPH free radical-scavenging and β-carotene/linoleic acid systems. Non-polar subfractions of the methanol extract did not show any antioxidant activity in both test systems. The polar subfractions showed antioxidant activity. In the β-carotene/linoleic acid systems polar extract of *Salvia sclarea* showed 63.5 per cent inhibition rate.

Kumar and Singh (2011) evaluated the antioxidant activities of ethanol extracts of Trans Himalayan plants. *Salvia sclarea* leaf showed maximum DPPH radical scavenging activity with IC_{50} value of 14.97µg/ml.

Salvia staminea Montbret et Aucher ex Benth. Family: Lamiaceae (Labiatae)

Tepe (2008) examined the *in vitro* antioxidant activities of rosmarinic acid levels of the methanol extracts of *Salvia staminea* and two other species of *Salvia*. The extracts were screened for their possible antioxidant activity by two complementary test systems, namely DPPH free radical scavenging and β-carotene/linoleic acid systems. *S. staminea* exhibited weak antioxidant activity in this test system of which IC_{50} value was 75.40µg/mg.

Salvia stenophylla Burch. ex Benth. Family: Lamiaceae (Labiatae)

The antioxidant and antiinflammatory activities of the methanol:chloroform (1:1) extracts of 16 *Salvia* species indigenous to South Africa were evaluated by Kamatou *et al.* (2010). Antioxidant activity was measured using ABTS and DPPH scavenging assays and compared to the control values obtained with Trolox. The solvent extract of *Salvia stenophylla* displayed antioxidant activity, with IC_{50} value of 14.9 µg/ml using DPPH whilst the IC_{50} value was 20.8µg/ml, when tested with ABTS. Total phenolic content was 161 mgGAE/g. Betulafolientriol oxide and rosmarinic acid were detected in all the species investigated and rosmarinic acid, carnosic acid, carnosol and oleanolic acid/ursolic acid were abundant in many species.

Salvia syriaca L. Family: Lamiaceae (Labiatae)

Firuzi *et al.* (2010) evaluated the antioxidant activity and total phenolic content of 24 Lamiaceae species growing in Iran. *Salvia syriaca* showed high antioxidant activity with 28.0µM/g DW FRAP value and IC_{50} = 371.7µg/ml of DPPH radical scavenging activity. It also showed high phenolic content of 13.0 mg CE/g. FRAP and DPPH assay results showed good correlations with the total phenolic contents of the plant measured by the Folin-Ciocalteu assay.

Salvia tomentosa Mill. Family: Lamiaceae (Labiatae)

Zupkó *et al.* (2001) evaluated the antioxidant activity of leaves of eleven *Salvia* species in enzyme-dependent and enzyme-independent systems of lipid peroxidation and their phenolic constituents. The 50 per cent aqueous methanolic extract of *Salvia tomentosa* displayed considerable concentration-dependent antioxidative effects that were comparable to those of *Salvia officinalis*.

Salvia triloba L. Family: Lamiaceae (Labiatae)

Protogeras *et al.* (1998) reported antioxidant activity of *Salvia triloba*. Yildirim *et al.* (2000) reported antioxidant activity of *Salvia triloba*. Antioxidant activity and reducing power activity was concentration dependent. Even in the presence of 50µg of the extract, the reducing power was significantly higher than that of the control in which there was no extract. They are of the opinion that although the reducing power of a substance may be an indicator of its potential antioxidant activity, there may not always be a linear correlation between these two activities.

Salvia verbenaca L. Family: Lamiaceae (Labiatae)

English Name: Clary

Khalil *et al.* (2007) studied the growth, phenolic compounds and antioxidant activity of clary grown under organic farming condition. Antioxidant activity of ethanol extract (inhibition per cent) was 77.11 at 200µl extract.

Tepe (2008) examined the *in vitro* antioxidant activities of rosmarinic acid levels of the methanol extracts of *Salvia verbenaca* and two other species of *Salvia*. The extracts were screened for their possible antioxidant activity by two complementary test systems, namely DPPH free radical scavenging and β-carotene/linoleic acid systems. *S. verbenaca* exhibited high antioxidant activity (14.30µg/mg). In β-carotene/linoleic acid system, *S. verbenaca* extract was superior to the other extracts studied (inhibition value is 77.03).

Salvia verticillata L. Family: Lamiaceae (Labiatae)

Souri *et al.* (2004) evaluated the antioxidative activity of *Salvia verticillata* by linoleic acid peroxidation test using 1,3-diethyl-2-thiobarbituric acid as the reagent. Methanolic extract showed 76.01 per cent inhibition at 40µg concentration with IC_{50} value of 6.38µg.

The radical scavenging activity of ethanolic extract of *Salvia verticillata* was evaluated *in vitro* by Nickavar *et al.* (2007) with the spectrophotometric method based on the reduction of the stable DPPH free radical. The extract exhibited concentration dependent radical scavenging activity and IC_{50} value was 23.53µg/ml. Total flavonoid content in rutin equivalent was 10.81µg/mg. However, a favourable correlation was not found between the radical scavenging potency and the total flavonoid content.

Matkowski *et al.* (2008) reported antioxidant activity in methanol extracts from roots and leaves of *Salvia verticillata*. The antioxidant activity correlates to the total polyphenol and, depending on the assay, to the hydroxycinnamic acids content.

Methanolic extracts of *Salvia verticillata* and seven other species of *Salvia* sampled from Eastern Anatolia in Turkey, were screened by Tosun *et al.* (2009) for their possible antioxidant activities by two complementary test systems, namely DPPH free radical scavenging and β-carotene/linoleic acid. Total phenolic content of the extracts of *Salvia* species were performed by Folin-Ciocalteu reagent and gallic acid used as standard. A wide variation has been observed among species in terms of antioxidant activity and total phenolic content. In both DPPH and β-carotene system, the most active plant was *Salvia verticillata* with a value of IC_{50} =18.3µg/ml and 75.8 per cent, respectively. This species also has the highest total phenolic content (167.1 mg GAE/g DW). A positive linear correlation was observed between total phenolic content and antioxidant activity of the extracts.

Salvia virgata Jacq. Family: Lamiaceae (Labiatae)

The radical scavenging activity of ethanolic extract of *Salvia virgata* was evaluated *in vitro* by Nickavar *et al.* (2007) with the spectrophotometric method based on the reduction of the stable DPPH free radical. The extract exhibited concentration dependent radical scavenging activity and IC_{50} value was 27.01μg/ml. Total flavonoid content in rutin equivalent was 8.54μg/mg. However, a favourable correlation was not found between the radical scavenging potency and the total flavonoid content.

Tepe (2008) examined the *in vitro* antioxidant activities of rosmarinic acid levels of the methanol extracts of *Salvia virgata*. The extracts were screened for their possible antioxidant activity by two complementary test systems, namely DPPH free radical scavenging and β-carotene/linoleic acid systems. *S. virgata* exhibited moderate antioxidant activity (65.70μg/mg).

Firuzi *et al.* (2010) evaluated the antioxidant activity and total phenolic content of 24 Lamiaceae species growing in Iran. *Salvia virgata* showed high antioxidant activity with 27.5μM/g DW FRAP value and IC_{50} = 644.8μg/ml of DPPH radical scavenging activity. It also showed high phenolic content of 7.5 mg CE/g. FRAP and DPPH assay results showed good correlations with the total phenolic contents of the plant measured by the Folin-Ciocalteu assay.

Salvia viridis L. Family: Lamiaceae (Labiatae)

English Name: Sage

Erdemoglu *et al.* (2006) evaluated the antioxidant activity of the extract by using DPPH as well as by flow-injection analysis-luminol chemiluminiscence (FIA-CL). Extracts were shown to possess a significant scavenger activity against DPPH free radical and an inhibitory effect on H_2O_2- or HOCl-luminol chemiluminiscence. The results concluded that the extracts have a potential source of antioxidants of natural origin.

Khalil *et al.* (2007) studied the growth, phenolic compounds and antioxidant activity of *Salvia viridis* grown under organic farming condition. Antioxidant activity of ethanol extract (inhibition per cent) was 91.34 at 200μl extract.

Sambucus caerulea Raf. Family: Caprifoliaceae

English Name: Blue Elderberry

The EtOAc extract obtained from *Sambucus caerulea* was tested in the DPPH free radical assay. High antioxidant activity was obtained from the extract of fruit (Acuna *et al.*, 2002).

Sambucus canadensis L. Family: Caprifoliaceae

English Name: American elderberry, Elderberry.

Ozgen *et al.* (2010) analyzed the total phenolic and total monomeric anthocyanin contents and their antioxidant capacity by ferric reducing antioxidant power (FRAP) and DPPH radical scavenging assays. Total phenolic and anthocyanin contents were measured using the Folin-Ciocalteu reagent and the pH differential methods, respectively. Overall, the phytonutrient contents and antioxidant capacity of elderberry accessions were similar to those typically reported for black raspberries, blackberries and other dark-fleshed small fruits. Variations among accessions was evident for TP, FRAP and DPPH values (CV 14.4, 21.7 and 26.8 per cent, respectively). TP and TMA values were highly correlated to antioxidant capacity values.

Sambucus ebulus L. Family: Caprifoliaceae

Kiselova *et al.* (2006) studied the antioxidant activity of *Sambucus ebulus*. Antioxidant activity was measured by ABTS cation radical decoloration assay and the total polyphenol content was measured according to the Folin-Ciocalteu method. *Sambucus ebulus* extract exhibited higher antioxidant activity than mate. A positive correlation between antioxidant activity and polyphenol content was found, suggesting that the antioxidant capacity of the aqueous plant extract was due to a greater extent to their polyphenols.

Extracts of *Sambucus ebulus* were investigated by Hosseinimehr *et al.* (2007) for their total flavonoids, phenol contents and their radical scavenging activity using DPPH assays. Quercetin and butylated hydroxytoluene were used as standard reference with well documented antioxidant activity. Methanolic extract sample showed free radical scavenging activity. A correlation between radical scavenging capacitiy of the extract with total phenolic compounds content was observed.

Sambucus nigra L. Family: Sambucaceae

English Name: Elderberry.

Youdin *et al.* (2004) demonstrated that elderberry anthocyanin incorporation into endothelial cells confers increased protection against oxidative stress. Murkovic *et al.* (2004) investigated the effects of elderberry juice on fasting and postprandial serum lipids and low-density lipoprotein oxidation in healthy volunteers. They reported that elderberry spray-dried extract at a low dose exerts a minor effect on serum lipids and antioxidant capacity. Higher, but nutritionally relevant doses might significantly reduce postprandial lipids.

Sanguisorba officinalis L. Family: Rosaceae

Cai *et al.* (2004, 2006) evaluated the antioxidant activity and phenolic compounds in traditional Chinese medicinal plants associated with anticancer. The improved ABTS.+ method was used to systematically assess the total antioxidant capacity (Trolox equivalent antioxidant capacity, TEAC) of the medicinal extracts. The TEAC values and total phenolic content for methanolic extract of root of *Sanguisorba officinalis* were 4446.6 µmol Trolox equivalent/100g dry weight (DW), and 15.87 g of gallic acid equivalent/100g DW respectively. A positive significant linear relationship between antioxidant activity and total phenolic content showed that phenolic compounds were the dominant antioxidant components in the tested herbs. Major types of phenolic compounds in *Sanguisorba officinalis* include phenolic acids (gallic acid), hydrolysable and condensed tannins, flavonols (flavan-3-ols; (+)-catechin, gallocatechin).

Gan *et al.* (2010) evaluated the antioxidant activity of 40 medicinal plants associated with prevention and treatment of cardiovascular and cerebrovascular diseases. A high correlation between antioxidant capacity and total phenolic content indicated that phenolic compounds could be the main contributor of the antioxidant activity of these plants. *Sanguisorba officinalis* showed highest antioxidant activity of 2025.33 μmol Fe(II)/g in FRAP assay and 1363.33μmol in TEAC assay. Total phenolic contet was also very high, 75.71 mg GAE/g.

Sanicula europaea L. Family: Apiaceae (Umbelliferae)

Le Claire *et al.* (2005) reported antioxidant activity in *Sanicula europaea*. The active constituent was found to be rosmarinic acid derivative.

Sanicula graveolens Poepp. ex DC. Family: Apiaceae (Umbelliferae)

The water-soluble extract of *Sanicula graveolens* showed radical scavenging activity in the DPPH discoloratioan assay (Viturro *et al.*, 1999). Bioassay-guided isolation led to caffeic acid derivatives and flavonoids as the main active compounds. After hydrolysis, caffeic acid and quercetin proved to be the bioactive principles.

Santalum album L. Family: Santalaceae

Scartezzini and Speroni (2000) were of the opinion that this plant contains antioxidant principles, that can explain and justify their use in traditional medicine in the past as well as the present.

Alanine, β-sitosterol, eugenol, palmitic acid, phenol are the constituents found in the plant that have antioxidative effect. Mainly the wood or the bark of the plant is used. It prevents premature ageing and rejuvenates the skin. Alanine, β-sitosterol, eugenol, palmitic acid and phenols are the antioxidative constituents in sandalwood (Mishra *et al.*, 2007).

Sapindus emarginatus Vahl Family: Sapindaceae

The antioxidant activities of methanolic extract of *Sapindus emarginatus* leaf were evaluated by Srikanth and Muralidharan (2010) and compared with α-tocopherol and BHT as synthetic antioxidants and ascorbic acid as natural based antioxidant. They studied *in vitro*, its antioxidant activities, radical-scavenging effects, Fe^{2+}-chelating ability and reducing power. The extract showed variable activities in all of these *in vitro* tests. The antioxidant effect of *S. emarginatus* was strongly dose dependent, increased with increasing leaf extract dose and then leveled off with further increase in extract dose. Compared to other antioxidants used in the study, *S. emarginatus* leaf extract showed less scavenging effect on DPPH radical and less reducing power on Fe^{3+} ferriccynide complex but better Fe^{2+}-chelating ability.

Sapindus mukorossi Gaertn. Family: Sapindaceae

The extracts of *Sapindus mukorossi* seeds using methanol, ethyl acetate or hexane as solvents were evaluated by Chen *et al.* (2010) for their tyrosinase inhibition, free radical scavenging, antimicrobial and anticancer properties. The anti-tryrosinase assay and antioxidant potentials were determined by *in vitro* mushroom tyrosinase assay and the radical scavenging methods. The results obtained from biological assays showed that *S. mukorossi* extracts possessed multiple bioactivities, including anti-tyrosinase, antioxidant, antimicroorganism and anticancer proliferation properties.

Sapium macrocarpum Muell.-Arg. Family: Euphorbiaceae

Ruiz-Terán *et al.* (2008) evaluated the antioxidant activity of aerial parts of *Sapium macrocarpum* using DPPH and β-carotene-linoleic acid bleaching assay. The extract displayed antioxidant activity two times more than the commercial antioxidant BHA. For the methanolic extract analysed, a clear relation between the total phenolic content of the extracts and their antioxidant activity was found.

Saposhnikovia divaricata Turcz. Family: Apiaceae (Umbelliferae)

Antioxidant capacities of 56 selected Chinese medicinal plants, including *Saposhnikovia divaricata* were evaluated by Song *et al.* (2010) using the TEAC and FRAP assays, and their total phenolic content was measured by the Folin-Ciocalteu method. *Saposhnikovia divaricata* showed low antioxidant activity with TEAC value of 6.39μmol Trolox/g, FRAP value of 14.22μmol Fe^{2+}/g and phenolic content was 2.31mg GAE/g.

Saraca asoca (Roxb.) de Wilde Family: Caesalpiniaceae

Kshirsagar and Upadhyay (2009) reported antioxidant activity of the methanol extract of stem and leaf. Their ability of scavenging radicals was measured by DPPH reduction spectrophotometric assay.

Saraca thaipingensis Prain Family: Caesalpiniaceae

Leelapornpisid *et al.* (2007) reported antioxidant activity of absolutes (obtained by solvent extraction) of *Saraca thaipingensis*.

Sarcolobus globosus Wall. Family: Asclepiadaceae

Wangensteen *et al.* (2006) identified two rotenoids, one isoflavone, and four phenolic glycosides. Extracts and compounds were evaluated for their DPPH radical scavenging and 15-lipoxygenase inhibitory activities. All tested rotenoids were found to inhibit 15-LO, while they lacked DPPH radical scavenging effect.

Sarcopoterium spinosum (L.) Spach. Family: Rosaceae

Al-Mustafa and Al-Thunibat (2008) evaluated the antioxidant activity of Jordanian medicinal plant *Sarcopoterium spinosum* root. The level of antioxidant activity was determined by DPPH and ABTS assay in relation to total phenolic contents of the medicinally used part. They concluded that *Sarcopoterium* root possess low antioxidant capacity (DPPH-TEAC <20mg/g).

Sarcostemma brevistigma Wight Family: Asclepiadaceae

Surveswaran *et al.* (2010) surveyed antioxidant properties and total phenolic and flavonoid contents of 15 samples, representing 12 Indian medicinal plant species. Total antioxidant assay was performed using ABTS and FRAP methods. The stems of *Sarcostemma brevistigma* exhibited the highest xanthine oxidase (XO) inhibitory activity. The highly significant and positive correlations between total antioxidant capacity parameters and total phenolic content indicated that the phenolic compounds contributed significantly to the antioxidant activity of the tested plant samples.

Sargentodoxa cuneata Rehd. et Wils. Family: Sargentodoxaceae

Li *et al.* (2008) evaluated the antioxidant capacities of 45 selected medicinal plants using ferric reducing antioxidant power (FRAP) and Trolox equivalent antioxidant capacity (TEAC) assays respectively, and the total phenolic contents of these plants were measured by the Folin-Ciocalteau method. It was found that *Sartentodoxa cuneata* possessed the highest antioxidant capacities with 453.53 µmol Fe(II)/g FRAP value, 265.43µmol Trolox/g of TEAC value and 52.35 mg GAE/g of Phenolic content. A strong correlation between TEAC values and those obtained from FRAP assay implied that antioxidants in these plants were capable of scavenging free radicals and reducing oxidants. A high correlation between antioxidant capacities and their total phenolic contents indicated that phenolic compounds were a major contributor of antioxidant activity of these plants.

Sasa borealis (Hack.) Makino et Shibata Family: Poaceae (Graminae)

The n-BuOH extract of *Sasa borealis* leaves exhibited significant antioxidant activity against the DPPH radical and a cytoprotective effect against oxidative damage in HepG2 cells (Park *et al.*, 2007). Isoorientin and Isoorientin 2-O-α-L-rhamnoside isolated from the extract showed potent free radical scavenging activity with IC_{50} values of 9.5 and 34.5 µM respectively, and strong cytoproetective effects against t-BOOH-induced oxidative damage in HepG2 cells at very low concentrations of 1.1 µM isoorientin and 0.8µM isoorientin2-O- α-L-rhamnoside (Park *et al.*, 2007).

Satureja cuneifolia Ten. Family: Lamiaceae (Labiatae)

English Name: Savory.

Kosar *et al.* (2003) prepared water extracts from steam distilled essential oil-extracted *Satureja cuneifolia*. The HPLC-DPPH on-line method was applied to the qualitative and quantitative analysis of this plant extract. There was a strong correlation between the scavenging (negative) peak area and the concentration of the radical scavenging reference substances used. The radical scavenging compounds within the extracts were determined as benzoic acid and hydroxycinnamic acid derivatives, flavonoids and diterpenoids according to their retention time and UV spectral data. Rosmarinic acid and carnosic acid were identified as the dominant radical scavengers in these extracts by this method.

Water-soluble extract from *Satureja cuneifolia* was screened by Dorman *et al.* (2004) for antioxidant properties in a battery of six *in vitro* assays. The extract demonstrated varying degrees of efficacy in each screen. The savory extract was the most effective at reducing iron (III), scavenging DPPH radicals, inhibiting ascorbate-iron(III)-catalyzed hydroxyl radical-mediated brain phospholipids peroxidation, and site-specific hydroxyl-mediated 2-deoxy-d-ribose degradation. The extract contained Folin-Ciocalteu reagent-reactive substances, which was confirmed by the presence of polar phenolic analytes (*i.e.*, hydroxybenzoates, hydroxycinnamates, and flavonoids).

Eminagaoglu *et al.* (2007) examined the *in vitro* antioxidant activities of the essential oil and methanol extracts of *Satureja cuneifolia* using DPPH and β-carotene-linoleic acid assays. In DPPH test system, free radical-scavenging activity of *S.cuneifolia* oil was determined to be $IC_{50} = 89.1$ µg ml. In the β-carotene-linoleic acid test system, antioxidant activity of the oil was 93.7 per cent.

The antioxidant capacity of the essential oil obtained by hydrodistillation from aerial parts of *Satureja cuneifolia* collected in three different maturation stages such as preflowering, flowering and postflowering were examined *in vitro*. The essential oils obtained from *S. cuneifolia* in three different stages and its main components were interacted with DPPH as a nitrogen-centred stable radical, resulting $IC_{50} = 1.6$-2.1 mg/ml (Kosar *et al.*, 2008). In addition, the effects on inhibition of lipid peroxidation on the essential oils were assayed using β-carotene bleaching method. All the tested oils inhibited the linoleic acid peroxidation at almost the same level as BHT (93.54-94.65 per cent). BHT and ascorbic acid were used as positive controls in the antioxidant assays.

Satureja forbesii (Benth.) Briq. Family: Lamiaceae (Labiatae)

The chemical composition and antioxidant properties of the essential oil from the air-dried aerial parts of *Satureja forbesii* were studied by Ortet *et al.* (2009). Using DPPH free-radical scavenging method and the *in vitro* assay for prevention of lipid peroxidation by thiobarbituric reactive species, significant activities were evidenced.

Satureja hortensis L. Family: Lamiaceae (Labiatae)

English Name: Savory.

Kosar *et al.* (2003) prepared water extracts from steam distilled essential oil-extracted *Satureja hortensis*. The HPLC-DPPH on-line method was applied to the qualitative and quantitative analysis of this plant extract. There was a strong correlation between the scavenging (negative) peak area and the concentration of the radical scavenging reference substances used. The radical scavenging compounds within the extracts were determined as benzoic acid and hydroxycinnamic acid derivatives, flavonoids and diterpenoids according to their retention time and UV spectral data. Rosmarinic acid and carnosic acid were identified as the dominant radical scavengers in these extracts by this method.

Souri *et al.* (2004) investigated the antioxidant activity of methanol extracts of savory used as vegetable in Iranian diet against linoleic acid peroxidation. The extracts showed good antioxidant activity. Savory showed exceptionally high antioxidant activity even higher than that of quercetin.

Mariutti *et al.* (2008) evaluated the antioxidant activity of savory employing ABTS and DPPH radical assay. DPPH EC_{50} value was 20.3g/kg while DPPH-TEAC value was 13.5mM/g. ABTS-TEAC value was 20.4mM/g.

Satureja intricata Lange Family: Lamiaceae (Labiatae)

Jordan *et al.* (2010) investigated the chemical composition and antiradical activity of *Satureja intricata*. The phenolic content resulted in its essential oil having greatest activity against the DPPH and ABTS radicals.

Satureja macrostoma (Moc.& Sesse ex Benth.) Briq. Family: Lamiaceae (Labiatae)

Gutierrez and Navarro (2010) evaluated the antioxidative potential of methanol extract of *Satureja macrostoma* using various antioxidant assays, including DPPH, superoxide, Nitric oxide, hydroxyl radical scavenging and iron-chelating activity. Total phenolic and flavonoid content was also determined. The extract exhibited powerful free radical scavenging, especially against DPPH, hydroxyl radical scavenging and iron-chelating activity as well as a moderate effect on NO and superoxide anions.

Satureja mutica Fisch. & C.A. Mey Family: Lamiaceae

Motamed and Naghibi (2010) evaluated the antioxidant properties *Satureja mutica* and *Spinacia turkestanica* by four different methods: free radical scavenging using DPPH, evaluation of xanthine-oxidase activity, inhibition of lipid peroxidation by the ferric thiocyanate method, and the deoxyribose degradation assay. The extracts showed antioxidant activity. The pro-oxidant activities were also assayed for this species and *S. mutica* showed significant pro-oxidant activity.

Satureja obovata Lag. Family: Lamiaceae (Labiatae)

Jordan *et al.* (2010) investigated the chemical composition and antiradical activity of *Satureja obovata*. The essential oil of *Satureja obovata* showed *no activity* against DPPH and ABTS radicals.

Satureja pilosa Velen Family: Lamiaceae (Labiatae)

Nikolova and Dzhurmanski (2009) evaluated the free radical scavenging activity of total methanol extract of *Satureja pilosa*. The extract exhibited strong antioxidant activity and its IC_{50} value for DPPH radical was 53.12 µg/ml.

Satureja spicigera (K.Koch) Boiss. Family: Lamiaceae (Labiatae)

Eminagaoglu *et al.* (2007) examined the *in vitro* antioxidant activities of the essential oil and methanol extracts of *Satureja spicigera* using DPPH and β-carotene-linoleic acid assays. In DPPH test system, free radical-scavenging activity of *S.spicigera* oil was determined to be IC_{50} = 127 µg ml. In the β-carotene-linoleic acid test system, antioxidant activity of the oil was 81.7 per cent.

Satureja x delpozoi Family: Lamiaceae (Labiatae)

Jordan *et al.* (2010) investigated the chemical composition and antiradical activity of hybrid *Satureja x delpozoi*. Both the hybrids showed low activity against the DPPH and ABTS radicals.

Saurauia oldhamii Hmse. Family: Saurauiaceae

Shyur *et al.* (2005) examined antioxidant activities of 26 medicinal herbal extracts that have been popularly used as folk medicines in Taiwan. The results of scavenging DPPH radical activity show that, among the 26 tested medicinal plants, *Saurauia oldhamii* exhibited strong activities and its IC_{50} value for superoxide anion scavenging activity was 124µg/ml.

Sauropus androgynous (L.) Merrill Family: Euphorbiaceae

Wong *et al.* (2006) investigated the antioxidant properties of 25 edible tropical plants including *Sauropus androgynous* using DPPH and FRAP assays. Total phenolic content was also estimated. *Sauropus androgynous* showed very high antioxidant activity. A strong correlation between TEAC values obtained for the DPPH assay and those for FRAP assay implied that compounds in the extracts were capable of scavenging DPPH radical and reducing ferric ions. A satisfactory correlation of TPC with TEAC suggested that polyphenols in the extracts were partly responsible for the antioxidant activities while its correlation with CCA was poor, indicating that polyphenols might not be the main cupric ion chelators.

Subhasree *et al.* (2009) carried out the antioxidant activity of leafy vegetable of *Sauropus androgynous* by measuring the ability of methanol extracts of this plant to scavenge radicals generated by *in vitro* systems and by their ability to inhibit lipid peroxidation. The levels of non-enzymatic antioxidants were also determined by standard spectrophotometric methods. Correlation and regression analysis established a positive correlation between some of these antioxidants and the *in vitro* free radical-scavenging activity of the plant extracts. The conclusions drawn from this study indicate that *in vivo* studies, isolation and analysis of individual bioactive compounds will reveal the crucial role that this plant may play in several therapeutic formulations.

Extracts from 11 vegetables of Indonesian origin were screened for flavonoid content, total phenolics, and antioxidant activity by Andarwulan *et al.* (2010). Flavonoid content in *Sauropus androgynous* was 143mg/100g fresh weight. The antioxidant activity was measured by DPPH and ABTS scavenging and inhibition of linoleic acid oxidation. They concluded that *S. androgynous* was rich rource of dietary flavonoids and antioxidants.

Narayanaswamy and Balakrishnan (2011) evaluated the antioxidant activity of 13 medicinal plants including leaf of *Sauropus androgynous*. The aqueous extract showed highest 82 per cent inhibition of DPPH radical while ethanolic extract showed 40 per cent inhibition of DPPH radical.

Saururus chinensis (Lour.) Baill. Family: Saururaceae

Chang *et al.* (2002) evaluated the antioxidant activity and free radical scavenging capacity of Korean medicinal plants using commonly accepted assays. They were extracted with dichloromethane, methanol or ethanol, respectively and selected for the best antioxidant results. Each sample under assay condition showed a dose dependent free radical scavenging effect of DPPH and a dose-dependent inhibitory effect of xanthine oxidase and lipid peroxidation. Among plant extracts the root bark of *Saururus chinensis* showed stronger SC_{50} and ID_{50} values than other plant extracts.

Flavonol glycosides with free radical-scavenging activity of *Saururus chinensis* were isolated by Kang *et al.* (2005). An activity-guided fractionation procedure was used to identify the antioxidative components of the aerial parts. The antioxidant activity was investigated with the DPPH radical and superoxide anion-scavenging assays.

Sauchinone, a lignan isolated from the roots of *Saururus chinensis* showed protective effect against oxidative injury (Jeong *et al.*, 2010).

Saussarea costus (Falc.) Lipschitz. Family: Asteraceae (Compositae)

Pandey *et al.* (2005) studied the antioxidant activity of *Saussarea costus* using its ability to scavenge DPPH, nitric oxide, superoxide radicals along with its ability to inhibit lipid peroxidation and GSH oxidation. The 1mg/ml extract had antioxidant activity with 85.2 per cent reduction of DPPH and a 72.7 per cent decrease in lipid peroxidation. It showed maximum inhibition of superoxide radical of 66.0 per cent, and 58.4 per cent inhibition of nitric oxide formation. The concentration of chlorogenic acid was 0.027 per cent in the extract.

Saussurea lappa (Decne) Sch.-Bip. Family: Asteraceae (Compositae)

Alcoholic extract of *Sassurea lappa* showed significant free radical scavenging activity *in vitro* (Achliya *et al.*, 2003). The phytochemical screening revealed presence of resins and glycosides.

Saxifraga stolonifera (L.) Meeb. Family: Saxifragaceae

Sohn *et al.* (2008) evaluated the antioxidant activity of fresh juice and methanol extract of aerial parts of *Saxifraga stolonifera*. The fresh juice showed ignorable DPPH scavenging activity but the methanol extract showed strong antioxidant activity (IC_{50} of 37.5µg/ml). Sequential organic solvent fractionation of methanol extract showed that IC_{50}s of ethyl acetate and the butanol fraction were 6.9 and 7.8µg/ml, respectively, that is comparable with vitamin C or butylated hydroxytoluene. Analysis of component in extract and fractionates suggested that the antioxidants in fractions are diverse and the active substances have glycosylated phenolic structure.

Scariola orientalis (Boiss.) Sojak. Family: Asteraceae (Compositae)

Souri *et al.* (2004) evaluated the antioxidative activity of *Scariola orientalis* by linoleic acid peroxidation test using 1,3-diethyl-2-thiobarbituric acid as the reagent. Methanolic extract showed 71.24 per cent inhibition at 40µg concentration with IC_{50} value of 2.71µg.

Schefflera actinophylla (Endl.) Harms. Family: Araliaceae

The chloroform and methanolic extracts of 124 Egyptian plant species belonging to 56 families were investigated by Moussa *et al.* (2011) and compared their antioxidant activity by DPPH scavenging assay. Among the 124 plant species tested 18 exhibited extremely high antiradical activity (>80 per cent inhibition). The IC_{50} value of DPPH radical scavenging of *Schefflera actinophylla* was 19.23µg/ml, while total phenolic and flavonoid contents were 32.56 mg Tannic acid equivalent/g extract and 220.86 mg Rutin equivalent/g extract, respectively. Correlation coefficient between DPPH radical scavenging activity and the total phenolic and flavonoid contents suggest that phenolics and flavonoids in the extracts were partly responsible for the antiradical activities.

Schefflera leucantha R.Vig. Family: Araliaceae

Kruawan and Kangsadalampai (2006) investigated the antioxidant activity and phenolic compound contents of water extract of *Schefflera leucantha*. The extract exhibited moderate scavenging activity against DPPH radicals (64.45 per cent) and their FRAP value was 523.36 µmol/g. The content of phenolic compound in the extracts was determined using the Folin-Ciocalteu reagent and calculated as gallic acid equivalent (GAE). The herbal extract gave a high phenolic compound content (GAE 129.86 mg/g).

Schima wallichii (DC) Korth. Family: Theaceae

The butanol fraction of *Schima wallichii* leaves have been shown to have antimutagenic and antioxidant activities (Didi Jauhari *et al.*, 2000). Kshirsagar and Upadhyay (2009) have found that leaf has antioxidant activity in methanol fraction and the stem has both ethyl acetate and methanol fractions. The leaf showed good activity in dose dependent study, *i.e.*, 50 per cent reduction of DPPH is achieved at the dose of 0.03mg ml that is comparable to the standards (which are pure molecules), *viz.* catechin and Trolox. Moreover, their activities were better than that of Curcumin for which 50 per cent reduction was achieved by 0.04 mg/ml. Successive ethyl acetate extracts have shown equal antioxidant activity as that of successive methanol.

Schinus molle L. Family: Anacardiaceae

Bendaoud *et al.* (2010) investigated the chemical composition and anticancer and antioxidant activities of *Schinus molle* berries essential oil by ABTS assay. The essential oil has IC_{50} value of 257 mg/l in ABTS assay.

Schinus terebinthifolia Engler. Family: Anacardiaceae

The antioxidant property of *Schinus terebinthifolia* was studied using free radical-generating systems (Velazquez *et al.*, 2003). The methanol extract protected against enzymatic and non-enzymatic lipid peroxidation in microsomal membranes of rat. The extract showed the highest scavenging activity on the superoxide and DPPH radicals.

Bendaoud *et al.* (2010) investigated the chemical composition and anticancer and antioxidant activities of *Schinus terebinthifolius* berries essential oil by ABTS assay. The essential oil has IC_{50} value of 24 mg/l in ABTS assay.

Schinus wenmannifolia Engler. Family: Anacardiaceae

The antioxidant property of *Schinus weinmannifolia* was studied using free radical-generating systems (Velazquez *et al.*, 2003). The methanol extract protected against enzymatic and non-enzymatic lipid peroxidation in microsomal membranes of rat. The extract showed the highest scavenging activity on the superoxide and DPPH radicals.

Schisandra chinensis (Turca) Baill. Family: Schisandraceae

Methanol aqueous extract of *Schisandra chinensis* was screened by Kim *et al.* (1997) for free radical scavenging activity using the DPPH free radical generating system. The extract showed free radical scavenging activity.

Cai *et al.* (2004) evaluated the antioxidant activity and phenolic compounds in traditional Chinese medicinal plants associated with anticancer. The improved ABTS.+ method was used to systematically assess the total antioxidant capacity

(Trolox equivalent antioxidant capacity, TEAC) of the medicinal extracts. The TEAC values and total phenolic content for methanolic extract of fruit of *Schisandra chinensis* were 253.6 µmol Trolox equivalent/100g dry weight (DW), and 1.05 g of gallic acid equivalent/100g DW respectively. A positive significant linear relationship between antioxidant activity and total phenolic content showed that phenolic compounds were the dominant antioxidant components in the tested herbs. Major types of phenolic compounds in *Schisandra chinensis* include lingnas (schizandrins, schizatherins, wulignan).

Gomisin J, a lignan compound, isolated from the fruits of *Schsandra chinensis* was evaluated for antioxidant activity. Trolox, one of the well-known antioxdidant, used as a positive control. Gomisin J showed protective effect with an EC_{50} value of 213µM (An *et al.*, 2006).

ESP-102, a combined extract of *Angelica gigas, Saururus chinensis* and *Schisandra chinensis* protects against glutamate-induced toxicity in primary cultures of rat cortical cells (Ma *et al.*, 2009). ESP-102 decreased the cellular calcium concentration increased by glutamate and inhibited the subsequent overproduction of cellular nitric oxide and ractive oxygen species to the level of control cells. It also preserved cellular activities of enzymes such as superoxide dismutase, glutathione peroxidase and glutathione reductase reduced in the glutamate-injured neuronal cells.

Schizonepeta ternnuifolia (Benth.) Briq. Family: Lamiaceae (Labiatae)

Antioxidant capacities of 56 selected Chinese medicinal plants, including *Schzonepeta ternnuifolia* were evaluated by Song *et al.* (2010) using the TEAC and FRAP assays, and their total phenolic content was measured by the Folin-Ciocalteu method. *Schizonepeta ternnuifolia* showed moderate antioxidant activity with TEAC value of 47.13µmol Trolox/g, FRAP value of 67.97µmol Fe^{2+}/g and phenolic content was 8.17mg GAE/g.

Schrankia leptocarpa DC. Family: Mimosaceae

Lagnika *et al.* (2011) evaluated the antioxidant activity of *Schrankia leptocarpa* using DPPH radical scavenging assay. The extract showed good radical scavenging activity with IC_{50} value of 1.35µg/ml while the standard L-Ascorbic acid had IC_{50} value of 1.1µg/ml.

Schrebera swietenioides Roxb. Family: Oleaceae

Ethanolic extract of *Schrebera swietenioides* root was screened by Manda *et al.* (2009) for its antioxidant activity, using DPPH, ABTS and lipid peroxidation assay methods. In antioxidant assay, per cent inhibition and IC_{50} was calculated. The IC^{50} was 1.7µg/ml in DPPH, 4.35µg in ABTS and 23.53µg in lipid peroxidation assay and showed more potent antioxidant activity compared to standard compound.

Sclerocarya birrea (A.Rich.) Hochst. Family: Anacardiaceae

Polyphenol content and antioxidant activity of fourteen wild edible fruits from Burkina Faso were investigated by Lamien-Meda *et al.* (2008). The data obtained show that the total phenolic and total flavonoid levels were significantly higher in the acetone than in the methanol extracts. *Sclerocarya birrea* fruit had higher flavonoid content and higher antioxidant activity.

Total phenolic content, proanthocyanidins, gallotannins, flavonoids and antioxidant activities of *Sclerocarya birrea* methanolic extracts were evaluated by Moyo *et al.* (2010) using *in vitro* assays. *S. birrea* young stem extract contained the highest levels of total phenolic content (14.15mg GAE/g), flavonoids (1.21 mg CE/g) and gallotannins (0.24 mg GAE/g). The EC_{50} values of the extracts in the DPPH free radical scavenging assay ranged from 4.26 to 6.92 µg/ml, compared to 6.86µg/ml for ascorbic acid. A dose dependent linear curve was obtained for all extracts in the ferric-reducing power assay. All extracts exhibited high antioxidant activity comparable to butylated hydroxytoulene based on the rate of β-carotene bleaching (84.1-93.9 per cent).

Scolopia braunii (Klotzsch) Sleumer Family: Flacourtiaceae

Mosaddik *et al.* (2004) screened some Australian Flacourtiaceae species for *in vitro* antioxidant activity.The total antioxidant activity has been assessed based on scavenging activity of stable ABTS free radicals. Mature stem extract of *Scolopia braunii* was found to have most antioxidant activity (IC_{50} = 2.9 µg/ml).

Scoparia dulcis L. Family: Scrophulariaceae

English Name: Broomweed.

Ratnasooriya *et al.* (2005) studied the antioxidative potentiality of plant. An aqueous extract showed marked antioxidative activity *in vitro*. These results suggest that at least some of its therapeutic indications claimed by traditional physicians may be mediated via its antioxidant activity.

Scrophularia auriculata L. Family: Scrophulariaceae

The antioxidant properties of twenty medicinal herbs used in the traditional Mediterranean and Chinese medicine including *Scrophularia auriculata* were studied by Schinella *et al.* (2002). Most of the extracts were weak scavengers of the hydroxyl radical.

Scrophularia ningpoensis Hemsl. Family: Asclepiadaceae

Li *et al.* (2008) evaluated the antioxidant properties of 45 medicinal plants using FRAP and TEAC assays and the total phenolic contents of these plants were measured. *Scrophularia nigpoensis* showed low antioxidant activity with 21.86 µmol Fe(II)/g FRAP value, 12.36µmol Trolox/g of TEAC value and 5.80 mg GAE/g of Phenolic content.

Scutellaria baicalensis Georgi Family: Lamiaceae (Labiatae)

The antioxidative effect of ganhuangenin (GHG), isolated from *Scutellaria baicalensis*, was examined by Lim *et al.* (1999) by measuring the capacity to suppress the formation of phospatidylcholine hydroperoxide (PCOOH). The results showed that a pretreatment with GHG effectively suppressed PCOOH formation, which was initiated by the peroxyl-generating oxidant, AAPH (2,2'-azobis-2-aminopropane hydrochloride). The protective action of GHG against the formation of PCOOH was observed in liver, lung and kidney. When compared with known antioxidants, they found that the antioxidative potency of GHG to be greater than that of α-tocopherol which strongly indicate that GHG is a powerful antioxidant against lipid peroxidation.

Free radical scavenging and antioxidant activities of baicalein, baicalin, wogonin and wogonoside, the four major flavonoids in the radix of *Scutellaria baicalensis*, were examined by Gao *et al.* (1999) in different systems. ESR results showed that baicalein and baicalin scavenged hydroxyl radical, DPPH radical and alkyl radical in a dose-dependent manner, while wogonin and wogonoside showed subtle or no effect on these radicals. Ten μmole/l of baicalein and baicalin effectively inhibited lipid peroxidation of rat brain cortex mitochondria induced by Fe(2+)-ascorbic acid, AAPH and NADPH, while wogonin and wogonoside showed significant effects only on NADPH-induced lipid peroxidation. In a study on cultured human neuroblastoma SH-SY5Y cell system, it was found that 10 μmol/l of baicalein and baicalin significantly protected cells against H_2O_2-induced injury. Baicalein was the most effective antioxidant among the four tested compounds in every system due to its o-tri-hydroxyl structure in the A ring. Compared with a well-known flavonoid, quercetin, the antioxidant activity of baicalein was lower in DPPH or AAPH system, but a little higher in those systems which might associate with iron ion.

The antioxidant properties of twenty medicinal herbs used in the traditional Mediterranean and Chinese medicine including *Scutellaria baicalensis* were studied by Schinella *et al.* (2002). Most of the extracts were weak scavengers of the hydroxyl radical. Although *S. baicalensis* inhibited the lipid peroxidation in rat liver microsomes and red blood cells, the extract showed inhibitory actions on aminopyrine N-demethylase and xanthine oxidase activities as well as a pro-oxidant effect observed in the Fe^{3+}-EDTA-H_2O system. The results of this work suggest that the anti-inflammatory activities of the same extracts could be explained at least in part by their antioxidant properties.

The neuroprotective effects of flavonoids from the stems and leaves of *Scutellaria baicalensis* against hydrogen peroxide-induced rat oheochromocytoma line PC12 injury wer evaluated by Shang *et al.* (2006) by cell lesion, free radicals and ATPase disorders. The results indicated that the extracts exert neuroprotective effects against H_2O_2 toxicity.

Li *et al.* (2008) evaluated the antioxidant capacities of 45 selected medicinal plants using ferric reducing antioxidant power (FRAP) and Trolox equivalent antioxidant capacity (TEAC) assays respectively, and the total phenolic contents of these plants were measured by the Folin-Ciocalteau method. It was found that *Scutellaria baicalensis* possessed the highest antioxidant capacities with 304.86 μmol Fe(II)/g FRAP value, 184.34μmol Trolox/g of TEAC value and 36.30 mg GAE/g of phenolic content. A strong correlation between TEAC values and those obtained from FRAP assay implied that antioxidants in these plants were capable of scavenging free radicals and reducing oxidants. A high correlation between antioxidant capacities and their total phenolic contents indicated that phenolic compounds were a major contributor of antioxidant activity of these plants.

Cai *et al.* (2004, 2006) evaluated the antioxidant activity and phenolic compounds in traditional Chinese medicinal plants associated with anticancer. The improved ABTS.+ method was used to systematically assess the total antioxidant capacity (Trolox equivalent antioxidant capacity, TEAC) of the medicinal extracts. The TEAC values and total phenolic content for methanolic extract of whole plant of *Scutellaria baicalensis* were 1323.5 µmol Trolox equivalent/100g dry weight (DW), and 8.17 g of gallic acid equivalent/100g DW respectively. A positive significant linear relationship between antioxidant activity and total phenolic content showed that phenolic compounds were the dominant antioxidant components in the tested herbs. Major types of phenolic compounds in *Scutellaria baicalensis* include flavones (baicalein, baicalin, chrysin, wogonin).

Scutellaria barbata D.Don Family: Lamiaceae (Labiatae)

English Name: Barbed Skulcap.

The tea made from this herb is fruitful in treating liver, lung and rectal cancer. It has also been found potent in reducing tumor growth. Gallic acid in *Scutellaria* is anticancerous in function. Cai *et al.* (2004) evaluated the antioxidant activity and phenolic compounds in traditional Chinese medicinal plants associated with anticancer. The improved ABTS.+ method was used to systematically assess the total antioxidant capacity (Trolox equivalent antioxidant capacity, TEAC) of the medicinal extracts. The TEAC values and total phenolic content for methanolic extract of whole plant of *Scutellaria barbata* were 155.3 µmol Trolox equivalent/100g dry weight (DW), and 1.37 g of gallic acid equivalent/100g DW respectively. A positive significant linear relationship between antioxidant activity and total phenolic content showed that phenolic compounds were the dominant antioxidant components in the tested herbs. Major types of phenolic compounds in *Scutellaria barbata* include flavones (baicalein, baicalin, apigenin).

Gan *et al.* (2010) evaluated the antioxidant activity of 40 medicinal plants. *Scutellaria barbata* showed very low antioxidant activity of 22.23 µmol Fe(II)/g in FRAP assay and 26.35µmol in TEAC assay. Total phenolic content was also very high, 2.12 mg GAE/g.

Scutellaria lateriflora L. Family: Lamiaceae (Labiatae)

Wojcikowski *et al.* (2007) evaluated the antioxidant activity of aerial parts of *Scutellaria lateriflora*. Oxygen radical absorbance capacity of the ethyl extract, methanol extract and aqueous extracts were 269.99, 123.17 and 634.09µmol Trolox equivalent per g of dried starting material, respectively.

Scutellaria multicaulis Boiss. subsp. *multicaulis* Family: Lamiaceae (Labiatae)

Souri *et al.* (2004) evaluated the antioxidative activity of *Scutellaria multicaulis* subsp. *multicaulis* by linoleic acid peroxidation test using 1,3-diethyl-2-thiobarbituric acid as the reagent. Methanolic extract showed 92.48 per cent inhibition at 40µg concentration with IC_{50} value of 2.83µg.

Firuzi *et al.* (2010) evaluated the antioxidant activity and total phenolic content of 24 Lamiaceae species growing in Iran. *Scutellaria multicaulis* showed low antioxidant activity with 9.4µM/g DW FRAP value and IC_{50} = 1215.9µg/ml of DPPH radical scavenging activity. It also showed high phenolic content of 4.2 mg CE/g.

Scutellaria pinnatifida Art. et Hamilt. Family: Lamiaceae (Labiatae)

Souri *et al.* (2004) evaluated the antioxidative activity of *Scutellaria pinnatifida* by linoleic acid peroxidation test using 1,3-diethyl-2-thiobarbituric acid as the reagent. Methanolic extract showed 94.87 per cent inhibition at 40µg concentration with IC_{50} value of 0.76µg which is approximately in the range of α-tocopherol (IC_{50}=0.60µg).

Scutellaria rivularis Benth. Family: Lamiaceae (Labiatae)

Shieh *et al.* (2000) evaluated the antioxidant activity of baicalein, baicalin and wogonin, isolated from *Scutellaria rivularis*. The results showed that the order of activity on xanthine oxidase inhibition was baicalein>wogonin>baicalin, IC_{50} = 3.12, 157.38 and 215.19 µM, respectively, whereas the activity on cytochrome C reduction was baicalin>wogonin >baicalein (IC_{50} = 224.12, 300.10 and 370.33 µM, respectively). Both baicalein and baicalin demonstrated a strong activity on eliminating the superoxide radical. The IC_{50} of baicalein was 2.8 fold higher than that of baicalin. However, they had no significant effect on scavenging hydroxyl radical.

Secale cereale L. Family: Poaceae (Graminae)

English Name: Rye.

The antioxidative activity of phenolic extracts from rye was examined by autoxidation of methyl linoleate by Kahkonen *et al.* (1999). Rye extract did not show remarkable antioxidant activity. The processing seems to be important when evaluating the antioxidant activity of cereals as brans were more active than other products from the same cereal species. This is obviously due to localization of the phenolics in the grain: the outer layers of the grain (husk, pericarp, testa, and aleurone cells) contain the greatest concentrations of total phenolic, whereas their concentrations were considerably lower in the endosperm layers. About 80 per cent of the trans-ferulic acid in rye was found in the bran (White and Xing, 1997).

Korycinska *et al.* (2009) evaluated the antioxidant properties of rye bran alkylresorcinols (C15:0-C25:0) and extracts from whole-grain cereal products using their radical-scavenging activity on DPPH and the chemiluminescence method (CL). DPPH radical reduction varied from ~10 to ~60 per cent for the alkylresorcinol homologues at concentrations from 5 to 300µM and was not dependent on the length of the alkyl side chain of the particular homologue. Differences in the EC_{50} values for the studied compounds were not statistically significant, the values varying from 157µM for homologue C23:0 to 195µM for homologue C15:0. Moreover, values of EC_{50} for all the alkylresorcinol homologues were significantly higher than those for Trolox and α-, δ-, γ-tocopherols, compounds with well-defined antioxidant activity and used as positive controls. CL inhibition was evaluated for all the tested alkylresorcinol homologues at concentrations of 5 and 10 µM and varied from ~27 to ~77 per cent. Similar to the DPPH method, the slight differences in CL inhibition suggest that the length of the alkyl side chain had no major impact on their antioxidant properties. The extracts from whole-grain products were added to the DPPH and CL reaction systems and their antioxidant activities were tested and compared with the total amount of alkylresorcinols evaluated in the extracts. DPPH radical and CL reduction for the whole-grain products varied from ~70 to ~43 per cent and from ~37 to ~91 per cent, respectively. A clear relationship between DPPH radical and CL reduction levels and the amount of total alkylresorcinols was obtained for whole-grain breakfast cereals, in which the reduction level decreased in the order rye>wheat>mixed>barley. Therefore, it may be considered that the antioxidant activity of alkylresorcinols could be of potential importance to the food industry, which is continuously searching for natural antioxidants for the production of food products during their processing and storage.

Secamone afzelii (Schult.) K.Schum. Family: Asclepiadaceae

The antioxidant activity of a methanol extract of *Secamone afzelii* stems was tested using the DPPH assay and the active compound was identified as α-tocopherol (Mensah *et al.*, 2004). The total extract also showed effective free radical scavenging activity in the assay for non-enzymatic lipid peroxidation in liposomes with an IC_{50} value of 90µg/ml, with α-tocopherol isolated from the plant having an IC_{50} of 15 µg/ml in the same system, thus demonstrating the presence of other antioxidants.

Extracts of *Secamone afzelii* have been shown to have anti-inflammatory and antioxidant properties due to flavonoids, triterpenoids, diterpenoids and caffeic acid derivatives (Houghton *et al.*, 2005).

Lagnika *et al.* (2011) evaluated the antioxidant activity of *Secamone afzelii* using DPPH radical scavenging aassay. The extract showed good radical scavenging activity with IC_{50} value of 1.74µg/ml while the standard L-Ascorbic acid had IC_{50} value of 1.1µg/ml.

Sechium edule Sw. Family: Cucurbitaceae

English Name: Chayote.

Ordonez *et al.* (2006) reported that *Sechium edule* leaf ethanolic and seed water extracts exerted prominent hydrogen-donating ability in the presence of DPPH stable radical (IC_{50} 2 µg/ml).

Securidaca longepedunculata Fres. Family: Polygalaceae

Akinmoladun *et al.* (2010) assessed the phytochemical constituents and antioxidant and free radical scavenging activities of *Securidaca longepedunculata* using seven different antioxidant assay methods. The results suggest that the methanolic extracts possess significant antioxidant and radical scavenging activity that may be due to the phytochemical content of the plant.

Securigera securidaca (L.) Dalla Torre et Sarntheim Family: Fabaceae (Papilionaceae)

Azad *et al.* (2009) investigated the effect of total extract of *Securigera securidaca* seeds on serum lipid profile and oxidative stresses caused by hyperocholesterolemia. Compared to ascorbic acid the extract had a weak antioxidant activity *in vitro* with no effect on total serum antioxidant and serum MDA level.

Securinega leucopyrus (Willd.) Muell. Family: Euphorbiaceae

Antioxidant activity of aerial parts of *Securinega leucopyrus* was studied by Vidyadhar *et al.* (2010) for its free radical scavenging property on different *in vitro* models *e.g.*, DPPH, nitric oxide and hydroxyl radical scavenging activity by using chloroform extract. The phytochemical screening has shown the presence of steroids, alkaloids, triterpenoids, flavonoids and tannins. *In vitro* antioxidant activity revealed that the chloroform extract showed more activity.

Securinega virosa Roxb. ex Willd. Family: Euphorbiaceae

One new flavonoid glycoside, 3-O-kaempferol 4-O-(galloyl)-β-D-glucoside, one new bergenin derivative, 11-O-caffeoylbergenin, along with other known flavonoids and phenolic derivatives, were isolated by Sanogo *et al.* (2009) from the leaves of *Securinega virosa*. Their structures were established on the basis of detailed spectral analysis. *In vitro* biological analysis of the isolated compounds showed that they were able to quench DPPH radicals and had a direct scavenging activity on superoxide anion. Kaempferol 3-O-(4-galloyl)-β-D-glucopyranoside (1), 11-O-caffeoylbergenin (2), and glucogallin (6) exhibited the highest antioxidant capacity, being also able to modulate hydroxyl radical formation more efficiently than the other compounds, acting as direct hydroxyl radical scavengers and chelating iron.

Sedum sarmentosum Bge. Family: Crassulaceae

Cai *et al.* (2004) evaluated the antioxidant activity and phenolic compounds in traditional Chinese medicinal plants associated with anticancer. The improved ABTS.+ method was used to systematically assess the total antioxidant capacity (Trolox equivalent antioxidant capacity, TEAC) of the medicinal extracts. The TEAC values and total phenolic content for methanolic extract of aerial parts of *Sedum sarmentosum* were 86.7 μmol Trolox equivalent/100g dry weight (DW), and 0.87 g of gallic acid equivalent/100g DW respectively. A positive significant linear relationship between antioxidant activity and total phenolic content showed that phenolic compounds were the dominant antioxidant components in the tested herbs.

Antioxidative activity of solvent fractions from *Sedum sarmentosum* were determined by Kim *et al.* (2006) using the DPPH radical assay, the ferric thiocyanate (FTC) method and the TBA method. Values were then compared with that of BHT (a synthetic antioxidant). Ethyl acetate and butanol fractions showed marked DPPH radical scavenging activity with values of 90.61 and 87.02 per cent, respectively. Total phenols contents in solvent fractions, determined using the Folin-Denis method, were in the order: ethyl acetate > butanol > ethanol > chloroform > water. There was a positive correlation between antioxidative activity and the total phenols content of fractions. Antioxidative activity in a linoleic acid system was determined using the FTC and TBA methods.

Kim *et al.* (2008) investigated growth characteristics, vitamin C content and antioxidative activity among local strains of *Sedum sarmentosum*.

Sedum sempervioides Fisch. ex Bieb. Family: Crassulaceae

Antioxidant and radical scavenging activities, reducing powers and the amount of total phenolic compounds of aqueous and methanolic extracts of *Sedum sempervioides* were studied by Mavi *et al.* (2004). The extract showed moderate antioxidant activity.

Semecarpus anacardium L. Family: Anacardiaceae

Semecarpus anacardium nut milk was administered orally (200 mg/kg body weight/day) by Premalatha *et al.* (1997) for 14 days to hepatocarcinoma bearing rats. The level of lipid peroxides and antioxidant enzymes' activity were determined in liver and kidney homogenates. Marked increase in lipid peroxide levels and a concomitant decrease in enzymatic antioxidant levels were observed in carcinoma induced rats, while drug treatment reversed the conditions to near normal levels.

Verma and Vinayak (2009) investigated antioxidant activity of the aqueous extract of nuts of *Semecarpus anacardium* in AKR mouse liver during the development of lymphoma. Antioxidant action was monitored by the activities of antioxidant enzymes catalase, superoxide dismutase and glutathione transferase. The effect of *S. anacardium* was also studied by observing the activity of LDH (lactate dehydrogenase), an enzyme of anaerobic metabolism. LDH activity serves as a tumour marker. The activities of antioxidant enzymes decreased gradually as lymphoma developed in mouse. However, LDH activity increased progressively. Administration of the aqueous extract of *S. anacardium* to lymphoma-transplanted mouse led to an increase in the activities of antioxidant enzymes, whereas LDH activity decreased significantly, indicating a decrease in carcinogenesis. The aqueous extract was found to be more effective than doxorubicin, a classical anticarcinogenic drug, with respect to its action on antioxidant enzymes and LDH in the liver of mice with developing lymphomas.

Sempervivum tectorum L. Family: Crassulaceae

English Name: House leek.

The antioxidant properties of components of leaf extracts of *Sempervivum tectorum* (ST), have been evaluated by Sentjurc *et al.* (2003) using UV irradiated liposomal systems containing the spin trap 5-(diethoxyphosphoryl)-5-methyl-pyrroline-*N*-oxide. Decreases in free radical activity in the liposomal systems as measured by electron paramagnetic resonance (EPR) spectroscopy demonstrate that the lipophilic ST juice components, kaempferol (KA) and kaempferol-3-glucoside (KG) contribute significantly to the antioxidant properties of the juice. EPR spectral simulation established the presence of oxygen and carbon centered free radical adducts. The mixtures with low pH, citric and malic acid, and ST juice reveal increased EPR signals from oxygen centered radicals in comparison to the control, pointing to the important role of pH in oxygen radical formation. Parallel assays that measured thiobarbituric acid related substances confirmed the antioxidant effects of KA and KG and explain the results of spin trapping experiments complicated by low pH's.

Liver protecting, lipid level lowering and HDL-cholesterol enhancing activities of a lyophilized extract of *Sempervivum tectorum* was examined in experimental hyperlipidaemia, in the pathomechanism of steatotic liver of rats by Blazovics *et al.* (1993). Diffuse hepatocellular degeneration by a fat rich diet was improved after treatment with this extract.

Senecio argunensis Turcz Family: Asteraceae (Compositae)

Zhou *et al.* (2008) evaluated the antioxidant activities of different extracts of *Senecio argunensis*.The antioxidant activities of *S. argunensis* extracts with acetoacetate, n-Butanol and water were detected by DPPH free radical-scavenging method and β-carotene/linoleic acid system.The acetoacetate, n-Butanol and water extracts from *S. argunensis* eliminated DPPH in dose-dependent manner, their EC_{50} values were 0.0198, 0.0219 and 0.092 mg/ml, respectively. The strength order of the antioxidant activities of the three parts in β-carotene/linoleic acid system was acetoacetate, n-Butanol and water extracts.The extracts of the three parts of *S. argunensis* all have antioxidant activities. Among these extracts, extracts with acetoacetate have the highest antioxidant activity.

Senecio cineraria DC. Family: Asteraceae (Compositae)

Souri *et al.* (2004) evaluated the antioxidative activity of *Senecio cineraria* by linoleic acid peroxidation test using 1,3-diethyl-2-thiobarbituric acid as the reagent. Methanolic extract showed 83.26 per cent inhibition at 40µg concentration with IC_{50} value of 7.29µg.

Senecio herzogui Family: Asteraceae (Compositae)

The aerial parts of 17 Bolivian plants including *Senecio herzegui* were screened by Rosas-Romero and Saavedra (2005) to determine antioxidant activity. A methanol extract of each plant was prepared and portioned sequentially with hexane, chloroform, and ethyl acetate, leaving an aqueous solution. All extracts and their 5 fractions for a total of 102 samples, were evaluated using two techniques: an adaptation of the β-carotene bleaching technique using an emulsion of linoleic acid in water as the oxidizable substrate, and the DPPH free radical trapping technique. The results of the β-carotene bleaching technique were more discriminating and better related to the rancidity process under normal conditions; with this assay, 11 species provided at least one fraction with highly promising antioxidant activity. All species gave good results under the DPPH technique, and in most cases they performed better than BHA, which was used as a reference antioxidant (Rosas-Romero and Saavedra, 2005).

Senna alata (L.) Roxb. (Syn.: *Cassia alata*) Family: Caesalpiniaceae

Chomnawang *et al.* (2007) reported that *Senna alata* had a moderate antioxidant activity. Antioxidant activity was determined by DPPH scavenging and NBT reduction assay.

Senna italica Mill. (Syn.: *Cassia italica* (Mill.) F.W. Andrews) Family: Caesalpiniaceae

Masoko *et al.* (2009) evaluated the antioxidant and antibacterial properties of acetone extract of the roots of *Senna italica*. The extract was assayed for the *in vitro* antioxidant activity using DPPH assay. The acetone extract of the roots of *Senna italica* showed free radical scavenging activity.

Sericostoma pauciflorum Stocks ex Wight Family: Boraginaceae

Singh *et al.* (2009) evaluated the antioxidant activity of dichloromethane and methanolic extracts of 12 arid zone medicinal plants. *Sericostoma pauciflorum* showed appreciable antioxidant activity and per cent of inhibition of DPPH where RC_{50} (µg/ml) was 60 (Dichloromethane extract) and 14.5 (MeOH extract). per cent of inhibition of DPPH was 94.4 per cent in methanol extract of *Sericostoma pauciflorum*. Key phytochemicals include α-, β-amyrin, sericostinyl acetate and triterpenes.

Serenoa repens (Bartram.) J.K.Small Family: Arecaceae (Palmae)

Wojcikowski *et al.* (2007) evaluated the antioxidant activity of aerial parts of *Serenoa repens*. Oxygen radical absorbance capacity of the ethyl extract, methanol extract and aqueous extracts were 0.91, 48.50 and 69.95μmol Trolox equivalent per g of dried starting material, respectively.

Serratula coronata L. Family: Asteraceae (Compositae)

The antioxidant effect of aqueous methanolic herb extracts of *Serratula coronata* was investigated by Báthori *et al.* (2004) using both enzyme-dependent and enzyme-independent systems. The extract displayed concentration-dependent inhibition of lipid peroxidation. Flavonoids and ecdysteroids present in the extracts were evaluated as antioxidant components. The flavonoid-containing fraction of the herb extract was more effective in lipid peroxidation than the ecdysteroid-containing fraction. Quercetin 3-O-methyl ether, apigenin, luteolin, quercetin, luteolin4'β-D-glucoside and quercetin4'β-D-glucoside were isolated from the extract.

Serratula strangulata Ilgin Family: Asteraceae (Compositae)

The antioxidative and free radical scavenging effects of four ecdysteroids, 20-hydroxyecdysone (E1), 25-deoxy-11,20-dihydroxyecdysone (E2), 24-(2-hydroxyethyl) 20-hydroxyecdysone (E3), and 20-hydroxyecdysone-20,22-monoacetonide (E4), isolated from *Serratula strangulata* have been investigated *in vitro* (Cai *et al.*, 2002). Thes ecdysteroids could protect human erythrocytes against oxidative hemolysis induced by a water soluble azo iniator AAPH. They could also inhibit the peroxidation of rat liver microsomes induced by hydroxyl radicals, as monitored by the formation of TBARS, and prevent radical-induced decrease of membrane fluidity as determined by fluorescence polarization. They reacted with galvinoxyl radicals in homogenous solution, and the pseudo-first-order rate constants were determined spectrophotometrically by following the disappearance of galvinoxyl radicals. Compounds E1 and (or) E3 were the most active in both antioxidative and radical-scavenging reactions.

Serratula tinctoria L. Family: Asteraceae (Compositae)

The antioxidant effect of aqueous methanolic herb extracts of *Serratula tinctoria* was investigated by Báthori *et al.* (2004) using both enzyme-dependent and enzye-independent systems. The extract displayed concentration-dependent inhibition of lipid peroxidation. Flavonoids and ecdysteroids present in the extracts were evaluated as antioxidant components.

Serratula wolffii Andrae Family: Asteraceae (Compositae)

The antioxidant effect of aqueous methanolic herb extracts of *Serratula wolffii* was investigated by Báthori *et al.* (2004) using both enzyme-dependent and enzye-independent systems. The extract displayed concentration-dependent inhibition of lipid peroxidation. Flavonoids and ecdysteroids present in the extracts were evaluated as antioxidant components.

Sesamum indicum L. Family: Pedaliaceae

English Name: Sesame.

The antioxidant activity of ethanolic extracts of sesame coat (EESC) was investigated by Chang *et al.* (2002). The antioxidant activity (91.4 per cent) of 1.0 mg EESC was equal to 1.0 mg tocopherol (90.5 per cent) but was weaker than 1.0 mg butylated hydroxyanisole (98.6 per cent) on peroxidation of linoleic acid. EESC showed an inhibitory effect against the formation of TBARS in a liposome model system. EESC at 10.0 mg exhibited a 94.9 per cent scavenging effect on DPPH radicals and marked reducing power, indicating that EESC acts as a primary antioxidant. The extracts, at a dose of 1.0 mg, showed a 50.0 per cent scavenging effect on the hydroxyl radical. EESC also exhibited a metal-binding ability. Sesamin and sesamolin, the lignan substances, were found in EESC, by HPLC analysis. In addition, chromatographic analysis demonstrated that phenolic compounds and tetranortriterpenoids, which had positive reactions with β-carotene, indicating antioxidant activity, are present in EESC. According to these results, termination of free radical reaction, metal-binding ability and quenching of reactive oxygen are suggested to be, in part, responsible for the antioxidant activity of EESC.

The free radical scavenging capacity of antioxidants from sesame cake extract was studied using DPPH radical on a kinetic model. Pure lignans and lignan glycosides isolated from methanolic extract by preparative HPLC were used in the study. To understand the kinetic behavior better and to determine the radical scavenging activity of sesame antioxidants, the second-order rate constant (k_2) was calculated for the quenching reaction with DPPH radical. The k_2 values for sesamol, sesamol dimer, sesamin, sesamolin, sesaminol triglucoside and sesaminol diglucoside were 4.00×10^{-5}, 0.50×10^{-5}, 0.36×10^{-5}, 0.13×10^{-5}, 0.33×10^{-5} and $0.08 \times 10^{-5} \mu M^{-1} s^{-1}$, respectively (Suja *et al.*, 2004).

Antioxidant activities of *Sesamum indicum* extracts derived from sesame seed by supercritical carbon dioxide extraction and by n-Hexane were determined by Hu *et al.* (2004) using DPPH radical scavenging and linoleic acid system methods. The extracts at 30MPa presented the highest antioxidant activities assessed in DPPH method.

The antioxidant activities of brown pigment, extract of n-Hexane and extract of supercritical carbon dioxide extraction of black sesame seeds were investigated by Xu *et al.* (2005). The results indicated that the brown pigment of sesame seed possessed excellent antioxidant activity.

Sesbania grandiflora (L.) Pers. Family: Fabaceae (Papilionaceae)

The antioxidant activity of 70 per cent acetone and 50 per cent ethanol extracts of *Sesbania grandiflora* were determined by DPPH radical scavenging method. Among the two different extracts acetone extract was found to be more efficient (Gowri and Vasantha, 2010).

The aqueous leaf extracts of *Sesbania grandiflora* have been analysed by Padmaja *et al.* (2011) for their free radical-scavenging activity, hydroxyl radical-scavenging activity in Fe^{3+}/ascorbate/EDTA/H_2O_2 system, inhibition of lipid peroxidation induced by $FeSO_4$ in egg yolk, metal chelating activity. The results indicated that *Sesbania grandiflora* showed significant antioxidant activity *in vitro*.

Ouattara *et al.* (2011) evaluated the antibacterial and antioxidant activities of *Sesbania grandiflora*. Aqueous, methanolic and hydro-acetone extractions were carried out on the leaves, stems, and granules, pods of fruit and roots of the plant. Gallic acid, caffeic acid, kaempferol, quercetin, rutin were isolated from the extracts. Antioxidant activity of the extracts was evaluated by the test of reduction of the stable radical DPPH and test for reduction of iron (FRAP). This activity was related to phenolic compounds contained in the extracts.

Sesbania pachycarpa DC. Family: Fabaceae (Papilionaceae)

Using Trolox assay, Cook *et al.* (1998) estimated the antioxidant activity of 17 wild edible plants of Niger Republic used for food and traditional medicine. They observed that *Sesbania pachycarpa* possessed strong antioxidant activity.

Seseli rigidum Waldst. et Kit. Family: Apiaceae (Umbelliferae)

The essential oil was isolated by hydrodistillation from the flowers of *Seseli rigidum*. The free radical scavenging capacities of oil were measured by DPPH assay. The 50 per cent of DPPH scavenging by oil was achieved at 24.5 µl/ml (Stojkovic *et al.*, 2009).

Sesuvium portulacastrum L. Family: Aizoaceae

Hydrodistillation was used to extract the essential oil from the fresh leaves of *Sesuvium portulacastrum* by Magwa *et al.* (2006). Using β-carotene, acetone and linoleic acid method for the antioxidant testing, the essential oil showed antioxidant activity threshold of 15.9 mm mean zone of color retention (Magwa *et al.*, 2006).

Vadlapudi and Naidu (2009) evaluated the antioxidant potential of *Sesuvium portulacastrum*, a mangrove plant. The SOD was 0.41U/mg, Catalase was 0.27U/mg, Ascorbic acid was 280mg/100g, DPPH radical inhibition was 34.5 per cent and FRAP units were 320.

Setaria italica (L.) Beauv. Family: Poaceae (Graminae)

English Name: Foxtail millet.

The less explored, commercially available foxtail millet-milled fractions like whole flour and bran rich fraction were studied by Suma and Urooj (2011) for its antioxidant potency. Phytochemicals like alkaloids, phenolics, reducing sugars and flavonoids were found only in methanolic and aqueous extracts, while tannins and terpenoids were present in all the solvent extracts of whole flour and bran rich fraction. Antioxidants were extracted using methanol, ethanol and water. Methanolic extracts of whole flour and bran rich fraction exhibited a significantly higher radical scavenging activity (44.62 per cent and 51.80 per cent respectively) using DPPH model system, and reducing power (0.381 and 0.455 respectively) at 2 mg, than the other solvents used for extraction. As bran rich fraction showed the highest antioxidant activity, suggesting the presence of antioxidant components in the bran layer.

Setaria viridis (L.) Beauv. Family: Poaceae (Graminae)

The EtOAc and-BuOH soluble fractions from the aerial part of *Setaria viridis* showed a strong free radical scavenging activity (Kwon *et al.*, 2002). Six major compounds were isolated from these fractions. They were identified by spectral data as tricin (*1*), *p*-hydroxycinnamic acid (*2*), vitexin 23-O-xyloside (*3*), orientin 23-O-xyloside (*4*), tricin-7-O-β-D-glucoside (*5*) and vitexin 23-O-glucoside (*6*). Among these compounds, 4 and 5 exhibited strong free radical scavenging activities on DPPH. They further studied the effects of these isolated compounds on the lipid peroxidation in rat liver microsomes induced by non-enzymatic method. As expected, 4 and 5 exhibited significant inhibition on ascorbic/Fe^{2+} induced lipid peroxidation in rat liver microsomes.

Shorea tumbaggaia Roxb. Family: Dipterocarpaceae

Ragini *et al.* (2011) evaluated the antidiabetic and antioxidant activity of *Shorea tumbaggaia*. In order to determine antioxidant activity of extract *in vivo* assays TBARS, SOD, GSH and CAT were performed for *in vivo* activity and DPPH scavenging method, reducing power method and hydrogen peroxide method were followed for *in vitro* assay. Oral administration of *S. tumbaggaia* for 28 days resulted in significant reduction in blood glucose levels. The lipid metabolites were significantly altered to near normal. And there was significant improvement in TBARS, SOD, GSH and CAT in liver tissues of alloxan induced diabetic rats when compared to untreated diabetic rats. For *in vitro* activity the increase in extract concentration increase the absorbance at various wavelengths. IC_{50} values were very much comparable to the standard drug ascorbic acid.

Sida acuta Burm.f. Family: Malvaceae

Karou *et al.* (2005) evaluated the polyphenol content and antioxidant activity of *Sida acuta* by ABTS assay. Polyphenols in the lyophilized extract of whole plant was 10.11 per cent. The antioxidant activity of whole plant was 1.20µmol Trolox/µg in the Phosphomolybdenum assay and 6.12 µmol Trolox/µg in ABTS assay. The total phenolic compounds were highly correlated with the antioxidant activities.

Pieme *et al.* (2010) evaluated the antioxidant activity of *Sida acuta* extracts. The results of the antioxidant properties showed that these extracts induced significantly increase SOD, CAT and GST activity after 48h.

Sida alba L. Family: Malvaceae

Polyphenol contents, antioxidant and antiinflammatory activities of *Sida alba* were investigated by Kiessoun *et al.* (2010). Total phenolic content was 32.53 mg GAE/100g extract, total flavonoid content was 1.02 quercetin equivalent/100mg extract, total flavonol content was 0.68mg quercetin equivalent/100 mg extract and total tannin content was 4.35 mg TAE/100 mg extract. Antioxidant activity in DPPH assay was 5.344 mmol Ascorbic acid equivalent (AAE)/g extract, in FRAP assay 3.69 mmol AAE/g extract and in ABTS assay 0.81 mmol AAE/g extract.

Sida cordifolia L. Family: Malvaceae

Dhalwal *et al.* (2005) studied the comparative antioxidant potential of ethanol extracts of *Sida cordifolia* leaf, stem, root and whole plant. Anti-lipid peroxidation, free radical scavenging, reducing power, nitric oxide scavenging, superoxide scavenging antioxidant assay and further estimation of total phenolic content and HPTLC studies were carried out. Various antioxidant activities were compared with standard antioxidant such as BHT, α-tocopherol, and ascorbic acid. Ethanol extracts were found to be good scavenger of DPPH radical in the order roots > stem > leaves > whole plant with values 76.62 per cent, 63.87 per cent, 58 per cent and 29 per cent at a dose of 1 mg respectively. All extracts of *Sida cordifolia* have effective reducing power and free-radical scavenging activity. Only the root extract exhibited superoxide-scavenging activity and inhibited lipid peroxidation in rat liver homogenate. All these antioxidant properties were concentration dependent. In addition Total phenolics content of all the extracts of *Sida cordifolia* were determined as gallic acid equivalents. The highest antioxidant activity was observed in the root extract.

Polyphenol contents, antioxidant and anti-inflaammatory activities of *Sida cordifolia* were investigated by Kiessoun *et al.* (2010). Total phenolic content was 10.25 mg GAE/100g extract, total flavonoid content was 1.92 quercetin equivalent/ 100mg extract, total flavonol content was 0.57mg quercetin equivalent/100 mg extract and total tannin content was 7.83 mg TAE/100 mg extract. Antioxidant activity in DPPH assay was 8.49 mmol Ascorbic acid equivalent (AAE)/g extract, in FRAP assay 3.47 mmol AAE/g extract and in ABTS assay 0.69 mmol AAE/g extract.

Sida retusa L. Family: Malvaceae

The methanolic extract of *Sida retusa* roots were found to inhibit lipid peroxidation, scavenge hydroxyl and superoxide radicals *in vitro* (Lissy *et al.*, 2006). IC_{50} of root extract of *Sida retusa* was 71.29 µg/ml, 1763.22µg/ml and 1130.24µg/ml for superoxide radical scavenging, hydroxyl radical scavenging and lipid peroxidation respectively.

Sida rhombifolia L. Family: Malvaceae

Polyphenol contents, antioxidant and anti-inflaammatory activities of *Sida rhombifolia* were investigated by Kiessoun *et al.* (2010). Total phenolic content was 5.75 mg GAE/100g extract, total flavonoid content was 1.02 quercetin equivalent/ 100mg extract, total flavonol content was 0.68mg quercetin equivalent/100 mg extract and total tannin content was 4.35 mg TAE/100 mg extract. Antioxidant activity in DPPH assay was 5.344 mmol Ascorbic acid equivalent (AAE)/g extract, in FRAP assay 3.69 mmol AAE/g extract and in ABTS assay 0.81 mmol AAE/g extract.

Sida rhombifolia L. ssp. *retusa* L. Family: Malvaceae

Dhalwal *et al.* (2007) studied the comparative antioxidant potentials of ethanol extracts of roots, stems, leaves and whole plant. Estimation of total polyphenolic content and high performance thin-layer chromatography profile were determined. Further inhibition of oxygen-derived free radicals, *viz.*, assays for free radical scavenging, reducing power, superoxide anion scavenging, nitric oxide scavenging and anti-lipid peroxidation, were performed. All the antioxidant activities were compared with standard antioxidants such as butylated hydroxyanisole and α-tocopherol acetate. Extracts were found to be good scavengers of the DPPH radical in the order root>leaves>whole plant>stem with 50 per cent inhibitory concentrations of 546.1, 852.8, 983.8 and 1,222.5 µg/ml, respectively. All the extracts of this plant showed effective free radical scavenging activity, reducing power, and superoxide scavenging activity. Only root extract inhibited lipid peroxidation in rat liver and brain homogenate. All these antioxidant properties were concentration dependent. In addition, total polyphenolic contents of all the extracts were determined as gallic acid equivalents. The highest antioxidant activity was observed in root extract.

Sida urens L. Family: Malvaceae

Polyphenol contents, antioxidant and anti-inflammatory activities of *Sida urens* were investigated by Kiessoun *et al.* (2010). Total phenolic content was 4.21 mg GAE/ 100g extract, total flavonoid content was 0.43 quercetin equivalent/100mg extract, total flavonol content was 0.31mg quercetin equivalent/100 mg extract and total tannin content was 2.28 mg TAE/100 mg extract. Antioxidant activity in DPPH assay was 5.083 mmol Ascorbic acid equivalent (AAE)/g extract, in FRAP assay 3.18 mmol AAE/g extract and in ABTS assay 0.57 mmol AAE/g extract.

Sideritis caesarea Duman *et al.* Family: Lamiaceae (Labiatae)

Sagdic *et al.* (2009) investigated the total phenolic, flavanol and flavonol compounds of the methanolic extracts of *Sideritis casearea* and their antioxidant effects. Antioxidant activity of the extracts tested by the DPPH radical scavenging system was 72.47 at 100 ppm concentration of the extracts.

Sideritis condensata Boiss. et Heldr. Family: Lamiaceae (Labiatae)

Ozkan *et al.* (2005) investigated the antioxidant properties of *Sideritis condensata*. Phenolic compounds concentration (as gallic acid equivalent) was 247.62 mg/g. Free radical scavenging activities (DPPH method) at 100 ppm concentration was 72.01 per cent. Antioxidative activity (phosphomolybdenum method) was 279.37 mg/g.

Sideritis erythrantha Boiss. et Heldr. Family: Lamiaceae (Labiatae)

Ozkan *et al.* (2005) investigated the antioxidant properties of *Sideritis erythrantha*. Phenolic compounds concentration (as gallic acid equivalent) was 217.61 mg/g. Free radical scavenging activities (DPPH method) at 100 ppm concentration was 71.48 per cent. Antioxidative activity (phosphomolybdenum method) was 146.11 mg/g.

Kose *et al.* (2010) investigated the chemical composition and antioxidant activities of the essential oils of *Sideritis erythrantha* var. *erythrantha* and var. *cedretorum*. Antioxidant activities of the essential oils were determined by DPPH, β-carotene/ linoleic acid and reducing power. Both the essential oils exhibited weak antioxidant activity.

Sideritis libanotica Labill. subsp. *linearis* (Benth.) Borm. Family: Lamiaceae (Labiatae)

Tepe *et al.* (2006) screened the antioxidative properties of the methanolic extracts of *Sideritis libanotica* subsp. *linearis* by two complementary test systems, namely DPPH free radical-scavenging and β-carotene/linoleic acid. The extract showed moderate antioxidant activity.

Sideritis montana L. Family: Lamiaceae (Labiatae)

English Name: Sideritis.

Koleva *et al.* (2003) determined the antioxidant activity of *Sideritis montana* by the β-carotene bleaching test (BCBT), DPPH radical scavenging method and static headspace gas chromatography. The highest antioxidant activity in the BCBT, close to that of BHT, was observed for the more apolar extracts. Extracts from Butanol and from ethyl acetate and the total methanol extracts showed a strong radical scavenging activity against DPPH, close to that of rosmarinic acid. The antioxidant activity of *Sideritis* extracts was attributed to the presence of flavonoid and phenylpropanoid glycosides.

Khalil *et al.* (2007) studied the growth, phenolic compounds and antioxidant activity of sideritis grown under organic farming condition. Antioxidant activity of ethanol extract (inhibition per cent) was 73.21 at 200μl extract.

Firuzi *et al.* (2010) evaluated the antioxidant activity and total phenolic content of 24 Lamiaceae species growing in Iran. *Siderites montana* showed high antioxidant activity with 14.3μM/g DW FRAP value and IC_{50} = 828.3μg/ml of DPPH radical scavenging activity. It also showed high phenolic content of 5.0 mg CE/g. FRAP and DPPH assay results showed good correlations with the total phenolic contents of the plant measured by the Folin-Ciocalteu assay.

Sideritis ozturkii Aytac et Aksoy Family: Lamiaceae (Labiatae)

Sagdic *et al.* (2009) investigated the total phenolic, flavanol and flavonol compounds of the methanolic extracts of *Sideritis ozturkii* and their antioxidant effects. Antioxidant activity of the extracts tested by the DPPH radical scavenging system was 72.47 at 100 ppm concentration of the extracts.

Sideritis raeseri Boiss. et Heldr. subsp. *raeseri* Family: Lamiaceae (Labiatae).

Nine 7-O-allosyl glucosides of 5,8-dihydroxy substituted flavones were isolated from a fraction of the methanol extract of the aerial parts of *Sideritis raeseri* subsp. *raeseri*. The antioxidant activity of this fraction, evaluated by the Co(II)EDTA-induced luminal chemiluminescence and by the DPPH scavenging activity was found to be moderate. The activity can be related with the presence of 5- and 8-O-substitued flavones (Gabrieli *et al.*, 2005).

Flower extracts contained phenolic compounds and their amounts varied from 15.3 to 34.1 mg GAE/g DW (Pljevljakusic *et al.*, 2011).

Sideritis scardica L. Family: Lamiaceae (Labiatae)

Koleva *et al.* (2003) determined the antioxidant activity of *Sideritis scardica* by the β-carotene bleaching test (BCBT), DPPH radical scavenging method and static headspace gas chromatography. The highest antioxidant activity in the BCBT, close to that of BHT, was observed for the more apolar extracts. The inhibition of hexanal formation in bulk safflower oil by most of *Sideritis scardica* extract was as effective as BHT but less so than rosmarinic acid. Extracts from Butanol and from ethyl acetate and the total methanol extracts showed a strong radical scavenging activity against

DPPH, close to that of rosmarinic acid. The antioxidant activity of *Sideritis* extracts was attributed to the presence of flavonoid and phenylpropanoid glycosides.

Nikolova and Dzhurmanski (2009) evaluated the free radical scavenging activity of total methanol extract of *Siderites scardica*. The extract exhibited strong antioxidant activity and its IC_{50} value for DPPH radical was 59.44 µg/ml.

Sideritis syriaca L. ssp. *syriaca* Family: Lamiaceae (Labiatae)

Koleva *et al.* (2003) determined the antioxidant activity of *Sideritis syriaca* by the β-carotene bleaching test (BCBT), DPPH radical scavenging method and static headspace gas chromatography. The highest antioxidant activity in the BCBT, close to that of BHT, was observed for the more apolar extracts. The inhibition of hexanal formation in bulk safflower oil by most of *Sideritis syriaca* extract was as effective as BHT but less so than rosmarinic acid. Extracts from Butanol and from ethyl acetate and the total methanol extracts showed a strong radical scavenging activity against DPPH, close to that of rosmarinic acid. The antioxidant activity of *Sideritis* extracts was attributed to the presence of flavonoid and phenylpropanoid glycosides.

Armata *et al.* (2008) investigated the constituents of *Sideritis syriaca* ssp. *syriaca* and their antioxidant activity. The antioxidant activities of the extracts were evaluated through *in vitro* model systems, such as DPPH and CO(II) EDTA-induced luminal chemiluminiscence. In both the model systems the ethyl acetate extract was the most effective.

Sideroxylon capiri (A.DC.) Pittier Family: Sapotaceae

Ruiz-Terán *et al.* (2008) evaluated the antioxidant activity of aerial parts of *Sideroxylon capiri* using DPPH and β-carotene-linoleic acid bleaching assay. The extract displayed antioxidant activity comparable to the commercial antioxidant BHA. For the methanolic extract analysed, a clear relation between the total phenolic content of the extracts and their antioxidant activity was found.

Sideroxylon obtusifolium T.D. Penn. Family: Sapotaceae

The *in vitro* antioxidant and free radical scavenging properties of bark extract of *Sideroxylon obtusifolium* used as anti-inflammatory agent in the Brazilian state of Bahis, was studied using different bioassays (Desmarchelier *et al.*, 1999). The total reactive antioxidant potential (TRAP) of the aqueous and methanolic extracts was determined by monitoring the intensity of luminal-enhanced chemiluminescence (CL), using 2,2'-azo-bis(2-amidinopropane) as a peroxyl radical source. The extract was active in this method. Extract also reduced hydroperoxide-initiated CL.

Siegesbeckia orientalis L. Family: Asteraceae (Compositae)

Gan *et al.* (2010) evaluated the antioxidant activity of *Siegesbeckia orientalis* using the FRAP and TEAC assays, and its total phenolic content was measured by the Folin-Ciocalteu method. For FRAP assay the value was 91.25 µmol Fe(II)/g dry weight. For TEAC assay, the value was 60.23µmol Trolox/g dry weight. Total phenolic content was 6.18 mg gallic acid equivalent/g dry weight of plant material.

Silene coronaria (L.) Clairv. Family: Caryophyllaceae

Souri *et al*. (2004) evaluated the antioxidative activity of *Silene coronaria* by linoleic acid peroxidation test using 1,3-diethyl-2-thiobarbituric acid as the reagent. Methanolic extract showed 89.11 per cent inhibition at 40µg concentration with IC_{50} value of 2.04µg.

Silene guntensis B. Fedtsch Family: Caryophyllaceae

Phytoecdysteroids from aerial parts of *Silene guntensis* were investigated and three phytoecdysteroids were isolated: 2,3-diacetate-22-benzoate-20-hydroxyecdysone (1), 2-deoxy-20-hydroxyecdysone (2), and 20-hydroxyecdysone (3) (Mamadalieva *et al*., 2011). The isolated compounds 1-3 and crude extracts were evaluated for their antiproliferative and antioxidant activities. Water and n-butanol extracts exhibited good antioxidant activities [IC_{50} values of (68.90 µg/ml and (69.12) µg/ml, respectively].

Silybum marianum Gaertn. Family: Asteraceae (Compositae)

English Name: Milk thistle

Silymarin and its active constituent, silybin, have been reported to work as antioxidants, scavenging free radicals and inhibiting lipid peroxidation. They protect against genoic injury, increase hepatocyte protein synthesis, decrease the activity of tumor promoters, stabilize mast cells, chelate iron and slow calcium metabolism (Berger and Kowdley, 2003). It influences the enzyme systems associated with glutathione and superoxide dismutase (Letteron *et al.*, 1990). A significant increase in the amount of the reduced glutathione (GSH) content was found in the liver, intestine and stomach after treatment with silibinin intravenously or silymarin intraperitoneally whereas there was no change in the lungs, spleen and kidneys of rats (Valenzula *et al.*, 1989). It may protect the brain from oxidative damage for its ability to prevent lipid peroxidation and replenishing the GSH levels (Nencini *et al.*, 2007).

A regioisomer of the known flavanolignan (-)-silandrin (3a), named (-)-isosilandrin (8a), was isolated from the fruits of a white-flowered variant of *Silybum marianum* populated in Hungary. Its structure was established both by spectroscopic methods and total synthesis, and its absolute configuration was determined by means of circular dichroism. This compound showed stronger inhibitory activity on the superoxide anion (O2*-) release by human polymorphonuclear leukocytes (PMNL) than (+)-silybin (1a,b) (Samu *et al.*, 2004).

Silibinin displays cytoprotective properties (Dehmlow *et al.*, 1996) and it may protect blood constituents from oxidative damage (Filipe *et al.*, 1997). The antioxidant properties were evaluated by studying the ability of this drug to react with relevant biological oxidants such as superoxide anion radical (O_2^{-}), hydrogen peroxide (H_2O_2), hydroxyl radical (HO) and hypochlorous acid (HOCl) (Mira *et al.*, 1994). Silibinin and silibinin dihemisuccinate (SDH) proved to be a strong scavenger of hypochlorous acid (Dehmlow *et al.*, 1996) but not of superoxide anion radical ((Dehmlow *et al.*, 1996; Mira *et al.*, 1994) produced by human granulocytes, but no reaction with H_2O_2 was detected (Mira *et al.*, 1994). However, SDH reacted rapidly with hydroxyl radical

(HO.) and appears to be a weak iron ion chelator. The studies on rat liver microsome lipid peroxidation induced by Fe(III)/ascorbate showed that SDH has an inhibitory effect, which is dependent on its concentration and the magnitude of lipid peroxidation (Mira *et al.*, 1994).

The antioxidant properties of silymarin and its flavanolignan components (silybin, silychristin and silydianin) were tested by Sersen *et al.* (2006). Silymarin, silychristin and silydianin exhibit relatively good antioxidant effectiveness against phenylglyoxylic ketyl radicals and DPPH. The most effective scavengers of phenylglyoxylic key radicals were silymarin and silychristin whereas silydianin was ca. 5-times less active than the first two compounds whereas silybin was ineffective. The scavenging properties of the studied compounds against DPPH radicals were in the same sequence silymarin>silychristin>silydianin>silybin. Silymarin and its flavanolignan components exhibit also antioxidant properties (Kvasnicka *et al.*, 2003).

An *in vitro* study of the protective effect of the flavonoid silydianin (an active constituent of *Silybum marianum*) against reactive oxygen species was carried out by Zielinska-Przyjemska and Wiktorowicz (2006). The results indicate a possible anti-inflammatory activity for silydianin, which regulates caspase activation, affects cell membranes and acts as a free radical scavenger.

Khalil *et al.* (2007) studied the growth, phenolic compounds and antioxidant activity of milk thistle grown under organic farming condition. Antioxidant activity of ethanol extract (inhibition per cent) was 68.69 at 200µl extract.

Wojcikowski *et al.* (2007) evaluated the antioxidant activity of aerial parts of *Silybum marianum*. Oxygen radical absorbance capacity of the ethyl extract, methanol extract and aqueous extracts were 516.86, 26.34 and 10.70µmol Trolox equivalent per g of dried starting material, respectively.

Hadaruga and Hadaruga (2009) presented the antioxidant activity evaluation of *Silybum marianum* extract in comparison with the commercial silymarin standard solution (2mg/ml) and the time-dependent antioxidant capacity by calculating the free radical DPPH reaction rate in the presence of these biosystems. The best result was obtained in the case of *Silybum marianum* extract with a double DPPH reaction rate after 1200s in comparison with the commercial silymarin standard solution.

Sinapis arvensis L. Family: Brassicaceae (Cruciferae)

Coruh *et al.* (2007) reported high antioxidant activity in *Sinapis arvensis*. The total phenolic content was 22.25 µgGAE/mg. There was a low correlation (R = 0.447) between total phenolic content and antioxidant activity in the plant samples.

Sinomonium acutum (Thunb.) Rehder et E.H.Wilson (Syn. *Cocculus heterophyllus* Hemsl. et E.H.Wilson) Family: Menispermaceae

Scopoletin was isolated from *Sinomonium acutum* and studied by Shaw *et al.* (2003) using four experimental models designed to assess antioxidant properties. The results indicated that scopoletin scavenged superoxide anion in the xanthine/ xanthine oxidase reaction system in concentration-dependent manner, but did not inhibit xanthine oxidase. Scopoletin may therefore be responsible for the superoxide anion scavenging activity seen in *Sinomonium acutum* extracts and may be of use in preventing superoxide anion-induced toxicity *in vivo*.

Gan *et al.* (2010) evaluated the antioxidant activity of *Sinomonium acutum* using the FRAP and TEAC assays, and its total phenolic content was measured by the Folin-Ciocalteu method. For FRAP assay the value was 245.94 µmol Fe(II)/g dry weight. For TEAC assay, the value was 122.24µmol Trolox/g dry weight. Total phenolic content was 16.21 mg gallic acid equivalent/g dry weight of plant material.

Siraitia grosvenori (Swingle) C.Jeffrey ex A.M.Lu et Z.Y.Zhang Family: Cucurbitaceae

Qi *et al.* (2008) reported that magrosides extract (MG) from fruits of *Siraitia grosvenori* scavenges free radicals *in vitro* and lowers oxidative stress, serum glucose, and lipid levels in alloxan-induced diabetic mice. Antioxidant capacity evaluated *in vitro* showed that MG and magroside V, which was the main component of MG, possessed strong oxygen free radical scavenging activities.

Sisymbrium thellungi O. Schnez Family: Brassicaceae

In a screening of South African indigenous food plants for antioxidant activity by testing for inhibition of lipid peroxidation the pot herb *Sisymbrium thellungi* had very high activity (Lindsey *et al.*, 2006).

Smallanthus sonchifolius (Poepp. et Endl.) H.Robinson Family: Asteraceae (Compositae)

Radical scavenging and anti-lipoperoxidative effects of two organic fractions and two aqueous extracts from the leaves of *Smallanthus sonchifolius* were determined by Valentova *et al.* (2005) using various *in vitro* models. The extracts' total phenolic content was 10.7-24.6 per cent. They exhibited DPPH (IC$_{50}$=16.14-33.39µg/ml). In the xanthine/xanthine oxidase superoxide radical generating system, the extracts' activities were 26.10-37.67 superoxide dismutase equivalents/mg. As one of the extracts displayed xanthine oxidase inhibitory activity, the effect of the extracts on a nonenzymatically generated superoxide was determined (IC$_{50}$ 7.36-21.01 µg/ml). The extracts inhibited t-butyl hydroperoxide-induced lipoperoxidation of microsomal and mitochondrial membranes (IC$_{50}$ 22.15-465.3 µg/ml) (Valentova *et al.*, 2005).

Smilax aspera L. Family: Smilacaceae

Demo *et al.* (1998) reported antioxidant α-tocopherol from the Hexane extract of *Smilax aspera*.

Smilax china L. Family: Smilacaceae

Lee *et al.* (2001) investigated free radical scavenging and antioxidant enzyme fortifying activities of extracts from *Smilax china* root. Methanol extract revealed the presence of high DPPH free radical scavenging activity (IC$_{50}$ 7.4µg/ml) and protective property of cell viability. Further fractionation with various solvent extraction and assay showed high levels of DPPH free radical scavenging activity in the ethyl acetate, butanol and water extracted fractions. In addition V79-4 cells treated with Methanol extract of *Smilax china* root induced an increase of superoxide dismutase, catalase and glutathione peroxidase activities in a dose dependent manner between 4-100µg/ml (Lee *et al.*, 2001).

Song *et al.* (2006) evaluated the antioxidant activity of aqueous and 25 per cent EtOH extract from *Smilax china* root and six different evaluation assay methods, *i.e.*, DPPH, NO and nitrite, reducing power and inhibitory effect on tyrosinase activity, were used.

Smilax excelsa Art. et Hamilt. Family: Smilacaceae

Souri *et al.* (2004) evaluated the antioxidative activity of *Smilax excelsa* by linoleic acid peroxidation test using 1,3-diethyl-2-thiobarbituric acid as the reagent. Methanolic extract showed 94.21 per cent inhibition at 40µg concentration with IC_{50} value of 2.00µg.

Different antioxidant tests were employed in order to evaluate the antioxidant activities of water, infusion, ethanol and ethyl acetate extracts of *S. excelsa* leaves (Ozsoy *et al.*, 2008). The extracts were found to have different levels of antioxidant properties in the test models used. All extracts had good total phenolic and flavonoid contents, inhibited lipid peroxidation, showed radical scavenging and iron-chelating activities.

The antioxidative activity of the 80 per cent ethanol extract obtained from eleven commonly consumed wild edible plants was determined according to the phosphomolybdenum method, reducing power, metal chelating, superoxide anion and free radical scavenging activity (Özen, 2010). *Smilax excelsa* had the highest antioxidant capacities. The results indicated that the antioxidant activity of vegetables seems to be due to the presence of polyphenols, flavonoid and anthocyanoside. Total phenolics in *Smilax excelsa* were 49.9mg pyrocatechol equivalent/g dw, Flavonoids were 12.6 mg quercetin equivalent/g dw and anthocyanins 2.5 mg cyanidin glycoside equivalent/g dw.

Smilax glabra Roxb. Family: Smilacaceae

Cai *et al.* (2004) evaluated the antioxidant activity and phenolic compounds in traditional Chinese medicinal plants associated with anticancer. The improved ABTS.+ method was used to systematically assess the total antioxidant capacity (Trolox equivalent antioxidant capacity, TEAC) of the medicinal extracts. The TEAC values and total phenolic content for methanolic extract of root of *Smilax glabra* were 137.3 µmol Trolox equivalent/100g dry weight (DW), and 0.74 g of gallic acid equivalent/100g DW respectively. A positive significant linear relationship between antioxidant activity and total phenolic content showed that phenolic compounds were the dominant antioxidant components in the tested herbs. Major types of phenolic compounds in *Smilax glabra* include flavan-3-ols (epicatechin), tannins.

Li *et al.* (2008) evaluated the antioxidant properties of 45 medicinal plants using FRAP and TEAC assays and the total phenolic contents of these plants were measured. *Smilax glabra* showed 103.25 µmol Fe(II)/g FRAP value, 61.81µmol Trolox/g of TEAC value and 14.24 mg GAE/g of phenolic content.

Smilax glyciphylla Sm. Family: Smilacaceae

A hot water extract of the Australian native sarsaparilla *Smilax glyciphylla* inhibited peroxidation of phosphatidylcholine liposomes initiated by Fe^{2+}/ascorbate (IC_{50}, 10µg/ml) and AAPH (IC_{50}, 33µg/ml) *in vitro* (Cox *et al.*, 2005). It also inhibited deoxyribose degradation and quenched chemically generated superoxide anion IC_{50}, 50µg/ml). Reactivity towards ABTS radical cation was equivalent to 48.4 mM Trolox, the water soluble α-tocopherol analogue. Cox *et al.* (2005) is of the opinion that consuming *Smilax glyciphylla* as a tea may be sufficient to reduce oxidative damage in the gastrointestinal tract.

Smilax zeylanica L. Family: Smilacaceae

Methanol and aqueous extract of leaves of *Smilax zeylanica* were evaluated by Murali *et al.* (2010) for *in vitro* antioxidant activity using DPPH, hydrogen peroxide, ABTS, nitric oxide and superoxide free radicals. The plant extracts exhibited dose dependent scavenging effects against the different free radicals tested. The methanol extract was subjected to *in vivo* antioxidant activity studies using CCl_4 induced hepatotoxicity model in Wistar albino rats. The extract exhibited significant increase in the levels of glutathione, tissue proteins and enzymes, vis., SOD, catalase and peroxidase at different dose levels. The extent of lipid peroxidation was significantly reduced in the extract treated groups. Results were comparable with that of standard antioxidant silymarin.

Solanum aculeastrum Dunal Family: Solanaceae

The antioxidant activity of the crude methanol, acetone and water extracts of berries of *Solanum aculeastrum* was examined by DPPH radical-scavenging (Koduru *et al.*, 2006). The methanol and water extracts had moderate antioxidant activity ranging between 53.1 to 65.5µg/ml, while the acetone extract did not demonstrate significant activity at the tested concentrations. The higher antioxidant activity of this plant exhibited by the water extract may be due to the presence of substantial amount of polar constituents from the plant material (Koduru *et al.*, 2006).

Solanum americanum L. Family: Solanaceae

Iwaleva *et al.* (2005) reported antioxidant and cytoprotective activities of boiled, cold and methanolic extracts of nine edible vegetables in South West Nigeria which were evaluated in the DPPH assay and hemeagglutination assay in bovine erythrocytes, respectively. *Solanum americanum* exhibited significant antioxidant activity.

Solanum brevifolium Dunal Family: Solanaceae

Mosquera *et al.* (2007) studied the antioxidant activity of the extracts of *Solanum brevifolium* using DPPH free-radical scavenging method. The percentage of the antioxidant activity of the n-Hexane fraction was 26.984, Dichloromethane fraction was 32.183 and methanol fraction was 43.9.

Solanum centrale J.M.Black Family: Solanaceae

English Name: Australian bush tomato.

The antioxidant capacities and phenolic composition in six native, commercially grown, Australian herbs and spices were investigated by Konczak *et al.* (2010). *Solanum centrale*, with Total phenolics content of 12.4 mg GAE/g DW and Total Reducing capacity of 206.2 µmol Fe^{+2}/g DW, resemble the Chinese Barbary Wolfberry fruit.

Solanum deflexiflorum Bitter Family: Solanaceae

Mosquera *et al.* (2007) studied the antioxidant activity of the extracts of *Solanum deflexiflorum* using DPPH free-radical scavenging method. The percentage of the antioxidant activity of the n-Hexane fraction was 25.85, Dichloromethane fraction was 34.253 and methanol fraction was 42.96.

Solanum fastigiatum Family: Solanaceae

Sabir and Rocha (2008) evaluated the potential antioxidant and hepatoprotective activity of aqueous extracts of leaves using *in vitro* and *in vivo* model using TBARS, DPPH radical and metal iron-chelating activities. The extract showed inhibition against TBARS, induced by 10microM FeSO(4) and 5microM sodium nitroprusside in rat liver, brain and phospholipid homogenates from egg yolk. The plant exhibited strong antioxidant activity in the DPPH (IC(50), 68.96µg/ml) assay. The aqueous extract also showed significant hepatoprotective activity that was evident by enzymatic examination and brought back the altered levels of TBARS, non-protein thiol and ascorbic acid to near the normal levels in a dose dependent manner.

Solanum indicum L. Family: Solanaceae

Narayanaswamy and Balakrishnan (2011) evaluated the antioxidant activity of 13 medicinal plants including leaf of *Solanum indicum*. The aqueous extract showed highest 90 per cent inhibition of DPPH radical while ethanolic extract showed 92 per cent inhibition of DPPH radical.

Solanum leucocarpum Dunal Family: Solanaceae

Mosquera *et al.* (2007) studied the antioxidant activity of the extracts *Solanum leucocarpum* using DPPH free-radical scavenging method. The percentage of the antioxidant activity of the n-Hexane fraction was 31.293, Dichloromethane fraction was 32.19 and methanol fraction was 30.05.

Solanum macrocarpon L. Family: Solanaceae

Oboh and Akindahunsi (2004) investigated the effect of preservation of green leafy vegetables of Nigeria on the antioxidant phytoconstituents (Vitamin C and Total phenol) and activity (reducing property and free radical scavenging ability) including *Solanum macrocarpon*. The result of the study revealed that sun-drying of green leafy vegetables cause a significant decrease in the vitamin C content (16.67-64.68 per cent loss). Conversely it leads to a significant increase in the total phenol content (6.45 -223.08 per cent grain) reducing property (16.00-362.50 per cent grain) and free radical scavenging ability (126.00-5757.00 per cent gain) of green leafy vegetables.

Odukoya *et al.* (2007) evaluated the antioxidant activity and total phenolic contents of Nigerian green leafy vegetables including *Solanum macrocarpon*.

Antioxidant activities of eight leafy vegetables of Ghana including *Solanum macrocarpon* were assessed by Morrison and Twumasi (2010). The total antioxidant capacity (TAC) and Total Phenol content (TPC) in the methanol extracts (METE) and hydro-ethanol extracts (HETE) were measured. The TPC of the methanol extract of *Solanum macrocarpon* was 0.267 mg/ml TAE and TAE was 0.261 mg/ml Ascorbic acid equivalent (AAE) while TPC of hydro-ethanol extract was 0.380 mg/ml TAE and TAE was 0.365 mg/ml Ascorbic acid equivalent (AAE). Radical scavenging activity of methanol extract was $EC_{50} = 0.1541$ mg/ml and HETE was 0.1380 mg/ml while Fe^{3+} reducing potential was $EC_{50} = 0.1380$ (METE) and 0.1380 (HETE).

Solanum melongena L. Family: Solanaceae

English Name: Egg plant, Brinjal; Japanese: Nasu.

Flavonoids isolatd from *Solanum melongena* showed potent antioxidant activity (Sudheesh *et al.*, 1999). Concentrations of malondialdehyde, hydroperoxides and conjugated dienes were lowered significantly. The activity of catalse was found to be significantly enhanced in the tissues of normal and cholesterol fed rats administered 1 mg flavonoid from brinjal/100g BW/day. The concentration of glutathione also showed elevated values in the experimental animals. The elevated levels of glutathione and significantly stimulated activity of catalase may be responsible for the antioxidant effect of these flavonoids.

Cai *et al.* (2004) evaluated the antioxidant activity and phenolic compounds in traditional Chinese medicinal plants and vegetables associated with anticancer. The improved ABTS.+ method was used to systematically assess the total antioxidant capacity (TEAC) of the medicinal extracts. The TEAC values and total phenolic content for methanolic extract of fruit of brinjal were 166.9 µmol Trolox equivalent/100g dry weight (DW), and 1.08 g of gallic acid equivalent/100g DW respectively.

Huang *et al.* (2004a) evaluated the antioxidant activities of various fruits and vegetables produced in Taiwan. At the level of 1 g fresh sample, low-density lipoprotein peroxidation was inhibited by at least 90 per cent by eggplant skin. The total phenolic content was significantly correlated with antioxidant activities measured by TBA and Iodometric assays.

Maeda *et al.* (2006) examined the phenolic contents of fruit of brinjal and its radical scavenging activities for DPPH radical. The polyphenol content was 80.5 mg-gallic acid equivalent/100 g) and DPPH scavenging activity was 403.4 µmol-Trolox eq./100 g.

Odukoya *et al.* (2007) evaluated the antioxidant activity and total phenolic contents of Nigerian green leafy vegetables including *Solanum melongena*.

Phenolic compounds and their antioxidant activities from fruits of eleven Thai eggplant varieties were determined by Chanasut and Rattanapanone (2008). DPPH assay, β-carotenoid bleaching assay, and ABTS assay were followed for the antioxidant activity. The three antioxidant assay methods gave different antioxidant activity trends. Phenol contents of the eggplant extracts were related to their antioxidant activities measured by ABTS and DPPH assay. In contrast, the results measured by the β-carotenoid bleaching assay were not related to the phenol contents of the extracts.

Nisha *et al.* (2009) evaluated the antioxidant activities of different varieties of *Solanum melongena* in terms of total phenolic content, DPPH, total reducing power, superoxide radical scavenging activity, metal chelating activity and total anthocyanin content. Extracts from purple colour small size egg plant fruit demonstrated better antioxidant activities than the other samples which may be attributed to the higher phenolic and anthocyanin content since a linear relation was observed between the TPC and the antioxidant parameters.

Antioxidant potential of selected vegetables commonly used in diet in Asian subcontinent, including *Solanum melongena* fruits has been evaluated by Gacche *et al.* (2010). The DPPH radical scavenging activity was 51.6 per cent, inhibition of lipid peroxidation was 16.4 per cent, FICA was 32.8, Vitamin C content was 20.48 mg/100g and total phenols was 102.41 mg/g.

Akanitapichat *et al.* (2010) evaluated the antioxidant and hepatoprotective activities of five eggplant varieties. Total phenolic content found in methanol extracts ranged from 739.36 to 1116.13 GAE mg/100g extract and total flavonoid content from 1991.29 to 3954.20 catechin equivalents mg/100g extract. White-green colored moderate size and long green varieties contained high total phenolic and flavonoid content and had better antioxidant activities than other varieties. Significant correlation was found between hepatoprotective activities and total phenolic/flavonoid content and antioxidant activities, indicating the contribution of the phenolic antioxidant present in eggplant to its hepatoprotective effect.

Solanum nigrum L. Family: Solanaceae

Sultana *et al.* (1995) reported that the presence of plant extract *Solanum nigrum* in the reaction mixture containing calf thymus DNA and free radical generating system protect DNA against oxidative damage to its deoxyribose moiety. The effect was dependent on the concentration of plant extracts.

Cai *et al.* (2004) evaluated the antioxidant activity and phenolic compounds in traditional Chinese medicinal plants associated with anticancer. The improved ABTS.+ method was used to systematically assess the total antioxidant capacity (Trolox equivalent antioxidant capacity, TEAC) of the medicinal extracts. The TEAC values and total phenolic content for methanolic extract of aerial parts of *Solanum nigrum* were 46.7 µmol Trolox equivalent/100g dry weight (DW), and 0.22 g of gallic acid equivalent/100g DW respectively. A positive significant linear relationship between antioxidant activity and total phenolic content showed that phenolic compounds were the dominant antioxidant components in the tested herbs. Major types of phenolic compounds in *Solanum nigrum* include phenolic acids (chlorogenic acid).

Abbas *et al.* (2006) evaluated the antioxidant activity of *Solanum nigrum* using DPPH assay. Methanolic extract showed DPPH radical scavenging activity.

Methanol extract of fruits of *Solanum nigrum* showed effective free radical scavenging activites in the DPPH assay (Al-Fatimi *et al.*, 2007). A study which utilizes six pretreatment methods before cooking on the peroxidase activity, chlorophyll and antioxidant status of *Solanum nigrum* showed that pretreatment methods have significant effects on the parameters measured. A sharp difference in the carotenoids, phenolics, flavonoids, and tannins contents has been reported, indicating the fragility of this antioxidant present in *Solanum nigrum* (Adebooye *et al.*, 2008). *Solanum nigrum* glycoprotein showed a dose-dependent radical scavenging activity on radicals, including DPPH radicals, hydroxyl radical and superoxide anion.

Although *Solanum nigrum* acts as anti-tumor, the *Solanum nigrum* glycoprotein may induce apoptosis through the inhibition of NF-kB activation, induced by oxidative stress in HT-29 cells (Heo *et al.*, 2004). A 50 per cent ethanol extract of the whole plant of *S. nigrum* also possess hydroxyl radical scavenging potential which is suggested as cytoprotective mechanism (Kumar *et al.*, 2001; Mohammed *et al.*, 2007). Evaluation of the antioxidant potential of *Solanum nigrum* leaves on the modulation of a 6 h restraint induced oxidative stress, which suggest that *Solanum nigrum* was better as an antioxidant with post-restraint treatment than with prerestraint administration.

The methanolic crude extracts of *Solanum nigrum* were screened by Veeru *et al.* (2009) for their free radical scavenging properties using ascorbic acid as standard antioxidant. Free radical scavenging activity was evaluated using DPPH free radical. The overall antioxidant activity of *Solanum nigrum* was found to be stronger with IC_{50} value of 0.12 mg/l. The ascorbic acid level was 8.43 mg/100g and the carotenoids content were 17.66 mg/100g in plant extracts. The total phenol content was 5.73 mg/g.

Maharana *et al.* (2010) evaluated the *in vitro* antioxidant activity of aqueous leaf extract of *Solanum nigrum*. The plant extract showed the total antioxidant activity of 54.16mg ascorbic acid equivalent/g as compared to 117.83 mg of the reference standard ascorbic acid. The reducing power of the extract was found to be significant and in a concentration dependent manner. The test extract showed marked antioxidant activity with IC_{50} vlue of 165μg/ml for DPPH radical, 472μg/ml for superoxide radical, 417μg/ml for H_2O_2 radical and 4,83μg/ml for nitric oxide radical.

Solanum ocharanthum Dunal Family: Solanaceae

Mosquera *et al.* (2007) studied the antioxidant activity of the extracts using DPPH free-radical scavenging method. The percentage of the antioxidant activity of the n-Hexane fraction was 33.79, Dichloromethane fraction was 39.081 and methanol fraction was 32.863.

Solanum ovalifolium Dunal Family: Solanaceae

Mosquera *et al.* (2007) studied the antioxidant activity of the extracts using DPPH free-radical scavenging method. The percentage of the antioxidant activity of the n-Hexane fraction was 27.44, Dichloromethane fraction was 37.012 and methanol fraction was 40.61.

Solanum pseudocapsicum L. Family: Solanaceae

In vitro antioxidant properties of *Solanum pseudocapsicum* leaf extracts were investigated by Badami *et al.* (2005). The crude methanolic extract showed potent antioxidant activity with 49.57 and 79 mg/ml for antioxidant and nitricoxide methods, respectively. The successive ethyl acetate extracts also exhibited potent antioxidant activity in the DPPH method with 91.50 and 101.50 mg/ml, respectively. The successive chloroform extract also exhibited antioxidant activity in nitric oxide method.

Solanum quitoense Lam. Family: Solanaceae

English Name: Naranjilla

Chemical characterization, antioxidant properties and volatile constituents of naranjilla was investigated by Acosta *et al.* (2009). H-ORAC value was 17μmol Trolox equivalent/g. Total polyphenolic content was 48 mg gallic acid equivalent/100g and ascorbic acid content was 12.5 mg/100g. Carotenoid content of the whole fruit and pulp was 33.3 and 7.2μg/g, respectively. The predominant carotenoid among the compounds identified in the whole fruit was β-carotene.

Solanum torvum Swartz. Family: Solanaceae

Sivapriya and Leela (2007) described the isolation and purification of a novel antioxidant protein from the water extract of dried, powdered *Solanum torvum* seeds. The purified protein, at 0.8μM, inhibited deoxyribose degradation induced by generation of hydroxyl radicals by 90 per cent and scavenged DPPH radicals by 76 per cent. The reducing power and chelating power of the purified protein at 0.8μM, were found to be 72 per cent and 85 per cent, respectively.

Phenolic compounds and their antioxidant activities from fruits of *Solamm torvum* were determined by Chanasut and Rattanapanone (2008). On the basis of DPPH results, *Solanum torvum* has both the highest total phenolic content and total antioxidant activity.

Kusirisin *et al.* (2009) investigated the effect of *Solanum torvum* (ST) extracts on the inhibition of CYP2E1 activity in human liver microsomes. ST extract was analyzed for antioxidant activity by the ABTS method. Polyphenolic compounds were measured by the total phenol content using the Folin-Ciocalteau reagent. One gram of concentrated ST extract had an antioxidant activity index of 3.68 mg of Trolox and 360.53 mg of ascorbic acid equivalent. Effects of free radical-scavenging, as measured by TBARS and superoxide anion, showed IC_{50} values of 20.60 and 10.26 μg/ml, respectively. Polyphenolic compounds found included phenol, flavonoid and tannin, measuring 160.30, 104.36 and 65.91 mg/g respectively.

Solanum trilobatum L. Family: Solanaceae

Sini and Devi (2004) investigated the antioxidant activities of chloroform extract of *Solanum trilobatum*. Phytochemical analysis showed the presence of simple phenols, phenolic acids, isoflavones, xanthones, and lignans. The antioxidant activity of alcohol chloroform extract was found to be concentration dependent to a certain extent and then leveled off with further increase in concentration. When compared to the reference antioxidant BHT, the extract exhibited less scavenging effect on DPPH radicals and reducing power but better superoxide radical scavenging effect. From comparison of the hydroxyl radical scavenging effect the extract with catechin, it seemed that the extract was four-times more effective than catechin. The extract also able to prevent the formation of OH-induced malondialdehyde in rat liver homogenate (Sini and Devi, 2004).

Solanum tuberosum L. Family: Solanaceae

English Name: Potato.

Antioxidant activity of potato compared with that of broccoli, onion, carrot and bell peppers was higher than all except broccoli. Patatin appeared to be the major water-soluble compound that showed antioxidant activity. The activity varied among potato cultivars, but was not related to flesh color or total phenolics. Antioxidant activity was evenly distributed within tuber parts and/or sections, except for skin tissue which had the greatest antioxidant activity and total phenolic content. Total phenolics varied among cultivars, with some containing two fold higher concentrations than other cultivars. Phenolic content differences were genotype dependent and not related to flesh color (Sai Khan *et al.*, 1995).

The antioxidative activity of phenolic extracts from fruits of potato was examined by autoxidation of methyl linoleate by Kahkonen *et al.* (1999). The peel of purple skinned potato vatiety Rosamunda showed remarkable antioxidant activity. The yellows-skinned potato variety Matilda had lower total phenolic content and antioxidant activity than Rosamunda. In a study where several red-fleshed potato varieties and breeding clones were analyzed the anthocyanin content of 21.7 mg in 100 g of skin (fresh weight) was detected and the dominant anthocyanin was identified as pelargonidin-3-rutinoside-5-glucoside. The major phenolics in the skin were chlorogenic and *p*-coumaric acids (Rodriguez-Saona *et al.*, 1998). The difference in antioxidant activity between potato varieties may result partly from the presence of anthocyanins, although pelargonin has been reported to show poor antioxidant activity in LDL and pro-oxidant activity in a lecithin liposome system (Satue-Garcia *et al.*, 1997). Methanolic extracts of purple potatoes showed strong antioxidant activities in a β-carotene bleaching method (Velioglu *et al.*, 1998).

Penarrieeta *et al.* (2005) measured the Total Antioxidant Capacity (TAC) in some Andean foods. Eight Andean foods, including *Solanum tuberosum* were analyzed by two methods ABTS and FRAP to assess TAC. TAC value of *Solanum tuberosum* was good in both water-soluble fraction and water-insoluble fraction.

Solidago canadensis L. Family: Asteraceae (Compositae)

McCune (1999) evaluated antioxidant activities of indigenous medicinal plants of Canadian boreal forest by three different assays (DPPH, NBT/xanthine oxidase and DCF/APPH). *Solidago canadensis* showed antioxidant activity similar to vitamin C, and green tea in all the assays.

Wojcikowski *et al.* (2007) evaluated the antioxidant activity of aerial parts of *Solidago canadensis*. Oxygen radical absorbance capacity of the ethyl extract, methanol extract and aqueous extracts were 1.55, 145.63 and 209.18μmol Trolox equivalent per g of dried starting material, respectively.

Solidago virgaurea L. Family: Asteraceae (Compositae)

English Name: Golden rod.

Aqueous-ethanolic extracts from *Fraxinus excelsior*, *Populus tremula* and *Solidago virgaurea* inhibit biochemical model reactions representing inflammatory situations to various extents. These model reactions include xanthine oxidase, diaphorase in the presence of the autoxidizable quinone juglone, lipoxygenase and photodynamic reactions driven by riboflavin or rose bengal. The tested extracts are the components of the phytomedicine Phytodolor N (abbreviated as PD) which possesses antipyretic, analgesic, antiinflammatory and antirheumatic activity. Since several reactive oxygen species produced by the mentioned model systems are also involved in inflammatory processes, the beneficial activities of the complete drug may at least in part be due to the reported antioxidative functions of the individual components (Meyer *et al.*, 1995).

The antioxidative activity of phenolic extracts from leaf and bark of golden rod was examined by autoxidation of methyl linoleate by Kahkonen *et al.* (1999). The total phenolics and antioxidant activity were low. Phenolics were 8.2 mg/g GAE and antioxidant activity was 28 per cent at 500 ppm level.

Sonchus arvensis L. Family: Asteraceae (Compositae)

English Name: Field milk thistle.

The antioxidative activity of phenolic extracts from herb of field milk thistle was examined by autoxidation of methyl linoleate by Kahkonen *et al.* (1999). The total phenol contents (5.6 mg/g GAE) and antioxidant activity (4 per cent at 500 ppm level) in *Sonchus arvensis* was very low.

Sonchus asper (L.) Hill Family: Asteraceae (Compositae)

Jimoh *et al.* (2011) evaluated the nutritive value, antioxidant and antibacterial activities of *Sonchus asper*. DPPH and ABTS radical scavenging assays were followed for antioxidant activity assay. With respect to the total phenol and flavonoids and acetone and methanol extracts had higher content of this constituent than the water extracts. The methanol and water extracts of *S. asper* had higher content of proanthocyanidins relative to acetone extract. The FRAP value for the plant showed that methanol extract was much higher than that of acetone while acetone in turn was higher than those of water extracts. The FRAP values for the methanol and actone extracts of the two plants were higher than those of BHT but lower than those of catechin, ascorbic acid and quercetin. At 1 mg/ml, the acetone extract of *S. asper* caused 97.8 per cent ABTS radical scavenging inhibition while the methanol, water and BHT caused inhibition at 98.0, 99 per cent and 99.3 per cent respectively. At 1 mg/ml the acetone, methanol, water and ascorbic acid caused DPPH radical scavenging activity at 85.6, 85.3, 81.8 and 99.8 per cent respectively for *S. asper*.

Sonchus oleraceus L. Family: Asteraceae (Compositae)

Yin *et al.* (2007) investigated *in vitro* antioxidant activity of *Sonchus oleraceus* by extraction solvent which were examined by reducing power, hydroxyl radical-scavenging activity and DPPH radical scavenging assays. 70 per cent MeOH extract had the greatest reducing power while EtOH extract had the greatest HRSA. The antioxidant activity of *S. oleraceus* extracts was concentration dependent and its IC_{50} values ranged from 47.1 to 210.5µg/ml and IC_{50} of 70 per cent MeOH, boiling water and 70 per cent EtOH extracts were 47.1, 52. and 56.5 µg/ml, respectively. 70 per cent MeOH extract of *S. oleraceus* contained the greatest amount of both phenolic and flavonoid contents. The extracts tested had greater nitrite scavenging effects at lower pH conditions.

Jimoh *et al*. (2011) evaluated the nutritive value, antioxidant and antibacterial activities of *Sonchus oleraceus*. DPPH and ABTS radical scavenging assays were followed for antioxidant activity assay. With respect to the total phenol and flavonoids and acetone and methanol extracts had higher content of this constituent than the water extracts. The acetone and methanol extracts had less proanthocyanidins than those of water extract. The FRAP value for the plant showed that methanol extract was much more than that of acetone while acetone is turn is higher than those of water extracts. The FRAP values for the methanol and acetone extracts of the two plants were higher than those of BHT but lower than those of catechin, ascorbic acid and quercetin. At 1 mg/ml, the acetone extract of *S. oleraceus* caused 99.4 per cent ABTS radical scavenging inhibition while the methanol, water and BHT caused inhibition at 95.7, 93.7 per cent and 99.3 per cent respectively. At 1 mg/ml the acetone, methanol, water and ascorbic acid caused DPPH radical scavenging activity at 56.1, 86.9, 92.9 and 100 per cent respectively for *S. oleraceus*.

Sonneratia apetala Buch.-Ham. Family: Sonneratiaceae

Vadlapudi and Naidu (2009) evaluated the antioxidant potential of *Sonneratia apetala*, a mangrove plant. The SOD was 1.64U/mg, Catalase was 0.68U/mg, Ascorbic acid was 360mg/100g, DPPH radical inhibition was 35.4 per cent and FRAP units were 620.

Sophora alopecuroides L. Family: Fabaceae (Leguminosae)

Souri *et al*. (2004) evaluated the antioxidative activity of *Sophora alopecuroides* by linoleic acid peroxidation test using 1,3-diethyl-2-thiobarbituric acid as the reagent. Methanolic extract showed 94.09 per cent inhibition at 40µg concentration with IC_{50} value of 4.70µg.

Sophora chrysophylla (Salisb.) Seem. Family: Fabaceae (Leguminosae)

The antioxidant activity of 6α-hydroxypterocarpon isolated from the root has been investigated *in vitro* by Toda and Shirataki (2005) on lipid peroxidation by interaction of haemoglobin and hydrogen peroxide. The results showed antioxidant activity of the chemical.

Sophora flavescens Air. Family: Fabaceae (Leguminosae)

Cai *et al*. (2004) evaluated the antioxidant activity and phenolic compounds in traditional Chinese medicinal plants associated with anticancer. The improved ABTS.+ method was used to systematically assess the total antioxidant capacity (Trolox equivalent antioxidant capacity, TEAC) of the medicinal extracts. The TEAC values and total phenolic content for methanolic extract of root of *Sophora flavescens* were 413.1 µmol Trolox equivalent/100g dry weight (DW), and 2.08 g of gallic acid equivalent/100g DW respectively. A positive significant linear relationship between antioxidant activity and total phenolic content showed that phenolic compounds were the dominant antioxidant components in the tested herbs. Major types of phenolic compounds in *Sophora flavescens* include flavonoids (kushenol, kurarinone, kuraridin, formononetin).

The methanol extract of *Sophora flavescens* showed a potent glycosidase inhibitory activity (Kim *et al.*, 2006). Active components were identified as known flavonoid antioxidant. All flavonoids were effective inhibitors of α-glucosidase and β-amylase. Results showed that 8-lavandulyl group in B-ring was a key factor of the glycosidase inhibitory activities.

Li *et al.* (2008) evaluated the antioxidant properties of 45 medicinal plants using FRAP and TEAC assays and the total phenolic contents of these plants were measured. *Sophora flavescens* showed low antioxidant activity with 26.36 µmol Fe(II)/g FRAP value, 71.45µmol Trolox/g of TEAC value and 17.60 mg GAE/g of phenolic content.

Sophora japonica L. Family: Fabaceae (Leguminosae)

Cai *et al.* (2004, 2006) evaluated the antioxidant activity and phenolic compounds in traditional Chinese medicinal plants associated with anticancer. The improved ABTS.+ method was used to systematically assess the total antioxidant capacity (Trolox equivalent antioxidant capacity, TEAC) of the medicinal extracts. The TEAC values and total phenolic content for methanolic extract of flower of *Sophora japonica* were 1850.3 µmol Trolox equivalent/100g dry weight (DW), and 5.95 g of gallic acid equivalent/100g DW respectively. A positive significant linear relationship between antioxidant activity and total phenolic content showed that phenolic compounds were the dominant antioxidant components in the tested herbs. Major types of phenolic compounds in *Sophora japonica* include flavonols (rutin, quercetin, kaempferol).

Gan *et al.* (2010) evaluated the antioxidant activity of 40 medicinal plants associated with prevention and treatment of cardiovascular and cerebrovascular diseases. A high correlation between antioxidant capacity and total phenolic content indicated that phenolic compounds could be the main contributor of the antioxidant activity of these plants. *Sophora japonica* showed very high antioxidant activity of 577.88 µmol Fe(II)/g in FRAP assay and 318.92µmol in TEAC assay. Total phenolic contet was also very high, 30.02 mg GAE/g.

Sophora mooracraftiana (Benth.) Baker Family: Fabaceae (Leguminosae)

The *in vitro* inhibitory effects were investigated by Toda and Shirataki (2004) for two resveratrol oligomers, sophorastilbene A and (+)-α-viniferin, isolated from *Sophora mooracraftiana* on lipid peroxidation by superoxide anion. They each inhibited lipid peroxidation by superoxide anion and generation of superoxide anion by the xanthine-xanthine oxidase system. The inhibitory effects of sophorastilbene A were stronger than those of (+)-α-viniferin on lipid peroxidation. Their effects were similar to superoxide dismutase as a superoxide anion scavenger. These results demonstrate that these resveratrol oligomers have the inhibitory effects on oxidative stress.

Sophora tonkinensis Gagnep. Family: Fabaceae (Leguminosae)

Cai *et al.* (2004, 2006) evaluated the antioxidant activity and phenolic compounds in traditional Chinese medicinal plants associated with anticancer. The improved ABTS.+ method was used to systematically assess the total antioxidant capacity (Trolox equivalent antioxidant capacity, TEAC) of the medicinal extracts. The TEAC values and total phenolic content for methanolic extract of root of *Sophora tonkinensis* were 137.7 µmol Trolox equivalent/100g dry weight (DW), and 1.55 g of gallic acid equivalent/100g DW respectively. A positive significant linear relationship between antioxidant activity and total phenolic content showed that phenolic compounds were the dominant antioxidant components in the tested herbs. Major types of phenolic compounds in *Sophora tonkinensis* include flavonoids (sophoraclin, genistein).

Sorbus aria (L.) Crantz. Family: Rosaceae

The antioxidant potential of 70 per cent methanolic extracts from the inflorescences, leaves and fruits of *Sorbus aria* was evaluated by Olszewska and Michel (2009) using three *in vitro* test systems (DPPH, TEAC and FRAP). High activities found were expressed as micromoles trolox equivalents per gram dry weight and were in the range of 86.9-956.2 for the DPPH test, 65.3-577.2 for the TEAC assay and 221.1-1915.2 for the FRAP method. Significant linear correlations between these values and the contents of the total phenolics, total flavonoids, total proanthocyanidins and chlorogenic acid isomers showed that the listed phenolic compounds are the determinants of the antioxidant capacity tested.

Sorbus aucuparia L. Family: Rosaceae

English Name: Rowanberry.

The antioxidative activity of phenolic extracts from fruits of rowanberry was examined by autoxidation of methyl linoleate by Kahkonen *et al.* (1999). The formation of MeLo-conjugated diene hydroperoxides was inhibited over 90 per cent by rowanberry extracts. The phenolic contents were low (18.7mg/g GAE) while the antioxidant activity was high (98 per cent at 500 ppm level). Rowanberry contains high amounts of ferulic acid.

The antioxidant potential of 70 per cent methanolic extracts from the inflorescences, leaves and fruits of *Sorbus aucuparia* was evaluated by Olszewska and Michel (2009) using three *in vitro* test systems (DPPH, TEAC and FRAP). High activities found were expressed as micromoles trolox equivalents per gram dry weight and were in the range of 86.9-956.2 for the DPPH test, 65.3-577.2 for the TEAC assay and 221.1-1915.2 for the FRAP method. Significant linear correlations between these values and the contents of the total phenolics, total flavonoids, total proanthocyanidins and chlorogenic acid isomers showed that the listed phenolic compounds were the determinants of the antioxidant capacity tested. As the superior activity and the highest phenolic levels (11.83 per cent of total phenolics calculated as gallic acid equivalents, 4.35 per cent chlorogenic acid isomers, 5.01 per cent of proanthocyanidins and 1.28 per cent of flavonoid aglycones) were found for *S. aucuparia* inflorescence, this plant material has the greatest potential as a source for natural health products.

Sorbus domestica L. Family: Rosaceae

Termentzi *et al.* (2006) reported the AE values (DPPH radical scavenging) of the methanol extracts of fruit pulp from ripe *Sorbus domestica* fruits as 0.682.

Sorbus intermedia (Ehrh.) Pers. Family: Rosaceae

The antioxidant potential of 70 per cent methanolic extracts from the inflorescences, leaves and fruits of *Sorbus intermedia* was evaluated by Olszewska and Michel (2009) using three *in vitro* test systems (DPPH, TEAC and FRAP). High activities found were expressed as micromoles trolox equivalents per gram dry weight and were in the range of 86.9-956.2 for the DPPH test, 65.3-577.2 for the TEAC assay and 221.1-1915.2 for the FRAP method. Significant linear correlations between these values and the contents of the total phenolics, total flavonoids, total proanthocyanidins and chlorogenic acid isomers showed that the listed phenolic compounds were the determinants of the antioxidant capacity tested.

Sorbus torminalis (L.) Crantz. var. *pinnatifida* Boiss. Family: Rosaceae

Serteser *et al.* (2009) studied the antioxidant activity of 50 per cent aqueous methanol extracts of fruits by various antioxidant assays, including free radical scavenging, hydrogen peroxide scavenging and metal (Fe^{2+}) chelating activities. The DPPH radical scavenging effect of fruit extract was 0.534. Fe^{2+} chelating activity (per cent) of fruit extract was 32.61 and Hydrogen peroxide inhibition activity of fruit extract was 34.32 per cent.

Sorbus torminalis (L.) Crantz. var. *torminalis* Family: Rosaceae

Serteser *et al.* (2009) studied the antioxidant activity of 50 per cent aqueous methanol extracts of fruits by various antioxidant assays, including free radical scavenging, hydrogen peroxide scavenging and metal (Fe^{2+}) chelating activities. The DPPH radical scavenging effect of fruit extract was 0.543. Fe^{2+} chelating activity (per cent) of fruit extract was 33.65 and Hydrogen peroxide inhibition activity of fruit extract was 38.94 per cent.

Sorghum bicolor (L.) Moench. (Syn.: *Sorghum vulgare* (L.) Pers.) Family: Poaceae (Graminae)

English Name: Jowar, Hindi: Jowar.

Specialty sorghums, their brans, and baked and extruded products were analyzed for antioxidant activity using three methods: Oxygen radical absorbance capacity (ORAC), ABTS and DPPH. Phenol contents of the sorghums correlated highly with their antioxidant activity measured by three methods. The ABTS and DPPH methods, which are more cost effective and simpler, were demonstrated to have similar predictive power as ORAC on sorghum antioxidant activity (Awika *et al.*, 2003).

The effects of plant color, pericarp thickness, pigmented testa, and spreader genes on phenols and antioxidant activity levels of 13 sorghum genotypes were evaluated by Dykes *et al.* (2005). Antioxidant activity levels using ABTS and DPPH assays were evaluated. Sorghums with pigmented testa and spreader genes had the highest levels of phenols and antioxidant activity. In addition, sorghums with purple/red plants and thick pericarp genes had increased levels of phenols and antioxidant activity. Sorghums with a black pericarp had higher levels of flavan-4-ols and anthocyanins than the other varieties.

Sreeramulu *et al.* (2009) evaluated the antioxidant activity of jowar by DPPH, scavenging assay, FRAP assay and reducing power. Phenolic content was 57.55 mg/100g. DPPH radical scavenging activity expressed as trolox equivalent, FRAP (µmol/g) and reducing power (mg/g) were 1.27, 67 and 0.82, respectively in jowar.

Soymida febrifuga (Roxb.) A.Juss. Family: Meliaceae

Srinivas Reddy *et al.* (2008a,b) evaluated the antioxidant and antimicrobial properties of hexane, methanol and aqueous extracts of *Soymida febrifuga* leaves, which is a traditional medicine in India. Antioxidant activity of different extracts was evaluated by free radical scavenging activity taking total phenolic content (TPC) as an index. The results showed that the methanol and aqueous extracts of leaf had a higher antioxidant activity and total phenolic content than the hexane extract. The antioxidant activity and TPC of the extracts were highly correlated. Methanolic and aqueous extracts of bark and leaf have shown higher total phenolic content, reducing power, metal chelating, superoxide, hydroxyl radical, hydrogen peroxide and nitric oxide (murine macrophage cells) scavenging activity than the root and root bark extracts.

Sparganium stoloniferum Buch. Family: Sparganiaceae

Gan *et al.* (2010) evaluated the antioxidant activity of 40 medicinal plants associated with prevention and treatment of cardiovascular and cerebrovascular diseases. *Sparganium stoloniferum* showed moderate antioxidant activity of 338.73 µmol Fe(II)/g in FRAP assay and 31.32µmol in TEAC assay. Total phenolic content was very low, 2.58 mg GAE/g.

Spartium junceum L. Family: Fabaceae (Leguminosae)

The possible superoxide dismutase-like activity of the extracts, fractions and constituents of *Spartium junceum* obtained through activity-guided fractionation were studied by using *in vitro* electron spin resonance spectrometry, in order to explain the role of antioxidant principles in the potent antiulcerogenic activity of the extract (Yesilada *et al.,* 2000). Despite the fact that the triterpene, spartitrioside, which was previously reported as the active antiulcerogenic constituent of the flowers was found almost inactive, the flavonoid-rich fractions showed potent antioxidant activity. Five flavonoid glycosides bearing catechol structure in ring B were isolated from the butanol extract and their structures were elucidated as isoquercitrin (quercetin 3β-glucoside) (1); luteolin 4′β-glucoside (2); quercetin 3, 42-diglucoside (3); azaleatin 3β-glucoside (quercetin 5-methylether 3β-glucoside) (4), quercetin 42β-glucoside (5). Flavonoids (2) and (4) showed the highest *in vitro* antioxidant activity with 22.59 and 19.08 U/ml, respectively.

Spathodea campanulata P.Beauv. Family: Bignoniaceae

Extracts of *Spathodea campanulata* have been shown to have anti-inflammatory and antioxidant properties due to flavonoids, triterpenoids, diterpenoids and caffeic acid derivatives (Houghton *et al.,* 2005).

Alcoholic extract of *Spathodea campanulata* aerial parts, and two of the isolated fractions from celite column showed strong antioxidant activity (92, 94 and 89 per cent RSA, Radical Scavenging Activity) (Nazif, 2007). Phytochemical investigation of chloroform/EtOAc faction of this column led to the isolation of phenolic acids, caffiec acid (1), and ferulic acid (2), fraction EtOAc/MeOH on further fractionation afforded 3 Flavonoids, kampferol 3-O-glucoside (3), quercetin 3-methyl ether (4) and 8-methoxy kampferol 3-O-glucoside (5).

Shanmukha *et al.* (2010) evaluated the antioxidant activity of ethanolic extract of *Spathodea campanulata*. Pretreatment of the extract orally was found to ameliorate the effects of gentamycin on lipid peroxide formation and showed a decrease in serum marker enzymes. It also prevented depletion of tissue GSH levels.

Spathodea nilotica Seem. Family: Bignoniaceae

The chloroform and methanolic extracts of 124 Egyptian plant species belonging to 56 families were investigated by Moussa *et al.* (2011) and compared their antioxidant activity by DPPH scavenging assay. Among the 124 plant species tested 18 exhibited extremely high antiradical activity (>80 per cent inhibition). The IC_{50} value of DPPH radical scavenging of *Spathodea nilotica* was 30.97µg/ml, while total phenolic and flavonoid contents were 129.96 mg tannic acid equivalent/g extract and 197.26 mg rutin equivalent/g extract, respectively. Correlation coefficient between DPPH radical scavenging activity and the total phenolic and flavonoid contents suggest that phenolics and flavonoids in the extracts were partly responsible for the antiradical activities.

Spatholobus suberectus Dunn Family: Fabaceae (Leguminosae)

Cai *et al.* (2004) evaluated the antioxidant activity and phenolic compounds in traditional Chinese medicinal plants associated with anticancer. The improved ABTS.+ method was used to systematically assess the total antioxidant capacity (Trolox equivalent antioxidant capacity, TEAC) of the medicinal extracts. The TEAC values and total phenolic content for methanolic extract of stem of *Spatholobus suberectus* were 1995.2 µmol Trolox equivalent/100g dry weight (DW), and 8.63 g of gallic acid equivalent/100g DW respectively. A positive significant linear relationship between antioxidant activity and total phenolic content showed that phenolic compounds were the dominant antioxidant components in the tested herbs. Major types of phenolic compounds in *Spatholobus suberectus* include flavonoids (formomonetin, daidzein, ononin, epicatechin) phenolic acid (protocatechuic acid).

Spergularia rubra Family: Caryophyllaceae

Antioxidant potential of *Spergularia rubra* was investigated by Vinholes *et al.* (2011). Thirty-six phenolic compounds were determined by HPLC-DAD, comprising non-acylated *C*-glycosyl flavones (38 per cent), *C*-glycosyl flavones acylated with aromatic acids (36 per cent), *C*-glycosyl flavones acylated with aliphatic acids (13 per cent) and 10 per cent corresponded to *C*-glycosyl flavones with a mixed acylation. Organic acids (oxalic, citric, malic, quinic and fumaric acids) and fatty acids (azelaic, myristic, palmitic, linoleic, linolenic and stearic acids) were described. Their determination by HPLC-UV and GC-IT–MS allowed finding concentrations of 192.15 and 34.87 g/kg, respectively. The extract showed a dose-dependent response against DPPH, superoxide and nitric oxide radicals.

Spermacoce articularis L. f. Family: Rubiaceae

Methanol extracts of seven Malaysian medicinal plants, including *Spermacoce articularis* were screened by Saha *et al.* (2004) for antioxidant and nitric oxide inhibitory activities. Antioxidant activity was measured by using FTC, TBA and DPPH free radical scavenging methods and Greiss assay was used for the measurement of nitric oxide inhibition in LPS and interferon-γ-treated RAW264.7 cells. All the extracts showed strong antioxidant activity comparable to or higher than that of α-tocopherol, BHT and quercetin in FTC and TBA methods. The extracts from *Spermacoce articularis* showed strong DPPH free radical scavenging activity comparable with quercetin, BHT and Vit C. In the Greiss assay *Spermacoce articularis* showed strong inhibitory activity on nitric oxide production in LPS and IFN-γ-induced RAW 264.7 cells.

Spermacoce exilis (L.O.Williams) C.D.Adams ex W.C.Burger and C.M.Taylo Family: Rubiaceae

Methanol exteracts of seven Malaysian medicinal plants, including *Spermacoce exilis* were screened by Saha *et al*. (2004) for antioxidant and nitric oxide inhibitory activities. Antioxidant activity was measured by using FTC, TBA and DPPH free radical scavenging methods and Greiss assay was used for the measurement of nitric oxide inhibition in LPS and interferon-γ-treated RAW264.7 cells. All the extracts showed strong antioxidant activity comparable to or higher than that of α-tocopherol, BHT and quercetin in FTC and TBA methods. In the DPPH radical scavenging assay the extract showed only moderate activity. Extracts from *Spermacoce exilis* also inhibited NO production but this was due to their cytotoxic effects upon cells during culture.

Ahmad *et al*. (2010) evaluated the antioxidant potential of 22 species of medicinal plants from Malaysian Rubiaceae including *Spermacoce exilis*. FTC, TBA, TPC and the DPPH assays were employed. The tested extracts showed strong antioxidant potential with percent inhibition of 98.2 per cent in the FTC, and 95.5 per cent in TBA assays. The TPC of the extract was 27.57mg GAE/g PE. A good correlation was observed between total phenolic content and radical-scavenging activities.

Spermacoce hispida L. Family: Rubiaceae

Antioxidant efficacy of flavonoid-rich fraction from *Spermacoce hispida* in hyperlipidemic rats was investigated by Kaviarasan *et al*. (2008). The flavonoid-rich fraction effectively scavenged DPPH and ABTS radicals *in vitro*. Further, the results showed elevated activities of free radical-scavenging enzymes (SOD, CAT and GPx) and increased levels of non-enzymatic antioxidants (GSH, vitamins C and E). TBARS and lipid hydroperoxides decreased significantly in flavonoid-rich fraction treated rats compared to HFD control. Among the doses used, 40 mg/kg BW dose showed maximum effect.

Spermacoce latifolia Aubl. Family: Rubiaceae

Ahmad *et al*. (2010) evaluated the antioxidant potential of 22 species of medicinal plants from Malaysian Rubiaceae including *Spermacoce latifolia*. FTC, TBA, TPC and the DPPH assays were employed. The tested extracts showed strong antioxidant potential with percent inhibition of 96.4 per cent in the FTC, and 77.3 per cent in TBA assays. The TPC of the extract was 17.00mg GAE/g PE. A good correlation was observed between total phenolic content and radical-scavenging activities.

Sphaeranthus indicus L. Family: Asteraceae (Compositae)

The free radical scavenging potential of *Sphaeranthus indicus* was studied by Shirwaikar *et al*. (2006). The ethanolic extract at 1000μg/ml showed maximum scavenging of the radical cation ABTS observed up to 41.99 per cent followed by the scavenging of the stable radical DPPH (33.27 per cent), superoxide dismutase (25.14 per cent) and nitric oxide radical (22.36 per cent) at the same concentration. However, the extract showed only moderate scavenging activity of iron chelation (14.2 per cent). Total antioxidant capacity of the extract was found to be 160.85nmol/g ascorbic acid.

Tiwari and Khosa (2009) evaluated the hepatoprotective and antioxidant effect of aqueous and methanolic extract of flower heads of *Sphaeranthus indicus*. Methanolic extract exhibited a significant effect showing increasing levels of SOD, CAT and GPx by reducing malondialdehyde levels.

Vadlamudi and Naidu (2010) evaluated the antioxidant activity of ethanolic extracts of *Sphaeranthus indicus* leaves. SOD activity was 1.3U/mg, Catalase activity was 1.9U/mg while vitamin C content was 70 mg/100g.

Sphallerocarpus gracilis (Besser ex Treviranus) Koso-Poljansky (Syn.: *Chaerophyllum gracile* Besser ex Treviranus) Family: Apiaceae (Umbelliferae)

Myagmar and Aniya (2000) evaluated the free radical scavenging action of some medicinal herbs growing in Mongolia. The aqueous extract of nine herbs including *Sphallerocarpus gracilis* were used. The free radical scavenging action was determined *in vitro* and *ex vivo* by using ESR spectrometer and chemiluminiscence analyzer. The results showed that *Sphallerocarpus gracilis* extract possess strong scavenging action of DPPH, superoxide and hydroxyl radicals.

Sphenocentrum jollyanum Pierre Family: Menispermaceae

Oke and Hamburger (2002) studied the antioxidant activity of ethanol extract of leaf and root using DPPH assay. The leaf extract showed moderate antioxidant activity while root showed high antioxidant activity.

Sphenostylis stenocarpa (Hochst. ex A.Rich.) Harms Family: Fabaceae (Papilionaceae)

English Name: African yam bean.

The antioxidant properties of some commonly consumed African yam bean were assessed with regard to its vitamin C, total phenol and phytate content, as well as antioxidant activity as typified by its reducing power and free radical scavenging ability (Oboh, 2006a). Vitamin C content of the African yam bean was 0.8 mg/100g. The phenol content of the African yam bean was 0.7 mg/g. The phytate content of yam bean was 2.4 per cent while the reducing power and free radical scavenging ability was 23.6 per cent.

The effect of fermentation on the polyphenol distribution and antioxidant activity of African yam bean was investigated by Oboh *et al.* (2009). The results of the study revealed that fermentation caused a significant decrease in the bound phenol content of the legume. Free soluble phenol from both the fermented and unfermented legumes had a significantly higher reducing power.

Spilanthes acmella Murr. Family: Asteraceae (Compositae)

Wongsawatkul *et al.* (2008) evaluated the antioxidant activity of *Spilanthes acmella*. Results showed that the extracts exert maximal vasorelaxations in a dose-dependent manner, but their effects were less than acetylcholine-induced nitric oxide (NO) vasorelaxation. This demonstrates that the extracts exhibit vasorelaxation *via* partially endothelium-induced NO and prostacyclin in a dose-dependent manner. Significantly, the ethyl acetate extract exerts immediate vasorelaxation (ED_{50} 76.1 ng/ml) and is the most potent antioxidant (DPPH assay). The chloroform extract shows the highest vasorelaxation and antioxidation (SOD assay). These reveal a potential source of vasodilators and antioxidants.

Spinacea oleracea L. Family: Chenopodiaceae

English Name: Spinach; Japanese: Hourensou.

Cai *et al.* (2004) evaluated the antioxidant activity and phenolic compounds in traditional Chinese medicinal plants and vegetables associated with anticancer. The improved ABTS.+ method was used to systematically assess the total antioxidant capacity (TEAC) of the medicinal extracts. The TEAC values and total phenolic content for methanolic extract of spinach were 167.1 µmol Trolox equivalent/100g dry weight (DW), and 0.90 g of gallic acid equivalent/100g DW respectively.

Maeda *et al.* (2006) examined the phenolic contents of Spinach and its radical scavenging activities for DPPH radical. The polyphenol content was 74.1 mg-gallic acid equivalent/100 g and DPPH scavenging activity was 149.6 µmol-Trolox eq./100 g.

Antioxidant potential of selected vegetables commonly used in diet in Asian subcontinent, including *Spinacia oleracea* leaves has been evaluated by Gacche *et al.* (2010). The DPPH radical scavenging activity was 20.4 per cent, inhibition of lipid peroxidation was 29.4 per cent and FICA was 43.9 per cent, Vitamin C content was 15.36 mg/100g and total phenols was 2.60 mg/g.

The hepatoprotective activity of the ethanolic extract of the leaves of spinach (EESO) was studied by Al-Dosani (2010) against carbon tetrachloride (CCl_4)-induced oxidative stress (OS) and liver injury in rats. Pretreatment of rats with EESO, at 250 and 500 mg/kg body weight for 21 consecutive days significantly prevented the CCl_4-induced hepatic damage as indicated by the serum marker enzymes (SGOT, SGPT, ALP and GGT) and bilirubin levels. Parallel to these changes, the leaves extract also prevented CCl_4-induced OS in rat liver by inhibiting lipid peroxidation (LPO) and restoring the levels of antioxidant non-enzymatic biomarker, such as total protein (TP) and non-protein sulfhydryl (NP-SH) in liver tissue. The biochemical changes were consistent with the histological findings of the liver tissue suggesting marked hepatoprotective effect of the leaves extract in a dose-dependent manner, besides, a significant reduction was also observed in pentobarbital-induced sleeping time in mice. The results of spinach extract were comparable to that of silymarin. The protective effect of the EESO against CCl_4-induced acute hepatotoxicity could be attributed to the potent antioxidant constituents of the spinach.

Evanjelene and Natarajan (2011) evaluated the free radical scavenging activity of *Spinacea oleracea* leaves and stem. The results highlighted that aqueous and methanol extracts of the leaves showed better antioxidant activity followed by the ethanol and ethyl acetate extracts.

Spinacia turkestanica Iliin Family: Chenopodiaceae

The antioxidant properties of 10 edible plants of the Turkmen Sahra region in Golestan province in northern Iran, including *Allium paradoxum, Allium rubellum, Foeniculum vulgare, Mentha longifolia, Origanum vulgare, Prunus divaricata, Rubus sanctus, Rumex tuberosus, Satureja mutica* and *Spinacia turkestanica* were evaluated by Motamed and Naghibi (2010) by four different methods: free radical scavenging using 2,2-diphenyl-1-picrylhydrazyl (DPPH), evaluation of xanthine-oxidase activity, inhibition of lipid peroxidation by the ferric thiocyanate method, and the deoxyribose degradation assay. All species tested except *A. paradoxum* and *A divaricata* showed antioxidant activity at least in one assay. The pro-oxidant activities were also assayed for these species and *S. mutica, M. longifolia,* and *O. vulgare* showed significant pro-oxidant activity.

Spiraea alba Du Roi Family: Rosaceae

English Name: Medow sweet

Borchardt *et al.* (2008) evaluated the antioxidant activity of methanol extracts of seeds of *Spiraea alba* using DPPH antioxidant assay.The antioxidant value reported in µM Trolox/100g (TE) from DPPH radical scavenging activity of crude seeds extract was 63,985.

Spiraea tomentosa L. Family: Rosaceae

English Name: Steeple bush.

Borchardt *et al.* (2008) evaluated the antioxidant activity of methanol extracts of seeds of *Spiraea tomentosa* using DPPH antioxidant assay.The antioxidant value reported in µM Trolox/100g (TE) from DPPH radical scavenging activity of crude seeds extract was 141,430.

Spirodela polyrrhiza (L.) Schleid. Family: Lemnaceae

Antioxidant capacities of 56 selected Chinese medicinal plants, including *Spirodela polyrrhiza* were evaluated by Song *et al.* (2010) using the TEAC and FRAP assays, and their total phenolic content was measured by the Folin-Ciocalteu method. *Spirodela polyrrhiza* showed moderate antioxidant activity with TEAC value of 54.84µmol Trolox/g, FRAP value of 89.55µmol Fe^{2+}/g and phenolic content was 10.53mg GAE/g.

Spondias mangifera Willd. Family: Anacardiaceae

Acharya *et al.* (2010) evaluated the antioxidant and antimicrobial activities of methanol and aqueous extracts of *Spondias mangifera* root. The principal bioactive compounds such as saponins, flavonoids and tannins were positive for both the extracts, alkaloids were detected only in methanol extract and were absent in aqueous extracts. The extracts were screened for their *in vitro* antioxidant potential inhibition of oxygen-derived free radicals *viz.*, assays for free radical scavenging by DPPH, reducing power ability and nitric oxide scavenging were performed. Both the extracts of this plant showed effective free radical scavenging activity, reducing power and nitric oxide scavenging activity.

Spondias mombin L. Family: Anacardiaceae

Akinmoladun *et al.* (2010) assessed the phytochemical constituents and antioxidant and free radical scavenging activities of *Spondias mombin* using seven different antioxidant assay methods. Percentage DPPH radical scavenging activity was highest (88.58 per cent) and compared with values obtained for ascorbic aicd and gallic acid. The extract had low nitric oxide radical scavenging activity. The results suggest that the methanolic extracts possess significant antioxidant and radical scavenging activity that may be due to the phytochemical content of the plant.

Spondias pinnata (L.f.) Kurz. Family: Anacardiaceae

Hazra *et al.* (2008) evaluated the antioxidant activity of *Spondias pinnata* stem bark. The extract showed total antioxidant activity with a trolox equivalent antioxidant concentration (TEAC) value of 0.78. The IC_{50} values for scavenging of free radicals were 112.18µg/ml, 13.46µg/ml, and 24.48µg/ml for hydroxyl, superoxide and nitric oxide, respectively. The IC_{50} for hydrogen peroxide scavenging was 44.74µg/ml. For the peroxynirite, singlet oxygen and hypochlorous acid scavenging activities the IC_{50} values were 716.32µg/ml, 58.07 µg/ml and 127.99µg/ml, respectively. The extract was found to be a potent iron chelator with IC_{50} = 66.54µg/ml. The reducing power was increased with increasing amounts of extract. The plant extract (100 mg) yielded 91.47mg/ml GAE equivalent phenolic content and 350.5 mg ml quercetin-equivalent flavonoid content.

Stachys acerosa Boiss. Family: Lamiaceae (Labiatae)

The antioxidant activity of the extracts of *Stachys acerosa* was measured by Salehi *et al.* (2007) by DPPH assay. The free radical scavenging activity of the n-butanol subfraction of methanol extract (IC_{50} =22.7µg/ml) was superior to all other extracts. The strong antioxidant activity could be related to its higher phenolic content.

Stachys alpina L. Family: Lamiaceae (Labiatae)

Haznagy-Radnai *et al.* (2006) investigated the antioxidant activity of *Stachys alpina* native to Hungary in an enzyme-independent lipid-peroxidation system. The methanolic extracts were found to be more effective than controls. Their study also revealed that there is relationship between the antioxidant activity and tannin (like pyrogallol) content as well as between the antioxidant potency and total phenol content expressed in gallic acid. No relationship was found between the antioxidant activity and either hydroxycinnamic acids – expressed in caffeic acid – or the flavonoid content.

Stachys alpina L. *ssp. dinarica* Family: Lamiaceae (Labiatae)

Methanol extract of aerial flowering parts of *Stachys alpina* ssp. *dinarica* was studied by Kukic *et al.* (2006) for total antioxidant activity, along with DPPH and OH radical scavenging activity and lipid peroxidation. High correlation between total phenolics content, total antioxidant activity and scavenging DPPH radical indicate that polyphenols were the main antioxidants. The extract exhibited high anti-DPPH activity, and scavenged OH radical. As per lipid peroxidation, IC_{50} value of the extract was 49.00µg/ml.

Stachys anisochila Vis. et Panc. Family: Lamiaceae (Labiatae)

Methanol extract of aerial flowering parts of *Stachys anisochila* was studied by Kukic *et al.* (2006) for total antioxidant activity, along with DPPH and OH radical scavenging activity and lipid peroxidation. High correlation between total phenolics content, total antioxidant activity and scavenging DPPH radical indicate that polyphenols were the main antioxidants. The extract exhibited high anti-DPPH activity, and scavenged OH radical.

Stachys annua L. Family: Lamiaceae (Labiatae)

Haznagy-Radnai *et al.* (2006) investigated the antioxidant activity of *Stachys annua* native to Hungary in an enzyme-independent lipid-peroxidation system. The methanolic extracts were found to be more effective than controls. Their study also revealed that there is relationship between the antioxidant activity and tannin (like pyrogallol) content as well as between the antioxidant potency and total phenol content expressed in gallic acid. No relationship was found between the antioxidant activity and either hydroxycinnamic acids – expressed in caffeic acid – or the flavonoid content.

Stachys beckeana Dorfl. et Hayek Family: Lamiaceae (Labiatae)

Methanol extract of aerial flowering parts of *Stachys beckeana* was studied by Kukic *et al.* (2006) for total antioxidant activity, along with DPPH and OH radical scavenging activity and lipid peroxidation. High correlation between total phenolics content, total antioxidant activity and scavenging DPPH radical indicate that polyphenols were the main antioxidants. The extract exhibited high anti-DPPH activity, and scavenged OH radical. As per lipid peroxidation, IC_{50} value of the extract was 25.07µg/ml

Stachys byzantha C. Koch Family: Lamiaceae (Labiatae)

Erdemoglu *et al.* (2006) evaluated the antioxidant activity of the *Stachys byzantha* extract by using DPPH as well as by flow-injection analysis-luminol chemiluminiscence (FIA-CL). Extracts were shown to possess a significant scavenger activity against DPPH free radical and an inhibitory effect on H_2O_2- or HOCl-luminol chemiluminiscence. The results concluded that the extracts have a potential source of antioxidants of natural origin.

The methanolic extracts of the aerial parts of *Stachys byzantina* and eight other species of *Stachys* were investigated by Khanavi *et al.* (2009) for their antioxidant activity and total phenolic content using FRAP and Folin-Ciocalteu assays respectively. *S.byzantina* had low antioxidant activity (9.3283 mmol FeII/100g) and total phenolic content (638.304 mg gallic acid/100g). There was a direct correlation between total phenol and antioxidant activity which indicates that polyphenols were the main antioxidants.

Stachys fruticulosa M.B. Family: Lamiaceae (Labiatae)

The methanolic extracts of the aerial parts of *Stachys fruticulosa* and eight other species of *Stachys* were investigated by Khanavi *et al.* (2009) for their antioxidant activity and total phenolic content using FRAP and Folin-Ciocalteu assays respectively. *S. fruticulosa* had the highest antioxidant activity (62.02 mmol FeII/100g) and total phenolic content (4550.368 mg gallic acid/100g). There was a direct correlation between total phenol and antioxidant activity which indicates that polyphenols were the main antioxidants.

Stachys iberica Family: Lamiaceae (Labiatae)

Tepe *et al.* (2011) evaluated the antioxidant activity of *Stachys iberica*.

Stachys inflata Benth. Family: Lamiaceae (Labiatae)

The methanolic extracts of the aerial parts of *Stachys inflata* and eight other species of *Stachys* were investigated by Khanavi *et al.* (2009) for their antioxidant activity and total phenolic content using FRAP and Folin-Ciocalteu assays respectively. *S. inflata* had good antioxidant activity (31.0787 mmol FeII/100g) and total phenolic content (1478.808 mg gallic acid/100g). There was a direct correlation between total phenol and antioxidant activity which indicates that polyphenols were the main antioxidants.

Composition and antioxidant activities of essential oil and methanol extract polar and nonpolar subfractions of *Stachys inflata* were determined by Ebrahimabadi *et al.* (2010). Essential oil and extracts were tested for their antioxidant activities using DPPH and β-carotene/linoleic acid assays. In the DPPH test, IC_{50} value for the polar subfraction was 89.50µg/ml, indicating an antioxidant activity of about 22 per cent of that of butylated hydroxytoulene (IC_{50} =19.72µg/ml) for this extract. In β-carotene/linoleic acid assay, the best inhibition belonged to the nonpolar subfraction (77.08 per cent). Total phenolic content of the polar and nonpolar extract subfractions was 5.4 and 2.8 per cent (w/w), respectively.

Stachys lavandulifolia Vahl. Family: Lamiaceae (Labiatae)

Souri *et al.* (2004) evaluated the antioxidative activity of *Stachys lavandulifolia* by linoleic acid peroxidation test using 1,3-diethyl-2-thiobarbituric acid as the reagent. Methanolic extract showed 91.56 per cent inhibition at 40µg concentration with IC_{50} value of 3.11µg.

Stachys laxa Boiss. et Buhse. Family: Lamiaceae (Labiatae)

The methanolic extracts of the aerial parts of *Stachys laxa* and eight other species of *Stachys* were investigated by Khanavi *et al.* (2009) for their antioxidant activity and total phenolic content using FRAP and Folin-Ciocalteu assays respectively. *S. laxa* had good antioxidant activity (35.0629 mmol FeII/100g) and total phenolic content (2089.992 mg gallic acid/100g). There was a direct correlation between total phenol and antioxidant activity which indicates that polyphenols were the main antioxidants.

Stachys macrantha C. Koch. Family: Lamiaceae (Labiatae)

Haznagy-Radnai *et al.* (2006) investigated the antioxidant activity of *Stachys macrantha* native to Hungary in an enzyme-independent lipid-peroxidation system.

The methanolic extracts were found to be more effective than controls. Their study also revealed that there is relationship between the antioxidant activity and tannin (like pyrogallol) content as well as between the antioxidant potency and total phenol content expressed in gallic acid. No relationship was found between the antioxidant activity and either hydroxycinnamic acids – expressed in caffeic acid – or the flavonoid content.

Stachys officinalis L. Family: Lamiaceae (Labiatae)

English Name: Wood betony.

Haznagy-Radnai *et al.* (2006) investigated the antioxidant activity of *Stachys officinalis* native to Hungary in an enzyme-independent lipid-peroxidation system. The methanolic extracts were found to be more effective than controls. Their study also revealed that there is relationship between the antioxidant activity and tannin (like pyrogallol) content as well as between the antioxidant potency and total phenol content expressed in gallic acid. No relationship was found between the antioxidant activity and either hydroxycinnamic acids – expressed in caffeic acid – or the flavonoid content.

Matkowski and Piotrowska (2006) studied the antioxidative effects of methanolic extract with the use of three *in vitro* assays (DPPH assay, phosphomolybdenum method and lipid peroxidation). The extract showed strong antioxidant activity.

Wojcikowski *et al.* (2007) evaluated the antioxidant activity of aerial parts of *Stachys officinalis*. Oxygen radical absorbance capacity of the ethyl extract, methanol extract and aqueous extracts were 2.68, 110.78 and 148.84µmol Trolox equivalent per g of dried starting material, respectively.

Stachys persica Gmel. Family: Lamiaceae (Labiatae)

The methanolic extracts of the aerial parts of *Stachys persica* and eight other species of *Stachys* were investigated by Khanavi *et al.* (2009) for their antioxidant activity and total phenolic content using FRAP and Folin-Ciocalteu assays respectively. *S. persica* had the highest antioxidant activity (61.42 mmol FeII/100g) and total phenolic content (3294.96 mg gallic acid/100g). There was a direct correlation between total phenol and antioxidant activity which indicates that polyphenols were the main antioxidants.

Stachys plumosa Griseb. Family: Lamiaceae (Labiatae)

Methanol extract of aerial flowering parts of *Stachys plumosa* was studied by Kukic *et al.* (2006) for total antioxidant activity, along with DPPH and OH radical scavenging activity and lipid peroxidation. High correlation between total phenolics content, total antioxidant activity and scavenging DPPH radical indicate that polyphenols were the main antioxidants. The extract showed high DPPH activity of 60.67 per cent at 100µg/ml.

Stachys prumeri Family: Lamiaceae (Labiatae)

Couladis *et al.* (2003) screened Greek aromatic plants from the Lamiaceae family for the antioxidant activity. Of the 21 plants they tested they found ethanol extracts prepared from *Stachys prumeri*, as well as 9 other plants exhibited the same antioxidant activity as α-tocopherol in their ability to inhibit bleomycin-Fe(II) complex-induced arachidonic acid superoxidation to MDA.

Stachys recta L. Family: Lamiaceae (Labiatae)

Haznagy-Radnai *et al*. (2006) investigated the antioxidant activity of *Stachys recta* native to Hungary in an enzyme-independent lipid-peroxidation system. The methanolic extracts were found to be more effective than controls. Their study also revealed that there is relationship between the antioxidant activity and tannin (like pyrogallol) content as well as between the antioxidant potency and total phenol content expressed in gallic acid. No relationship was found between the antioxidant activity and either hydroxycinnamic acids – expressed in caffeic acid – or the flavonoid content.

Stachys setifera C.A.Mey. Family: Lamiaceae (Labiatae)

The methanolic extracts of the aerial parts of *Stachys setifera* and eight other species of *Stachys* were investigated by Khanavi *et al*. (2009) for their antioxidant activity and total phenolic content using FRAP and Folin-Ciocalteu assays respectively. *S.setifera* had moderate antioxidant activity (11.3923 mmol FeII/100g) and total phenolic content (708.744 mg gallic acid/100g). There was a direct correlation between total phenol and antioxidant activity which indicates that polyphenols were the main antioxidants.

Stachys subaphylla Rech.f. Family: Lamiaceae (Labiatae)

The methanolic extracts of the aerial parts of *Stachys subaphylla* and eight other species of *Stachys* were investigated by Khanavi *et al*. (2009) for their antioxidant activity and total phenolic content using FRAP and Folin-Ciocalteu assays respectively. *S.subaphylla* had good antioxidant activity (17.1142 mmol FeII/100g) and total phenolic content (1016.04 mg gallic acid/100g). There was a direct correlation between total phenol and antioxidant activity which indicates that polyphenols were the main antioxidants.

Stachys sylvatica L. Family: Lamiaceae (Labiatae)

Haznagy-Radnai *et al*. (2006) investigated the antioxidant activity of *Stachys sylvatica* native to Hungary in an enzyme-independent lipid-peroxidation system. The methanolic extracts were found to be more effective than controls. Their study also revealed that there is relationship between the antioxidant activity and tannin (like pyrogallol) content as well as between the antioxidant potency and total phenol content expressed in gallic acid. No relationship was found between the antioxidant activity and either hydroxycinnamic acids – expressed in caffeic acid – or the flavonoid content.

Stachys trinervis Aitch. et Hemsl. Family: Lamiaceae (Labiatae)

The methanolic extracts of the aerial parts of *Stachys trinervis* and eight other species of *Stachys* were investigated by Khanavi *et al*. (2009) for their antioxidant activity and total phenolic content using FRAP and Folin-Ciocalteu assays respectively. *S. trinervis* had low antioxidant activity (9.1092 mmol FeII/100g) and total phenolic content (430.584 mg gallic acid/100g).

Stachys turcomanica Trautv. Family: Lamiaceae (Labiatae)

The methanolic extracts of the aerial parts of *Stachys turcomanic* and eight other species of *Stachys* were investigated by Khanavi *et al.* (2009) for their antioxidant activity and total phenolic content using FRAP and Folin-Ciocalteu assays respectively. *S. turcomanica* had good antioxidant activity (22.5698 mmol FeII/100g) and total phenolic content (1313.568 mg gallic acid/100g). There was a direct correlation between total phenol and antioxidant activity which indicates that polyphenols were the main antioxidants.

Stachytarpheta angustifolia (Mill.) Vahl Family: Verbenaceae

The free radical scavenging activity of an ethanol extract of the leaves of *Stachytarpheta angustifolia* was assessed by measuring its capability for scavenging DPPH radical, superoxide anion radical, hydroxyl radical, nitric oxide radicals as well as its ability to inhibit lipid peroxidation, using appropriate assay systems (Awah *et al.*, 2010). This extract showed a potent antioxidant activity in the DPPH radical-scavenging assay (EC_{50}= 9.65µg/ml), superoxide anion radical (IC_{50}= 64.68 µg/ml), non-enzymatic lipid peroxidation (IC_{50}= 282.91µg/ml), and accumulation of nitrite *in vitro*.

Stachytarpheta cayennensis (Rih.) Vahl Family: Verbenaceae

De Souza (2011) evaluated the antioxidant capacity of verbascoside, martinoside, betulinic acid from the *Stachytarpheta cayennensis* and quercetin by an *in vitro* assay with isolated mitochondria from mice's brain The results showed that all compounds tested exhibited a scavenger effect on the ROS generated by the isolated mitochondria, which displayed a dependent dose increase.

Stachytarpheta jamaicensis (L.) Vahl (Syn.: *Verbena jamaicensis* L.)

Family: Verbenaceae

The antioxidant effects of ethyl acetate and n-Hexane extracts of dried leaves of *Stachytarpheta jamaicensis* on the reactive oxygen species generating during the respiratory burst of rat peritoneal macrophages were investigated by Alvarez *et al.* (2004). Only Ethyl acetate extract, at concentration between 0.4 and 40 µg/ml inhibited the extacellular release of oxygen radicals by resident peritoneal macrophages stimulated with phorbol-12-myristate 13-acetate. At concentrations above 40µg/ml, ethyl acetate extract inhibited the production of nitric oxide in macrophages stimulated *in vivo* with sodium thioglycollate then *in vitro* with lipopolysaccharide and gamma-interferon (IFN-gamma), n-hexane extracts at concentrations between 0.4 and 40µg/ml did not scavenge oxygen radical generated enzymatically by hypoxanthine/xanthine oxidase (HX/XO) system, but ethyl acetate extract at the same concentrations showed potent oxygen radical-scavenging activity. At 40µg/ml, ethyl acetate extract also inhibited XO activity (Alvarez *et al.*, 2004).

Stellaria chamaejasme L. Family: Caryophyllaceae

Myagmar and Aniya (2000) evaluated the free radical scavenging action of some medicinal herbs growing in Mongolia. The aqueous extract of nine herbs including *Stellaria chamaejasme* were used. The free radical scavenging action was determined *in vitro* and *ex vivo* by using ESR spectrometer and chemiluminiscence analyzer. The results showed that *Stellaria chamaejasme* extract possess strong scavenging action of DPPH, superoxide and hydroxyl radicals. *Stellaria chamaejasme* also depressed reactive oxygen production from polymorphonuclear leukocytes stimulated by phorbol-12-myristate *ex vivo*.

Stellaria dichotoma Bge. Family: Caryophyllaceae

Li *et al.* (2008) evaluated the antioxidant properties of 45 medicinal plants using FRAP and TEAC assays and the total phenolic contents of these plants were measured. *Stellaria dichotoma* showed low antioxidant activity with 32.04 μmol Fe(II)/g FRAP value, 16.49μmol Trolox/g of TEAC value and 5.99 mg GAE/g of Phenolic content.

The antioxidant properties of enzymatic extracts from *Stellaria dichotoma* were evaluated by Lim *et al.* (2008) using seven carbohydrates and five proteases for DPPH radical, hydroxyl radical and alkyl radical scavenging activity using electron spin resonance spectrometer. The DPPH radical scavenging activities of pancreatic trypsin and Celluclast extracts from *S. dichotoma* were the highest among various protease and carbohydrate extracts, and the 50 per cent inhibitory concentration (IC$_{50}$) values were 10.45 and 13.80 μg/ml, respectively. The *S. dichotoma* enzymatic extracts also exhibited alkyl radical scavenging activity in a dose dependent manner.

Stellaria media (L.) Vill. Family: Caryophyllaceae

A total of 27 extracts from non-cultivated and weedy vegetables traditionally consumed by ethnic Albanians in southern Italy were tested for their free radical scavenging activity (FRSA) in the DPPH screening assay, for their *in vitro* non-enzymatic inhibition of bovine brain lipid peroxidation and for their inhibition of xanthine oxidase. Extracts from *Stellaria media* showed strong *in vitro* inhibition of xanthine oxidase, with an activity higher than that of a reference extract from *Ledum groenlandicum* (Pieroni *et al.*, 2002).

Stemmadenia bella Miers Family: Apocynaceae

Ruiz-Terán *et al.* (2008) evaluated the antioxidant activity of aerial parts of *Stemmadenia bella* using DPPH and β-carotene-linoleic acid bleaching assay. The extract displayed antioxidant activity comparable to the commercial antioxidant BHA. For the methanolic extract analysed, a clear relation between the total phenolic content of the extracts and their antioxidant activity was found.

Stemona species Family: Stemonaceae

From the roots of various *Stemona* species four new dehydrotocopherols (chromenols) were isolated and their structures and stereochemistry elucidated by Brem *et al.* (2004). Various C-methylations of the aromatic ring reflect differences in methyltransferase activities and agreed with the current species delimitations showing an exclusive accumulation of dehydro-δ-tocopherol for the *Stemona tuberosa* group, whereas different provenances of *Stemona curtisii* were characterized by dehydro-γ-tocopherol accompanied by small amounts of dehydro-α-tocopherol. From *Stemona collinsae* all four tocopherols were isolated with a clear preponderance of dehydro-δ-tocopherol accompanied by smaller amounts of the rare dehydro-β-tocopherol. *Stemona burkillii* and a group of unidentified species showed a weak accumulation trend towards dehydro-α-tocopherol, whereas *Stemona cochinchinensis* and especially *Stemona kerrii* clearly differed by a preponderance of chromanol derivatives. In *Stemona* cf. *pierrei* no tocopherols could be detected at all. Based on TLC tests and microplate assays with the free radical 2,2-diphenyl-1-picrylhydrazyl (DPPH) the antioxidant capacities of all chromenol derivatives were comparable with that of α-tocopherol showing no significant differences among each other, except for a more rapid kinetic behaviour of the 5,7,8-methylated dehydro-α-tocopherol. The species studied by Brem *et al.* (2004) include *Stemona tuberosa, Stemona collinsae, Stemona curtisii, Stemona cochinchinensis, Stemona kerrii, Stemona burkilli* and *Stemona pierrei*.

Stemona burkillii Prain Family: Stemonaceae

From the roots of various *Stemona* species four new dehydrotocopherols (chromenols) were isolated and their structures and stereochemistry elucidated by spectroscopic methods. *Stemona burkillii* and a group of unidentified species showed a weak accumulation trend towards dehydro-α-tocopherol. Based on TLC tests and microplate assays with the free radical DPPH the antioxidant capacities of all chromenol derivatives were comparable with that of α-tocopherol (Brem *et al.*, 2004).

Stemona cochinchinensis Gagnep. Family: Stemonaceae

From the roots of various *Stemona* species four new dehydrotocopherols (chromenols) were isolated and their structures and stereochemistry elucidated by spectroscopic methods. *S. cochinchinensis* clearly differed by a preponderance of chromanol derivatives. Based on TLC tests and microplate assays with the free radical DPPH the antioxidant capacities of all chromenol derivatives were comparable with that of α-tocopherol (Brem *et al.*, 2004).

Stemona collinasae Family: Stemonaceae

From the roots of various *Stemona* species four new dehydrotocopherols (chromenols) were isolated and their structures and stereochemistry elucidated by spectroscopic methods. The double bond between C-3 and C-4 proved to be a typical chemical character of the genus found in most of the species. From *Stemona collinasae* all four tocopherols were isolated with a clear preponderance of dehydro-delta-tocopherol accompanied by smaller amounts of the rare dehydro-α-tocopherol. Based on TLC tests and microplate assays with the free radical DPPH the antioxidant capacities of all chromenol derivatives were comparable with that of α-tocopherol (Brem *et al.*, 2004).

Stemona kerrii Craib. Family: Stemonaceae

From the roots of various *Stemona* species four new dehydrotocopherols (chromenols) were isolated and their structures and stereochemistry elucidated by spectroscopic methods. *Stemona kerrii* clearly differed by a preponderance of chromanol derivatives. Based on TLC tests and microplate assays with the free radical DPPH the antioxidant capacities of all chromenol derivatives were comparable with that of α-tocopherol (Brem *et al.*, 2004).

Stemona sessilifolia (Miq.) Franch. Family: Stemonaceae

Cai *et al.* (2004) evaluated the antioxidant activity and phenolic compounds in traditional Chinese medicinal plants associated with anticancer. The improved ABTS.+ method was used to systematically assess the total antioxidant capacity (Trolox equivalent antioxidant capacity, TEAC) of the medicinal extracts. The TEAC values and total phenolic content for methanolic extract of root of *Stemona sessiliflora* were 139.4 µmol Trolox equivalent/100g dry weight (DW), and 1.26g of gallic acid equivalent/100g DW respectively. A positive significant linear relationship between antioxidant activity and total phenolic content showed that phenolic compounds were the dominant antioxidant components in the tested herbs.

Antioxidant capacities of 56 selected Chinese medicinal plants, including *Stemona sessilifolia* were evaluated by Song *et al.* (2010) using the TEAC and FRAP assays, and their total phenolic content was measured by the Folin-Ciocalteu method. *Stemona sessilifolia* showed low antioxidant activity with TEAC value of 12.21µmol Trolox/g, FRAP value of 22.87µmol Fe^{2+}/g and phenolic content was 5.55mg GAE/g.

Stephania erecta L. Family: Menispermaceae

Patnibul *et al.* (2008) screened 18 vegetables for antioxidant activity using silica gel Thin Layer chromatography followed by spraying with DPPH. *Stephania erecta* exhibited strong antioxidant activity.

Stephania hernandifolia (Willd.) Walp. Family: Menispermaceae

Sharma *et al.* (2010) investigated the antidiabetic and antioxidant potential of the powdered corm of *Stephania hernandifolia*. This was tested in normal and Streptozotocin (STZ)-induced diabetic rats, using oral administration of ethanol and an aqueous extract (400 mg/kg body weight) of *Stephania hernandifolia* corm. The experimental data revealed that both extracts has significant antihyperglycemic and antioxidant activity in Streptozotocin-induced rats compared to the standard drug. The antioxidant activity *in vitro* was measured by means of the DPPH and Superoxide-free radical scavenging assay. The extracts of ethanol and aqueous strongly scavenged DPPH radicals, with IC_{50} being 265.33 and 217.90 µg/ml, respectively. The extracts of ethanol and aqueous moderately scavenged the superoxide radical with IC_{50} values of 526.87 and 440.89 µg/ml.

Stephania japonica (Thunb.) Miers Family: Menispermaceae

Rahman *et al.* (2011) investigated the antioxidant and analgesic potentiality of crude methanolic extract of *Stephania japonica* leaf. The extract showed antioxidant activity in DPPH radical scavenging activity, nitric ocide scavenging activity and reducing power assays. In both DPPH radical and NO scavenging assay, the extract exhibited moderate antioxidant activity and the IC_{50} values in DPPH radical scavenging and NO scavenging assays were found to be 105.55 and 129.12 µg/ml, respectively while the IC_{50} values of ascorbic acid were 12.30 and 18.64 µg/ml, respectively. Reducing power activity of the extract increased in a dose dependent manner.

Stephania rotunda Lour. Family: Menispermaceae

Gülçin *et al.* (2010) determined the antioxidant activity of cepharanthine and fangchinoline from *Stephania rotunda* by performing different *in vitro* antioxidant assays, including DPPH free radical scavenging, ABTS radical scavenging, N,N-dimethyl-p-phenylenediamine dihydrochloride (DMPD) radical scavenging, superoxide anion (O_2^{*-}) radical scavenging, hydrogen peroxide scavenging, total antioxidant activity, reducing power, and ferrous ion (Fe^{2+}) chelating activities. Cepharanthine and fangchinoline showed 94.6 and 93.3 per cent inhibition on lipid peroxidation of linoleic acid emulsion at 30 µg/ml concentration, respectively. On the other hand, butylated hydroxyanisole (BHA), butylated hydroxytoluene (BHT), α-tocopherol, and trolox indicated inhibitions of 83.3, 92.2, 72.4, and 81.3 per cent on peroxidation of linoleic acid emulsion at the same concentration (30 µg/ml), respectively.

Stephania venosa (Blume) Spreng. Family: Menispermaceae

Ethanol and water extracts of the *Stephania venosa* were investigated by Leewanich *et al.* (2011) for their cytotoxic and antioxidative activities. Both extracts were shown to be antioxidants. The ethanol extract exhibited the strongest superoxide and radical scavenging activities.

Sterculia diversifolia G. Don Family: Sterculiaceae

The chloroform and methanolic extracts of 124 Egyptian plant species belonging to 56 families were investigated by Moussa *et al.* (2011) and compared their antioxidant activity by DPPH scavenging assay. Among the 124 plant species tested 18 exhibited extremely high antiradical activity (>80 per cent inhibition). The IC_{50} value of DPPH radical scavenging of *Bombax malabaricum* was 27.32µg/ml, while total phenolic and flavonoid contents were 111.06 mg Tannic acid equivalent/g extract and 171.66 mg Rutin equivalent/g extract, respectively. Correlation coefficient between DPPH radical scavenging activity and the total phenolic and flavonoid contents suggest that phenolics and flavonoids in the extracts were partly responsible for the antiradical activities.

Sterculia lychnophora Hance Family: Sterculiaceae

Lam *et al.* (2011) reported that *Sterculia lynchnophera* possessed high free radical scavenging activity with IC_{50} = 11.02µM.

Sterculia scaphigera Wall. Family: Sterculiaceae

Antioxidant capacities of 56 selected Chinese medicinal plants, including *Sterculia scaphigera* were evaluated by Song *et al.* (2010) using the TEAC and FRAP assays, and their total phenolic content was measured by the Folin-Ciocalteu method. Bark of root of *Sterculia scaphigera* showed low antioxidant activity with TEAC value of 52.26µmol Trolox/g, FRAP value of 57.28µmol Fe^{2+}/g and phenolic content was 5.49mg GAE/g.

Sterculia striata St. Hil. et Naudin Family: Sterculiaceae

Costa *et al.* (2010) investigated the chemical constituents, total phenolics and antioxidant activity of *Sterculia striata*. The phytochemical investigation of the stem bark by chromatographic methods led to the isolation of sitosterol, stigmasterol and sitosterol-3-O-β-D-glucopyranoside, besides pentacyclic triterpenoids, lupeol, 3-β-O-acyl-lupeol, lupenone and betulinic acid (Costa *et al.*, 2010). For determining of the phenolic content of the ethanolic extract of *Sterculia striata* they used the Folin Ciocalteu reagent, and for the evaluation of antioxidant activity, they utilized the DPPH free radical.

Sterculia tragacantha Lindl. Family: Sterculiaceae

The anti-nociceptive, anti-inflammatory and anti-oxidant activities of *Sterculia tragacantha* were evaluated by Udegbunam *et al.* (2011). The extract showed significant anti-oxidant activity. Phytochemical screening showed that the extract contained alkaloids, flavonoids, tannins, glycosides and saponins.

Stereospermum colais Mabb. Family: Bignoniaceae

Vijaya Bharathi *et al.* (2010) evaluated the antioxidant and wound healing properties of different extracts of *Stereospermum colais* leaf. Chloroform extract showed maximum antioxidant activity with an IC_{50} value of 35µg/ml.

Stereospermum kunthianum Cham. Family: Bignoniaceae

Compaore *et al.* (2011) evaluated the antioxidant, diuretic activities and polyphenol content of *Stereospermum kunthianum*. In biological studies, antioxidant activities were investigated with water, methanol and aqueous acetone extracts. Furthermore, the xanthine oxidase inhibitory activity and the diuretic activity of an aqueous acetone extract were evaluated. The DPPH, FRAP and ABTS methods have shown that the aqueous acetone extract presents the best antioxidant activities. This aqueous acetone extract was further proven to have interesting xanthine oxidase inhibitory activity, but only a weak diuretic activity. This aqueous acetone extract also possessed the highest phenolic and flavonoid contents. HPLC-MS analysis allowed identifying and quantifying, rutin, isoquercitrin, quercetin, hyperoside, quercitrin and luteolin and the glycosides of ferulic, sinapic *p*-coumaric acids and kaempferol, apigenin in aqueous-acetone extract.

Stereospermum personatum (Hassk.) D.Chatterjee Family: Bignoniaceae

Bio-assay guided fractionation of different extracts of both stem and stem bark of *Stereospermum personatum* led to the isolation of free radical scavenging and xanthine oxidase inhibitory molecules along with three new anthraquinones, sterequinones F-H, a new naphthaquinone, two new phenyl esters together with known compounds. The antioxidant and xanthine oxidase inhibitory potentials of the isolated compounds were reported by Sampathkumar *et al.* (2005).

Stereospermum suaveolens DC. Family: Bignoniaceae

Chandrasekhar *et al.* (2010) evaluated the hepatoprotective activity of *Stereospermum suaveolens* against carbon tetrachloride (CCl_4)-induced liver damage in albino rats. The degree of protection in this activity has been measured by using biochemical parameters such as serum glutamate oxaloacetate transaminase (SGOT), serum glutamate pyruvate transaminase (SGPT), alkaline phosphatase (ALP), total bilirubin, LDL-cholesterol and SOD, CAT, GSH, total thiols, NO, and lipid peroxidation in liver tissue homogenate. The results suggest that the methanol stem bark extract of *Stereospermum suaveolens* at the doses 125, 250, and 500 mg/kg and reference standard Liv-52 treated group produced significant hepatoprotection against CCl_4-induced liver damage by decreasing the activities of serum enzymes, bilirubin and lipid peroxidation. The extract significantly increased levels of SOD, CAT, GSH and total thiols, as compared to control group.

The ethanol extract of *Stereospermum suaveolens* (EESS) bark was evaluated by Balasubramanian and Chatterjee (2010) for its hepatoprotective and antioxidant activities against carbon tetrachloride (CCl_4)-induced liver damage, in wistar albino rats. The ethanol extract of *Sterespermum suaveolens* (EESS) bark (200 and 400 mg/kg body weight, p.o.) was administered to the experimental rats for 14 days. The hepatoprotective activity was assessed using various serum biochemical parameters as glutamate oxaloacetate transaminase (SGOT), glutamate pyruvate transaminase (SGPT), alkaline phosphatase (ALP), bilirubin, and total proteins. Malondialdehyde (MDA) level as well as the activities of superoxide dismutase (SOD), reduced glutathione (GSH) and catalase (CAT) were determined to explain the possible mechanism of activity. The substantially elevated levels of serum GOT, GPT, ALP and total bilirubin, due to CCl_4 treatment, were restored towards near normal by EESS, in a dose dependent manner. EESS also increased the serum total proteins of CCl_4-intoxicated rats. Reduced enzymatic and non-enzymatic antioxidant levels and elevated lipid peroxide levels were restored towards near normal, by administration of EESS.

Stevia rebaudiana (Bert.) Bertoni Family: Asteraceae (Compositae)

Stevia rebaudiana is used in many parts of the world as a non-caloric sweetener. Stevioside, along with steviobioside, isosteviol and steviol caused inhibition of oxidative phosphorylation in rat liver mitochondria (Kelmer *et al.*, 1985). The effect of several natural products extracted from the leaves of *S. rebaudiana* on rat liver mitochondria was investigated. They inhibited oxidative phosphorylation including on ATPase, NADH-oxidase, succinic-oxidase, succinic dehydrogenase and glutamate dehydrogenase activity. The ADP/O ratio decreased and substrate respiration increased at low concentrations and at higher concentrations there was complete inhibition. It was concluded that, in addition to the inhibitory effects, *S. rebaudiana* natural products may also act as uncouplers of oxidative phosphorylation (Bracht *et al.*, 1985).

Four steviol glycosides, Stevioside, rebaudiosides A and C, and dulcoside A, showed strong inhibitory activity against 12-O-tetradecanoylphorbol-13-acetate (TPA)-induced inflammation in mice. The 50 per cent inhibitory dose of these compounds markedly inhibited the promoting effect of TPA (1µg/mouse) on skin tumor formation initiated with 7, 12-dimethylbenz{a}anthracene (Yasukawa *et al.*, 2002).

Leaf extract of *S. rebaudiana* promotes effects on certain physiological systems such as the cardiovascular and renal systems and influences hypertension and hyperglycaemia. Since these activities may be correlated with the presence of antioxidant compounds, leaf and callus extracts of *S. rebaudiana* were evaluated for their total phenols, flavonoids content and total antioxidant capacity. Total phenols and flavonoids were analyzed according to the Folin-ciocalteu methods and total antioxidant activity of water and methanolic extracts of *Stevia* leaves and callus was assessed by ferric reducing/antioxidant power (FRAP) assay as well as DPPH assay. The total phenolic compounds were found to be 25.18 mg/g for *Stevia* leaves and 35.86 mg/g for callus on dry weight basis. The flavonoids content was found to be 21.73 and 31.99 mg/g in the leaf and callus, respectively. The total antioxidant activity was expressed as mg equivalent of gallic acid, ascorbic acid, BHA and trolox per gram on dry weight basis. Total antioxidant activity was reported in the range of 9.66 to 38.24 mg and 11.03 to 36.40 mg equivalent to different standards in water and methanolic extracts of *Stevia* leaves, respectively. In case of *Stevia* callus, it was found to be 9.44 to 37.36 for water extract and 10.14 to 34.37 mg equivalent to standards for methanolic extract. The concentrations required for 50 per cent inhibition (IC_{50}) of DPPH radicals were 11.04, 41.04 and 57.14 µg/ml for gallic acid, trolox and butylated hydroxyanisole (BHA), respectively. The per cent inhibition of DPPH radical of various extracts of *Stevia* leaves and callus ranged from 33.17 to 56.82 per cent. The highest per cent inhibition was observed in methanolic extract of callus (Tadhani *et al.*, 2007).

At 0.1 mg/ml, the ethyl acetate extract (EAE) of the crude 85 per cent methanolic extract (CAE) of *Stevia rebaudiana* leaves exhibited preventive activity against DNA strand scission by ˙OH generated in Fenton's reaction on pBluescript II SK (–) DNA. Its efficacy was better than that of quercetin. The radical scavenging capacity of CAE was evaluated by the DPPH test (IC_{50} = 47.66 µg/ml). EAE was derived from CAE scavenged DPPH (IC_{50} = 9.26 µg/ml), ABTS⁺ (IC_{50} = 3.04 µg/ml) and ˙OH (IC_{50} = 3.08 µg/ml). Additionally, inhibition of lipid peroxidation induced with 25 mM $FeSO_4$ on rat liver homogenate as a lipid source was noted with CAE (IC_{50} = 2.1 mg/ml). The total polyphenols and total flavonoids of EAE were 0.86 mg gallic acid equivalents/ mg and 0.83 mg of quercetin equivalents/mg, respectively. Flavonoids, isolated from EAE, were characterized as quercetin-3-O-arabinoside, quercitrin, apigenin, apigenin-4-O-glucoside, luteolin, and kaempferol-3-O-rhamnoside by LC-MS and NMR analysis(Ghanta *et al.*, 2007).

Phansawan and Poungbangpho (2007) reported antioxidant capacities of *Stevia rebaudiana* by ABTS method. Ethanol was the best solvent for extraction to give the highest antioxidant capacity, followed by acetone, methanol, distilled water and acetic acid.

Shukla *et al.* (2009) assessed the *in vitro* potential of ethanolic leaf extract of *Stevia rebaudiana* as a natural antioxidant. The DPPH activity of the extract (20, 40, 50, 100 and 200 µg/ml) was increased in a dose dependent manner, which was found in the range of 36.93-68.76 per cent as compared to ascorbic acid 64.26-82.58 per cent. The IC(50) values of ethanolic extract and ascorbic acid in DPPH radical scavenging assay were obtained to be 93.46 and 26.75 µg/ml, respectively. The ethanolic extract was also found to scavenge the superoxide generated by EDTA/NBT system.

Measurement of total phenolic content of the ethanolic extract of *S. rebaudiana* was achieved using Folin-Ciocalteau reagent containing 61.50 mg/g of phenolic content, which was found significantly higher when compared to reference standard gallic acid. The ethanolic extract also inhibited the hydroxyl radical, nitric oxide, superoxide anions with IC_{50} values of 93.46, 132.05 and 81.08 µg/ml, respectively. However, the IC_{50} values for the standard ascorbic acid were noted to be 26.75, 66.01 and 71.41 µg/ml respectively.

Kim *et al.* (2011) investigated the antioxidant activity and the bioactive compounds found in water extracts taken from *Stevia rebaudiana* (SR) leaves and calli. Analysis of vitamins in the leaves showed that folic acid (52.18 mg/100 g) was a major component, followed by vitamin C. The total phenolic and flavonoid contents were found to be 130.76 µg catechin and 15.64 µg quercetin for leaves and 43.99 µg catechin and 1.57 µg quercetin for callus at mg of water extracts, respectively. Pyrogallol was the major material among the phenolic compounds in both leaf and callus extracts. Furthermore, their results showed that the leaf extracts contained higher amounts of free radicals, hydroxyl radicals and superoxide anion radical scavenging activities than those of the callus extracts.

DPPH-scavenging antioxidant potential of *in vitro* regenerated tissues of *Stevia rebaudiana* was carried out by Ahmad *et al.* (2011). Their study revealed that significantly higher activity (87.7 per cent) was observed in callus of *Stevia rebaudiana* followed by regenerated shoots (86.3 per cent) and regenerated plantlets (83.5 per cent).

Streblus asper Lour. Family: Moraceae

The methanolic extract from leaves of *Streblus asper* was screened by Kanti *et al.* (2009) for *in vitro* antioxidant activity, and for anti-diabetic activity. The *in vitro* antioxidant activity was evaluated using DPPH and H_2O_2 for their radical scavenging property. The data were expressed as IC_{50} and compared with that of ascorbic acid and tocoferol that served as reference standard. The leaf extract showed an IC_{50} of 1.01mg/ml and 700 µg/ml as compared to the standards ascorbic acid and tocoferol that showed an IC_{50} of 4nm and 215 µg/ml respectively.

Narayanaswamy and Balakrishnan (2011) evaluated the antioxidant activity of 13 medicinal plants including leaf of *Streblus asper*. The aqueous extract showed 25 per cent inhibition of DPPH radical while ethanolic extract showed 38 per cent inhibition of DPPH radical.

The leaf extract of *Streblus asper* (SAPE) was evaluated by Choudhury *et al.* (2011) for its hypoglycemic property as well *in-vivo* antioxidant activity in diabetic rats at cellular level. The Petroleum ether extract of SAPE exhibited anti diabetic property as well as increased the levels of enzymatic and non enzymatic antioxidant entities along with reduced MDA levels.

Striga hermonthica (Del.) Benth. Family: Scrophulariaceae

The aqueous acetone extract of *Striga hermonthica* was further separated into aqueous and ethyl acetate fractions and assayed for their *in vitro* antioxidant properties using DPPH. The crude extract exhibited a weak antioxidant activity (IC_{50} of 95.27

µg/ml). Lutolin was isolated and identified as the main DPPH radical scavenger of the ethyl acetate fraction, exhibiting an IC_{50} value of 6.80µg/ml (Kiendrebeogo *et al.*, 2005).

Striga orobanchoides Benth. Family: Scrophulariaceae

The ethanolic extract of *S. orobanchoides* possesses strong antioxidant properties as evidenced by the significant increase in the level of catalase, SOD and ascorbic acid and decrease in the levels of thio-barbituric acid reactive substances (TBARS). Badami *et al.* (2003) confirmed antioxidant property of this plant by DPPH and nitric oxide radical inhibition assays. A large number of flavonoids including apigenin and luteolin are known to possess strong antioxidant properties (Raj and Shalini, 1999; Badami *et al.*, 2003).

Strobilanthus crispus (L.) Bremek. (Syn: *Saricocalyx crispus*). Family: Acanthaceae

Ismail *et al.* (2000) investigated the components present in and the total antioxidant activity of leaves of *Strobilanthes crispus*. Proximate analyses and total antioxidant activity using ferric thiocyanate and thiobarbituric acid methods were employed. High content of water-soluble vitamins (C, B(1), and B(2)) contributed to the high antioxidant activity of the leaves. The leaves also contained a moderate amount of other proximate composition as well as other compounds such as catechins, alkaloids, caffeine, and tannin, contributing further to the total antioxidant activity. Catechins of *Strobilanthes crispus* leaves showed highest antioxidant activity when compared to Yerbamate and vitamin E.

Abu Bakar *et al.* (2004) reported antioxidant activity in *Strobilanthes crispus*.

Strophanthus hispidus DC. Family: Apocynaceae

Ayoola *et al.* (2008) carried out phytochemical and antioxidant screening of *Strophanthus hispidus*. The extract tested positive for tannins, flavonoids and glycosides. DPPH free radical scavenging assay was followed and the extracts showed good antioxidant activity.

Struchium sparganophora (L.) O.Ktze. Family: Asteraceae (Compositae)

Oboh and Akindahunsi (2004) investigated the effect of preservation of green leafy vegetables of Nigeria on the antioxidant phytoconstituents (Vitamin C and Total phenol) and activity (reducing property and free radical scavenging ability) including *Struchium sparganophora*. The result of the study revealed that sun-drying of green leafy vegetables cause a significant decrease in the vitamin C content (16.67-64.68 per cent loss). Conversely it leads to a significant increase in the total phenol content (6.45-223.08 per cent grain) reducing property (16.00-362.50 per cent grain) and free radical scavenging ability (126.00-5757.00 per cent gain) of green leafy vegetables.

The ethanolic extract of the leaf of *Struchium sparganophora* contains alkaloids, tannin, saponins, phlobatannin, anthraquinone and glycosides (Oboh, 2006). Further more the extract had high antioxidant activity as typified by its high total phenol content (5.4g/100g), reducing power (2.50 OD (700)), and free radical scavenging ability (65.2 per cent).

Odukoya *et al.* (2007) evaluated the antioxidant activity and total phenolic contents of Nigerian green leafy vegetables including *Struchium sparganophora*.

Strychnos henningsii Gilg. Family: Loganiaceae

The antioxidant and free radical scavenging activity of aqueous extract of bark of *Strychnos henningsii* was investigated by Oyedemi *et al.* (2010) both *in-vivo* and – *vitro* using spectroscopic method against DPPH, superoxide, hydrogen peroxide, nitric oxide, ABTS and the ferric reducing agent. Free radical scavenging activity of the plant extract against H_2O_2, ABTS and NO was concentration dependent with IC_{50} value of 0.023, 0.089 and 0.49 mg/ml respectively. However, *Strychnos henningsii* exhibited lower inhibitory activity against DPPH with IC_{50} value of 0.739 mg/ml. The reducing power of the extract was found to be concentration dependent. The administration of the aqueous extract at 250, 500 and 1000mg/kg body weight to male wistar rats significantly increased the percentage inhibition of reduced glutathione (GSH), superoxide dismutase (SOD) and catalase (CT). Whereas, lipid peroxidation level in hepatotoxic rats decreased significantly at the dose of 500 and 1000mg/kg body weight at the end of 7 days. Th extract yielded high phenol content (48 mg/g tannic acid equivalent) followed by proanthocyanidin (8.7mg/g catechin equivalent) respectively. A positive correlation was observed between these polyphenols and the free radical scavenging activities.

Strychnos minor Dennst. Family: Loganiaceae

Patnibul *et al.* (2008) screened 18 vegetables for antioxidant activity using silica gel Thin Layer chromatography followed by spraying with DPPH. *Strychnos minor* exhibited strong antioxidant activity.

Strychnos potatorum L. Family: Loganiaceae

Strychnos potatorum seed powder (SPP) and aqueous extract at the doses of 100 and 200 mg/kg, p.o. offered significant hepatoprotective action (Sanmugapriya and Venkataraman, 2006). Reduced enzymatic and nonenzymatic antioxidant levels and elevated lipid peroxide levels were restored to normal by administration of SPP and SPE.

The methanolic extract of seeds of *Strychnoss potatorum* were found to show moderate antioxidant potential (Dhobal *et al.*, 2008).

Strychnos spinosa Lam. Family: Loganiaceae

The phenolic content and antioxidant capacities of *Parinari curatelifolia*, *Strychnos spinosa* and *Adansonia digitata* were determined and compared to orange juice and baobab nectar, a commercial beverage. Methanolic extracts were investigated for their ability to scavenge free radicals by the 1, 1-diphenyl-2 picrylhydrazyl radical and superoxide radical scavenging assays whilst the β-Carotene Linoleic Acid Model System and inhibition of phosholipid peroxidation were used as model systems. The reducing power assay was used to determine the reducing potential of the extracts. Results showed that the beverages in this investigation were capable of acting as antioxidant sources as they displayed significant radical scavenging properties. The total phenolic, ascorbic acid, proantocyanidin and flavonoid contents ranged between 12 and 58 mg GAE/100 ml, 0.00 to 51.26 mg/100 ml, 0.35–1.071 per cent and 18.3–124 mg/100 ml, respectively. There was a positive correlation between antioxidant activities and phenolic compounds content but there was no clear relationship between proanthocyanidin content and antioxidant activity.

Strychnos vanprukii Craib Family: Loganiaceae

From the stem of *Strychnos vanprukii*, two new lignan glucosides, vanprukoside and strychnoside were isolated together with the known lignan glucoside, (+)-lyoniresinol-3α-*O*-β-glucopyranoside (Thongphasuk *et al.*, 2004). All three compounds exhibited antioxidant activity.

Stryphnodendron obovatum Benth. Family: Fabaceae (Leguminosae)

The antioxidant activity of stem bark extracts from *Stryphnodendron obovatum*, including fractions and isolated compounds, was evaluated by Sanches *et al.* (2005) by DPPH in thin-layer chromatography. All the fractions and isolated compounds showed antioxidant activity. Chemical isolation of the ethyl-acetate fraction resulted in the identification of three compounds: epigallocatechin, gallocatechin and epigallocatechin-(4β→8)-gallocatechin.

Stryphnodendron polyphyllum Mart. Family: Fabaceae (Leguminosae)

Stem bark of *Stryphnodendron obovatum* and *Stryphnodendron polyphyllum* was investigated by Lopes *et al.* (2005) for wound healing, antibacterial and antioxidant activity. The antioxidant activity through reduction of the DPPH radical in TLC, confirmed the anti-radical properties of these extracts in both species. CE and EAF of both species showed a radical scavenging activity (RSA) and protected DPPH from discolouration, already at 0.032 microg/ml. The extract from *Stryphnodendron polyphyllum* were more effective than those *Stryphnodendron obovatum*, although the former had a lower tannin content.

Styrax formosana Matsum. Family: Styracaceae

Hou *et al.* (2003) evaluated the antioxidant activity of 70 per cent aqueous acetone extract of *Styrax formosana* by various assays including DPPH, hydroxyl radicals and reducing power assay. The extract exhibited moderate activity against DPPH radicals with IC_{50} value of 31.5 µg/ml and inhibited formation of OH generated in the Fenton reaction system with IC_{50} value of 0.3 µg/ml). The amount of phenolics was 2.7 mg of GAE/g.

Styrax japonica Sieb. et Zucc. Family: Styracaceae

Min *et al.* (2004) reported new furofuran and butyrolactone lignans with antioxidant activity from the stem bark of *Styrax japonica*. Styraxlignolide C, Styraxlignolide D, Styraxlignolide E, and (-)-pinoresinol glucoside exhibited weak free radical-scavenging activity in the DPPH assay with IC_{50} values of 380, 278, 194 and 260 micro M, respectively.

Succisa pratensis Moench. Family: Dipsacaceae

English Name: Devil's bit scabious.

The antioxidative activity of phenolic extracts from leaf and bark of *Succisaa pratensis* was examined by autoxidation of methyl linoleate by Kahkonen *et al.* (1999). The phenol contents and antioxidant activity was low.

Sutherlandia frutescens R.Br. subsp. *microphylla* Family: Fabaceae (Leguminosae)

Fernandes *et al.* (2004) analyzed the antioxidant potential of *Sutherlandia frutescens* subsp. *microphylla* – one of the best known multipurpose medicinal plant used in the treatment of cancer, viral diseases and inflammatory conditions, etc in South Africa. Their result indicated that hot water extract of *S. frutescens* possess superoxide as well as hydrogen peroxide scavenging activities at low concentrations (10µg/ml).

Katerere and Eloff (2005) evaluated the antioxidant activity of leaves of *Sutherlandia frutescens* using DPPH free-radical scavenging assay.

Swartzia langsdorffii Desf. Family: Fabaceae (Leguminosae)

Santos *et al.* (2009) evaluated the antioxidant activity of extracts from nine plant species belonging to the Brazilian flora acting on the free radicals DPPH and TEMPOL. Results showed that the extracts of *Swartzia langsdorffii* display moderate antioxidant activities with DPPH scavenging IC_{50} value of 1.532 mg/ml while Hydroxyl scavenging IC_{50} value was 5.408 mg/ml.

Swertia chirata Buch.-Ham. Family: Gentianaceae

Scartezzini and Speroni (2000) are of the opinion that this herb contains antioxidant principles, that can explain and justify their use in traditional medicine in the past as well as the present.

Chen *et al.* (2011) investigated the *in vitro* and *in vivo* antioxidant effects of *Swertia chirayita* extracts (SCE). Antioxidant ability of *Swertia chirayita* was investigated by employing several established *in vitro* methods. *In vivo* antioxidant activity was tested against CCl(4)-induced toxicity in mice. The levels and activities of malondialdehyde (MDA) and antioxidant enzymes, including superoxide dismutase (SOD), catalase (CAT), and glutathione (GSH), were then assayed using standard procedures. SCE exhibited strong antioxidant ability *in vitro*. The liver and kidney of CCl(4)-intoxicated animals exhibited a significant decrease in SOD, CAT, and GSH levels. Additionally, these organs exhibited a significant increase in MDA level. CCl(4) did not exhibit toxicity on mice treated with SCE and Vitamin E. The effects of *Swertia chirayita* (three dosages) were comparable to those of Vitamin E, except in MDA level in the liver and GSH level in the kidney.

Swertia corymbosa (Griseb.) Wight ex Clarke Family: Gentianaceae

Selvameena *et al.* (2010) investigated the antioxidant potential of methanolic extract of *Swertia corymbosa* by analyzing the enzymes of lipid peroxidation like catalase, superoxide dismutase and glutathione peroxidase from the samples of alloxan induced diabetic rats. It was found tat the extract exhibited significant antioxidant potentials.

Swertia decussata Nimmo ex C.B. Clarke Family: Gentianaceae

The lipid peroxidation inhibitory activities of four hydroxyxanthones, isolated from the whole plant *Swertia decussata,* were evaluated by Patro *et al.* (2005). The most promising antioxidant among the xanthones, 1,7,8-trihydroxy-3-methoxyxanthone (swertianine, 4) was also tested for its scavenging potential against DPPH and superoxide radicals. The data clearly revealed good antioxidant activity of the xanthones, especially 4 which also showed strong protection against γ-ray induced pBR322 DNA damage. A comparison of the radioprotecting activities of the monomethylated tetrahydroxyxanthone 4 with that of its congener, 1,3,7-trihydroxy-8-methoxyxanthone (5) revealed that the radioprotecting activity was not affected by the position of methylation.

Swertia japonica Makino Family: Gentianaceae

Substances with antioxidative properties were obtained from an ether extract of *Swertia japonica.* Six active components of the extract were isolated and identified as methylbellidifolin, methylswertianin, swertianin, bellidifolin, norswertianin and desmethylbellidifolin. These six xanthone derivatives were shown to possess different antioxidant activities by chemiluminescent assay. The antioxidative activities of bellidifolin, norswertianin and desmethylbellidifolin were higher than those of butylated hydroxytoluene (BHT) and α-tocopherol. On autoxidation of methyl linoleate, bellidifolin had activity similar to that of BHT (Ashida *et al.,* 1994).

Swertia speciosa Wall. Family: Gentianaceae

A new compound, 1-O-primeverosyl-6-hydroxy-3,5-dimethoxyxanthone together with eight known compounds have been identified on the basis of assay-guided isolation of different extractives of the rhizomes of *Swertia speciosa* by Rawat *et al.* (2004). *In vitro* acetone and methanol extracts of the rhizomes showed profound antioxidant activities in linoleic acid antioxidation, xanthine/xanthine oxidase superoxide and DPPH assays. Ursolic acid and 1,5,8-trihydroxyxanthone also showed moderate to high antioxidant activity (Rawat *et al.,* 2004).

From the rhizomes of *Swertia speciosa* a new constituent, 6-hydroxy-3,5-dimethoxy-1-[(6-O-β-D-xylopyranosyl-β-D-glucopyranosyl)oxy]-9H-xanthen-9-one (8), was isolated by Rana and Rawat (2005), together with nine known compounds. Both the acetone and MeOH extracts, as well as the known constituents ursolic acid (5) and 1,5,8-trihydroxy-3-methoxy-9H-xanthen-9-one (6) exhibited potent antioxidant activities in different biological assays.

Swietenia humilis Zucc. Family: Meliaceae

Ruiz-Terán *et al.* (2008) evaluated the antioxidant activity of aerial parts of *Swietenia humilis* using DPPH and β-carotene-linoleic acid bleaching assay. The extract displayed antioxidant activity comparable to the commercial antioxidant BHA. For the methanolic extract analysed, a clear relation between the total phenolic content of the extracts and their antioxidant activity was found.

Swietenia macrophylla King Family: Meliaceae

Chemical constituents of the bark of *Swietenia macrophylla* was investigated by Falah *et al.* (2008). A new phenylpropanoid-substituted catechin, namely, swietemacrophyllanin [(2R*,3S*,7"R*)-catechin-8,7"-7,2"-epoxy-(methyl 4",5"-dihydroxyphenylpropanoate)] (1) was isolated from the bark of *S. macrophylla* together with two known compounds, catechin (2) and epicatechin (3). The structure of 1 was elucidated by spectroscopic data and by comparison of the NMR data with those of catiguanins A and B, phenylpropanoid-substituted epicatechins. The DPPH free radical scavenging activity of the isolated compounds indicated that all of the three compounds have strong activity compared with trolox as a reference. Swietemacrophyllanin (1) had the strongest activity with a 50 per cent inhibitory concentration (IC_{50}) value of 56 µg/ml.

Tan *et al*. (2009) evaluated the antioxidant activity of methanol, dichloromethane and n-Hexane extracts of *Swietenia macrophylla* leaves using DPPH radical scavenging assay in which all the leaf extracts showed remarkable activities. At the concentration of 320µg/ml, the methanol extract, the dichloromethane extract and the n-hexane extract showed 90.61 per cent, 84.17 per cent and 89.46 per cent scavenging activities respectively. The extracts were found to contain a sizable amount of total phenolic, total tannin and total flavonoid contents.

Swietenia mahagoni (L.) Jacq. Family: Meliaceae

Sahgal *et al*. (2009) examined the *in vitro* antioxidant activities of the methanol extract of *Swietenia mahagoni* seeds (SMCM) using DPPH, xanthine oxidase inhibition (XOI), hydrogen peroxide scavenging activity (HPSA) and FRAP assays. The extract exhibited antioxidant activity of 23.29 per cent with an IC^{50} value of 2.3 mg/ml in the DPPH scavenging method, 47.2 per cent in the XOI assay, 49.5 per cent by the HPSA method, and 0.728 mmol/Fe(II)g in the FRAP method at the concentration tested. The amount of total phenolics and flavonoid contents was 70.83 mg gallic acid equivalent and 2.5 mg of catechin equivalent per gram of dry extract, respectively.

Panda *et al*. (2010) evaluated the antidiabetic and antioxidant potential of the methanol extract of *Swietenia mahagoni* (MESM) bark. Antioxidant effects were assayed in diabetic rats by estimating thiobarbituric acid reactive substances (TBARS), glutathione (GSH), and catalase (CAT) levels.Oral administration of MESM at the doses of 25 and 50 mg/kg b.w. resulted in a significant reduction in blood glucose levels in diabetic rats. Body weights were significantly reduced in STZ-induced diabetic rats when compared to normal rats, while the extract significantly restored body weight. They evaluated the antioxidant activity of MESM in STZ-induced diabetic rats. Decreased levels of TBARS and increased levels of GSH and CAT activity indicated a reduction in free radical formation in tissues such as the liver and kidney of diabetic rats.

Hajra *et al*. (2011) evaluated the antioxidant and antidiabetic potential of ethanolic extract of *Swietenia mahagoni* seeds. The extract showed DPPH radical scavenging activity 4.8, 6.74, 7.67, 9.18 and 17.25 per cent at respective concentrations of 10, 50, 100, 250 and 400 µg/ml. The extract also showed significant hydroxyl radical scavenging activity of 25.00, 33.52, 40.45, 50.57 and 88.03 per cent at respective concentrations. The extract significantly inhibited nitric oxide radical and ferric reducing power in a concentration dependent manner. The total phenolic content of seeds extract was found to be 1µg/mg of catechol equivalent.

Symphonia globulifera L. f. Family: Clusiaceae (Guttiferae)

Guttiferone A isolated from the seed shells of *Symphonia globulifera* showed a potent free radical scavenging activity compared to the well-known antioxidant (Ngouele *et al.*, 2006).

Symphytum asperum Lepech. Family: Boraginaceae

A water-soluble hydroxycinnamate-derived polymer (>1000 kDa) from *Symphytum asperum* strongly reduced the DPPH radical (IC(50) approximately 0.7 microg/ml) and inhibited the nonenzymatic lipid peroxidation of bovine brain extracts (IC(50) approximately 10 ng) (Barthomeuf *et al.*, 2001). This polymer exhibited only a low hydroxyl radical scavenging effect in the Fe(3+)-EDTA-H(2)O(2) deoxyribose system (IC(50) > 100 microg/ml) but strongly decreased superoxide anion generation in either the reaction of phenazine methosulfate with NADH and molecular oxygen (IC(50) approximately 13.4 microg/ml) or in rat PMA-activated leukocytes (IC(50) approximately 5 microg/ml). The ability to inhibit both degranulation of azurophil granules and superoxide generation in primed leukocytes indicates that the NADPH oxidase responsible for this later effect was inhibited, pointing to the *Symphytum asperum* polymer as a potent antiinflammatory and vasoprotective agent.

Two high-molecular water-soluble preparations with high anticomplement and antioxidant activity were isolated from the roots of *Symphytum asperum*. Their main chemical constituent was found to be poly[oxy-1-carboxy-2-(3,4-dihydroxyphenyl)ethylene] according to IR and NMR spectroscopy (Barbakadze *et al.*, 2005; Merlani *et al.*, 2010).

Symphytum caucasicum Bieb. Family: Boraginaceae

Two high-molecular water-soluble preparations with high anticomplement and antioxidant activity were isolated from the roots of *S. caucasicum*. Their main chemical constituent was found to be poly[oxy-1-carboxy-2-(3,4-dihydroxyphenyl)ethylene] according to IR and NMR spectroscopy(Barbakadze *et al.*, 2005; Merlani *et al.*, 2010).

Symphytum officinale L. Family: Boraginaceae

Two high-molecular-weight water soluble biopolymers, the main components of which was poly3(3,4-dihydrophenyl)glyceric acid or polyoxyl-carboxy-2-(3,4-dihydroxyphenyl)ethylene, were isolated from the roots of compfrey *Symphytum officinale* by Barbakadze *et al.* (2009). They exhibit antioxidant activity as expressed in decrease of active oxygen species (AOS) by interfering directly in their formation process by polymorphonuclear neutrophils (PMN) and biding directly AOS.

Symplocos cochinchinensis S. Moore Family: Symplocaceae

Methanol extract of *Symplocos cochinchinensis* leaves was evaluated by Sunil and Ignacimuthu (2011) for its *in vitro* and *in vivo* antioxidant activity. The total phenolic content of the extract was 230 mg of gallic acid equivalents/g extract. The extract showed very good scavenging activity on DPPH (IC$_{50}$ 620.30 µg/ml), hydroxyl (IC$_{50}$ 730.21 µg/ml), nitric oxide (IC$_{50}$ 870.31 µg/ml) radicals, as well as high reducing power. The extract also showed strong suppressive effect on lipid peroxidation. In *in vivo* study CCl_4 induced oxidative stress produced significant increase in SGOT, SGPT and LDH levels along with reduction in liver SOD, CAT, GSH and GPx levels. Pre-treatment of rats with the extract (250 and 500 mg/kg) for 7 days showed significant reduction in the levels of SGOT, SGPT and LDH compared to CCl_4 treated rats. SOD, CAT, GSH and GPx levels were increased significantly due to treatment with the extract. The activity of the extract was comparable to the standard drug, silymarin (25 mg/kg).

Symplocos racemosa Roxb. Family: Symplocaceae

Vijayabaskaran *et al.* (2010) investigated the antitumor activity and antioxidant status of ethanol extract of *Symplocos racemosa* (EESR). Treatment with EESR decreased the level of lipid peroxidation and increased the levels of catalase.

Synedrella nodiflora (L.) Gaertn. Family: Asteraceae (Compositae)

The hexane, ethyl acetate, ethanol and water extracts of *Synedrella nodiflora* were assessed by Wijaya *et al.* (2011) for their antibacterial and antioxidant capacities. The antioxidant capacities were evaluated using the FRAP and β-carotene bleaching assays. The ethanol extract displayed significant antioxidant capacities both in the FRAP and β-carotene bleaching assays. Folin-Ciocalteu and aluminium chloride spectrometry assays indicated the presence of phenolic compounds, including flavonoids in the ethanol extract.

Syringa pubescens Turcz. Family: Oleaceae

Xu *et al.* (2010) evaluated the antioxidant activity of differents extracts of *Syringa pubescens* using ABTS and DPPH radical scavenging and reducing reaction. The results showed good dose dependent activity. The butanol fraction from the skin showed the most potent antioxidant activities in each assay. With the concentration of 0.62 mg/ml showed 96.28 per cent ABTS radical scavenging and 96.16 per cent in the DPPH radicl scavenging. Phenolics content have a greater correlation with antioxidant activity.

Syzygium anisatum (Vickery) Craven et Biffin Family: Myrtaceae

English Name: Anise myrtle

The antioxidant capacities and phenolic composition in six native, commercially grown, Australian herbs and spices were investigated by Konczak *et al.* (2010). Anise myrtle contained 55.9mg gallic acid equivalent (GAE)/g dry weight. Anise myrtle exhibited good oxygen radical absorbance capacity and highest total reducing capacity [TRC; Ferric Reducing Antioxidant Power (FRAP)].

Syzygium aqueum (Burm.f.) Alston Family: Myrtaceae

The antioxidant activity of fresh and dried plant extracts of *Syzygium aqueum* were studied using β-carotene bleaching and the ABTS radical cation assay. The percentage of antioxidant activity for all extract samples using both assays was between 58 and 80 per cent. The fresh samples had higher antioxidant activity than the dried samples. The results of β-carotene bleaching assay were correlated with those of the ABTS assay.

Palanisamy *et al.* (2011) showed *S. aqueum* leaf extracts to have a significant composition of phenolic compounds, protective activity against free radicals as well as low pro-oxidant capability. Its ethanolic extract, in particular, is characterized by its excellent radical scavenging activity of EC_{50} of 133 µg/ml DPPH, 65 µg/ml ABTS and 71 µg/ml (Galvinoxyl), low pro-oxidant capabilities and a phenolic content of 585–670 mg GAE/g extract.

Syzygium aromaticum (L.) Merr. et Perry Family: Myrtaceae

English Name: Cloves

Odukoya *et al.* (2005) evaluated the antioxidant activity of Nigerian dietary spices including *Syzygium aromaticum*. The antioxidant activity (expressed as per cent inhibition of oxidation) was 68.65 per cent, reducing power was 11.37 per cent and Total phenolic content was 216.38 mg/100g Tannic acid Equivalent. Antioxidant activity correlated significantly and positively with total phenolics while there was no linear correlation between total antioxidant activity and reducing power neither between reducing power and total phenolic content.

Wojdylo *et al.* (2007) investigated Total equivalent antioxidant capacities (TEAC) and phenolic contents of 32 species extracts from 21 botanical families grown in Poland. The total antioxidant capacity was estimated by ABTS, DPPH and FRAP expressed as TEAC. The herbs with highest TEAC values were *Syzygium aromaticum* and *Epilobium hirsutum*.

Using a multiple-method approach, the antioxidant activity of the essential oils from *Syzygium aromaticum* was tested by Viuda-Martos *et al.* (2010). The total phenolic compound content was also determined. The clove essential oil had the highest amount of total phenols (898.89 mg/l GAE) and showed the highest inhibition of DPPH radicals (98.74 per cent) and the highest FRAP value (1.47 TEAC).

Niwano *et al.* (2011) summarized their research for herbal extracts with potent antioxidant activity obtained from a large scale screening based on superoxide radical. Experiments with the Fenton reaction and photolysis of H_2O_2 induced by UV irradiation demonstrated that extract of *Syzygium aromaticum* (bud) have potent ability to directly scavenge OH radicals.

Syzygium cerasoides (Roxb.) Raiz. Family: Myrtaceae

Kshirsagar and Upadhyay (2009) found antioxidant activity in methanolic fraction of stem. Methanol extracts of twig showed good activity in dose dependent study, *i.e.*, 50 per cent reduction of DPPH is achieved at the dose of 0.03mg/ml that is comparable to the standards (which are pure molecules), *viz.*, catechin and Trolox (Kshirsagar and Upadhyay, 2009). Moreover, their activities were better than that of curcumin for which 50 per cent reduction is achieved by 0.04 mg/ml.

Syzygium cumini (L.) Skeels (Syn.: *Syzygium jambolanum* (Lam.) DC.) Family: Myrtaceae

English: Black plum.

Muruganandan *et al.* (2001) reported antioxidant and anti-inflammatory activities with the ethanolic extract of stem bark. Banerjee *et al.* (2005) reported IC_{50}

values for DPPH radical scavenging activity in fruit skin as 168µg/ml, superoxide radical 260µg/ml, hydroxyl radical 428µg/ml and prevention of lipid peroxidation was 222µg/ml.

The effect of black plum seed extract on lipid peroxidation in alloxan induced diabetic rats was studied. Alloxan (150mg.kg body weight) increased significantly the glucose level in blood and induced lipid peroxidation in liver. The antioxidant enzymes catalase, glutathione, glutathione peroxidase, superoxide dismutase in liver were decreased. Oral administration of the extract for 15 days to alloxan treated animals showed remarkable increase in the level of antioxidant enxymes and reduced the level of lipid peroxidation activity and blood glucose. The results suggested that seed extract of the plant is an antioxidant and acute hyperglycaemic drug and might be used in the regulation of lipid peroxidation without detectable adverse side effects.

Bajpai *et al.* (2005) investigated the phenolic contents and antioxidant activity of some food and medicinal plants including *Syzygium cumini*. The seeds were found to have high phenolic contents and antioxidant activity. The TPC of seeds of *Syzygium cumini* was 108.7 mg/g GAE. The antioxidant activity of seeds of *Syzygium cumini*, as assayed by auto-oxidation of β-carotene and linoleic acid and expressed as the per cent of inhibition relative to the control was 85.4 per cent.

Ruan *et al.* (2008) reported that methanol extract and ethyl acetate fraction from *Syzygium cumini* leaves showed significant free radical scavenging activity.

A major underlying factor of peptic ulcer disorder is the generation of free radicals in addition to increased and prolonged acid secretion. Udaparkar *et al.* (2009) investigated the antioxidant activity of *Syzygium cumini* seeds on aspirin induced peptic ulcer in normal and diabetic rats. The level of lipid peroxide was elevated and the activities of antioxidant enzymes were decreased by administration of aspirin in both groups. The hydroalcoholic extract of the *Syzyygium cumini* seeds showed decreased level of lipid peroxides and increased activity of antioxidant enzymes in both groups. The results obtained in this study indicate that the antioxidant activity of the seed extract was more significant in aspirin induced peptic ulcer in normal rats as compared to diabetic rats.

Cuminoside, isolated from the seeds of *Syzygium cumini*, was studied for its hypoglycemic and antioxidant potential. Cuminoside caused a significant decrease in FBS level, lipid peroxidation level, and improvement in the levels of antioxidant enzymes (reduced glutathione, superoxide dismutase and catalase) in diabetic rats. A considerable decrease in lipid peroxidation and improvement in the antioxidant enzymes level in NIDDM rats indicated that cuminoside has antioxidant potential with antidiabetic activity (Farswan *et al.*, 2009).

Hydromethanol extracts of 15 Bangladeshi medicinal plants, including *Syzygium cumini*, were evaluated by Hasan *et al.* (2009) for antioxidant potential using DPPH radical scavenging assay. *Syzygium cumini* seeds exhibited radical scavenging activity with an IC_{50} value of 4.25 µg/ml compared to the IC_{50} value of 5.15µg/ml as shown by the reference antioxidant ascorbic acid, in a dose dependent fashion.

Syzygium currani (C.B. Robinson) Merr. Family: Myrtaceae

Quantitative analysis of antiradical phenolic constituents from fourteen edible Myrtaceae fruits was carried out by Reynertson *et al*. (2008). Results of DPPH assay was IC_{50} = 33.4µg/ml, Total antioxidant capacity of *Syzygium currani* was 12.1mg C3G/g dry weight and Total Phenolic content was 39.6mg GAE/g dry weight.

Syzygium formosanum (Hayata) Moti Family: Myrtaceae

Lee *et al*. (2006) investigated the antioxidant activity of ethanol extract of leaves using DPPH, hydroxyl and superoxide radical scavenging and reducing power activities as well as the induction of heme oxygenase-1 (HO-1). The IC_{50} values against DPPH radical were 5.8µg/ml, OH radical 0.42µg, superoxide radical 12.9µg/ml and total phenolics was 36.1 mg of GAE/g.

Extracts of *Syzygium formosanum* leaf showed remarkable free radical scavenging activity with IC_{50} value of 10.8µg/ml and significant inhibitory effects on lipid peroxidation with 8258 per cent inhibition at the concentration of 50ng/ml (Thoung *et al.*, 2006).

Syzygium gratum (Wt.) S.N.Mitra Family: Myrtaceae

Kukongviriyapan *et al*. (2007) reported antioxidant, antiinflammatory and vascular protectivities in the water extract of leaf of *Syzygium gratum*.

Syzygium jambos (L.) Alston (Syn.: Eugenia jambos L.) Family: Myrtaceae

Quantitative analysis of antiradical phenolic constituents from fourteen edible Myrtaceae fruits was carried out by Reynertson *et al*. (2008). Results of DPPH assay was IC_{50} = 92.0µg/ml, Total Phenolic content was 8.69mg GAE/g dry weight.

Syzygium javanicum Miq. Family: Myrtaceae

Quantitative analysis of antiradical phenolic constituents from fourteen edible Myrtaceae fruits was carried out by Reynertson *et al*. (2008). Results of DPPH assay was IC_{50} = 81.4µg/ml, Total antioxidant capacity of *Syzygium javanicum* was 0.09mg C3G/g dry weight and Total Phenolic content was 3.57mg GAE/g dry weight.

Syzygium malaccense (L.) Merr. et Perry Family: Myrtaceae

Lako *et al*. (2006) analysed phytochemical flavonols, carotenoids and the antioxidant properties of a wide selection Fijian fruits, vegetables and other readily available foods. Total antioxidant capacity of *Syzygium malaccense* was 11mg/100g of TEAC, Total polyphenol content was 32mg GAE/100g.

Quantitative analysis of antiradical phenolic constituents from fourteen edible Myrtaceae fruits was carried out by Reynertson *et al*. (2008). Results of DPPH assay for *Syzygium malaccense* was IC_{50} = 269µg/ml, and Total Phenolic content was 8.58mg GAE/g dry weight.

Syzygium samarangense (Blume) Merr. et Perry Family: Myrtaceae

Chen *et al*. (2009) studied the antioxidant activity of leaves extract towards xanthine oxidase, lipoxygenase and tyrosinase activities. The inhibitory effect of acetone extract on xanthine oxidase activity was nil, tyrosine activity was about 60 per cent and lipoxygenase activity was about 20 per cent.

Tabebuia impetiginosa Martius ex DC. Family: Bignoniaceae

Volatiles were isolated from the dried inner bark of *Tabebuia impetiginosa* using steam distillation under reduced pressure followed by continuous liquid-liquid extraction. The extract was analyzed by gas chromatography and gas chromatography-mass spectrometry. The major volatile constituents of *T. impetiginosa* were 4-methoxybenzaldehyde (52.84 microg/g), 4-methoxyphenol (38.91 microg/g), 5-allyl-1,2,3-trimethoxybenzene (elemicin; 34.15 microg/g), 1-methoxy-4-(1E)-1-propenylbenzene (trans-anethole; 33.75 microg/g), and 4-methoxybenzyl alcohol (30.29 microg/g). The antioxidant activity of the volatiles was evaluated using two different assays. The extract exhibited a potent inhibitory effect on the formation of conjugated diene hydroperoxides (from methyl linoleate) at a concentration of 1000 µg/ml. The extract also inhibited the oxidation of hexanal for 40 days at a level of 5 µg/ml. The antioxidative activity of *T. impetiginosa* volatiles was comparable with that of the well-known antioxidants, α-tocopherol, and butylated hydroxytoluene (Park *et al.*, 2003).

Tabebuia rosea (Bertol.) DC. Family: Bignoniaceae

Antioxidant potential of *Tabebuia rosea* leaves was tested by Ramalakshmi and Muthuchelian (2010) by DPPH free radical scavenging assay. The DPPH scavenging effect of the extract was increased with the increasing dosage and the IC_{50} estimations for the extract and the ascorbic acid standard were found to be 32.34µg and 1.51µg respectively. Moreover the total phenolics concentration equivalent to gallic acid was found to be 170mg/g of plant extract, which correlated with antioxidant activity.

Tabernaemontana divaricata (L.) R.Br. ex Roem. et Schult. Family: Apocynaceae

Successive soxhlet extraction of leaves of *Tabernaemontana divaricata* was done with petroleum ether, ethanol and water. All the extracts were screened for antioxidant activity in NBT model at different concentrations (5-100 µg/ml). The antioxidant activity was dose and time dependent. Phytochemical studies have shown presence of phenolic compounds, flavonoids, alkaloids and glycosides in ethanolic and aqueous extracts (Jain *et al.*, 2007).

Tacca leontopetaloides (L.) O.Kuntze Family: Taccaceae

Habila *et al.* (2011) evaluated the phytochemicals and antioxidant activity of four Nigerian medicinal plants, including *Tacca leotopetaloides*. The percentage of antioxidant activity (AA per cent) of the plants extract showed a dose dependent increase. *Tacca leontopetaloides* showed highest activity at 125 µg/ml (86 per cent). The results of reducing potential and Total Phenolic contents expressed in terms of Gallic acid equivalent showed that *Tacca leontopetaloides* possessed the least reducing potential (0.217nm, 6.90mg) when compared with Gallic acid standard (1.268nm).

Tachigalia paniculata Aubl. Family: Caesalpiniaceae

Myricetin 7-O-β-D-glucopyranosyl-(1→6) β-D-glucopyranoside (1) and myricetin 7-O-α-L-rhamnopyranosy-(1→6) β-D-glucopyranoside (2) isolated from the leaves of

Tachigalia paniculata showed antioxidant activity in the TEAC test, and xanthine oxidase test (Cioffi *et al.*, 2002).

Tagetes erecta L. Family: Asteraceae (Compositae)

English Name: Marigold.

The different cultivars of marigold showed marked variations in total phenols and flavonoids, as well as antioxidant and radical scavenging activities. The ethanolic extracts exhibited the highest antioxidant and radical scavenging activities. The cultivar Xinhong had the highest activity, as well as one of the highest lutein contents and antioxidant activities (Li *et al.*, 2007).

Tagetes lucida Cav. Family: Asteraceae (Compositae)

Aquino *et al.* (2002) evaluated the antioxidant activity of methanolic extract and some of its constituents of *Tagetes lucida*. The extract and some of its constituents showed a significant free-radical-scavenging effect in comparison to α-tocopherol and standard flavonols.

Tagetes maxima Kuntze Family: Asteraceae (Compositae)

Extracts and fractions from *Tagetes maxima* among the nine Bolivian plants were attributed to the phenolic compounds present in this bioactive species (Parejo *et al.*, 2005).

Tagetes mendocina Phil. Family: Asteraceae (Compositae)

Schmeda-Hirschmann *et al.* (2004) evaluated the antioxidant activity of methanol extract of the aerial parts of *Tagetes mendocina* using DPPH and superoxide anion scavenging assay. Assay-guided isolation led to 4'-hydroxyacetophenone (1), protocatechuic acid (2), syringic acid (3), patuletin 4), quercetagetin 7-O-β-D-glucoside (5), patuletin 7-O-β-D-glucoside (6) and axillarin7-O-β-D-glucoside (7) as the free radical scavengers and antioxidant compounds from *Tagetes mendocina*. Quercetagetin 7-O-β-D-glucoside proved to be the main free radical scavenger of the extracts measured by the DPPH decoloration test as well as for quenching the superoxide anion and inhibition of lipoperoxidation in erythrocytes.

Tagetes multiflora Kunth. Family: Asteraceae (Compositae)

The aerial parts of 17 Bolivian plants including *Tagetes multiflora* were screened by Rosas-Romero and Saavedra (2005) to determine antioxidant activity. A methanol extract of each plant was prepared and portioned sequentially with hexane, chloroform, and ethyl acetate, leaving an aqueous solution. All extracts and their 5 fractions for a total of 102 samples, were evaluated using two techniques: an adaptation of the β-carotene bleaching technique using an emulsion of linoleic acid in water as the oxidizable substrate, and the DPPH free radical trapping technique. The results of the β-carotene bleaching technique were more discriminating and better related to the rancidity process under normal conditions; with this assay, 11 species provided at least one fraction with highly promising antioxidant activity. All species gave good results under the DPPH technique, and in most cases they performed better than BHA, which was used as a reference antioxidant (Rosas-Romero and Saavedra, 2005).

Talinum portulacifolium (Forsk.) Aschers et Schweinf. Family: Portulacaceae

Babu *et al*. (2009) evaluated the antihyperglycemic and antioxidant effects of hexane, ethanolic and aqueous extracts of *Talinum portulacifolium* leaves. All the extracts exhibited dose dependent scavenging activities against DPPH radicals, nitric oxide radicals and hydrogen peroxide. Further, all extracts had relatively lower reducing power, compared to that of ascorbic acid. The total phenolic content of hexane, ethanol and aqueous extracts were found to be 61, 100 and 114 mg/gm of the dry extract respectively. TLC of the above extracts using the DPPH as a spraying reagent revealed yellow spots against purple background indicating the presence of potent antioxidant compounds.

Talinum triangulare (Jacq.) Willd. Family: Portulacaceae

English Name: Javanese Ginseng.

Antioxidant activity of the javanese ginseng root was investigated by Estiasih and Kurniawan (2006). The root extracts were prepared by solvent extraction using methanol, ethanol (96 per cent), ethanol (70 per cent), acetone, and hexane.Total antioxidant activity of the extracts was measured by ferric thiocyanate method, whereas radical scavenging capacity and reducing power were measured by the DPPH and the reducing potential methods, respectively. The result showed that the highest total antioxidant activity was observed in acetone and methanol extracts. It appeared that the ability of these extracts for partitioning at the interface of the emulsion in the tested oxidation system was the highest among other extracts, therefore it had the best activity to inhibit oxidation.The highest radical scavenging capacity measured by EC_{50} was observed in acetone extract. The type of phenolic compounds of this extract appeared to be responsible for the highest radical scavenging capacity. Different phenomena occurred for reducing power. Methanol extract had the highest reducing power and the least were found with the hexane and acetone extract. It was suggested that each extracts comprised different types of phenolic based on different polarity of solvents used for extraction. The antioxidant compounds of javanese ginseng root extracts were primary antioxidant based on the ability to scavenge free radical. It could be concluded that acetone was the best solvent for antioxidant extraction of the javanese ginseng root. However, all tested antioxidant mechanisms in this research showed that vitamin E (1000 ppm) used as control had better activity than javanese ginseng root extracts (1000 ppm) for all types of solvent. Javanese ginseng extracts might contain other compounds which were not responsible for antioxidant activity, therefore at the same concentration the activity were lower than that of vitamin E.

Odukoya *et al*. (2007) evaluated the antioxidant activity and total phenolic contents of Nigerian green leafy vegetables including *Talinum triangulare*. Phenol content was 21.83 mg/100 g dry weight.

Antioxidant activities of eight leafy vegetables of Ghana including *Talinum triangulare* were assessed by Morrison and Twumasi (2010). The total antioxidant capacity (TAC) and Total Phenol content (TPC) in the methanol extracts (METE) and hydro-ethanol extracts (HETE) were measured. The TPC of the methanol extract of

Talinum triangulare was 0.336 mg/ml TAE and TAE was 0.340 mg/ml Ascorbic acid equivalent (AAE) while TPC of hydro-ethanol extract was 0.494 mg/ml TAE and TAE was 0.462 mg/ml Ascorbic acid equivalent (AAE). Radical scavenging activity of methanol extract was EC_{50} = 0.1656mg/ml and HETE was 0.1799 mg/ml while Fe^{3+} reducing potential was EC_{50} = 0.1799 (METE) and 0.1799 (HETE).

Adefegha and Oboh (2011) reported that cooking decreases the vitamin C contents in the leafy vegetable *Talinum triangulare*, while it increased the phenolic content and antioxidant activities.

Liang *et al.* (2011) evaluated the antioxidant and hepatoprotective activity of polysaccharides from *T. triangulare* (TTP). The antioxidant activities of 40 per cent, 60 per cent, 80 per cent and crude TTP were evaluated using three different models *in vitro*, including reducing power, hydroxyl radicals, superoxide anion. The levels of aspartate aminotransferase (AST), alanine aminotransferase (ALT) in serum, and glutathione (GSH), superoxide dismutase (SOD), malondialdehyde (MDA) in liver tissues were measured. *In vitro* assays, TTP showed remarkably different degrees of antioxidant activities in dose-dependent manners. The crude TTP demonstrated a relatively strong antioxidant activity, while the 40 per cent TTP showed the strongest antioxidant activity, and the 60 per cent TTP had the weakest antioxidant ability. *In vivo* assay, pretreatment with TTP had significantly decreased the levels of AST, ALT and MDA against CCl(4) injures, and restored the activities of defense antioxidant substances SOD and GSH towards normalization.

Tamarindus indica L. Family: Caesalpiniaceae

English Name: Tamarind

Antioxidant activity of tamarind seeds was investigated by Tsuda *et al.* (1994). An ethanol extract prepared from the seed coat exhibited antioxidative activity as measured by the thiocyanate and TBA method, but there is no activity in the extract prepared from the germ. The ethyl acetate extract prepared from the seed coat had strong antioxidative activity. 2-hydroxy-3',4'-dihydroxyacetophenone (TA0), methyl 3,4-dihydroxybenzoate (TA1), 3,4-dihydroxyphenyl acetate (TA2) and (-)-epicatchin (TA3) were isolated from the ethyl acetate extract. TA0, TA1 and TA2 had strong antioxidative activity in the linoleic acid autoxidation system as measured by the thiocyanate and TBA method as well as α-tocopherol.

Chanwitheesuk *et al.* (2005) investigated the antioxidant activity of shoot tips of *Tamarindus indica* using a β-carotene bleaching method. The contents of plant chemicals such as vitamin C, Vitamin E, carotenoids, tannin and total phenolics, were also determined. Methanolic extract showed antioxidant activity. *Tamarindus indica* has antioxidant index of 1.10, Vitamin C 4.96 mg per cent, Vitamin E 0.0015mg per cent, Total carotenes 0.64 mg per cent, Total xanthophylls 1.05mg per cent, Tannins 77mg per cent and Total phenolics of 121mg per cent.

Polyphenol content and antioxidant activity of fourteen wild edible fruits from Burkina Faso were investigated by Lamien-Meda *et al.* (2008). The data obtained show that the total phenolic and total flavonoid levels were significantly higher in the acetone than in the methanol extracts. *Tamarindus indica* fruit had higher flavonoid content and higher antioxidant activity.

Povichit *et al.* (2010) investigated phenolic content and free radical scavenging effect of seed coat of *Tamarindus indica* by DPPH and ABTS assays, anti-lipid peroxidation activity by TBARS and for antiglycation activity. The results revealed that the total phenolic content showed good correlation with free radical scavenging by ABTS and anti-lipid peroxidation by TBARS, but showed no correlation with antiglycation. The extract of *Tamarindus indica* demonstrated a significant antioxidant effect and also showed a promising antiglycation effect. The IC_{50} were 0.09mg/ml for the DPPH method, TEAC value of 2.79 mg/Trolox/mg sample for the ABTS method, IC_{50} of 0.64 mg/ml for the TBARS method and IC_{50} of 0.01 for the antiglycation method.

Tamarix aphylla (L.) Karsten Family: Tamaricaceae

Vadlapudi and Naidu (2009) evaluated the antioxidant potential of *Tamarix aphylla*, a mangrove plant. The SOD was 0.65U/mg, Catalase was 1.6U/mg, Ascorbic acid was 290mg/100g, DPPH radical inhibition was 55.4 per cent and FRAP units were 790.

Tamarix aralensis Bge. Family: Tamaricaceae

Souri *et al.* (2004) evaluated the antioxidative activity of *Tamarix aralensis* by linoleic acid peroxidation test using 1,3-diethyl-2-thiobarbituric acid as the reagent. Methanolic extract showed 54.30 per cent inhibition at 40µg concentration with IC_{50} value of 34.70µg.

Tamarix hispida Willd. Family: Tamaricaceae

Sulanova *et al.* (2004) isolated a new pentacyclic triterpenoid 3-α-(3″,4″-dihydroxy-trans-cinnamoyloxy)-D-friedoolean-14-en-28-oic acid along with two known compounds, rhamnocitrin and isorhamnocitrin from the aerial parts of *Tamarix hispida*. All the free compounds showed significant antioxidant activity.

Tamarix nilotica (Ehrenb.) Bunge Family: Tamaricaceae

Antioxidant activity of a selection of commonly occurring wild plants growing in Beni-Sueif governorate, Upper Egypt, has been tested for their antioxidant actgivity. The plants selected are *Tamarix nilotica*, *Ambrosia maritima*, *Zygophyllum coccenium*, *Conyza dioscoridis*, *Chenopodium ambrosioides*, and *Calotropis procera*. The *in vitro* antioxidant assays used in this study were DPPH radical scavenging activity, superoxide anion scavenging activity and iron chelating activity. Extracts prepared from the leaves and flowers of *Tamarix nilotica* have shown the highest antioxidant activity in the three kinds of assay.

Tanacetum balsamita L. Family: Asteraceae (Compositae)

Nikolova and Dzhurmanski (2009) evaluated the free radical scavenging activity of total methanol extract of *Tanacetum balsamita*. The extract exhibited strong antioxidant activity and its IC_{50} value for DPPH radical was 42.73 µg/ml.

Tanacetum budjnurdense (Rech.f.) Tzvelev. Family: Asteraceae (Compositae)

The antioxidant activities of ethanolic extract of *Tanacetum budjnurdense* from Iran was examined by Esmaeili *et al.* (2010) using DPPH assay.

Tanacetum chiliophyllum (Fisch. et Mey.) Sch.-Bip. Family: Asteraceae (Compositae)

The antioxidant activities of ethanolic extract of *Tanacetum chiliophyllum* from Iran was examined by Esmaeili *et al.* (2010) using DPPH assay.

Tanacetum densum (Lab.) Schultz Bip. Family: Asteraceae (Compositae)

Methanolic extracts of three different *Tanacetum* subspecies [*Tanacetum densum* (Lab.) Schultz Bip. subsp. *sivasicum* Hub-Mor et Grierson, *Tanacetum densum* (Lab.) Schultz Bip. subsp. *eginense* Heywood and *Tanacetum densum* (Lab.) Schultz Bip. subsp. *amani* Heywood] which are endemic to Turkish flora were screened by Tepe and Sokment (2006) for their possible antioxidant activities by two complementary test systems namely DPPH free radical scavenging and β-carotene/linoleic acid. In DPPH system, the most active plant was *T. densum* subsp. *amani* with an IC_{50} value of 69.30 µg/ml. On the other hand, *T. densum* subsp. *sivasicum* exerted greater antioxidant activity than those of other subspecies in β-carotene/linoleic acid system (79.10 per cent). Antioxidant activities of BHT, curcumine and ascorbic acid were also determined as positive controls in parallel experiments. Total phenolic constituents of the extracts of *Tanacetum* subspecies were performed employing the literature methods involving Folin–Ciocalteu reagent and gallic acid as standard. The amount of total phenolics was highest in subsp. *sivasicum* (162.33µg/mg), followed by subsp. *amani* (158.44 µg/mg). Especially, a positive correlation was observed between total phenolic content and antioxidant activity of the extracts.

Tanacetum hololeucum (Bornm.) Podlech Family: Asteraceae (Compositae)

The antioxidant activities of ethanolic extract of *Tanacetum hololeucum* from Iran was examined by Esmaeili *et al.* (2010) using DPPH assay. The extract displayed antioxidant activity with IC_{50} value of 59.55 µg/ml using DPPH assay.

Tanacetum kotschyi (Boiss.) Grierson Family: Asteraceae (Compositae)

The antioxidant activities of ethanolic extract of *Tanacetum kotschyi* from Iran was examined by Esmaeili *et al.* (2010) using DPPH assay.

Tanacetum macrophyllum (Waldst. et Kitam.) Sch-Bip. Family: Asteraceae (Compositae)

The essential oil of *Tanacetum macrophyllum* growing wild in Turkey was obtained by hydrodistillation. Twenty-eight components, representing 95.6 per cent of the total oil, were identified. The main component of the oil was β-eudesmol (89.5 per cent). The antioxidant activity of the essential oil was measured by the FRAP assay (Javidnia *et al.*, 2010).

Tanacetum parthenium L. Family: Asteraceae (Compositae)

Bioactive ingredients showing antioxidant activity were isolated from *Tanacetum parthenium* by Koganov (2009). Nikolova and Dzhurmanski (2009) evaluated the free radical scavenging activity of total methanol extract of *Tanacetum parthenium*. The extract exhibited strong antioxidant activity and its IC_{50} value for DPPH radical was 39.23 µg/ml.

Tanacetum polycephalum Sch.-Bip. Family: Asteraceae (Compositae)

The chemical composition of the essential oil obtained by hydrodistillation from *Tanacetum polycephalum* was analysed by GC and GC/MS and 39 compounds constituting 94.02 per cent of the oil were identified, the major components being borneol (28.30 per cent), β-pinene (10.10 per cent), α-pinene (6.5 per cent), camphene (6 per cent), α-terpineol (5.16 per cent) and 1,8-cineol (5.10 per cent). The antioxidant activity of the essential oil was evaluated using the DPPH test and 5-lipoxygenase test. The antioxidant activity show that, in reduction of the stable radical DPPH, the highest activity was obtained with the essential oil and with Trolox (IC_{50} = 12.4 and 8.9 µg/ml, respectively (Hamzeh Amiri. 2007).

Tanacetum tabrisianum (Boiss.) Sosn. Family: Asteraceae (Compositae)

The antioxidant activities of ethanolic extract of *Tanacetum tabrisianum* from Iran was examined by Esmaeili *et al.* (2010) using DPPH assay.

Tanacetum vulgare L. Family: Asteraceae (Compositae)

English Name: Tansy

The antioxidative activity of phenolic extracts from herb of Tansy was examined by autoxidation of methyl linoleate by Kahkonen *et al.* (1999). The phenol content (14.2 mg/g GAeE) and antioxidant activity (44 per cent at 5ppm level) were moderate.

The methanolic extract of aerial parts of *Tanacetum vulgare* and its fractions were investigated by Juan-Badaturuqe *et al.* (2009) for antioxidant activity. The crude extract displayed DPPH radical scavenging effects with an EC_{50} value of 37 microg/ml. Activity-guided fractionations of the crude extract resulted in the isolation of three antioxidant compounds; 3,5-O-dicaffeoylquinic acid (3,5-DCQA), axillarin and luteolin. 3,5-DCQA was the major constituent with antioxidant activity (IC_{50} = 9.7 microM) comparable with that of the standard quercetin (IC_{50} = 8.8 microM).

Tapinanthus globiferus (A.Rich.) Tieghem Family: Amaryllidaceae

Using Trolox assay, Cook *et al.* (1998) estimated the antioxidant activity of 17 wild edible plants of Niger Republic used for food and traditional medicine. They observed that *Tapinanthus globiferus* possessed strong antioxidant activity.

Taraxacum mongolicum Hand.-Maz. Family: Asteraceae (Compositae)

Cai *et al.* (2004) evaluated the antioxidant activity and phenolic compounds in traditional Chinese medicinal plants associated with anticancer. The improved ABTS.+ method was used to systematically assess the total antioxidant capacity (Trolox equivalent antioxidant capacity, TEAC) of the medicinal extracts. The TEAC values and total phenolic content for methanolic extract of whole plant of *Taraxacum mongolicumi* were 142 µmol Trolox equivalent/100g dry weight (DW), and 0.50 g of gallic acid equivalent/100g DW respectively. A positive significant linear relationship between antioxidant activity and total phenolic content showed that phenolic compounds were the dominant antioxidant components in the tested herbs. Major types of phenolic compounds in *Taraxacum mongolicum* include flavonols (rutin).

Li *et al.* (2008) evaluated the antioxidant properties of 45 medicinal plants using FRAP and TEAC assays and the total phenolic contents of these plants were measured. *Taraxacum mongolicum* showed low antioxidant activity with 54.91 µmol Fe(II)/g FRAP value, 31.03µmol Trolox/g of TEAC value and 8.97 mg GAE/g of Phenolic content.

Taraxacum officinale Webber ex Wiggers Family: Asteraceae (Compositae)

English Name: Dandelion.

Hu and Kitts (2005) evaluated the antioxidant activity of flower extract of *Taraxacum officinale*. Flvaonoids and coumaric acid derivatives were identified from dandelion flower. Characteristics of chain breaking antioxidants, such as extended lag phase and reduced propagated rate, were observed in oxidation of linoleic acid emulsion with the addition of dandelion flower extract (DFE). DFE suppressed both superoxide and hydroxyl radical, while the latter was further distinguished by both site specific and non-specific hydroxyl radical inhibition. DPPH-radical-scavenging activity and a synergistic effect with α-tocopherol were attributed to the reducing activity derived from the phenolic content of DFE. A significant and concentration dependent, reduced nitric oxide production from bacterial-lipopolysaccharide stimulated mouse macrophage RAW 264.7 cells was observed with the addition of DFE. Moreover, peroxyl-radical-induced intracellular oxidation of RAW 264.7 cells was inhibited significantly by the addition of DFE over a range of concentrations.

Wojcikowski *et al.* (2007) evaluated the antioxidant activity of aerial parts of *Taraxacum officinale*. Oxygen radical absorbance capacity of the ethyl extract, methanol extract and aqueous extracts of leaf were 13.33, 40.50 and 42.83µmol Trolox equivalent per g of dried starting material, respectively. Oxygen radical absorbance capacity of the ethyl extract, methanol extract and aqueous extracts of root were 48.65, 49.15 and 54.20µmol Trolox equivalent per g of dried starting material, respectively.

Crude extract from *Taraxacum officinale* was screened by Sengul *et al.* (2009) for its *in vitro* antioxidant and antimicrobial properties. Total phenolic content of extract of this plant was also determined. β-carotene bleaching assay and Folin-Ciocalteu reagent were used to determine total antioxidant activity and total phenols of plant extract. Antioxidant activity was 43.05 per cent and the total phenolic content was 15.50 mgGAE/g DW).

Choi *et al.* (2010) investigated the hypolipidemic and antioxidant effects of root and leaf of *Taraxacum officinale*. After the treatment with dandelion root and leaf positively changed the plasma antioxidant enzymes.

Tarenna attanuata (Hook.f.) Hutch. Family: Rubiaceae

Iridoid constituents of *Tarenna attanuata* were evaluated for antioxidant activity, but none of the 10 compounds showed positive activity (Yang *et al.*, 2006).

Tasmannia lanceolata (Poir.) A.C.Sm. Family: Magnoliaceae

English Name: Tasmannia pepper.

The antioxidant capacities and phenolic composition in six native, commercially grown, Australian herbs and spices were investigated by Konczak *et al.* (2010). Tasmannia pepper leaf contained the highest level of total phenolics 102.1 mg gallic acid equivalent (GAE)/g dry weight. Tasmanni pepper leaf exhibited the highest oxygen radical absorbance capacity (ORAC assay). The TP content of Tasmannia pepper berry (16.86 mg GAE/g DW) was similar to that of black pepper, but its TRC was 25 per cent lower.

Tecoma radicans (L.) Juss. Family: Bignoniaceae

Successive extracts of aerial parts of *Tecoma radicans*, showed free radical scavenging activity. Ethyl acetate extract produced maximum inhibition (97.8 per cent) of DPPH and recorded by the electron spin resonance occurred with ethyl acetate extract at 97.8 per cent. Activity-guided fraction led to the isolation of five flavonoids. The five flavonoids were identified as luteolin, quercetin 3-methyl ether, apigenin, 6-hydroxyluteolin and chrysoeriol. The most potent free radical scavenger of the five identified flavonoids was quercetin 3-methyl ether with 68.9 per cent inhibition followed by luteolin with 39.4 per cent inhibition (Hashem, 2008).

Tecoma stans (L.) Juss. ex Kunth Family: Bignoniaceae

The ethanol, methanol and water extracts of *Tecoma stans* possessed strong radical scavenging activity from FRAP and DPPH assays. It was ranged from 1433.75 to 3841 g/ml. Phytochemical analysis revealed the presence of alkaloids, flavonoids, saponins, phenols, steroids, anthraquinones and tannins. The three extract fractions have showed highest Total Phenolic contet (177-216 mg gallic acid equivalent/g).

Tecomaria capensis (Thunb.) Lindl. Family: Bignoniaceae

The chloroform and methanolic extracts of 124 Egyptian plant species belonging to 56 families were investigated by Moussa *et al.* (2011) and compared their antioxidant activity by DPPH scavenging assay. Among the 124 plant species tested 18 exhibited extremely high antiradical activity (>80 per cent inhibition). The IC_{50} value of DPPH radical scavenging of *Tecomaria capensis* was 22.55µg/ml, while total phenolic and flavonoid contents were 111.06 mg Tannic acid equivalent/g extract and 226.40 mg Rutin equivalent/g extract, respectively. Correlation coefficient between DPPH radical scavenging activity and the total phenolic and flavonoid contents suggest that phenolics and flavonoids in the extracts were partly responsible for the antiradical activities.

Tectona grandis L. Family: Verbenaceae

Krishna and Nair (2010, 2011) investigated the antioxidant potential of different extracts from leaf, bark and wood of *Tectona grandis* using DPPH and ABTS assays. Ethyl acetate extract of wood showed very high activity with 98.6 per cent inhibition against DPPH and ABTS free radicals. The value was higher than standard compounds used for the study. Compounds isolated from the leaves showed low antioxidant activity.

Telfaria occidentalis Hook. Family: Cucurbitaceae

Oboh and Akindahunsi (2004) investigated the effect of preservation of green leafy vegetables of Nigeria on the antioxidant phytoconstituents (Vitamin C and Total phenol) and activity (reducing property and free radical scavenging ability) including *Telfaria occidentalis*. The result of the study revealed that sun-drying of green leafy vegetables cause a significant decrease in the vitamin C content (16.67-64.68 per cent loss). Conversely it leads to a significant increase in the total phenol content (6.45 -223.08 per cent gain) reducing property (16.00-362.50 per cent gain) and free radical scavenging ability (126.00-5757.00 per cent gain) of green leafy vegetables.

Nwanna and Oboh (2007) investigated the antioxidant and hepatoprotective properties of free and bound polyphenols from *Telfaria occidentalis* leaves. It was inferred that both soluble and free and bound polyphenols extracts of *T. occidentalis* leaf have antioxidant and hepatoprotective properties, however soluble free polyphenols had significantly higher antioxidant and hepatoprotective properties than the bound polyphenols.

Odukoya *et al.* (2007) evaluated the antioxidant activity and total phenolic contents of Nigerian green leafy vegetables including *Telferia occidentalis*.

Adefegha and Oboh (2011) reported that cooking decreases the vitamin C contents in the leafy vegetable *Telfaria occidentalis*, while it increased the phenolic content and antioxidant activities.

Telosma minor Craib. Family: Asclepiadaceae

Chanwitheesuk *et al.* (2005) investigated the antioxidant activity of *Telosma minor* using a β-carotene bleaching method. The contents of plant chemicals such as vitamin C, Vitamin E, carotenoids, tannin and total phenolics, were also determined. Methanolic extract showed antioxidant activity. *Telosma minor* has antioxidant index of 2.85, Vitamin C 23.6 mg per cent, Vitamin E 0.0070mg per cent, Total carotenes 1.29 mg per cent, Total xanthophylls 4.24mg per cent, Tannins 17.7mg per cent and Total phenolics of 98.4 mg per cent.

Tephrosia egregia Sandw. Family: Fabaceae (Papilionaceae)

The antioxidant and larvicidal activities of *Tephrosia egregia* extracts and its major component, dehydrorotenone, were studied by Arriaga *et al.* (2009). High antioxidant activity was found for dehydrorotenone and methanol and ethyl acetate extracts from roots and stems, respectively.

Tephrosia purpurea Pers. Family: Fabaceae (Papilionaceae)

The ethanol extract of *Tephrosia purpurea* was found to significantly inhibit the carbon tetrachloride-induced lipid peroxidation *in vivo* and superoxide generation *in vivo* (Soni *et al.*, 2006). The ethyl acetate fraction of the same extract was studied for free radical scavenging and antilipid peroxidation activity. The IC_{50} values in both of these *in vitro* assays were found to be significantly reduced for ethyl acetate fraction compared with the ethanolic extract of the plant. The observation was further supported by comparing the *in vivo* antioxidant activity for both the ethanolic extract and its ethyl acetate fraction.

Gunjegaonkar *et al.* (2010) evaluated the hepatorpotective and antioxidant activity of *Tephrosia purpurea* whole plant extract. IC_{50} DPPH scavenging activity of extract of *Tephrosia purpurea* was 400µg/ml while standard ascorbic acid was 174.79µg/ml.

Patel *et al.* (2010) investigated the *in vitro* antioxidant activity of aqueous and ethanolic extracts of the leaves of *Tephrosia purpurea*. The total phenolic content of aqueous and ethanolic extracts showed the content values of 9.44 per cent w/w and 18.44 per cent w/w, respectively, and total flavonoid estimation of aqueous and ethanolic extracts showed the content values of 0.91 per cent w/w and 1.56 per cent w/w, respectively, for quercetin and 1.85 per cent w/w and 2.54 per cent w/w, respectively, for rutin. Further investigations were carried out for *in vitro* antioxidant activity and radical scavenging activity by calculating its percentage inhibition by means of IC_{50} values, all the extracts' concentrations were adjusted to fall under the linearity range and here many reference standards like tannic acid, gallic acid, quercetin, ascorbic acid were taken for the method suitability. The results revealed that leaves of this plant have antioxidant potential. The results also showed the ethanolic extract to be more potent than the aqueous decoction which is claimed traditionally.

Root extracts of *Tephrosia purpurea* were screened by Nile and Khobragade (2011) for *in vitro* antioxidant and xanthine oxidase inhibitory activity. Antioxidant activity was measured using ABTS, DPPH, FRAP and OTAC methods. The root extract and phytochemicals, obtained in distilled water, inhibited bovine milk XO in a concentration-dependent manner, with an additional superoxide scavenging capacity, the average antioxidant activity of *T.purpurea* root extract in a concentration range of 100-200 µg/ml. The reacting system revealed significant antioxidant activity, viz., 42.2 (ABTS), 28.7 (DPPH), 36.5 (FRAP) and 25.6 per cent by ORAC assay. Screening of xanthine oxidase inhibitory activity by extract in terms of kinetic parameters revealed noncompetitive mode of inhibition, where the K_m and V_{max} value in presence of (25 to 100 µg/ml) *T. purpurea* root extract is 0.18 µg and 0.040, 0.037, 0.034 and 0.030 (µg/min) while for control K_m and V_{max} is 0.21 µg and 0.043 (µg/min), respectively. The phytochemical analysis revealed presence of significant amount of polyphenols and flavonoids (90 per cent and 80 per cent, respectively).

Tephrosia tinctoria Pers. Family: Fabaceae (Papilionaceae)

Rajaram and Surtesh Kumar (2011) carried out the *in vitro* antioxidant and antidiabetic activity of the different parts (leaf, stem and root) of *Tephrosia tinctoria* extracted with various solvents from non polar to polar basis (hexane, chloroform, ethyl acetate and ethanol). Among the various fractions tested, the ethyl acetate fraction of stem of *T. tinctoria* exhibited maximum antioxidant (DPPH with IC_{50} value 94.33µg/ml) and antidiabetic (α-Glucosidase inhibition IC_{50} value 33.54µg/ml) activity. Further the ethyl acetate fraction of stem extract was assayed for antioxidative assays such as reducing power, lipid peroxidation inhibition activity, total phenols and total flavonoids.

828 | *Encyclopaedia of Herbal Antioxidants*

Terminalia arjuna (Roxb. ex DC.) Wight et Arn. Family: Combretaceae

Nine plants that are components of Ayurvedic formulations used for the therapy of cardiovascular diseases were investigated by Munasinghe *et al.* (2001) to determine whether antioxidant activity is one of the mechanisms by which these plants exert cardioprotection. Antioxidant activity was measured both *in vitro* (DPPH radical scavenging and deoxyribose protection assays) and *in vivo* (by effects on lipid peroxidation). *Terminalia arjuna* showed significant DPPH radical scavenging activity with ED(50) 8.3microg/ml (similar to L-ascorbic acid). Of all the nine plants tested, *Terminalia arjuna* demonstrated the highest antioxidant activity.

Using rat liver mitochondria Tilak *et al.* (2003) examined the antioxidant effect of *Terminalia arjuna*. Among the various extracts from arjunsal the ones from organic solvents were found to be the most effective in DPPH, ABTS and FRAP assays. In *T. arjuna* bark, the methanolic extract showed the highest antioxidant activity. Baicalien showed antioxidant and radioprotective activity. The results of Tilak *et al.* (2003) indicate that various preparations from *T. arjuna* and its component baicalein have significant antioxidant activity.

Antidyslipidemic and antioxidant activities of different fractions of *Terminalia arjuna* stem bark was investigated by Chander *et al.* (2004). *In vitro* experiment of *T. arjuna* fractions at tested concentrations (50-500 microg/ml) inhibited the oxidative degradation of lipids in human low density lipoprotein and rat liver microsomes induced by metal ions.

Oral administration of *Terminalia arjuna* bark for 12 weeks in rabbits caused augmentation of myocardial antioxiants: superoxide dismutase (SOD), catalase (CAT), and glutathione (GSH) along with induction of heat shock protein 72 (HSP 72). *In vivo* ischemic-reperfusion injury induced oxidative stress, tissue injury of heart and haemodynamic effects were prevented in the *Terminalia arjuna* bark treated rabbit hearts (Gauthaman *et al.*, 2005).

Bajpai *et al.* (2005) investigated the phenolic contents and antioxidant activity of some food and medicinal plants including *Terminalia arjuna*. The leaves, fruits and bark were found to have high phenolic contents and antioxidant activity. The TPC of fruits, leaves and bark of *T.arjuna* were 133.9, 143.6 and 145.9 mg/g GAE, respectively. The antioxidant activity of *T.arjuna*, as assayed by auto-oxidation of β-carotene and linoleic acid and expressed as the per cent of inhibition relative to the control were, 83.2 per cent, 85.1 per cent and 89.7 per cent for fruits, leaves and bark, respectively.

Bark extract of *Terminalia arjuna* inhibited the induction of hydroxyl, superoxide, DPPH, ABTS radicals in a dose-dependent manner (Koti Reddy *et al.*, 2008).

Arjunolic acid (AA), a triterpenoid saponin isolated from *Terminalia arjuna*, was screened for antioxidant activity using four different *in vitro* methods, namely reducing power assay, DPPH radical scavenging assay, β-carotene bleaching assay and hydroxyl radical scavenging assay. The results thus obtained were compared with standard antioxidant compounds such as Ascorbic acid, BHT and Catechin. The results of all four *in vitro* antioxidant assays exhibited that Arjunolic acid possess antioxidant property and comparative anlysis showed that AA has relatively low antioxidant property than standards.

Biswas *et al.* (2010) evaluated the antioxidant and free radical potency of petroleum ether, chloroform and methanol extracts of *Terminalia arjuna* leaf. The different antioxidant assays, including DPPH radical scavenging, total reductive activity, super oxide anion radical, nitric oxide scavenging, lipid peroxidation and total phenolic content were estimated. Ascorbic acid was used as the reference. All the extracts exhibited potent *in vitro* antioxidant activity that increase with extract concentration.

Terminalia belerica (Gaertn.) Roxb. Family: Combretaceae

English Name: Beleric myrobalan.

The aqueous extract of the fruits of *Terminalia belerica* and their equiproportional mixture triphala were evaluated by Vani *et al.* (1997) for their *in vitro* antioxidant activity. They also inhibited radiation induced lipid peroxidation in rat liver microsomes effectively with IC_{50} values less than 15µg/ml. The extracts were found to possess ability to scavenge free radicals such as DPPH and superoxide. The phenolic contents present in the extract were also determined. Thus their mixture, triphala, is expected to be more efficient due to the combined activity of the individual component.

Saleem *et al.* (2001) extracted fruits of *Terminalia belerica* with 70 per cent acetone and analyzed for their total phenolic concentration and antioxidant potential. The IC_{50} results indicated that the extract is stronger antioxidant than α-tocopherol. Total phenolics concentration, expressed as gallic acid equivalents, showed close correlation with the antioxidant activity. HPLC analysis with diode array detection at 280 nm of the extract indicated the presence of hydroxybenzoic acid derivatives, hydroxycinnamic acid derivatives, flavonol aglycones and their glycosides as main phenolics compounds.

Terminalia belerica is one of the three constituents of the famous Indian preparation Triphala, the other two being *Phyllanthus emblica* and *Terminalia chebula*. The methanolic extract of 5 commercial Triphala was evaluated by Jayajothi *et al.* (2004) for antioxidant activity by DPPH free radical scavenging method, total phenolic content by Folin-Ciocalteu method and gallic acid equivalents (GAE) by HPTLC method. All extracts exhibited antioxidant activity significantly. A clear correlation between IC_{50} and content of GAE nor the total phenolic content could be observed. Consumption of Triphala has been suggested to exert several beneficial effects by virtue of its antioxidant activity.

Bajpai *et al.* (2005) investigated the phenolic contents and antioxidant activity of some food and medicinal plants including *Terminalia bellerica*. The leaves, fruits and bark were found to have high phenolic contents and antioxidant activity. The TPC of fruits, leaves and bark of *T.bellerica* were 164.5, 110.9 and 72.0 mg/g GAE, respectively. The antioxidant activity of *T.bellerica*, as assayed by auto-oxidation of β-carotene and linoleic acid and expressed as the per cent of inhibition relative to the control were, 90.6 per cent, 72.5 per cent and 69.6 per cent for fruits, leaves and bark, respectively.

Effect of continuous administration of dried 75 per cent methanolic extract of fruits of *Terminalia belerica* suspended in water was studied in alloxan induced hyperglycemia and antioxidant defense mechanism in rats. Oxidative stress produced by alloxan was found to be significantly lowered by the administration of *Terminalia belerica* fruit extract. The results suggest that *Terminalia belerica* fruit extract possessed anti-diabetic and anti-oxidant activity and these activities may be interrelated. Jain and Agrawal (2009) reported antioxidant activity of methanolic extract of *Terminalia belerica*.

Potential *in vitro* antioxidant activities of various fractions of crude methanol extract of *Terminalia belerica* fruits were evaluated by Sherin *et al*. (2010). Ethyl acetate fraction possess potent DPPH radical scavenging activity, which was better than crude methanol extract and reference antioxidant, Trolox. Maximum amounts of total phenolics and flavonoids were found in ethyl acetate fraction which correlates well with its antioxidant activity.

Povichit *et al*. (2010) investigated phenolic content and free radical scavenging effect of *Terminalia belerica* by DPPH and ABTS assays, anti-lipid peroxidation activity by TBARS and for antiglycation activity. The results revealed that the total phenolic content showed good correlation with free radical scavenging by ABTS and anti-lipid peroxidation by TBARS, but showed no correlation with antiglycation. The extract of *Terminalia belerica* demonstrated a moderate antioxidant effect and also showed a moderate antiglycation effect. The IC_{50} were 0.12mg/ml for the DPPH method, TEAC value of 1.20 mg/Trolox/mg sample for the ABTS method, IC_{50} of 0.87 mg/ml for the TBARS method and IC_{50} of 0.02 for the antiglycation method.

Terminalia brachystemma Welw. ex Hiern. Family: Combretaceae

Masoko and Eloff (2007) evaluated the antioxidant activity of leaves of six species of *Terminalia*. Acetone extract of *Terminalia brachystelma* showed strong antioxidant activity.

Terminalia catappa L. Family: Combretaceae

The antioxidant and hepatoprotective actions of *Terminalia catappa* collected from Okinawa island were evaluated by Kinoshita *et al*. (2007) *in vitro* and *in vivo* using leaves extract and isolated antioxidants. A water extract of the leaves of *T. catappa* showed a strong radical scavenging action for DPPH and superoxide anion. Chebulagic acid and corilagin were isolated as the active components from *T.catappa*. Both antioxidants showed a stong scavenging action for superoxide and peroxyl radicals and also inhibited reactive oxygen species production from leukocytes stimulated by phorbol-12-myristate acetate. Galactosamine (GaIN, 600mg/kg, s.c.,) and lipopolysaccharide (LPS, 0.5microg/kg, i.p.)-induced hepatotoxicity of rats as seen by an elevation of serum alanine aminotransferase, aspartate aminotransferase and glutathione S-transferase (GST) activities was significantly reduced when the herb extract or corilagin was given intraperitoneally to rats prior to GaIN/LPS treatment were also decreased by pretreatment with the herb/corilagin. In addition, apoptotic events such as DNA fragmentation and the increase in caspase-3 activity in the liver observed with GaIN/LPS treatment were prevented by the pretreatment with the herb/corilagin. These results showed that the extract of *T. catappa* and its antioxidant, corilagin are protective against GaIN/LPS-induced liver injury through suppression of oxidative stress and apoptosis.

Terminalia chebula Retz. Family: Combretaceae

Methanol aqueous extract of *Terminalia chebula* was screened by Kim *et al.* (1997) for free radical scavenging activity using the DPPH free radical generating system. The extract showed free radical scavenging activity.

Saleem *et al.* (2001) extracted fruits of *Terminalia chebula* with 70 per cent acetone and analyzed for their total phenolic concentration and antioxidant potential. The IC_{50} results indicate that the extract is stronger antioxidant than α-tocopherol. Total phenolics concentration, expressed as gallic acid equivalents, showed close correlation with the antioxidant activity. HPLC analysis with diode array detection at 280 nm of the extract indicated the presence of hydroxybenzoic acid derivatives, hydroxycinnamic acid derivatives, flavonol aglycones and their glycosides as main phenolics compounds.

Naik *et al.* (2003) evaluated the antioxidant activity of aqueous extract of *Acacia catechu* for its potential as antioxidant. The antioxidant activity of these extracts was tested by studying the inhibition of radiation induced lipid peroxidation in liver microsomes at different doses in the rages of 100-600 Gy as estimated by TBARS.

Cai *et al.* (2004, 2006) evaluated the antioxidant activity and phenolic compounds in traditional Chinese medicinal plants associated with anticancer. The improved ABTS.+ method was used to systematically assess the total antioxidant capacity (Trolox equivalent antioxidant capacity, TEAC) of the medicinal extracts. The TEAC values and total phenolic content for methanolic extract of fruit of *Terminalia chebula* were 4403.0 µmol Trolox equivalent/100g dry weight (DW), and 13.16 g of gallic acid equivalent/100g DW respectively. A positive significant linear relationship between antioxidant activity and total phenolic content showed that phenolic compounds were the dominant antioxidant components in the tested herbs. Major types of phenolic compounds in *Terminalia chebula* include tannins (ellagitannins), phenolic acids (gallic acid).

Terminalia chebula is one of the three constituents of the famous Indian preparation Triphala, the other two being *Phyllanthus emblica* and *Terminalia belerica*. The methanolic extract of 5 commercial Triphala was evaluated by Jayajothi *et al.* (2004) for antioxidant activity by DPPH free radical scavenging method, total phenolic content by Folin-Ciocalteu method and gallic acid equivalents (GAE) by HPTLC method. All extracts exhibited antioxidant activity significantly. A clear correlation between IC_{50} and content of GAE nor the total phenolic content could be observed. Consumption of Triphala has been suggested to exert several beneficial effects by virtue of its antioxidant activity.

Aqueous extract of *Terminalia chebula* was tested by Naik *et al.* (2004) for potential antioxidant activity by examining its ability to inhibit gamma-radiation-induced lipid peroxidation in rat liver microsomes and damage to superoxide dismutase enzyme in rat liver mitochondria. The extracts inhibited xanthine/xanthine oxidase activity and is also an excellent scavenger of DPPH radicals.

Bajpai *et al.* (2005) investigated the phenolic contents and antioxidant activity of some food and medicinal plants including *Terminalia chebula*. The leaves, fruits and

Encyclopaedia of Herbal Antioxidants

bark were found to have high phenolic contents and antioxidant activity. The TPC of fruits, leaves and bark of *T.chebula* were 144.7, 162.1 and 150.7 mg/g GAE, respectively. The antioxidant activity of *T.chebula*, as assayed by auto-oxidation of β-carotene and linoleic acid and expressed as the per cent of inhibition relative to the control were, 79.8 per cent, 80.1 per cent and 85.2 per cent for fruits, leaves and bark, respectively.

Aqueous extract of fruit of *Terminalia chebula* exhibited *in vitro* ferric-reducing antioxidant activity and DPPH free radical scavenging activities (Lee *et al.*, 2005). Jain and Agrawal (2009) reported antioxidant activity of methanolic extract of *Terminalia chebula*.

Terminalia crenulata Roth Family: Combretaceae

Jain *et al.* (2010) evaluated the antioxidant activity of aqueous and alcoholic extract of bark of *Terminalia crenulata*. The DPPH and nitric oxide radical inhibiting activity were used to detect oxidative activity. The results of DPPH method showed 50 per cent inhibition rate at the 116.62μg/ml and 153.06μg/ml with alcoholic and aqueous extract, respectively. Nitric oxide scavenging inhibition showed 50 per cent inhibition rate at the 232.85μg/ml and 474.80μg/ml using alcoholic and aqueous extract, respectively.

Terminalia gazensis Bak.f. Family: Combretaceae

Masoko and Eloff (2007) evaluated the antioxidant activity of leaves of six species of *Terminalia*. Acetone and Methanol extracts of *Terminalia gazensis* showed strong antioxidant activity.

Terminalia mollis Laws. Family: Combretaceae

Masoko and Eloff (2007) evaluated the antioxidant activity of leaves of six species of *Terminalia*. Acetone, Hexane and methanol extracts of *Terminalia mollis* showed strong antioxidant activity.

Terminalia muelleri Benth. Family: Combretaceae

Bajpai *et al.* (2005) investigated the phenolic contents and antioxidant activity of some food and medicinal plants including *Terminalia muelleri*. The leaves, fruits and bark were found to have high phenolic contents and antioxidant activity. The TPC of fruits, leaves and bark of *T.muelleri* were 167.2, 75.7 and 144.7 mg/g GAE, respectively. The antioxidant activity, as assayed by auto-oxidation of β-carotene and linoleic acid and expressed as the per cent of inhibition relative to the control were, 83.5 per cent, 81.5 per cent and 74.8 per cent for fruits, leaves and bark, respectively.

Terminalia pallida Brandis Family: Combretaceae

Ethanol extract of *Terminalia pallida* was evaluated for its antiulcer activity against various models of ulcers, such as drug-induced ulcers in Swiss albino rats. The extract at the doses of 250 and 500 mg/kg per os exhibited significant antioxidant protection against ulcers produced by indomethacin, histamine and the effect was comparable to that of the reference drug famotidine (30 mg/kg b.w.) orally). The extract significantly lowered the elevated lipid peroxide level (TBARS) and restored the altered glutathione level in ethanol-induced gastric ulceration (Gupta *et al.*, 2005).

Terminalia prunoides Welw. ex Hiern. Family: Combretaceae

Masoko and Eloff (2007) evaluated the antioxidant activity of leaves of six species of *Terminalia*. Acetone and methanol extracts of *Terminalia prunoides* showed strong antioxidant activity.

Terminalia sambesica Engl. et Diels. Family: Combretaceae

Masoko and Eloff (2007) evaluated the antioxidant activity of leaves of six species of *Terminalia*. Acetone extract of *Terminalia sambesiaca* showed strong antioxidant activity.

Terminalia sericea Burch. ex DC. Family: Combretaceae

Masoko and Eloff (2007) evaluated the antioxidant activity of leaves of six species of *Terminalia*. Acetone and methanol extracts of *Terminalia sericea* showed strong antioxidant activity.

Tetracera loureiri (Fin. et Gagnep.) Pierre ex Craib. Family: Dilleniaceae

Kukongviriyapan *et al.* (2003) described *in vitro* and *in vivo* antioxidant and hepatoprotective activities of *Tetracera loureiri*. The ethanol extract of *T. loureiri* possessed potent antioxidant and strong free radical scavenging properties assayed using FRAP and DPPH, respectively. The cytoprotective effects of *T. loureiri* were demonstrated in ethanolic extracts of freshly isolated rat hepatocytes against the chemical toxicants paracetamol and tertiary-butylhydroperoxide. The cells pretreated with the extract maintained the GSH/GSSG ratio and suppressed lipid peroxidation in a dose dependent manner. Pretreating rats with the ethanol extract orally, one hour prior to intraperitoneal injection of toxic doses of paracetamol, prevented elevations of plasma ALT and AST.

Tetracera potatoria G. Don Family: Dilleniaceae

Oluwole *et al.* (2008) investigated the mechanism underlying antiulcer activity and endogenous antioxidants of the methanolic extract of the root of *Tetracera potatoria* (MeTp). The MeTp treated animals showed significant increase in the activity of SOD with concurrent decrease in the level of malondialdehyde with respect to control.

Tetracera volubilis L. Family: Dilleniaceae

Lock *et al.* (2005) evaluated the *in vitro* antioxidant activity of 40 Peruvian plants using DPPH assay. *Tetracera volubilis* showed highest antioxidant activity with EC_{50} value of 5.29µg/ml. The crude ethanolic extract of *Tetracera volubilis* has better antioxidant activity *in vitro* than the pure natural antioxidant rutin (EC_{50} = 7.6µg/ml)

Tetrorchidium andinum Muell.-Arg. Family: Euphorbiaceae

Mosquera *et al.* (2007) studied the antioxidant activity of the extracts using DPPH free-radical scavenging method. The percentage of the antioxidant activity of the n-Hexane fraction was 24.04, Dichloromethane fraction was 28.97 and methanol fraction was 27.7.

Teucrium chamaedrys C.Koch Family: Lamiaceae (Labiatae)

English Name: Wall germander.

Kadifkova-Panovska *et al.* (2005) investigated the antioxidant activity of extracts of *Teucrium chamaedrys* prepared using different organic solvents namely diethyl ether, ethyl acetate and n-butanol. Using such extracts they assessed their antioxidant abilities in a series of assays that included inhibition of DPPH inhibition of hydroxyl radical (D-ribose assay) and inhibition of β-carotene oxidation. They found that a 0.4mg/ml extract of *T. chamaedrys* was very effective in inhibiting β-carotene oxidation.

The antioxidant properties of pure metabolites, as well as of crude organic extracts of the plant, have been analyzed by Pacifico *et al.* (2009) on the basis of their DPPH radical scavenging capability. The antioxidant capacity in cell-free systems of the isolated metabolites was carried out by measuring their capabilities to inhibit the synthesis of thiobarbituric acid reactive species in assay media using as oxidable substrates, a vegetable fat and a pentose sugar 2-deoxyribose and to prevent oxidative damage of the hydrosoluble bovine serum albumin (BSA) protein. Phenylethanoid glycosides resulted efficacious DPPH radical, while iridiod glycosides prevent massively the 2-deoxyribose and BSA oxidation in assay media.

Gursoy and Tepe (2009) examined the antioxidant activities and the amount of total phenolics of the methanol extracts of *Teucrium chamaedrys*. The subfractions were screened for their antioxidant activities by two complementary tests, namely DPPH free radical-scavenging and β-carotene/linoleic acid assays. Non polar extracts of the plant species remained inactive in both test systems. On the other hand polar extracts showed remarkable antioxidant activities. In DPPH system free radical scavenging effect of *T.chamaedrys* was measured as 18.00μg/mg. Polar subfraction was found to be as effective as the positive control BHT. Non-polar subfraction was found to have the highest total phenolic amount (97.12μg/mg).

Teucrium marum L. subsp. *marum* Family: Lamiaceae (Labiatae)

English Name: Cat thyme

Ricci *et al.* (2005) evaluated the chemical composition, antimicrobial and antioxidant activity of the essential oil of *Teucrium marum* subsp. *marum*. The antioxidant activity of the essential oil was evaluated using the DPPH test, 5-lipoxygenase test and luminal/xanthine/xanthine oxidase chemiluminiscence assay.

Teucrium montanum L. Family: Lamiaceae (Labiatae)

English Name: Mountain germander.

Kadifkova-Panovska *et al.* (2005) investigated the antioxidant activity of extracts of *Teucrium montanum* prepared using different organic solvents namely diethyl ether, ethyl acetate and n-butanol. Using such extracts they assessed their antioxidant abilities in a series of assays that included inhibition of DPPH inhibition of hydroxyl radical (D-ribose assay) and inhibition of β-carotene oxidation. They found that a 0.4mg/ml extract of *T.montanum* was very effective in inhibiting β-carotene oxidation.

The free radical scavenging activities of petroleum ether, chloroform, ethyl acetate, n-butanol and water extracts of *Teucrium montanum* were investigated by Djilas *et al.*

(2006). n-Butanol extract possessed potent DPPH free radical scavenging activity. Diterpenoids, flavonoids and phenolic acids have been reported in the plant.

Teucrium polium L. Family: Lamiaceae (Labiatae)

Couladis *et al.* (2003) screened Greek aromatic plants from the Lamiaceae family for the antioxidant activity. Of the 21 plants they tested they found ethanol extracts prepared from *Teucrium polium*, as well as 9 other plants exhibited the same antioxidant activity as α-tocopherol in their ability to inhibit bleomycin-Fe(II) complex-induced arichidonic acid superoxidation to MDA.

Suboh *et al.* (2004) investigated the antioxidant activity effects of various concentrations of methanol extracts of *Teucrium polium* and six other medicinal plants as high as 1 mg/ml. They reported that the extract significantly reduced 10mM hydrogen peroxide-induced lipid peroxidation in human erythrocytes. In contrast, the extract failed to protect erythrocytes against protein oxidation and loss of cell deformability of the oxidatively stressed erythrocytes.

Kadifkova-Panovska *et al.* (2005) investigated the antioxidant activity of extracts of *Teucrium polium* prepared using different organic solvents namely diethyl ether, ethyl acetate and n-butanol. Using such extracts they assessed their antioxidant abilities in a series of assays that included inhibition of DPPH inhibition of hydroxyl radical (D-ribose assay) and inhibition of β-carotene oxidation. They found that a 0.4mg/ml extract of *T.polium* was very effective in inhibiting β-carotene oxidation. Aerial parts of *T.polium* are rich in flavonoids.

Ljubuncic *et al.* (2006) investigated the antioxidant action of the extract of *Teucrium polium*. They assessed (i) its ability to inhibit (a) oxidation of β-carotene, (b) AAPH-induced plasma oxidation, (c) iron induced lipid-peroxidation in rat liver homogenates; (ii) to scavenge superoxide radical and the hydroxyl radical (iii) its effect on the enzyme xanthine oxidase activity, (iv) its capacity to bind iron and (v) its effect on cell glutathione (GSH) homeostasis in cultured HepG2 cells. They found that the extract (i) inhibited oxidation of β-carotene, (b) AAPH-induced plasma oxidation, (c) Fe^{2+}-induced lipid peroxidation in rat liver homogenates ($IC_{50} = 7\mu g/ml$) scavenged superoxide radical ($IC_{50} = 127\mu g/ml$) and hydroxyl radical ($IC_{50} = 667\mu g/ml$), (iii) binds iron ($IC_{50} = 797\mu g/ml$) and (iv) tended to increase intracellular GSH levels resulting in a decrease in the GSSG/GSH ratio.

Al-Mustafa and Al-Thunibat (2008) evaluated the antioxidant activity of Jordanian medicinal plant *Teucrium polium* shoot. The level of antioxidant activity was determined by DPPH and ABTS assay in relation to total phenolic contents of the medicinally used part. They concluded that *Teucrium polium* shoot possess low antioxidant capacity (DPPH-TEAC range 20-80mg/g).

Esmaeili *et al.* (2009) reported antioxidant and protective effects of major flavonoids from *Teucrium polium* on β-cell destruction in a model of streptozotocin-induced diabetes.

The crude extracts and isolated compounds were screened for their antioxidant and free radical scavenging activities using DPPH radical-scavenging, β-carotene/linoleic acid and ammonium thiocyanate methods (Sharififar *et al.*, 2009). Methanol extract, rutin and apigenin were found to be the most active fractions as radical-scavengers with IC_{50} values of 20.1, 23.7 and 30.3µg/ml, respectively. The samples with the highest inhibition of oxidation of β-carotene and lipid peroxidation in ammonium thiocyanate methods were also found to be methanol extract, rutin and apigenin. Methylated flavonoids exhibited a lesser antioxidant activity.

Tepe *et al.* (2011) reported that water extracts of *Teucrium polium* exhibited excellent antioxidant activity.

Thalictrum minus L. Family: Ranunculaceae

English Name: Low meadow-rue.

Karyagina *et al.* (2011) studied the antioxidant complex of *Thalictrum minus* in cell extracts and culture medium. In these model systems the inhibition of lipid oxidation and DPPH reduction (EC_{50} = 12-15µg/ml) were observed. At the phenolic compound concentration of 8-15µg/ml, the reducing capacity of cell extracts was equivalent to 1.5mM ascorbic acid. At the same time, berberine, a major alkaloid synthesized by the culture manifested a low antioxidant activity. The analysis of phenolic acid composition in low meadow-rue showed that one of the main components of its antioxidant system were caffeic, gallic, chlorogenic and ferulic acids.

Thelesperma megapotamicum (Spreng.) Herter Family: Asteraceae (Compositae)

Borneo *et al.* (2008) evaluated the antioxidant activity of ethanolic extract of *Thelesperma megapotamicum*. The FRAP value was 435.2µmol of Fe(II)/g and DPPH radical scavenging activity was EC_{50} = 1607.1.

Theobroma cacao L. Family: Sterculiaceae

English Name: Cocoa

Because rapid methods to evaluate and compare food products for antioxidant benefits are needed, a new assay based on liquid chromatography-mass spectrometry (LC-MS) was developed for the identification and quantitative analysis of antioxidants in complex natural product samples such as food extracts. This assay is based on the comparison of electrospray LC-MS profiles of sample extracts before and after treatment with reactive oxygen species such as hydrogen peroxide or DPPH. Using this assay, methanolic extracts of cocoa powder were analyzed, and procyanidins were found to be the most potent antioxidant species. These species were identified using LC-MS, LC-MS/MS, accurate mass measurement, and comparison with reference standards. Furthermore, LC-MS was used to determine the levels of these species in cocoa samples. Catechin and epicatechin were the most abundant antioxidants followed by their dimers and trimers. The most potent antioxidants in cocoa were trimers and dimers of catechin and epicatechin, such as procyanidin B2, followed by catechin and epicatechin. This new LC-MS assay facilitates the rapid identification and then the determination of the relative antioxidant activities of individual antioxidant species in complex natural product samples and food products such as cocoa.

Lee *et al.* (2003) compared the phenolic and flavonoid contents and total antioxidant capacities of cocoa, black tea, green tea, and red wine. Cocoa contained much higher levels of total phenolics (611 mg of gallic acid equivalents, GAE) and flavonoids (564 mg of epicatechin equivalents, ECE) per serving than black tea (124 mg of GAE and 34 mg ECE, respectively), green tea (165 mg of GAE and 47 mg of ECE), and red wine (340 mg of GAE and 163 of ECE). Total antioxidant activities were measured using ABTS and DPPH radical scavenging assays and are expressed as vitamin C equivalent antioxidant capacities (VCEACs). Cocoa exhibited the highest antioxidant activity among the samples in ABTS and DPPH assays, with VCEACs of 1128 and 836 mg/serving, respectively. The total antioxidant capacities from ABTS and DPPH assays were highly correlated with phenolic content and flavonoid content.

Osman *et al.* (2004) investigated the components of cocoa leaves. The main catechin-polyphenols in extracts were epicatechin (EC), epigallocatechin, gallate (EGCG), epigallocatechin (EGC), gallic acid (GA) and epicatechin gallate (ECG). The antioxidation properties of the polyphenol extracts were tested, using ferric chloride reduction, and compared against a synthetic antioxidant (BHA). The polyphenol extracts (CS and CL) showed similar antioxidation powers to GT and BHA throughout the entire concentration range (100–2000 ppm). In the oil-based test medium; the antioxidative performance of polyphenol extracts were better than BHA at 50 ppm. At 200 ppm, the performance is quite similar to BHA. At higher concentration (400 ppm) the antioxidant activities are much better than BHA. In the presence of Cu2þ prooxidant (20 ppm), BHA (200 ppm) and all the extracts (200 pmm) showed similar performances.

Arlorio *et al.* (2005) reported antioxidant activity of phenolic pigments from *Theobroma cacao* hulls extracted with supercritical CO_2.

Othman *et al.* (2010) investigated the relationship between antioxidant potential and epicatechin content of raw cocoa beans from different countries, namely Malaysia, Ghana, Cote d'Ivoire and Sulawesi (Indonesia). The epicatechin content of raw cocoa beans was in the range of 270-1235 mg/100g cocoa beans. Based on the TEAC and FRAP assays, Sulawesian beans exhibited the highest antioxidant capacity followed by Malaysian, Ghanaian and Cote d'Ivoirian beans. Both ethanolic and water extracts of cocoa beans showed a significant positive and high correlation between epicatechin and TEAC value. Similarly FRAP assay also showed a positive and high correlation with epicatechin for both ethanolic and water extracts.

Theobroma grandiflorum (Willd. ex Spreng.) Schum. Family: Sterculiaceae

English Name: Cupuacu.

Activity-guided fractionation of *Theobroma grandiflorum* seeds by Yang *et al.* (2003) resulted in the identification of two new sulfated flavonoid glycosides, theograndins I and II. In addition, nine known flavonoid antioxidants (+)-catechin, (-)-epicatechin, isoscuterllarin 8-O-β-D-glucoronide, quercetin-3-O- β-D-glucoronide, quercetin-3-O- β-D-glucoronide 6″-methyl ester, quercetin, kaempferol, and isoscutellarin 8-O- β-D-glucoronide 6″-methyl ester were identified. Theograndin II displayed antioxidant activity (IC_{50}=120.2 µM) in the DPPH free-radical assay with

IC_{50} value of 143µM. While theograndin I was less active as an antioxidant than theograndin II, the known compounds were more potent in the DPPH assay (IC_{50} range 39.7-89.7µM).

Thespesia populnea (L.) Soland. ex Corr. Family: Malvaceae

Antioxidant activity of the aqueous and methanolic extracts of the *Thespesia populnea* bark was investigated by Ilavarasan *et al.* (2003) in rats by inducing liver injury with carbon tetrachloride : Olive oil (1:1). Bark extracts at a dose level of 500 mg/kg showed significant antioxidant activity against carbon tetrachloride-induced liver injury in rats.

Thespesia populneoides (Roxb.) Kostel Family: Malvaceae

Vadlapudi and Naidu (2009) evaluated the antioxidant potential of *Thespesia populneoides*, a mangrove plant. The SOD was 1.65U/mg, Catalase was 0.24U/mg, Ascorbic acid was 170mg/100g, DPPH radical inhibition was 64.1 per cent and FRAP units were 1038.

Thevetia neriifolia Juss. ex Steud Family: Apocynaceae

Ayoola *et al.* (2008) carried out phytochemical and antioxidant screening of *Thevetia neriifolia*. The extract tested positive for tannins, flavonoids, glycosides and anthraquinone. DPPH free radical scavenging assay was followed and the extracts showed good antioxidant activity.

Thonningia sanguinea Vahl Family: Balanophoraceae

Work on some Ghananian medicinal plants revealed that *Thonningia sanguinea* possess free-radical scavenging capacity and strong hepatoprotective activity inhibiting hydrogen peroxide-induced lipid peroxidation, galactosamine-induced hepatitis and carbon tetrachloride-induced hepatotoxicity (Gyamfi *et al.*, 1999). Another report on *T. sanguinea* harvested from elsewhere in Africa confirmed its antioxidant potentials and free-radical scavenging activity (Ohtani *et al.*, 2000). In fact these authors isolated two ellagitannins, thonningianins A (1) and thonningianins A (2) as the major antioxidant principle in this plant.

Thunbergia laurifolia Lindl. Family: Acanthaceae

Different drying treatments were tested by Chan and Lim (2006)on *Thunbergia laurifolia* leaves to produce a tea with total phenolic content (TPC) and antioxidant activity (AOA) comparable to fresh leaves and with acceptable aroma. TPC was measured using the Folin-Ciocalteu method while AOA was measured using DPPH, ferric reducing antioxidant power (FRAP) and ferrous ion chelating (FIC) -assays. Using a household microwave oven for drying, *Thunbergia* tea with superior antioxidant activity could be produced. Unlike oven-dried and sun-dried leaves, TPC and AOA of microwave-dried *Thunbergia* leaves were higher than fresh leaves. The microwave dried *Thunbergia* tea was superior in terms of TPC and AOA compared with other commercial herbal teas tested.

Antioxidant activities and total phenolic content of *Thunbergia laurifolia* extracts were evaluated using free radical scavenging, FRAP and the Folin-Ciocalteu method.

It was found that water extraction of phenolic compounds was the most efficient (2433.9mg GAE/100g) compared to ethanol and acetone extraction which had phenolic contents 565 and 142.1 mg GAE/100g, respectively. In addition *Thunbergia laurifolia* water extract possessed the highest antioxidant activities using free radical scavenging at the EC_{50} values of 0.13 mg GAE/ml, whereas ethanol and acetone extract showed EC_{50} at 0.26 and 0.61 mg GAE/ml respectively. Finally, the water extract also showed the highest total antioxidant activity using FRAP assay at 0.93 mmol/g followed by ethanol and acetone extracts (0.18 and 0.04 mmol/g).

Palipoch *et al.* (2010) evaluated the antioxidant properties of leaf extracts of *Thunbergia laurifolia* using DPPH assay. The extracts exhibited strongly dose dependent and highly reducing power activity and scavenging abilities but less activity than positive control, ascorbic acid.

Thymbra spicata L. Family: Lamiaceae (Labiatae)

Kosar *et al.* (2003) prepared water extracts from steam distilled essential oil-extracted *Thymbra spicata*. The HPLC-DPPH on-line method was applied to the qualitative and quantitative analysis of this plant extract. There was a strong correlation between the scavenging (negative) peak area and the concentration of the radical scavenging reference substances used. The radical scavenging compounds within the extracts were determined as benzoic acid and hydroxycinnamic acid derivatives, flavonoids and diterpenoids according to their retention time and UV spectral data. Rosmarinic acid and carnosic acid were identified as the dominant radical scavengers in these extracts by this method.

Water-soluble extract from *Thymbra spicata* was screened by Dorman *et al.* (2004) for antioxidant properties in a battery of six *in vitro* assays. The extract demonstrated varying degrees of efficacy in each screen. The extract contained Folin-Ciocalteu reagent-reactive substances, which was confirmed by the presence of polar phenolic analytes (*i.e.*, hydroxybenzoates, hydroxycinnamates, and flavonoids).

Avci *et al.* (2006) evaluated the hypercholesterolaemic and antioxidant activities of *Thymbra spicata* var. *spicata*. The ethanol extract increased the serum HDL concentration and decreased serum LDL-C concentration. The ethanolic extracts significantly decreased MDA level in mice.

Thymophylla pentachaeta (DC.) Small var. *belenidium* (DC.) Strother Family: Asteraceae (Compositae)

Borneo *et al.* (2008) evaluated the antioxidant activity of ethanolic extract of *Thymophylla pentachaeta* var. *belenidium*. The FRAP value was 551.4μmol of Fe(II)/g and DPPH radical scavenging activity was $EC_{50} = 458.7$.

Thymus boveii Benth. Family: Lamiaceae (Labiatae)

Tepe *et al.* (2011) examined the chemical composition, radical scavenging and antimicrobial activity of the Essential oils of *Thymus boveii*. The scavenging ability of the essential oils showed a concentration-dependent activity profile. The strongest free radical scavenging activity in DPPH assay was exhibited by *T. bovei* essential oil (82.75 per cent at 0.050 mg/ml). In β-carotene-linoleic acid system the essential oil of

T.boveii at 2.0 mg/ml concentration, antioxidant activity of the oil was measured as 97.51 per cent. Reducing power of the sample at 1.0 mg/ml concentration was 0.565 nm.

Thymus bracteosus Vis. ex Benth. Family: Lamiaceae (Labiatae)

The composition of the essential oil and the antioxidant properties of the dried herbs of *Thymus bracteosus* were investigated by Brantner *et al.* (2005). The essential oil of *Thymus bracteosus* did not show any antioxidant activity in the test systems.

Thymus caespitosus Brot. Family: Lamiaceae (Labiatae)

Dandlen *et al.* (2010) evaluated the antioxidant activity of essential oil of six species of *Thymus* including *Thymus caespitosus*. The oil showed good activity in preventing lipid peroxidation and scavenging free radicals and hydroxyl and superoxide anions. The oil in which thymol or carvacrol predominated did not present significant activities, which according to the authors indicated that these phenolic compounds were not determinant in the ability for scavenging hydroxyl radicals.

Thymus camphoratus Hoffmans. et Link Family: Lamiaceae (Labiatae)

Dandlen *et al.* (2010) evaluated the antioxidant activity of essential oil of six species of *Thymus* including *Thymus camphoratus*. The oil showed good activity in preventing lipid peroxidation and scavenging free radicals and hydroxyl and superoxide anions. The oil in which thymol or carvacrol predominated did not present significant activities, which according to the authors indicated that these phenolic compounds were not determinant in the ability for scavenging hydroxyl radicals.

Thymus capitatus (L.) Hoffman et Link Family: Lamiaceae (Labiatae)

Al-Mustafa and Al-Thunibat (2008) evaluated the antioxidant activity of Jordanian medicinal plant *Thymus capitatus* shoot. The level of antioxidant activity was determined by DPPH and ABTS assay in relation to total phenolic contents of the medicinally used part. They concluded that *Thymus capitatus* shoot possesses moderate antioxidant capacity (DPPH-TEAC range 20-80mg/g).

Thymus capitellatus Hoffmans. et Link Family: Lamiaceae (Labiatae)

Dandlen *et al.* (2010) evaluated the antioxidant activity of essential oil of six species of *Thymus* including *Thymus capitellatus*. The oil showed good activity in preventing lipid peroxidation and scavenging free radicals and hydroxyl and superoxide anions.

Thymus caramanicus Jalas Family: Lamiaceae (Labiatae)

Chemical composition of the essential oil, antioxidant activity (DPPH and β-carotene/linoleic acid assays), and total phenolic content (Folin-Ciocalteu assay) of aerial parts of *Thymus caramanicus* were determined by Safaei-Ghomi *et al.* (2009). The highest radical scavenging activity (DPPH test) was shown by the polar subfraction of the methanol extract (IC_{50} = 43µg/ml) which was also higher than that of BHT (IC_{50} = 9.7 µg/ml). However it was the nonpolar subfraction of the methanol extract that showed the highest inhibition (84.4 per cent) as assessed by the β-carotene/linoleic acid assay, which was only slightly lower than that shown by BHT (93.3 per cent). Total phenolic content of the polar subfraction, as GAE, was 124.3µg/mg.

Thymus carnosus Boiss. Family: Lamiaceae (Labiatae)

Dandlen *et al.* (2010) evaluated the antioxidant activity of essential oil of six species of *Thymus* including *Thymus carnosus*. The oil showed good activity in preventing lipid peroxidation and scavenging free radicals and hydroxyl and superoxide anions.

Thymus daenensis Celak. Family: Lamiaceae (Labiatae)

Firuzi *et al.* (2010) evaluated the antioxidant activity and total phenolic content of 24 Lamiaceae species growing in Iran. *Thymus daenensis* showed high antioxidant activity with 25.1µM/g DW FRAP value and IC_{50} = 506.3µg/ml of DPPH radical scavenging activity. It also showed high phenolic content of 10.6 mg CE/g. FRAP and DPPH assay results showed good correlations with the total phenolic contents of the plant measured by the Folin-Ciocalteu assay.

Thymus eigii M.Zohary et P.H.Davis Family: Lamiaceae (Labiatae)

Essential oils, nonpolar and polar extracts were subjected to a screening for their possible antioxidant activity using DPPH and β-carotene –linoleic acid assays (Tepe *et al.,* 2004). In the former case, the polar subfraction of the methanol extract was found to be superior to all extracts tested, only 16.8 micro g/ml of which provided 50 per cent inhibition, whereas all extracts, particularly the polar ones seem to inhibit the oxidation of linoleic acid in the latter case.

Thymus guyonii Noe Family: Lamiaceae (Labiatae)

Hazzit *et al.* (2006) reported that essential oil of *Thymus guyonii* possessed antioxidant activity.

Thymus hirtus ssp. algeriensis Family: Lamiaceae (Labiatae)

Ahmed *et al.* (2011) evaluated the antioxidant activities of 10 plants from Tunisia by the DPPH method. The essential oil of *Thymus hirtus* ssp. *algeriensis* presented a moderate DPPH scavenging effect. The essential oil was rich in monoterpenoids especially linalool at 17.62 per cent and camphor at 13.82 per cent

Thymus hyemalis L. Family: Lamiaceae (Labiatae)

Tepe *et al.* (2011) examined the chemical composition, radical scavenging and antimicrobial activity of the Essential oils of *Thymus hyemalis*. The scavenging ability of the essential oils showed a concentration-dependent activity profile. The strongest free radical scavenging activity in DPPH assay was exhibited by *T. hyemalis* essential oil (73.48 per cent at 0.050 mg/ml). In β-carotene-linoleic acid system the essential oil of *T. hyemalis* at 2.0 mg/ml concentration, antioxidant activity of the oil was measured as 89.33 per cent. Reducing power of the sample at 1.0 mg/ml concentration was 0.548 nm.

Thymus longicaulis C. Presl. subsp. *longicaulis* var. *longicaulis* Family: Lamiaceae (Labiatae)

Sarikurkcu *et al.* (2010) examined the chemical composition and *in vitro* antioxidant activity of the hydrodistillated essential oil and various extracts obtained from *Thymus longicaulis* subsp. *longicaulis* var. *longicaulis*. Antioxidant activities of the samples were determined by four different test systems namely β-carotene/linoleic acid, DPPH, reducing power and chelating effect. Essential oil showed the highest antioxidant activity in β-carotene/linoleic acid, system. In the case of other test systems, in general, methanol and water extracts exhibited the strongest activity profiles. Especially, reducing power of water extract was found superior than those of synthetic antioxidants.

Thymus marschallinus Willd. Family: Lamiaceae (Labiatae)

A study was carried out by Budinevi *et al.* (1995) on the antioxidative activity of *Thymus marschallinus*. Using the ethanolic extracts of plant material the best results were obtained. Antioxidative activity of powdered plant was studied on the Rancimat apparatus at 100°C, and increased induction periods were observed. The extract of *Thymus marschallinus* was also in the emulsion of linoleic acid and the course of oxidation was followed by measuring the decoloration rate of β-carotene emulsion.

The ability to scavenge hydroxyl radicals of the essential oils of *Thymus marschallinus* and *Thymus proximus* were evaluated using the safranine method and the authors reported that the latter oil was more effective than the former. The activity of both the oils were dose-dependent and they were mainly constituted by thymol, *p*-cymene and γ-terpinene (Jia *et al.*, 2010).

Thymus migricus Klokov et Desj-Shost. Family: Lamiaceae (Labiatae)

Souri *et al.* (2004) evaluated the antioxidative activity of *Thymus migricus* by linoleic acid peroxidation test using 1,3-diethyl-2-thiobarbituric acid as the reagent. Methanolic extract showed 97.00 per cent inhibition at 40µg concentration with IC_{50} value of 0.77µg which is approximately in the range of α-tocopherol (IC_{50}=0.60µg).

Thymus munbyanus Boiss. et Reut. Family: Lamiaceae (Labiatae)

Hazzit *et al.* (2006) reported that essential oil of *Thymus munbyanus* possessed antioxidant activity.

Thymus numidicus Poir. Family: Lamiaceae (Labiatae)

Hazzit *et al.* (2006) reported that essential oil of *Thymus numidicus* possessed antioxidant activity.

Thymus pallescens Noe Family: Lamiaceae (Labiatae)

Hazzit *et al.* (2006) reported that essential oil of *Thymus pallescens* possessed antioxidant activity.

Thymus pectinatus Fisch. et Mey. var. *pectinatus* Family: Lamiaceae (Labiatae)

The essential oil and water-soluble (polar) and water-insoluble (nonpolar) subfractions of the methanol extract of *Thymus pectinatus* var. *pectinatus* were assayed for their antimicrobial and antioxidant properties. The essential oil, in particular the polar subfraction of the methanol extract showed antioxidant activity. Thymol and carvacrol were individually found to possess weaker antioxidant activity than the crude oil itself, indicating that other constituents of the essential oil may contribute to the antioxidant activity observed (Vardar-Unlu *et al.*, 2003).

Thymus praecox Opiz. subsp. *caucasicus* var. *caucasicus* Family: Lamiaceae (Labiatae)

The dichloromethane, ethyl acetate, ethanol and aqueous extracts of *Thymus praecox* subsp. *caucasicus* var. *caucasicus* along with the essential oils were assessed by Orhan *et al.* (2009) for their antioxidant activities. Antioxidant activity was evaluated by DPPH radical scavenging and ferrous iron-chelating tests. Ferric-reducing antioxidant power (FRAP) was also tested. The ethyl acetate and ethanol extracts exerted significant DPPH scavenger effect. The water extracts displayed the highest ferrous iron-chelating effect. The leaf and flower essential oils had the best FRAP.

Thymus proximus Sergl. Family: Lamiaceae (Labiatae)

The ability to scavenge hydroxyl radicals of the essential oils of *Thymus marschallinus* and *Thymus proximus* were evaluated using the safranine method and the authors reported that the latter oil was more effective than the former. The activity of both the oils were dose-dependent and they were mainly constituted by thymol, *p*-cymene and γ-terpinene (Jia *et al.*, 2010).

Thymus pubescens Boiss. et Kotschy ex Celak Family: Lamiaceae (Labiatae)

Souri *et al.* (2004) evaluated the antioxidative activity of *Thymus pubescens* by linoleic acid peroxidation test using 1,3-diethyl-2-thiobarbituric acid as the reagent. Methanolic extract showed 97.42 per cent inhibition at 40μg concentration with IC_{50} value of 0.84μg which is approximately in the range of α-tocopherol (IC_{50}=0.60μg).

Thymus pulegoides L. Family: Lamiaceae (Labiatae)

Dandlen *et al.* (2010) evaluated the antioxidant activity of essential oil of six species of *Thymus* including *Thymus pulegoides*. The oil showed good activity in preventing lipid peroxidation and scavenging free radicals and hydroxyl and superoxide anions.

Thymus satureioides Coss et Bal. Family: Lamiaceae (Labiatae)

The methanol extract of *Thymus satureioides* showed significant radical-scavenging effect (SC = 14.54 microg) (Ismaili *et al.* (2004).

Thymus sipyleus Boiss. subsp. *sipyleus* var. *sipyleus* Family: Lamiaceae

Tepe *et al.* (2005) evaluated the antioxidant activity of *Thymus sipyleus* subsp. *sipyleus* var. *sipyleus* and var. *rosulanus*. Essential oil of both varieties showed good antioxidant activity and in some cases comparable to synthetic antioxidant BHT.

One triterpenic acid (ursolic acid), one phenolic acid (rosmarinic acid), and four flavonoids (luteolin, luteolin 7-O-(6"-feruloyl)-β-glucopyranoside, and luteolin 7-O-β-glucuronide) were isolated from the aerial parts of *Thymus sipyleus* subsp. *sipyleus* var. *sipyleus*. *In vitro* lipid peroxidation inhibition effects of the compounds were determined using TBA test method in a bovine braine liposome system. All compounds inhibited lipid peroxidation in various degrees except for ursolic acid. The activity order of the compounds was completely different in DPPH radical-scavenging activity. None of the compounds showed Fe^{2+} chelating activity (Özgen *et al.*, 2011).

Thymus spathulifolius Hausskn. et Velen. Family: Lamiaceae (Labiatae)

Sokmen *et al.* (2004) evaluated the *in vitro* antimicrobial and antioxidant properties of essential oil and methanol extracts from *Thymus spathulifolius*. The antioxidant potential of the samples was evaluated using two separate methods, inhibition of free radical DPPH and β-carotene-linoleic acid systems. The polar subfraction of the methanol extract was able to reduce the stable free radical DPPH with an IC_{50} of 16.15µg/ml, which was lower than that of synthetic antioxidant BHT (19.8µg/ml). Inhibition values of linoleic acid oxidation were calculated as 92 per cent and 89 per cent for the oil and the polar subfraction, respectively. Gallic acid equivalent total phenolic constituent of the polar subfraction was 141.00µg/mg (14.1 per cent w/w).

Thymus vulgaris L. Family: Lamiaceae (Labiatae)

English Name: Thyme.

A biphenyl compound, 3,4,3',4'-tetrahydroxy-5,5'-diisopropyl-2,2'-dimethylbiphenyl (1), and a flavonoid, eriodicytol (2), were isolated as antioxidative components from the leaves of *Thymus vulgaris* by bioassay-directed fractionation. These compounds inhibited superoxide anion production in the xanthin/xanthine oxidase system. Mitochondrial and microsomal lipid peroxidation induced by Fe(III)-ADP/NADH of Fe(III)-ADP/NADPH were also inhibited by these compounds. Compound 1 was an extremely potent antioxidant; complete inhibition was observed at 1 microM against both microsomal and mitochondrial peroxidation (Haraguchi *et al.*, 1996).

The antioxidative activity of phenolic extracts from thyme was examined by autoxidation of methyl linoleate by Kahkonen *et al.* (1999). Thyme possessed strong antioxidant activity in this study as expected on the basis of previous studies (Takacsova *et al.*, 1995; Nakatani, 1997; Hirasa and Takemasa, 1998). Thymol and carvacrol were major aroma components of essential oil of thyme, and both showed high antioxidant and antimicrobial activity. Bi-phenyl compounds, dimerization products of thymol and carvacrol and a flavonoid (eriodicytol) have also been isolated as efficient antioxidants inhibiting superoxide anion production in the xanthine/xanthine oxidase system and mitochondrial and microsomal lipid peroxidation (Haraguchi *et al.*, 1996). Highly methylated flavonoids with antioxidant activity in linoleic acid oxidation system have been found in the less polar fraction (Miura and Nagatani, 1989).

Antioxidant activity of 4 flavonoids isolated from thyme was evaluated by the oil stability index method using a model substrate oil including methyl lineolate in silicone oil at 90°C (Miura *et al.*, 2002). Carnosol, rosamarol, epirosmarol, isorosmarol, galdosol and carnosic acid exhibited remarkably strong activity which was comparable to that of α-tocopherol. The activity of miltirone, atuntzenin A, luteolin, 7-O-methyl luteolin, and eupafolin was comparable to that of BHT. The activity of these compounds was mainly due to the presence of ortho-dihydroxy groups. The DPPH radical scavenging activity of these compounds showed the similar result.

DPPH scavenging-guided fractionation of a leaf extract of *Thymus vulgaris* led to the isolation of the radical scavengers rosmarinic acid 1, eriodictyol, taxifolin, luteolin 7-glucoronide, *p*-cymene 2,3-diol, *p*-cymene 2,3-diol-dimer, carvacrol, thymol and a new compound 2. Phenylpropanoid trimer 2 was a weaker and stronger radical scavenger than rosmarinic acid 1 in off-line TEAC and DPPH assays, respectively (Dapkevicius *et al.*, 2002).

Agbor *et al.* (2005) evaluated the antioxidant capacity of 14 herbs from Cameroon. Methanol and HCl in methanol extracts were analyzed using two different antioxidant assay methods (Folin-Ciocalteu method and FRAP method). The 1.2M HCl in methanol extract of *Thymus vulgaris* had significantly higher antioxidant capacity than the methanolic extract. The FRAP antioxidant values were significantly higher than the Folin antioxidant values. *Thymus vulgaris* showed good activity in Folin-Ciocalteu method. Irrespective of the assay methods used, the samples were rich in antioxidants.

The entire plant possesses antioxidative properties. It enhances blood circulation. 4-terpenol, alanine, β-carotene, caffeic acid, camphene, carvacrol, γ-terpinene, lycopene, myrcene and palmitic acid are the active antioxidative constituents in thyme (Mishra *et al.*, 2007).

Mariutti *et al.* (2008) evaluated the antioxidant activity of Thyme employing ABTS and DPPH radical assay. DPPH EC_{50} value was 6.4g/kg while DPPH-TEAC value was 3.3mM/g. ABTS-TEAC value was 3.8mM/g.

Using a multiple-method approach, the antioxidant activity of the essential oils from *Thymus vulgaris* was tested by Viuda-Martos *et al.* (2010). The thyme essential oil produced the highest percentage of inhibition of TBARS (89.84 per cent).

Twenty-five essential oils, including Thyme oil, were tested by Wei and Shibamoto (2010) for antioxidant activities using a conjugated diene assay, the aldehyde/carboxylic acid assay, the DPPH free radical scavenging assay, and the malonaldehyde/gas chromatography (MA/GC) assay. They were also tested for lipoxygenase inhibitory activities using the lipoxygenase inhibitor-screening assay. Thyme oil exhibited the greatest antioxidant effect in all assays (80-100 per cent) except DPPH assay (60 per cent). Wei *et al.* (2010) reported that Essential oil of *Thymus vulgaris* had appreciable antioxidant activities comparable to that of α-tocopherol, the reference chosen by the authors. In the thyme oil *p*-cymene and thymol predominated. Antioxidant activity of essential oils of five spice plants, including *Thymus vulgaris* used in the Mediterranean diet was evaluated by Viuda-Martos *et al.* (2010). All the essential oils tested had antioxidant activity, *Thymus vulgaris* oil presented the best activity, close to that verified for BHT. Composition and antioxidant activity of *Thymus vulgaris* volatiles was investigated by Grosso *et al.* (2010).

Spiridon *et al.* (2011) evaluated the total phenolic content and antioxidant activity of plants used in Romanian herbal medicine. Total phenolic concentration was determined using Folin-Ciocalteu phenol reagent method, while total flavonoids were measured using the aluminium chloride colorimetric method. *Thymus vulgaris* extracts showed good antioxidant activity.

Thymus zygis L. Family: Lamiaceae (Labiatae)

The antioxidant activities of methanol and ethyl ether extracts obtained from *Thymus zygis*, collected during flowering or non-flowering period, were evaluated and compared by Soares *et al.* (1997). Although methanol extracts reduce DPPH radicals more efficiently than ethyl ether extracts, suggesting a potent radical scavenger activity, the ethyl ether extracts were found to be most active in inhibiting lipid peroxidation in sarcoplasmic reticulum (SR) membranes. In addition, both extracts present peroxyl and superoxide radical scavenging activities. Methanolic extracts were more potent as scavengers of peroxyl and superoxide radicals than the ethyl ether extracts. Apparently, there is a relationship between antioxidant potency and the total phenolic groups content in each extract.

Dandlen *et al.* (2010) evaluated the antioxidant activity of essential oil of six species of *Thymus* including *Thymus zygis* subsp. *zygis* and *Thymus zygis* subsp. *sylvestris*. *Thymus zygis* subsp. *zygis* oil possessed the best capacity for preventing lipid peroxidation (IC_{50} = 0.030 mg/ml), followed by *Thymus zygis* subsp. *sylvestris*.

Tibouchia grossa (L.f.) Cogn. Family: Melastomataceae

Mosquera *et al.* (2009) evaluated the antioxidant activity of forty-six methanol plant extracts using DPPH radical scavenging assay. The plant extracts that showed greatest antioxidant activity were *Tiboucha grossa* with 47.0 per cent of antioxidant activity.

Tilia argentea Desf. ex DC. Family: Tiliaceae

Yildirim *et al.* (2000) reported antioxidant activity of the water extract of of *Tilia argentea* by the thiocyanate method. The antioxidant activity of the water extract increased with the increasing amount of lyophilized extract (50-400µg) added into the linoleic acid emulsion. Statistically significant effect was determined in 100µg and higher amounts. Antioxidant activity and reducing power activity was concentration dependent. Even in the presence of 50µg of the extract, the reducing power was significantly higher than that of the control in which there was no extract. They are of the opinion that although the reducing power of a substance may be an indicator of its potential antioxidant activity, there may not always be a linear correlation between these two activities.

Demiray *et al.* (2009) evaluated phenolic profiles and antioxidant activities of Turkish medicinal plant *Tilia argentea* flowers and leaves. 70 per cent acetone was found as the most efficient solvent targeting total phenolic content. Catechin, naringenin, quercetin, protocatechuic, *p*-coumaric, chlorogenic, ferulic and caffeic acids were determined in all *T. argentea* extracts, however gallic acid was found only in water extracts. The antioxidant capacities of *T. argentea* extracts were in the following descending order: acetone extract (14.7 mg AAE/g dw plant), pure water extract (6.1 mg AAE/g dw plant), and methanol extract (407 mg AAE/g dw plant). Although the methanol extract had higher total phenolic content than water extract, it did not exhibit higher antioxidant activity than the water extract (Demiray *et al.*, 2009).

Tinospora cordifolia (Willd.) Miers ex Hook.f. et Thoms. Family: Menispermaceae

Antioxidant properties of various components from *Tinospora cordifolia* have been reported. Oral administration of an aqueous *T. cordifola* root extract for 6 weeks resulted in a decrease in the levels of plasma lipid peroxidation, ceruloplasmin and α-tocopherol in alloxan diabetic rats (Prince and Menon, 1999). The root extract also caused an increase in the levels of glutathione and vitamin C in alloxan diabetes. Transina, an herbal preparation containing *T. cordifolia* induced a dose-related decrease in streptozotocin (STZ) induced hyperglycemia and attenuation of STZ induced decrease in islet SOD activity (Mathew and Kuttan, 1997). This indicates that the anti-hyperglycemic effect of transina may be due to free radical scavenging activity. The hyperglycemic activity of STZ may be the consequence of decrease in islet SOD activity leading to the accumulation of free radicals in islet beta-cells.

Extract of *Tinospora cordifolia* has been shown to inhibit the lipid peroxidation and formation of superoxide, and hydroxyl radicals *in vitro* (Bhattacharya *et al.*, 1997). Administration of the extract partially reduced the elevated lipid peroxides in serum. *Tinospora* extract may be useful in reducing the chemotoxicity induced by free radical forming chemicals like cyclophosphamide. Several glycosides with potential antioxidant activity were also isolated as polyacetates, from the n-BuOH fraction of the *T. cordifolia* stems. The structures of three new norditerpene, furan glycosides cordifoliside A,B and C have been established by ID and 2D NMR spectroscopy (Gagan *et al.*, 1994).

The antioxidant activity of arabinogalactan polysaccharide (TSP) from *Tinospora cordifolia* was studied by Subramanian *et al.* (2002). The polysaccharide showed good protection against iron-mediated lipid peroxidation of rat brain homogenate as revealed by the TBARS and lipid hydroperoxide assays. TSP also provided significant protection to protein against gamma-ray induced damage.

Alcohlic *Tinospora cordifolia* root extract administered at a dose of 100mg/kg body weight to diabetic rats orally for six weeks normalized the antioxidant activity status of liver and kidney (Stanley *et al.*, 2004). The effect of *Tinospora cordifolia* root extract was more potent than glibenclamide (600 µg/kg body weight).

Jadhao *et al.* (2009) evaluated the antioxidant in *Tinospora cordifolia*. Fresh leaves contain 360 mg ascorbic acid, 10.43 mg lycopene, 5.24 mg carotene, 20.296 mg anthocyanin, 400 mg phenol and dry leaves contain 290 mg ascorbic acid, 125 mg iron, 21.234 mg anthocyanin, 1240 mg phenol/100g. Fresh fruits show 40 mg ascorbic acid, 0.6709 mg lycopene, 6.36 mg carotene, 19.355 mg anthocyanin, 560 mg phenol and dry fruit shows 3660 mg ascorbic acid, 37.5 mg iron, 19.061 mg anthocyanin, 2600 mg phenol/100g.

Hydromethanol extracts of 15 Bangladeshi medicinal plants, including *Tinospora cordifolia*, were evaluated by Hasan *et al.* (2009) for antioxidant potential using DPPH radical scavenging assay. *Tinospora cordifolia* aerial parts exhibited radical scavenging activity with an IC_{50} value of 29.87 µg/ml compared to the IC_{50} value of 5.15µg/ml as shown by the reference antioxidant ascorbic acid, in a dose dependent fashion.

Lavanya *et al.* (2010) reported that the ethanolic extract of stem of *Tinospora cordifolia* at 1000microg/ml showed maximum scavenging of the radical cation DPPH, nitric oxide radical and superoxide radicals at the same concentration.The IC_{50} values of ethanol extract for superoxide radical scavenging activity, DPPH radical scavenging activity, nitric oxide radical and lipid peroxide inhibitory activity was found to be 0.8, 0.7, 0.7 and 0.9 mg/ml, respectively.

Sadiq *et al.* (2010) reported that intake of *Tinospora cordifolia* stem extract at the dose of 500 mg/kg body weight increase the antioxidant activity.

Tinospora crispa Miers Family: Menispermaceae

TLC autographic assays revealed in the CH_2Cl_2 extract of *Tinospora crispa* the presence of three compounds exhibiting antioxidant and radical scavenging properties towards β-carotene and DPPH radical (Cavin *et al.*, 1998). They were isolated and identified as N-CIS-feruloyltyramine, N-trans-feruolyltyramine and secosiolariciresinol. When tested in dilution assays on the reduction of the DPPH radical, these 3 compounds proved to be more active than the synthetic antioxidant BHT. Further investigation of the CH_2Cl_2 extract led to the isolation ov vanillin, syringin, the alkaloid N-formylnornuciferin and the diterpene derivatives borapetosides B and C.

Amom *et al.* (2009) evaluated the nutritional composition, antioxidant ability and flavonoid content of *Tinospora crispa*. The DPPH assay was utilized to evaluate the antioxidant ability to scavenge free radicals. The test demonstrated the percentage of inhibition of up tot 86 per cent in the *T.crispa* extract. FRAP assay on the extract produced a FRAP value of 0.89 mmol/l. The flavonoids detected in *T. crispa* are catechin, luteolin, morin and rutin. Morin has certain biological activities including antioxidant activities. Luteolin has strong scavenging properties for superoxide radicals and poses as a potent physical quencher of singlet oxygen. The total phenolic content of the stem aqueous extract was measured using the Folin-Ciocalteu assay. The result produced the value of 0.29 GAE/100g of fresh sample.

Toddalia aculeata Lam. Family: Rutaceae

The antioxidant activity of the crude methanolic extract of the seeds of *Entada pursaetha*, the stem of *Toddalia aculeata*, and the fruit of *Ziziphus mauritiana* was investigated by Pakutharivu and Suriyavadhana (2010). The total phenolic composition of methanolic extract was calculated to be 5.5mg catechol equivalents/g of sample. Antioxidant activity of the extract was evaluated on the basis of ability of scavenging free radical and hydroxyl radical with the IC_{50} values 2.12mg/ml and 1.034mg/ml respectively. Total antioxidant capacity of crude plant extract was found to be 1.43mg ascorbic acid equivalents at 250µg/ml extract concentration. The reducing power of the extract increased dose dependently and the extract reduced the most Fe^{3+} ions to the extent less than the standard ascorbic acid.

Toddalia asiatica (L.) Lam. Family: Rutaceae

Decoction and 75 per cent ethanolic extracts from leaves of *Toddalia asiatica* were examined by Sithisarn and Jarikem (2010) for antioxidant activity and phenolics content. Both the tested samples showed strong DPPH scavenging activity (EC_{50}<50µg/ml). Total phenolic content was 19.96g per cent CAE (decoction extract) and 14.29(ethanolic extract), flavonoid content was 0.78g per cent Rutin equivalent (Decoction extract) and 1.91 g per cent Rutin equivalent (Ethanolic extract).

Madhavan *et al.* (2010) investigated the *in vitro* and *in vivo* antioxidant activity of alcoholic and aqueous extracts of the roots of *Toddalia asiatica*. The tested extracts exhibited potential scavenging effects on DPPH, hydrogen peroxide and nitric oxide free radicals. The *in vivo* antioxidant activity was investigated using wistar albino rats. *Toddalia asiatica* extracts significantly increased the hepatic levels of reduced glutathione, proteins, antioxidant enzymes and decreased lipid peroxidation.

Balasubramaniam *et al.* (2012) investigated the anti-inflammatory and free radical scavenging potential of 50 per cent ethanolic extract of stem bark of *Toddalia asiatica*. The antioxidant activity of the same was assessed by Free radical scavenging activity using DPPH, Nitric oxide scavenging activity, Hydroxyl radical scavenging activity and Chelation of Ferrous (Fe^{2+}) ions. *Toddalia asiatica* showed good radical scavenging activity at various concentrations (200-1000 µg/ml) against DPPH with IC_{50} value of 240.07 µg/ml, OH radical with IC_{50} value of 432.17 µg/ml, NO radical with IC_{50} value of 324.81 µg/ml and the same showed moderate scavenging activity against Chelation of Fe^{2+} ions with IC_{50} value of 483.21 µg/ml.

Toona sinensis (A.Juss.) M.Roem. Family: Meliaceae

English Name: Chinese Toon.

Wang *et al.* (2007) evaluated the antioxidant activity of phenolics from fresh young leaves of *Toona sinensis*. The 80 per cent acetone extract of Chinese Toon exhibited considerable antioxidant activity in the DPPH radical-scavenging assay. Compounds isolated from the extract were examined for their antioxidant activities by the DPPH method, and the results showed that gallic acid and its derivatives, gallotannins and flavonoids were the main constituents contributing to the antioxidant activity of Chinese Toon.

Hseu *et al.* (2008) investigated the antioxidant activity of aqueous extracts of *Toona sinensis* (TS; 0-100 microg/ml) and gallic acid (0-50 microg/ml), with the purified natural phenolic components evaluated using different antioxidant models. It was found that the TS extracts and gallic acid possess effective antioxidant activity against various oxidative systems *in vitro*, including the scavenging of free and superoxide anion radicals, reducing power, and metal chelation. However, antioxidant activity in terms of metal chelation was not observed for the gallic acid. Moreover, TS extracts and gallic acid appear to possess powerful antioxidant properties with respect to oxidative modification of human LDL induced by $CuSO_4$, AAPH or sodium nitroprusside, as assessed by the relative electrophoretic mobility, TBARS formation, and cholesterol degradation of oxidized LDL. Furthermore, AAPH-induced oxidative hemolysis, lipid peroxidation, and decline in superoxide dismutase (SOD) activity in human erythrocytes were prevented by both the TS extracts and the gallic acid.

The methanolic crude extract and four subfractions (hexane fraction, ethyl acetate fraction, n-butanol fraction and aqueous fraction) of old leaves of *Toona sinensis* (L.T.S.) were prepared by sequential partitioning, and their antioxidant activities were evaluated by Jiang *et al.* (2009) using DPPH free radical-scavenging assay, ferric reducing ability of plasma (FRAP) assay, β-carotene bleaching method and stabilizing soybean oil. The results indicated that old L.T.S. possess powerful antioxidant activity and could prevent oxidation of soybean oil. The ethyl acetate-soluble fraction had the strongest antioxidant activity, with high levels of total flavonoids content (108.57 mg rutin equivalents/g), total phenol contents (262.09 mg gallic acid equivalents/g), reductive activity (467.53 mg ascorbic acid [Vc] equivalents/g), FRAP value (10578.71 μM FRAP/g) and strong DPPH radical scavenging capacity, respectively.

Yang *et al.* (2011) evaluated the protective effects of non-cytotoxic concentrations of aqueous leaf extracts of *Toona sinensis* (TS extracts; 50-100μg/ml) and gallic acid (5μg/ml), a major component of these extracts, against AAPH-induced oxidative cell damage in human umbilical vein endothelial cells (ECs). Exposure of ECs to AAPH (15mM) decreased cell viability from 100 per cent to 43 per cent. However, ECs were pre-incubated with TS extracts prior to AAPH induction resulted in increased resistance to oxidative stress and cell viability in a dose-dependent manner. An increase in ECs-derived PGI(2) and IL-1β in response to AAPH exposure was positively correlated with cytotoxicity and negatively with TS extracts concentrations. In addition, gallic acid also suppressed PGI(2) and IL-1β production in AAPH-induced ECs. Notably, TS extracts/gallic acid treatment significantly inhibited ROS generation, MDA formation, SOD/catalase activity, and Bax/Bcl-2 dyregulation in AAPH-stimulated ECs. Pretreatment of ECs with TS extracts/gallic acid also suppressed AAPH-induced cell surface expression and secretion of VCAM-1, ICAM-1 and E-selectin, which was associated with abridged adhesion of U937 leukocytes to ECs. Moreover, TS extracts/gallic acid treatment significantly inhibited the AAPH-mediated up regulation of PAI-1 and down regulation of t-PA in ECs, which may decrease fibrinolytic activity.

Toona sureni (Blume) Merr. Family: Meliaceae

An antioxidant compound has been isolated from the leaves *Toona sureni* by Ekaprasada *et al.* (2009). The structure was determined to be methyl 3,4,5-trihydroxybenzoate (methyl gallate), based on UV-vis, FTIR, NMR and MS spectra. The isolated compound exhibited potent antioxidant activity in the α,α-diphenyl-β-picrylhydrazyl (DPPH) radical scavenging test, with IC_{50} value 1.02 μg/ml.

Tordylium apulum L. Family: Apiaceae (Umbelliferae)

A total of 27 extracts from non-cultivated and weedy vegetables traditionally consumed by ethnic Albanians in southern Italy were tested by Pieroni *et al.* (2002) for their free radical scavenging activity (FRSA) in the DPPH screening assay, for their *in vitro* non-enzymatic inhibition of bovine brain lipid peroxidation and for their inhibition of xanthine oxidase. In the lipid peroxidation assay, extracts from leaves of *Tordylium apulum* showed a remarkable inhibitory activity (>50 per cent).

Torenia concolor Lindl. Family: Scrophulariaceae

Two new phenylethanol glycosides, phenylethyl-O-alpha-l-rhamnopyranosyl-(1 2)-beta-D-glucopyranoside (torenoside A, 1) and 2'-O-3,4-dihydroxy-beta-phenylethoxy-O-alpha-L-rhamnopyranosyl-(1'' 3')-(4'-O-caffeoyl)-beta-D-glucopyranoside (torenoside B, 2), along with the 17 known compounds (3-19) were isolated from *Torenia concolor* by Chou *et al.* (2009). Moreover, phenylethanol glycosides 3-6 exhibited significant antioxidant activities in DPPH radical scavenging assay.

Torenia fournieri Linden ex E. Fourn Family: Scrophulariaceae

The edible flower of *Torenia fournieri* was found to possess potent antioxidative activity in a rat brain homogenate model. Bioassay-guided isolation of the active compounds from a CH_2Cl_2–MeOH (1:1) extract led to the isolation of acteoside (*1*), luteolin-7-*O*-β-glucoside (*2*), apigenin-7-*O*-α-rhamnosyl-(1→6)-β-glucoside (*3*), and apigetrin (*4*) (Shindo *et al.*, 2008).

Torreya grandis Fort. ex Lindl. Family: Scrophulariaceae

Chen *et al.* (2006) investigated the antioxidative activity of the ethanolic extract of seeds. Exposure of human dermal fibroblasts to the extract at 50 and 250 μg/ml showed significant protective effect against Hydrogen peroxide.

Torreya grandis cv. Merrillii is an endemic tree species in China, seeds of which are used as a popular snack, possessing beneficial effects on preventing angiosclerosis and coronary heart diseases. Antioxidant activity and chemical constituents of *T. grandis* cv. Merrillii seed (TGMS) were investigated by Shi *et al.* (2009). The antioxidant activity of different fractions and the ethanol extract was evaluated using DPPH radical scavenging capacity, hydroxyl radical scavenging activity and lipid peroxidation assays. The oil, CH_2Cl_2 and n-BuOH fractions, and ethanol extract of TGMS all showed antioxidant activities in these models, especially the DPPH one. By GC-MS analysis, twenty-seven constituents were identified from the oil fraction of TGMS. The total content of phenolic compounds in the CH_2Cl_2 and n-BuOH fractions and ethanol extract was also determined by the Folin-Ciocalteau method as 17.6, 21.6 and 12.9 microg/mg, respectively. In addition, analysis of the CH_2Cl_2 fraction yielded four phenolic compounds: 4-hydroxybenzaldehyde, 4-methoxy pyrocatechol, coniferyl aldehyde, 4-hydroxy cinnamaldehyde, and two steroids, beta-sitosterol and daucosterol.

Torreya nucifera (L.) Siebold. et Zucc. Family: Scrophulariaceae

Investigation on antioxidant compounds from the ethanolic extracts of *Torreya nucifera* leaves by Lee *et al.* (2006) resulted in the isolation of abietane diterpenoids, a known 18-methylesterferruginol (*1*) and a new 18-dimethoxyferruginol (*2*). Compounds *1* and *2* inhibited the Cu^{2+}-mediated, 2,2'-azobis(2-amidinopropane)hydrochloride-mediated and 3-morpholinosydnonimine-1-mediated low-density lipoprotein (LDL) oxidation in the thiobarbituric acid-reactive substances assay as well as the macrophage-mediated LDL oxidation. Compounds *1* and *2* exhibited the potent antioxidant activities in the conjugated diene production, relative electrophoretic mobility, and apoB-100 fragmentation on copper-mediated LDL oxidation. Compound 1 also suppressed nitric oxide production and inducible nitric oxide synthase expression in lipopolysaccharide-stimulated RAW264.7 cells.

Tournefortia bicolor Sw. Family: Boraginaceae

The total phenolic content and antioxidant activity of extracts and four flavonoids isolated from leaves of *Tournefortia bicolor* were evaluated by da Silva *et al.* (2010) using Folin-Ciocalteu reagent, DPPH free radical scavenging and inhibition of peroxidation of linoleic acid by FTC method. *T. bicolor* showed higher phenolic content 68.8 to >1000 mg/g and also scavenged radicals (IC$_{50}$ 12.8 to 437 mg/l) and inhibited lipid peroxide formation was IC$_{50}$ = 51.2 to 89 mg/l. For these extracts a good correlation between the phenolic content and antioxidant activity was observed, suggesting that *T. bicolor* is richer in phenolic compounds and that it could serve as a new source of natural antioxidants or nutraceuticals with potential applications. Seven compounds, trans-phytol (1), taraxerol (2), 3,7,4″-trimethoxyflavone (3), 5,3′-dihydroxy-3,7,4′-trimethoxyflavone (4), quercetin (5), tiliroside (6), and rutin (7) were isolated from the extract. Compounds (4-7) were also evaluated and were effective as DPPH quenching (IC$_{50}$ 7.7 to 79.3 mg/l) and as inhibition of lipid peroxidation (IC$_{50}$ 80.1 to 88.7 mg/l).

Trachelospermum jasminoides Lem. Family: Apocynaceae

Gan *et al.* (2010) evaluated the antioxidant activity of *Trachelospermum jasminoides* using the FRAP and TEAC assays, and its total phenolic content was measured by the Folin-Ciocalteu method. For FRAP assay the value was 113.22 µmol Fe(II)/g dry weight. For TEAC assay, the value was 81.21µmol Trolox/g dry weight. Total phenolic content was 9.19 mg gallic acid equivalent/g dry weight of plant material.

Trachyspermum ammi (L.) Sprague ex Turril (Syn.: *Carum copticum*) Family: Apiaceae (Umbelliferae)

English Name: Bishop' weed; Hindi: Ajwan.

Souri *et al.* (2008) evaluated the antioxidant activity of fruit against linoleic acid peroxidation using 1,3-diethyl-2-thiobarbituric acid as reagent. Antioxidant activity (IC$_{50}$) against peroxidation of linoleic acid (2mg/ml) was 14.36 and phenolic content was 919.12 mg/100g dry weight. The results of this study showed that there is no significant correlation between antioxidant activity and phenolic content of the studied plant materials and phenolic content could not be a good indicator of antioxidant capacity.

Seeds of *Trachyspermum ammi* exhibited antioxidant activity with trolox equivalent antioxidant capacity using ABTS and DPPH assay methods, FRAP and total phenolic content. Seeds of *Trachyspermum ammi* showed antioxidant activity in all these models (Surveswaran *et al.*, 2007). Ethanolic extract of ajwan showed activity against hexachlorocyclohexane (HCH) induced lipid peroxidation. Prefeeding of ajwan extract decreased hepatic levels of lipid peroxides and increased GSH, GSH-peroxidase, G-6-PDH, SOD, Catalase and glutathione S-transferase activities. At the same time there was a significant reduction in hepatic levels of HCH-induced raise in lipid peroxides as a result of the prefeeding of the extract (Anilakumar *et al.*, 2009).

Radical scavenging activity (RSA), antioxidant activity (AA), reducing power, total polyphenol (TP) and flavonoid contents (TFC) of oregano (*Oreganum vulgare*), ajowan (*Trachyspermum ammi*) and Indian borage (*Plectranthus amboinicus*) extracts were evaluated by Khanum *et al.* (2011). Oregano exhibited maximum radical scavenging activity (88.2 per cent, 82.3 per cent) for aqueous and ethanolic extracts at 50 ppm concentration respectively, followed by ajowan (86.9 per cent, 68.4 per cent) and Indian borage (30.5 per cent, 30.4 per cent). Extracts of oregano and ajowan showed better antioxidant activity and reducing power than that of Indian borage. Aqueous extracts of oregano had highest TP (Gallic acid equivalents) and TFC (Catechin equivalents) of 27.7 per cent and 50.6 per cent respectively compared to ajowan (6.7 per cent, 24.4 per cent) and Indian borage (4.2 per cent, 5.5 per cent). Synergistic studies showed that the addition of oregano extract appreciably enhanced the RSA of ajowan and Indian borage extracts even at 50 ppm concentration.

Trachyspermum copticum (L.) Link. Family: Apiacae (Umbelliferae)

Antioxidant activities of ethanol extracts from seven Apiaceae fruits have been studied by Nickavar and Abolhasani (2009) by the DPPH radical scavenging test. All the studied extracts, including *Trachyspermum copticum*, showed antioxidant capability. For *Trachyspermum copticum* IC_{50} value of the DPPH scavenging activity was 126.40µg/ml and total flavonoid content was 40.18µg/mg. A positive correlation was found between the antioxidant potency and flavonoid content of the fractions.

Trachystemon orientalis L. Family: Boraginaceae

The antioxidative activity of the 80 per cent ethanol extract obtained from eleven commonly consumed wild edible plants was determined according to the phosphomolybdenum method, reducing power, metal chelating, superoxide anion and free radical scavenging activity (Ozen, 2010). *Trachystemon orientalis* had the highest antioxidant capacities; Total antioxidant activity was 18730 µg α-tocopherol equivalent/g, higher than BHT but lower than BHA. The results indicated that the antioxidant activity of vegetables seems to be due to the presence of polyphenols, flavonoid and anthocyanoside. Total phenolics in *Trachystemon orientalis* were 82.1mg pyrocatechol equivalent/g dw, Flavonoids were 3.63 mg quercetin equivalent/g dw and anthocyanins 15.2 mg cyanidin glycoside equivalent/g dw.

Tragia pungens (Forssk.) Muell.-Arg. Family: Euphorbiaceae

Methanolic extracts of *Tragia pungens* showed high free radical scavenging activity (Mothana *et al.*, 2010).

Tragopogon aureus Boiss. Family: Asteraceae (Compositae)

Coruh *et al.* (2007) reported high antioxidant activity in *Tragopogon aureus*. The total phenolic content was 12.56 µgGAE/mg. There was a low correlation (R = 0.447) between total phenolic content and antioxidant activity in the plant samples.

Trapa bispinosa Roxb. Family: Trapaceae

Methanol aqueous extract of *Trapa bispinosa* was screened by Kim *et al.* (1997) for anti-oxidative activity using Fenton's reagent/ethyl lineolate system. The extract showed antioxidant activity.

Trapa natans L. Family: Trapaceae

The *in vitro* antioxidant potential of aqueous extract of *Trapa natans* fruits rind was investigated by Malaviya *et al.* (2010). The extract was found to contain a large amount of polyphenols and also exhibited an immense reducing ability. The total content of phenolic, flavonoid and tannin compounds was estimated as 63.81mg of GAE/g dry material, 21.34 mg of rutin equivalent/g dry material and 17.11mg of total tannin equivalent/g of dry material, respectively. IC_{50} values for different antioxidant models were calculated as 128.86µg/ml for DPPH radicals, 97.65µg/ml for superoxide, 148.32 µg/ml for Hydrogen peroxide and 123.01µg/ml for NO, respectively.

Trapa pseudoincisa Nakai Family: Trapaceae

cis-Hinokiresinol, also known as (+)-nyasol, was isolated from *Trapa pseudoincisa*. cis-Hinokiresinol was also found to exhibit antioxidant and antiatherogenic activities. The IC_{50} values for the scavenging activities of cis-hinokiresinol on ABTS cation and superoxide anion radicals were 45.6 and 40.5 microM, respectively. The IC_{50} values for the inhibitory effects on Lp-PLA2, hACAT1, hACAT2 and LDL-oxidation were 284.7, 280.6, 398.9 and 5.6 microM, respectively.

Trapa taiwanensis Nakai Family: Trapaceae

English Name: Water caltrop.

Chiang and Ciou (2009) investigated the effect of pulverization treatment on the antioxidant properties of water caltrop pericarps. The results showed that crude fiber was the largest proportional constituent of water caltrop pericarps, comprising 72.3 g/100 g (dry basis). After 10 min milling, five fractions were obtained 1–5 different particle sizes (285, 211, 138, 62 and 41 µm, respectively). Comparing the effect of pulverization on the fiber component of each fraction, the fraction 1 exhibited higher hemi-cellulose but lower lignin than smaller particle size fraction except fraction 5. At a dosage of 250 µg/ml methanol extracts, the highest free radical-scavenging ability was obtained in the fraction 5, which is found to be 78.3 g/100 g, followed by the fractions 1–4. The same tendency was observed in the reducing power and cupric ions chelating ability. The fractions 1 and 5 contained significantly higher total phenolics and total flavonoids than fraction 2–4.

Trewia nudiflora L. Family: Euphorbiaceae

Total phenol, antioxidant and free radical scavenging activities of *Trewia nudiflora* were studied by Prakash *et al.* (2007). Leaves and fruits of *Trewia nudiflora* were found to have high TPC (107.8 mg/g) and high Antioxidant activity (96.9 per cent). Leaves and fruits showed very low inhibitory concentration of FRSA and Reducing power.

Trewia polycarpa Benth. Family: Euphorbiaceae

The alcoholic extract of *Trewia polycarpa* roots, which exhibited significant anti-inflammatory activity, was evaluated for possible mode of action by studying its antioxidant potential in adjuvant-induced arthritic rats (Chamundeswari *et al.*, 2003). The biological defence system constituting the superoxide dismutase, glutathione peroxidase, ascorbic acid showed a significant increase while the lipid peroxide content was found to decrease to a large extent on the treatment with *Trewia polycarpa* roots thereby indicating the extracts free radical scavenging property (Chamundeswari *et al.*, 2003).

Trianthema decandra L. Family: Aizoaceae

Singh *et al.* (2009) evaluated the antioxidant activity of dichloromethane and methanolic extracts of 12 arid zone medicinal plants. *Trianthema decandra* showed appreciable antioxidant activity and per cent of inhibition of DPPH where RC_{50} (µg/ ml) was 69 (Dichloromethane extract) and 9 (MeOH extract). per cent of inhibition of DPPH was 90.65 per cent in methanol extract of *Trianthema decandra*.

Trianthema portulacastrum L. Family: Aizoaceae

Mandal *et al.* (1996) reported restoration of antioxidant balance by *Trianthema portulacastrum* in carbon tetrachloride-induced hepatocellular injury in mice.

Tribulus terrestris L. Family: Zygophyllaceae

Crude extract from *Tribulus terrestris* was screened by Sengul *et al.* (2009) for its *in vitro* antioxidant and antimicrobial properties. Total phenolic content of extract of this plant was also determined. β-carotene bleaching assay and Folin-Ciocalteu reagent were used to determine total antioxidant activity and total phenols of plant extract. Antioxidant activity was 69.25 per cent and the total phenolic content was 14.43 mgGAE/g DW).

Singh *et al.* (2009) evaluated the antioxidant activity of dichloromethane and methanolic extracts of 12 arid zone medicinal plants. *Tribulus terrestris* showed appreciable antioxidant activity and per cent of inhibition of DPPH where RC_{50} (µg/ ml) was 8 (Dichloromethane extract) and 7.5 (MeOH extract). Per cent of inhibition of DPPH was 82.87 per cent in methanol extract of *Tribulus terrestris*. Key phytochemicals include sapogenin, diosgenin, harmine, saponins and tannins.

Trichilia emetica Vahl Family: Meliaceae

Germano *et al.* (2006) evaluated the antioxidant properties and bioavailability of free and bound phenolic acids from *Trichilia emetica*. Extracts submitted to alkaline hydrolysis showed high antioxidant properties in two *in vitro* assays: autooxidation of methyl lineolate and ascorbate/Fe^{2+} -mediated lipid peroxidation in rat microsomes. An *in vivo* study was also performed to investigate the intestinal absorption of phenolic acids after oral administration of *Trichilia emetica* root extracts. Results showed high levels of phenolic acids, free or conjugated to glucuronide, in the plasma of rats treated with the hydrolyzed extract. Due to the absence of chlorogenic acid in plasma samples, the presence of caffeic acid seems to be derived from hydrolysis of chlorogenic acid in the gastrointestinal tract (Germano *et al.*, 2006).

Trichilia prieuriana A. Juss. Family: Meliaceae

Kuglerova *et al.* (2008) investigated the antioxidant activity of seven Ugandan tree species. The crude ethanolic extract of bark of *Trichilia prieuriana* exhibited DPPH scavenging activity with IC_{50} value 377µg/ml.

Trichilia roka Chiov. Family: Meliaceae

Germano *et al.* (2001) have reported the hepatoprotective properties of root decoction of *Trichilia roka*, a plant used in Mali folk medicine, against carbon tetrachloride-induced hepatotoxicity and correlated this effect to the polyphenol antioxidant component of the fraction.

Tricholepis glaberrima DC. Family: Asteraceae (Compositae)

Antioxidant activity of aqueous, chloroform and methanol extract of the aerial parts of plant of *Tricholepis glaberrima* were evaluated by Naphade *et al.* (2009) for the antioxidant activity by the FTC and TBA methods. Methanolic extract showed higher antioxidant activity than the chloroform and aqueous extract. The results obtained in this study indicate that the plants of *Tricholepis glaberrima* are a potent source of natural antioxidants.

Trichopus zeylanicus Gaertn. Family: Dioscoreaceae

The antioxidant *properties* of *T. zeylanicus* were established on free radicals (DPPH and ABTS), its ability to reduce iron, lipoxygenase activity and hydrogen peroxide-induced lipid peroxidation. *T. zeylanicus* contains polyphenols and sulfhydryl compounds which have the ability to scavenge reactive oxygen species (Tharakan *et al.*, 2005).

Trichosanthes cucumeriana L. Family: Cucurbitaceae

Adebooye (2007) ensures from his study that Nigerian *Trichosanthes cucumeriana* possesses valuable antioxidant and nutraceutical properties. They reported Ascorbic acid 24.8-25.7 per cent, lycopene 16.1-18 per cent, carotenoids 15.6-18.4 mg, α-carotene 10.3-10.7 mg and β-carotene contents were 2.4-2.8 mg/100g fresh weight along with phenols 46.8 per cent, flavonoids 78 per cent and ferric reducing antioxidant power 25.2 per cent. In the seed oil punicic acid, oleic, linolenic, eleosteric, palmitic, stearic, arachidic, meso-inositol were found to be present (Khare, 2007).

Trichosanthes kirilowii Maxim. Family: Cucurbitaceae

Cai *et al.* (2004) evaluated the antioxidant activity and phenolic compounds in traditional Chinese medicinal plants associated with anticancer. The improved ABTS.+ method was used to systematically assess the total antioxidant capacity (Trolox equivalent antioxidant capacity, TEAC) of the medicinal extracts. The TEAC values and total phenolic content for methanolic extract of peel of *Trichosanthes kirilowii* were 69.8 µmol Trolox equivalent/100g dry weight (DW), and 0.50 g of gallic acid equivalent/100g DW respectively. A positive significant linear relationship between antioxidant activity and total phenolic content showed that phenolic compounds were the dominant antioxidant components in the tested herbs.

Li *et al.* (2008) evaluated the antioxidant properties of 45 medicinal plants using FRAP and TEAC assays and the total phenolic contents of these plants were measured. *Trochosanthes kirilowii* showed very low antioxidant activity with 1.85 µmol Fe(II)/g FRAP value, 1.05µmol Trolox/g of TEAC value and 2.13 mg GAE/g of Phenolic content.

Antioxidant capacities of 56 selected Chinese medicinal plants, including *Trichosanthes kirilowii* were evaluated by Song *et al.* (2010) using the TEAC and FRAP assays, and their total phenolic content was measured by the Folin-Ciocalteu method. *Trichosanthes kirilowii* showed low antioxidant activity with TEAC value of 11.01μmol Trolox/g, FRAP value of 9.53μmol Fe^{2+}/g and phenolic content was 1.66mg GAE/g.

Tridax procumbens L. Family: Asteraceae (Compositae)

The effect of *Tridax procumbens* on liver antioxidant defense system using lipopoysaccharide-induced hepatitis in D-galactosamine sensitized rats was studied by Ravikumar *et al.* (2005).

Nwanjo (2008) evaluated the effect of aqueous extract of *Tridax procumbens* leaves against free radical hepatocellular damage. An assessment of hepatospecific marker enzymes and antioxidant defense enzymes in rat plasma and liver homogenate indicated the extract could significantly inhibit lipid peroxide production in the plasma and liver induced in rats by chloroquine. While chloroquine treatment caused a significant increase in lipid peroxide end-product malondialdehyde, aspartate aminotransaminase in superoxide dismutase and catalase activities and reduced glutathione, vitamin C and vitamin E levels, aqueous extract of *Tridax* procumbens treatment prevented these changes in antioxidative activity.

Tridax procumbens was analyzed by Habila *et al.* (2010) for reducing power ability as an antioxidant using the DPPH assay and for total phenolics using the Folin-Ciocalteu method. The results of the analysis show that *T.procumbens* has a percentage antioxidant activity (AA per cent) of 96.70 which was observed to be higher than those of gallic (92.92 per cent) and ascorbic acids (94.81 per cent) used as standards. The reductive potential determination shows that *T. procumbens* has a very significant reductive potential of 0.89 nm at the same concentration with gallic acid whose reductive potential was 0.99 nm. The total phenolic determination shows that *T. procumbens* has a phenolic content of 12mg/g GAE.

Trifolium hybridum L. Family: Fabaceae (Papilionaceae)

English Name: Alsike clover.

The antioxidative activity of phenolic extracts from herb of alsike clover was examined by autoxidation of methyl linoleate by Kahkonen *et al.* (1999). The total phenols (5mg/g GAE) and antioxidant activity (4 per cent at ppm level) was low.

Trifolium pannonicum Jacq. Family: Fabaceae (Papilionaceae)

Godevac *et al.* (2007) evaluated the antioxidant activity of nine Fabaceae including *Trifolium pannonicum* using DPPH radical scavenging capacity, TEAC values by ABTS radical cation and inhibition of liposome peroxidation. The plant exhibited strong antioxidant capacity in all the tested methods.

Trifolium pratense L. Family: Fabaceae (Papilionaceae)

English Name: Red clover.

The antioxidative activity of phenolic extracts from herb of red clover was examined by autoxidation of methyl linoleate by Kahkonen *et al.* (1999). The total phenols (7.8 mg/g GAE) and antioxidant activity (19 per cent) was low.

Trifolium repens L. Family: Fabaceae (Papilionaceae)

English Name: White clover

The antioxidative activity of phenolic extracts from leaf and bark of white clover was examined by autoxidation of methyl linoleate by Kahkonen *et al.* (1999). The total phenols (2.9 mg/g GAE) and antioxidant activity (10 per cent) was low.

Trigonella foenum-graecum L. Family: Fabaceae (Papilionaceae)

English Name: Fenugreek.

Sabu and Kuttan (2003) investigated the antioxidant activity of Indian herbal drugs, including *Trigonella foenum-graecum*, in rats with alloxan-induced diabetes. The extract of *Trigonella foenum-graecum* showed antidiabetic and antioxidant activity.

Dixit *et al.* (2005) evaluated the antioxidant properties of germinated fenugreek seeds. Different fractions of the germinated seeds used to determine their antioxidant potential at different levels. The assays employed were FRAP, DPPH, ferromyoglobin/ 2,2'-azobis-3-ethylbenzthiazoline-6-sulfonic acid, pulse radiolysis, oxygen radical absorbance capacity and inhibition of lipid peroxidation in mitochondrial preparations from rat liver. An aqueous fraction of fenugreek exhibited the highest antioxidant activity compared with other fractions.

The free radical scavenging potential of the seed of *Trigonella foenum-graecum* was studied by Choudhary (2006) by using different antioxidant models of screening. The ethanolic extract at 500μg/ml showed maximum scavenging of the radical cation ABTS observed up to 97.56 per cent. It also scavenged stable radical DPPH (88.12 per cent) and nitric oxide radical (72.86 per cent) at the same concentration.

Syndrex® is a formulated herbal antidiabetic preparation containing powder of germinated fenugreek seeds. Dixit *et al.* (2007) assessed the antioxidant potential of this drug. Syndrex® was fractionated by soxhlet apparatus and fractions were used to determine their antioxidant potential at different levels. *In vitro* activity was assessed by FRAP assay, DPPH, ferromyoglobin/ABTS and pulse radiolysis. Methanolic fraction of Syndrex® exhibited the highest antioxidant activity as compared to other fractions. This fraction showed maximum phenolic and flavonoid contents.

Khalaf *et al.* (2008) reported antioxidant activity in the methanolic crude extracts of seeds of fenugreek using DPPH assay.

Subhasree *et al.* (2009) carried out the antioxidant activity of leafy vegetable of *Trigonella foenum-graecum* by measuring the ability of methanol extracts of this plant to scavenge radicals generated by *in vitro* systems and by their ability to inhibit lipid peroxidation. The levels of non-enzymatic antioxidants were also determined by standard spectrophotometric methods. Correlation and regression analysis established a positive correlation between some of these antioxidants and the *in vitro* free radical-scavenging activity of the plant extracts. The conclusions drawn from this study indicate that *in vivo* studies, isolation and analysis of individual bioactive compounds will reveal the crucial role that this plant may play in several therapeutic formulations.

Souri *et al.* (2008) evaluated the antioxidant activity of seed against linoleic acid peroxidation using 1,3-diethyl-2-thiobarbituric acid as reagent. Antioxidant activity (IC_{50}) against peroxidation of linoleic acid (2mg/ml) was 91.66 and phenolic content was 194.63 mg/100g dry weight. The results of this study showed that there is no significant correlation between antioxidant activity and phenolic content of the studied plant materials and phenolic content could not be a good indicator of antioxidant capacity.

The significant hepatoprotective effects were obtained by ethanolic extract of leaves of *Trigonella foenum-graecum* against liver damage induced by H_2O_2 and CCl_4 as evidenced by decreased levels of antioxidant enzymes (Meera *et al.*, 2009). The extract also showed significant anti lipid peroxidation effects *in vitro*, besides exhibiting significant activity in superoxide radical and nitric oxide radical scavenging, indicating higher potent antioxidant effects.

Antioxidant potential of selected vegetables commonly used in diet in Asian subcontinent, including *Trigonella foenum-graecum* leaves has been evaluated by Gacche *et al.* (2010). The DPPH radical scavenging activity was 25.7 per cent, inhibition of lipid peroxidation was 22.0 per cent and FICA was 16.8 per cent, Vitamin C content was 20.70 mg/100g and total phenols was 6.36 mg/g.

Tripleurospermum disciforme (C.A.Mey) Schultz.-Bip. Family: Asteraceae (Compositae)

Extract of *Tripleurospermum disciforme* with different solvents and a new isolated compound from its chloroform extract were tested by linoleic acid peroxidation for antioxidant activity using 1,3-diethyl-2-thiobarbituric acid as reagent (Souri *et al.*, 2006). The chloroform extract was found to be the most active one ($IC_{50}=10.75\mu g/ml$) comparable to α-tocopherol ($IC_{50}=14.75\mu g/ml$) as positive control. The isolated dioxaspiran derivative showed IC_{50} value of 185.50$\mu g/ml$ with the same condition. Since this is higher than IC_{50} value of the crude chloroform extract, other effective compounds may be responsible for such a difference (Souri *et al.*, 2006).

Tripleurosprmum inodorum L. Family: Asteraceae (Compositae)

The antioxidative activity of phenolic extracts from herb of *Tripleurospermum inodorum* was examined by autoxidation of methyl linoleate by Kahkonen *et al.* (1999). The total phenols (3.9 mg/g GAE) and antioxidant activity (6 per cent at 500 ppm level) was low.

Tripleurospermum oreades Tchihatchewii var. *oreades* Family: Asteraceae (Compositae)

Antioxidant and DPPH radical scavenging activities, reducing powers and the amount of total phenolic compounds of some medicinal Asteraceae species were studied by Özgen *et al.* (2004). Water extract of *Tripleurospermum oreades* var. *oreades* showed good antioxidant activity with 31.7 per cent inhibition at 500$\mu g/ml$ concentration. The DPPH radical scavenging activity was 49. per cent at 500$\mu g/ml$ concentration and reducing power was 12.1 $\mu g/ml$ Ascorbic acid equivalent. Total phenolic amount was 14.3$\mu g/ml$ Gallic acid equivalent.

Tripterygium wilfordii Hook.f. Family: Celastraceae

Cai *et al.* (2004) evaluated the antioxidant activity and phenolic compounds in traditional Chinese medicinal plants associated with anticancer. The improved ABTS.+ method was used to systematically assess the total antioxidant capacity (Trolox equivalent antioxidant capacity, TEAC) of the medicinal extracts. The TEAC values and total phenolic content for methanolic extract of flora bud of *Tripterygium wilfordii* were 1858.3 µmol Trolox equivalent/100g dry weight (DW), and 6.58 g of gallic acid equivalent/100g DW respectively. A positive significant linear relationship between antioxidant activity and total phenolic content showed that phenolic compounds were the dominant antioxidant components in the tested herbs. Major types of phenolic compounds in *Tripterygium wilfordii* include phenolic terpenes (triptophenolide, triptophenolide methyl ether).

Gan *et al.* (2010) evaluated the antioxidant activity of *Tripterygium wilfordii* using the FRAP and TEAC assays, and its total phenolic content was measured by the Folin-Ciocalteu method. For FRAP assay the value was 217.94 µmol Fe(II)/g dry weight. For TEAC assay, the value was 181.98µmol Trolox/g dry weight. Total phenolic content was 17.51 mg gallic acid equivalent/g dry weight of plant material.

Triticum aestivum L. Family: Poaceae (Graminae)

English Name: Wheat.

The antioxidative activity of phenolic extracts from grains of wheat was examined by autoxidation of methyl linoleate by Kahkonen *et al.* (1999). The total phenols were very low. The antioxidant activities of wheat extracts were very low. Wheat bran extract reached 35 per cent inhibition. Ferulic acid is the dominant form of phenolic acid in wheat. Ferulic acid showed moderate antioxidant activity in a study by Terao *et al.* (1993), where the activity was measured by autooxidation of bulk MeLo. The processing seems to be important when evaluating the antioxidant activity in wheat as brans were more active than other products from the grains. This is obviously due to localization of the phenolics in the grain: the outer layers of the grain (husk, pericarp, testa and aleurone cells) contain the greatest concentrations of total phenolic, whereas their concentrations were considerably lower in the endosperm layers. About 80 per cent of the trans-ferulic acid in wheat grain was found in the bran (White and Xing, 1997).

Several wheat brans (wheat bran alone, wheat bran powder, wheat bran with malt flour, bran breakfast cereal, table of bran and tablet of bran with cellulose) and oat brans (crunchy oat bran, oat bran alone, and oat breakfast cereal) used as dietary fiber supplements by consumers were evaluated as alternative antioxidant sources (i) in the normal human consumer, preventing disease and promoting health, and (ii) in food processing, preserving oxidative alterations (Martinez-Tome *et al.*, 2004). Products containing wheat bran exhibited higher peroxyl free radical scavenging effectiveness than those with oat bran. Wheat bran powder was the best hydroxyl radical (OH) scavenger. In terms of hydrogen peroxide (H_2O_2) scavenging, wheat bran alone was the most effective, while crunchy oat bran, oat bran alone, and oat breakfast cereal did not scavenge H_2O_2. The products made with oat bran showed lower Trolox equivalent antioxidant capacity (TEAC) values. In general, avenathramide showed a higher antioxidant level than each of the following typical cereal components: ferulic acid, genistic acid, p-hydroxy-benzoic acid, protocatechuic acid, syringic acid, vanillic acid, vanillin, and phytic acid (Martinez-Tome *et al.*, 2004).

Two pigmented wheat (*Triticum* spp.) genotypes (blue and purple) were fractionated in bran and flour fractions, examined, and compared for their free radical scavenging properties against TEAC, FRAP, TPC, phenolic acid composition, carotenoid composition and total anthocyanin content. Bran fractions had the highest antioxidant activities (1.9-2.3 mmol TEAC/100g) in both the grain genotypes and were 3-5-fold higher than the respective flour fractions (0.4-0.7 mmol TEAC/100g). Ferulic acid was the predominant phenolic acid in wheat genotypes (bran fractions). The highest contents of anthocyanins were found in the shorts of blue and purple wheat (Siebenhandl *et al.*, 2007).

Wheat is one of the major cereals in the world because of the universal use of wheat for a wide range of products such as bread, noodles, cakes, biscuits, cookies, etc. The graded flour fractions, which were milled from whole wheat grain from outer to inner parts without removal of germ and bran, are rich in dietary fibres and minerals, the source of nutrition for human beings. The whole waxy wheat was milled into five fractions using the gradual milling method and the phenolic contents and antioxidant capacity of these flours were investigated by Huang *et al.* (2009). The total phenolic and flavonoid contents of the free and bound phenolic extracts gradually increased in the order from the inner to the outer fractions. The flours milled from the outer parts of grain contained significantly higher amount of phenolics and exhibited significantly higher antioxidant capacity than did the whole grain. Likewise, the inner flour fractions milled from mostly endosperm part had significantly higher amount of phenolics and exhibited significantly higher antioxidant capacity than did the white flour, which was milled by a conventional milling method. Thus, the graded flour from whole waxy wheat should by encouraged to be used for processing whole-grain foods to improve both qualities of end-use products and health benefits.

Korycinska *et al.* (2009) evaluated the antioxidant properties of rye bran alkylresorcinols (C15:0-C25:0) and extracts from whole-grain cereal products using their radical-scavenging activity on DPPH and the chemiluminescence method (CL). DPPH radical reduction varied from ~10 to ~60 per cent for the alkylresorcinol homologues at concentrations from 5 to 300 µM and was not dependent on the length of the alkyl side chain of the particular homologue. Differences in the EC_{50} values for the studied compounds were not statistically significant, the values varying from 157µM for homologue C23:0 to 195µM for homologue C15:0. Moreover, values of EC_{50} for all the alkylresorcinol homologues were significantly higher than those for Trolox and α-, δ-, γ-tocopherols, compounds with well-defined antioxidant activity and used as positive controls. CL inhibition was evaluated for all the tested alkylresorcinol homologues at concentrations of 5 and 10 µM and varied from ~27 to ~77 per cent.

Similar to the DPPH method, the slight differences in CL inhibition suggest that the length of the alkyl side chain had no major impact on their antioxidant properties. The extracts from whole-grain products were added to the DPPH and CL reaction systems and their antioxidant activities were tested and compared with the total amount of alkylresorcinols evaluated in the extracts. DPPH radical and CL reduction for the whole-grain products varied from ~70 to ~43 per cent and from ~37 to ~91 per cent, respectively. A clear relationship between DPPH radical and CL reduction levels and the amount of total alkylresorcinols was obtained for whole-grain breakfast cereals, in which the reduction level decreased in the order rye>wheat>mixed>barley. Therefore, it may be considered that the antioxidant activity of alkylresorcinols could be of potential importance to the food industry, which is continuously searching for natural antioxidants for the production of food products during their processing and storage.

Sreeramulu *et al.* (2009) evaluated the antioxidant activity of wheat by DPPH, scavenging assay, FRAP assay and reducing power. Phenolic content was 109.34 mg/100g. DPPH radical scavenging activity expressed as trolox equivalent, FRAP (μmol/g) and reducing power (mg/g) were 0.24, 33 and 0.61, respectively in wheat.

No correlation among total phenolic content, DPPH and ABTS was reported in the wheat extracts (Lingi Yu *et al.*, 2002).

Triticum durum Desf. Family: Poaceae (Graminae)

English Name: Durum wheat.

The processing seems to be important when evaluating the antioxidant activity of cereals as brans were more active than other products from the same cereal species. This is obviously due to localization of the phenolics in the grain: the outer layers of the grain (husk, pericarp, testa, and aleurone cells) contain the greatest concentrations of total phenolic, whereas their concentrations are considerably lower in the endosperm layers. Onyeneho and Hettiarachchy (1992) observed that the freeze-dried fraction from durum wheat bran exhibited stronger antioxidant activity than extracts from other milling fractions.

Triumfetta pilosa Roth Family: Tiliaceae

Ramakrishna *et al.* (2011) evaluated the antioxidant activity of *Triumfetta pilosa*. The ethanolic extract of *Triumfetta pilosa* was screened for its free radical, hydroxyl radical, superoxide and nitric oxide scavenging activity. Total antioxidant activities of ethanolic extract was compared with standard antioxidants. Results indicated that the ethanolic extract exhibited antioxidant potential of *in vitro* screening methods.

Triumfetta rhomboidea Jacq. Family: Tiliaceae

The methanolic extract of *Triumfetta rhomboidea* roots were found to inhibit lipid peroxidation, scavenge hydroxyl and superoxide radicals *in vitro* (Lissy *et al.*, 2006). IC_{50} of root extract of *Triumfetta rhomboidea* was 336.65 μg/ml, 1346.03μg/ml and 1004.22μg/ml for superoxide radical scavenging, hydroxyl radical scavenging and lipid peroxidation respectively.

Sivakumar *et al*. (2008) investigated the possible antitumor effect and antioxidant role of methanol extract of *Triumfetta rhomboidea* (METR) leaves against Dalton's ascites lymphoma (DAL) bearing Swiss albino mice. The effects of METR on the growth of murine tumor, life span of DAL bearing mice were studied. Hematological profile and liver biochemical parameters (lipid peroxidation, antioxidant enzymes) were also estimated. Treatment with METR decreased the tumor volume and viable cell count there by increasing the life span of DAL bearing mice. METR brought back the hematological parameter more or less normal level. The effect of METR also decreased the levels of lipid peroxidation and increased the levels of glutathione (GSH), superoxide dismutase (SOD) and catalase (CAT).

Trollius chinensis Bunge Family: Ranunculaceae

English Name: Flos Trolli.

Microwave-assisted extraction techniques of flavone from *Trollius chinensis* were studied by Zhao *et al*. (2009), then the property of antioxidation of the flavonone was also investigated through comparing the eliminating effect to the free-radical of DPPH with VC and BHT.The optimum conditions were achieved and demonstrated as follows: ethanol solution (volume fraction as 60 per cent) as extraction solution, the ratio of materials to solution as 1:20(g/ml), microwave power as 600 W and 2 min extraction time. By the optimum conditions, the yield of flavone from *Trollius chinensis* was obtained at 83.7 per cent.The flavone from *Trollius chinensis* can eliminate DPPH efficiently, this indicated that the property of antioxidant of flavone from *Trollius chinensis* was higher, when compared with VC and BHT: VC flavone from *Trollius chinensis* BHT. Moreover, the flavone from *Trollius chinensis* exhibited synergistic effect with VC and BHT.

Flos trollii is considered as functional tea, as well as a traditional medicinal herb, in China. Total phenolic and flavonoid contents of Flos trollii were determined by Li *et al*. (2011) by a colorimetric method. The antioxidative potential of the hydroalcoholic extract of Flos trollii (FTE, extracted by alcohol:water, 80:20) was also evaluated by various antioxidant assays. Chemiluminescence technique was used to determine the radical scavenging activities of FTE toward different reactive oxygen species, including superoxide anion ($O(-2)$), hydroxyl radical (OH), lipid-derived radicals (R), and singlet oxygen ($(1)O(2)$). FTE could effectively scavenge $O(-2)$, OH , R , and (1)O(2) at an efficient concentration (EC(50)) of 46, 5.64, 5.19, and 3.97 mg/ml, respectively. Moreover, the radical scavenging activities of FTE were higher than those of ascorbic acid. Further, FTE had higher DPPH radical-scavenging activity with EC(50) 44 mg/ml, compared with BHTsynthetic antioxidant with EC(50) 52 mg/ml.

Trollius ledebouri Reichn. Family: Ranunculaceae

The protective effects of an ethanolic extract of the flower of *Trollius ledebouri* on DNA damage caused by potassium dichromate in mouse peripheral lymphocytes were investigated by Yang *et al*. (2008) using cell gel electrophoresis. The extract prevented potassium dichromate-induced reduction in plasma superoxide dismutase (SOD) and glutathione peroxidase activities. Moreover the extract showed potent scavenging activity of H_2O_2/Fe^{2+}-induced OH *in vitro*.

Tropaeolum majus L. Family: Tropaeolaceae

Major anthocyanins, ascorbic acid content, total phenolic content, and the radical scavenging activity against ABTS and DPPH radicals in petals of orange Nasturtium flowers (*Tropaeolum majus*), were investigated by Garzon and Wrolstad (2009). Anthocyanin (ACN) content in the petals was 72 mg/100 g FW and pelargonidin 3-sophoroside represented 91 per cent of the total ACN content. The ascorbic acid content was 71.5 mg/100 g and the total phenolic content as determined by the Folin–Ciocalteu method was 406 mg GAE/100 g FW. The radical scavenging activities against ABTS and DPPH radicals were 458 and 91.87 µm trolox eq/g FW, respectively.

Tropaeolum tuberosum Ruiz. et Pavon Family: Tropaeolaceae

English Name: Mashua.

The ORAC antioxidant activity contribution in the tubers related to the type of phenolic compounds present in *Tropaeolum tuberosum* was evaluated by Chirinos *et al.* (2008). Phenolic compounds were analysed by separating them into four main fractions: fraction I obtained by means of a liquid–liquid partition with ethyl acetate and fractions II, III and IV obtained by elution on a Sephadex LH-20 column. Fraction I revealed the presence of gallic acid, gallocatechin, procyanidin B_2 and epigallocatechin. Other phenolic compounds such as hydroxycinnamic and hydroxybenzoic acid derivatives, rutin and/or myricetin derivatives were also present in fraction I. Fraction II was mainly composed of epicatechin, hydroxycinnamic and hydroxybenzoic acid derivatives. Fraction III presented mainly anthocyanins for the purple coloured mashua tubers and rutin, hydroxycinnamic acid and hydroxybenzoic acid derivatives for the yellow coloured genotype. Fraction IV was composed of proanthocyanidins. Alkaline and acid hydrolysis of the different fractions revealed the presence of gallocatechin, epicatechin, *p*-coumaric acid, *o*-coumaric acid, cinnamic acid, protocatechuic acid, rutin and quercetin as the main phenolic moieties present. The proanthocyanidin fractions were the major contributors to the ORAC antioxidant activity of the mashua tubers for two of the three genotypes (34.7–39.2 per cent).

Campos *et al.* (2006) and Salluca *et al.* (2008) evaluated the phenolic compounds and antioxidant capacity of tuber of *Tropaeolum tuberosum*. Total antioxidant capacity was measured using ABTS and FRAP methods. Highest TAC value was observed in *Tropaeolum tuberosum* tubers with 0.35µMol Trolox equivalent/g dry sample and phenolics 0.02 µMol gallic acid equiv./g dry sample. Data from FRAP displayed a linear correlation with TPH data.

Tulbaghia violacea Harv. Family: Liliaceae

Four plant extracts with antioxidant activity were screened by Naidoo *et al.* (2008) for their anticoccidial activity *in vivo* with toltrazuril as the positive control.

Combretum woodii (160 mg/kg) proved to be extremely toxic to the birds, while treatment with *Tulbaghia violacea* (35 mg/kg), *Vitis vinifera* (75 mg/kg) and *Artemisia afra* (150 mg/kg) resulted in feed conversion ratios similar to toltrazuril, and higher than the untreated control. *T. violacea* also significantly decreased the oocyst production in the birds. From this study they concluded that antioxidant-rich plant extracts have potential benefits in treating coccidial infections. The promising results obtained with *T. violacea* justify further studies on the potential value of the plant as a therapeutic or prophylactic anticoccidial agent.

Turnera diffusa Willd. Family: Turneraceae

Wojcikowski *et al.* (2007) evaluated the antioxidant activity of aerial parts of *Turnera diffusa*. Oxygen radical absorbance capacity of the ethyl extract, methanol extract and aqueous extracts of leaf were 47.32, 783.58 and 92.84µmol Trolox equivalent per g of dried starting material, respectively.

Garza-Juarez *et al.* (2011) investigated the correlation between the antioxidant activity (obtained by DPPH assay) and chromatographic profiles of *Turnera diffusa* extracts.

Turnera ulmifolia L. Family: Turneraceae

Nascimento *et al.* (2006) evaluated the leaf extract of *Turnera ulmifolia* for its radical scavenging capacity (RSC). The *in vitro* RSC of a 50 per cent hydroethanolic (HE) extract was evaluated by β-carotene/linoleic acid coupled oxidation system for the inhibition of oxidation and the lipid peroxidation inhibition in rat brain homogenates, using thiobarbituric acid reactive substances (TBARS) and chemiluminescence (CL). Results indicated, through peroxidation suppression, that this extract exhibited greater antioxidative activity (77.4 per cent ± 10 per cent) than α-tocopherol (58.4 per cent ± 3.7 per cent). TBARS and CL inhibition was concentration-dependent and $Q_{1/2}$ values were 8.2 and 6.0 µg/ml for TBARS and CL, respectively. For α-tocopherol these values were 7.1 µg/ml (TBARS) and 9.8 µg/ml (CL). Phenolic compounds may be responsible for this antioxidant capacity.

Tussilago farfara L. Family: Asteraceae (Compositae)

Cho *et al.* (2005) investigated the antioxidant activities of ethyl acetate fraction by cell-free bioassays. It was shown to inhibit lipid peroxidation initiated by Fe(2+)/ascorbic acid in rat brain homogenates, and scavenge DPPH radicals.

A bioassay-guided fractionation of the ethylacetate soluble fraction from the flower buds of *Tussilago farfara* yielded two flavonoids, quercetin 3-O-β-L-arabinopyranoside and quercetin 3-O- β-D-glucopyranoside. These two sugar conjugates of quercetin exhibited higher antioxidative activity than their aglycone, quercetin by NBT superoxide scavenging assay.

Antioxidant capacities of 56 selected Chinese medicinal plants, including *Tussilago farfara* were evaluated by Song *et al.* (2010) using the TEAC and FRAP assays, and their total phenolic content was measured by the Folin-Ciocalteu method.

Tussilago farfara showed very high antioxidant activity with TEAC value of 217.62μmol Trolox/g, FRAP value of 455.64μmol Fe^{2+}/g and phenolic content was 30.03mg GAE/g.

Tylophora indica (Burm.f.) Merril Family: Asclepiadaceae

Gupta *et al.* (2011) carried out *in vitro* evaluation of antioxidant activity and total phenolic content of *Tylophora indica*. The total phenolic content of methanolic leaves extract was estimated by Folin-ciocalteu assay method and was found to be 0.160mg/CE/g. The DPPH radical scavenging activity of methanolic extract of *Tylophora indica* was found to be highest at 100μl concentration which was 30.74 per cent. Nevertless, per cent DPPH scavenging activity of standard ascorbic acid at same concentration was found to be 45.43 per cent. The per cent of DPPH scavenging activity increased with the increasing concentration. The concentration of *Tylophora indica* needed for 50 per cent inhibition (IC_{50}) was found to be 199.58μg/ml whereas 194.58μg/ml needed for ascorbic acid.

Typha capensis (Rohrb.) N.E.Br. Family: Typhaceae

English Name: Bulrush.

Henkel *et al.* (2011) investigated the antioxidant activity of *Typha capensis*. The extracts from leaves and rhizomes revealed dose dependent inhibitory activity for collagenase and free radical formation.

Typha latifolia L. Family: Typhaceae

English Name: Bulrush.

The antioxidative activity of phenolic extracts from leaf and stalk of bulrush was examined by autoxidation of methyl linoleate by Kahkonen *et al.* (1999). The total phenols of the leaf was 8.2 mg/g GAE and stalk 5.7 mg/g GAE. The antioxidant activities of the extracts in MeLo were 35 per cent for leaf and 80 per cent for stalk at 500 ppm level.

Typhonium flagelliforme (L.) Blume Family: Araceae

The antibacterial and antioxidant activity of different extracts from of *Typhonium flagelliforme* tuber, commonly called 'Rodent Tuber' was assessed by Mohan *et al.* (2008) towards selected bacteria as well as in different antioxidant models. Two complementary test systems, namely DPPH free radical scavenging and total phenolic compounds, were used for the antioxidant analysis. All the extracts were subjected to screening for their possible antioxidant activity. The DPPH assay showed that the inhibitory activity of ethyl acetate (77.6±0.9 per cent) and dichloromethane (70.5±1.7 per cent) extracts were having comparatively admirable inhibition capacity when compared to the positive control BHT (95.3±1.3 per cent). Total phenolic content of all extracts was also evaluated, and dichloromethane extracts (5.21±0.82 GAE mg/g extract) was superior to all other extracts, followed by hexane and ethyl acetate.

www.ingramcontent.com/pod-product-compliance
Lightning Source LLC
Chambersburg PA
CBHW020216290326
41948CB00001B/66